Advances in High-Pressure Mineralogy

edited by

Eiji Ohtani
Institute of Mineralogy, Petrology, and Economic Geology
Faculty of Science
Tohoku University
Sendai 980-8578
Japan

THE GEOLOGICAL SOCIETY OF AMERICA®

Special Paper 421

3300 Penrose Place, P.O. Box 9140 ▪ Boulder, Colorado 80301-9140 USA

2007

Copyright © 2007, The Geological Society of America, Inc. (GSA). All rights reserved. GSA grants permission to individual scientists to make unlimited photocopies of one or more items from this volume for noncommercial purposes advancing science or education, including classroom use. For permission to make photocopies of any item in this volume for other noncommercial, nonprofit purposes, contact the Geological Society of America. Written permission is required from GSA for all other forms of capture or reproduction of any item in the volume including, but not limited to, all types of electronic or digital scanning or other digital or manual transformation of articles or any portion thereof, such as abstracts, into computer-readable and/or transmittable form for personal or corporate use, either noncommercial or commercial, for-profit or otherwise. Send permission requests to GSA Copyright Permissions, 3300 Penrose Place, P.O. Box 9140, Boulder, Colorado 80301-9140, USA.

Copyright is not claimed on any material prepared wholly by government employees within the scope of their employment.

Published by The Geological Society of America, Inc.
3300 Penrose Place, P.O. Box 9140, Boulder, Colorado 80301-9140, USA
www.geosociety.org

Printed in U.S.A.

GSA Books Science Editor: Marion E. Bickford and Abhijit Basu

Library of Congress Cataloging-in-Publication Data

Advances in high-pressure mineralogy / edited by Eiji Ohtani.
    p. cm. (Special paper ; 421)
  Includes bibliographical references.
  ISBN-13 978-0-8137-2421-8 (pbk.)
    1. Mineralogy. 2. Materials at high pressures. 3. Materials at high temperatures. 4. Phase transformations (Statistical physics). I. Ohtani, Eiji. Special papers (Geological Society of America) ; 421.
QE364.2.H54A38 2007
549--dc22   2006052632

Cover, top: Image shows the global circulation of the materials in Earth; blue shows influx of materials from the surface to the core, whereas yellow shows outflow from the interior of Earth. The image was originally designed by Shio Watanabe and Shigenori Maruyama, and modified by Eiji Ohtani. (Published in Ohtani, E., 2005, Water in the Mantle, Elements, v. 1, no. 1, p. 25–30.)
Bottom left: Side view of diamond anvil cell with a metal gasket. Photograph courtesy of Takeshi Sakai. **Bottom right:** Top view of DIA-type high-pressure apparatus. Photograph courtesy of Hidenori Terasaki.

10 9 8 7 6 5 4 3 2 1

Contents

Preface v

Part 1. Phase Transitions of Earth and Planetary Materials at High Pressure

1. *Phase transitions of minerals in the transition zone and upper part of the lower mantle* 1
   M. Akaogi

2. *Properties of lower mantle Al-(Mg,Fe)SiO$_3$ perovskite* 15
   D. Andrault

3. *Discovery of post-perovskite phase transition and implications
   for the nature of the D″ layer of the mantle* 37
   K. Hirose and K. Kawamura

4. *High-pressure phase transformations in the system FeO-MgO* 47
   I.Yu. Kantor, A.P. Kantor, L.S. Dubrovinsky, and C.A. McCammon

5. *High-pressure mineral assemblages in shocked meteorites and
   shocked terrestrial rocks: Mechanisms of phase transformations
   and constraints to pressure and temperature histories* 57
   P. Gillet, A. El Goresy, P. Beck, and M. Chen

Part 2. Role of Volatiles in the Earth

6. *High-pressure mineralogy of diamond genesis* 83
   Y.A. Litvin

7. *Melting of ice VII and new high-pressure, high-temperature amorphous ice* 105
   L. Dubrovinsky and N. Dubrovinskaia

8. *Effect of water on the phase relations in Earth's mantle and deep water cycle* 115
   K.D. Litasov and E. Ohtani

Part 3. New Techniques in Mineral Physics

9. *Geophysical applications of nuclear resonant spectroscopy* 157
   W. Sturhahn and J.M. Jackson

10. *Single-crystal structure and electron-density analyses of Earth's interior under
    high-pressure and high-temperature conditions using synchrotron radiation* 175
    T. Yamanaka

11. *Phase-relation studies of mantle minerals by in situ X-ray diffraction
    using multianvil apparatus* 189
    T. Katsura

12. *Multianvil techniques in conjunction with synchrotron radiation at Deutsches
    ElektronenSYnchrotron (DESY) – HAmburger SYnchrotron LABor (HASYLAB)* 207
    H.J. Mueller, F.R. Schilling, and C. Lathe

13. *X-ray microtomography under high pressure* 227
    T. Uchida, Y. Wang, F. Westferro, M.L. Rivers, J. Gebhardt, and S.R. Sutton

Index 239

Preface

Recent advances in high-pressure mineralogy are remarkable. As demonstrated by the topics covered in this volume, various new techniques combined with an intensive X-ray from synchrotoron radiation have been introduced in this field. Appling these techniques, several new observations have become available, such as the post-perovsikite phase and spin transitions in the lower mantle phases at pressures above 100 GPa. This book provides a broad view of technical developments in high-pressure mineral physics and recent new results which have changed our view of Earth's deep interior.

This book is originated from the general symposium G-12 of the 19th International Geological Congress in Florence, Italy, in 2004. The symposium on high-pressure mineralogy was a great success, with many authors presenting papers. Although the session concentrated on presenting new results related to mineral physics, this volume aims to present rather a broad view of the current status of high-pressure mineralogy, based on the presentations made at the session. We also invited people who are actively conducting high-pressure X-ray spectroscopy and ultrahigh-pressure generation to explore the lower mantle and the core.

This volume's 13 papers, from several different fields of high-pressure mineralogy, will be of interest to readers from a broad range of earth science disciplines.

Part 1 contains topics on phase transitions in silicates. Five review papers are included here: four on phase transitions in the mantle, based on high-pressure and high-temperature experiments, and one on high-pressure phases observed in nature, especially in meteorites, which are the most common occurrence of the high-pressure phases on the surface of the Earth. Akaogi presents a review on phase transitions in the upper mantle and transition zone. These phase transitions are the physical basis for explaining seismological observations on major discontinuities in the mantle. Andrault presents a review on silicate perovskite, which is a major constituent of the lower mantle. Although there are many studies on thermoelastic properties of this phase, controversies still exist on the compression behavior, due to the complex chemistry of this phase that can accommodate various cations including aluminum, and ferric and ferrous iron. In order to formualte a better understanding of the lower mantle, it is essential to clarify compositional dependence on thermoelastic properties of this phase. Hirose and Kawamura review the post-perovskite phase transition, which is an exciting new observation in mantle phase relations. The post-perovskite phase may be the major component in the D″ layer at the base of the lower mantle. Kantor et al. report high-pressure properties of FeO, which is the end component of one of the most important lower mantle minerals, ferropericlase. The phase transition in ferropericlase may contribute affect seismic properties in the lower mantle. Gillet et al. review the natural occurrence of high-pressure phases in shocked meteorites. Recent technological advances in transmission electron microscopy (TEM) have made it possible to discover various types of high-pressure polymorphs such as α-PbO_2–type SiO_2 (seifertite) and ilmenite-type $MgSiO_3$ (akimotoite). The shock-induced textures are also useful in estimating the conditions of collision of the planetesimals in the early solar system.

Part 2 contains three papers relating volatiles in the mantle. Litvin reviews recent advances in studies of diamond genesis in the mantle. He emphasizes the importance of carbonatitie magmas for generation of natural diamond and the syngenetic primary inclusions in diamonds. Dubrovinsky and Dubrovinskaia briefly review the phase relations in ice that are currently available and report new and interesting observations on the amorphous phases near the melting temperature of ice at high pressure. Water circulation in the deep earth is currently one of the hotly debated issues. Litasov and Ohtani present a comprehensive review of recent studies on water in the deep mantle including the topics on water content in nominally anhydrous minerals, dense hydrous magnesium silicates (DHMS), and the effects of water on the phase boundaries in the mantle.

High-pressure research in mineral physics is one of the rapidly advancing fields in mineralogy, with several basic techniques developed in this decade. Further developments are expected in this particular field of mineralogy. Part 3 consists of five papers on the recent advances in experimental and theoretical techniques in high-pressure mineral physics. Sturhahn and Jackson review the application of nuclear resonant spectroscopy, such as nuclear resonant inelastic X-ray scattering and synchrotron Mössbauer spectroscopy, to high-pressure minerals. The spectroscopy combined with the diamond-anvil cell technique provides a powerful tool to clarify phonon density of states of minerals and permits determination of sound velocities and Grüneisen parameters, as well as important information on valence, spin state, and magnetic ordering at high pressure and temperature. Yamanaka reviews recent developments on the single-crystal X-ray diffraction studies at high pressure. Improvements of the diamond anvil cell design together with advanced methods for data reduction have made it possible to analyze single-crystal structure and electron density of high-pressure minerals. Katsura summarizes recent advances in the multianvil technique combined with the synchrotron X-ray radiation, especially its implications for the determination of the phase relations in the mantle. He provides a critical evaluation of the pressure scales and the reliable boundaries of the phase transitions of mantle minerals, which are essential in discussing the origin of the transition zone. Mueller et al. report the recent advances in the multianvil technology combined with several in situ measurements with a special emphasis on ultrasonic interferometry and in situ X-ray diffraction at high pressure and temperature. They also discuss future possibilities of this technique by showing their new high-pressure system under development. Uchida et al. present an in situ X-ray microtomogaphy technique using their Drickamer cell combined with the intense X-rays from the synchrotron radiation. They present basic information on the X-ray imaging technique at high pressure, which has a great potential for application to geology and mineral physics. X-ray imaging at high pressure and temperature may provide fruitful contributions to earth sciences.

I would like to acknowledge reviewers for spending time on critical reviews of the chapters for this volume. I would like to thank Professor A. Basu for providing us a chance to prepare this book on high-pressure mineralogy, and Mr. Sujoy Ghosh and Ms. Uliana Litasova for their technical assistance in the preparation of this book.

Eiji Ohtani
Tohoku University, Sendai
June 2006

Acknowledgments

The editor gratefully acknowledges the following individuals for their time and effort in reviewing chapters for this volume: M. Akaogi, D. Andrault, M. Arima, J. Badro, J. Bass, J. Chen, A. Cogne, L. Dubrovinsky, G. Fiquet, D. Frost, K. Funakoshi, P. Gillet, N. Hirao, K. Hirose, T. Inoue, T. Katsura, H. Kobayashi, T. Kondo, K. Litasov, H. Muller, J.B. Parise, C. Sanloup, T. Sharp, G. Shen, A. Suzuki, H. Terasaki, C. Vanpeteghem, Y. Wang, H. Yang.

Phase transitions of minerals in the transition zone and upper part of the lower mantle

M. Akaogi[†]
Department of Chemistry, Gakushuin University, 1-5-1 Mejiro, Toshima-ku, Tokyo 171-8588, Japan

ABSTRACT

High-pressure experiments on phase transitions of mantle-constituent minerals and bulk rocks provide indispensable data that clarify the mineralogical constitution of the deep mantle. This paper reviews the results of high-pressure experimental studies carried out in recent years. Phase relations of olivine-wadsleyite-ringwoodite transitions in pyrolite have been precisely determined to compare with seismological observations of the 410 and 520 km discontinuities. Results on the postspinel transition to perovskite + magnesiowüstite, which corresponds to the 660 km discontinuity, still have some controversies in transition pressure as well as the boundary slope. In pyrolite mantle, Ca-poor and Ca-rich pyroxenes are dissolved into garnet to form majorite in the transition zone. Recent studies have indicated that majorite transforms directly to aluminous perovskite in the normal mantle, but that it may transform first to aluminous ilmenite and then to perovskite at relatively low temperatures, such as in subducting slabs. Phase transitions in diopside and wollastonite have recently been examined in detail. The Ca component in majorite is exsolved as $CaSiO_3$-perovskite in the transition zone of the pyrolite mantle. Mg-rich perovskite in the lower mantle contains both Fe and Al components, in which Fe may be present in both ferrous and ferric states. The aluminum in Mg-rich perovskite introduces some vacancies in oxygen sites that may considerably affect elastic properties and possibly incorporate water in the structure. In basalt, Mg-rich perovskite becomes stable at higher pressure than that of the 660 km depth. Because basalt and continental crust materials have higher contents of Al and Si, several aluminous silicate phases that do not appear in pyrolite are stable in lower-mantle conditions. They are calcium ferrite– and hollandite-structured phases and a new hexagonal aluminous phase that can host Na and K in the deep lower mantle. Recent studies on phase transitions in SiO_2 have indicated that stishovite transforms to a $CaCl_2$-type phase, which further changes to an α-PbO_2-type phase. The Na- and K-hollandites and α-PbO_2-type SiO_2 were found in shocked meteorites.

Keywords: phase transition, high-pressure mineral, upper mantle, lower mantle, transition zone.

[†]E-mail: masaki.akaogi@gakushuin.ac.jp.

INTRODUCTION

Phase transformation studies of mantle minerals at high pressures and high temperatures are indispensable in the effort to clarify the nature of seismic discontinuities and mineralogical constitutions and dynamics of the deep mantle. For these purposes, current high-pressure experimental studies are carried out using a multianvil apparatus and laser-heated diamond anvil cell. Both of the experimental techniques developed rapidly in the past decades and have different features. Using the diamond anvil cell, pressure and temperature corresponding to the deep mantle and the core can be applied to a very small sample on a microgram scale. The multianvil apparatus is used for experiments to compress a much larger sample on the order of one to hundred milligrams with precisely controlled pressure and temperature that are relatively lower than those with the diamond anvil cell. Since the middle of 1980s, multianvil experiments coupled with synchrotron radiation have made it possible to accurately perform in situ X-ray observations at high pressure and high temperature. The combination of diamond anvil cell with synchrotron X-ray has also been used quite extensively. Currently, the in situ X-ray observation is widely used for accurate determination of compressibility, transformation kinetics, equilibrium phase relations, etc., at high pressure and high temperature. Another new improvement in the multianvil experiments is to adopt sintered diamond for anvil materials, instead of tungsten carbide widely used in high-pressure laboratories. Using the sintered diamond anvils in the multianvil system, pressures up to ~50 GPa have been generated at high temperature (Kubo et al., 2003).

A generally accepted petrological model for the upper mantle is pyrolite, which consists of olivine, Ca-poor orthopyroxene, Ca-rich clinopyroxene, and pyrope-rich garnet (Ringwood, 1975). Among them, olivine is the most abundant mineral (~60% in volume) in pyrolite, and it has an approximate composition of $(Mg_{0.9},Fe_{0.1})_2SiO_4$. The seismic discontinuities at 410 and 660 km depths that distinguish the transition zone from the upper and lower mantle are globally observed. The seismic discontinuities provide important clues that clarify the physical and chemical nature of the transition zone. The 410 km discontinuity is attributed to the transition of olivine (α phase) to wadsleyite (modified spinel, β phase). It is generally accepted that the 660 km discontinuity is caused by the postspinel transition of ringwoodite (spinel, γ phase) to perovskite + magnesiowüstite. Therefore, high-pressure phase relations in the system Mg_2SiO_4-Fe_2SiO_4 have been intensively studied. Recent studies have focused on details of the match between seismic observations of the discontinuities and accurate high-pressure phase relations. The issues include topics on sharpness of the 410 km discontinuity in terms of pressure interval of the olivine-wadsleyite transition and on the difference between the observed postspinel transition pressure and the 660 km discontinuity depth.

Ca-poor orthopyroxene and Ca-rich clinopyroxene are dissolved into pyrope-rich garnet to form a garnet solid solution called majorite at depths in the transition zone. In the lower mantle, Mg-perovskite, Ca-perovskite, and magnesiowüstite are major constituent minerals. The Mg-perovskite contains considerable amount of Al. Recent studies reveal that the Al^{3+} affects structure and properties of Mg-perovskite in the lower mantle, probably due to formation of defects in oxygen sites of the Mg-perovskite structure.

A typical oceanic lithosphere consists of an upper thin layer of basaltic crust, a middle layer of harzburgite, and a lower layer of depleted pyrolite. Subduction of oceanic lithospheres creates chemical heterogeneity in the deep mantle because of different chemistry of basaltic crust and harzburgite from pyrolite. Because the basaltic crust contains higher concentrations of Si, Al, Ca, Fe, and alkali elements than pyrolite, several aluminous phases other than aluminous perovskite exist in subducting basaltic materials in lower-mantle conditions. Structures, phase relations, and properties of the aluminous phases have been studied recently.

The purpose of this paper is to review the experimental studies on phase transitions of mantle minerals and bulk mantle rocks that have been carried out in the past several years. Most of the results discussed here were obtained by multianvil high-pressure experiments. First, we describe recent studies on the simple systems Mg_2SiO_4-Fe_2SiO_4 and $Mg_4Si_4O_{12}$-$Mg_3Al_2Si_3O_{12}$ as models of the olivine and pyroxene-garnet system, respectively. Next, a brief review on phase transitions of pyrolite and basalt is given. Perovskite in the lower mantle is particularly focused on defect chemistry. Several aluminous phases stable in lower-mantle conditions are discussed with respect to candidates of host phases of Al, Na, and K. Finally, some perspectives on important open questions for mantle mineralogy are given.

Phase Transitions in the System Mg_2SiO_4-Fe_2SiO_4

The high-pressure phase relations in the system Mg_2SiO_4-Fe_2SiO_4 among olivine, wadsleyite, and ringwoodite have been intensively studied since 1960s. By the end of 1980s, the α-β-γ phase relations in the whole compositional range of the Mg_2SiO_4-Fe_2SiO_4 system had been examined in detail by high-pressure experiments and by thermodynamic studies. Figure 1 shows equilibrium phase boundaries at 1400 °C for the α-β-γ transitions calculated using measured thermodynamic data (Akaogi et al., 1989) and transition pressures based on quench experiments (Katsura and Ito, 1989). Using the high-pressure phase relations for the α-β-γ transitions of the Mg_2SiO_4-Fe_2SiO_4 system and elastic properties of the three phases, seismic discontinuities and chemical compositions of the upper mantle and transition zone have been discussed (see Agee, 1999, for review).

In recent years, several attempts have been made to determine more precisely the phase relations using in situ X-ray observations on the basis of internal pressure standards. The α-β and β-γ transition boundaries in Mg_2SiO_4 were determined by Morishima et al. (1994) and Suzuki et al. (2000), respectively, at temperature below ~1000–1200 °C using a NaCl pressure scale. The determined boundaries have somewhat more positive pressure-temperature (dP/dT) slopes than those by Katsura and Ito

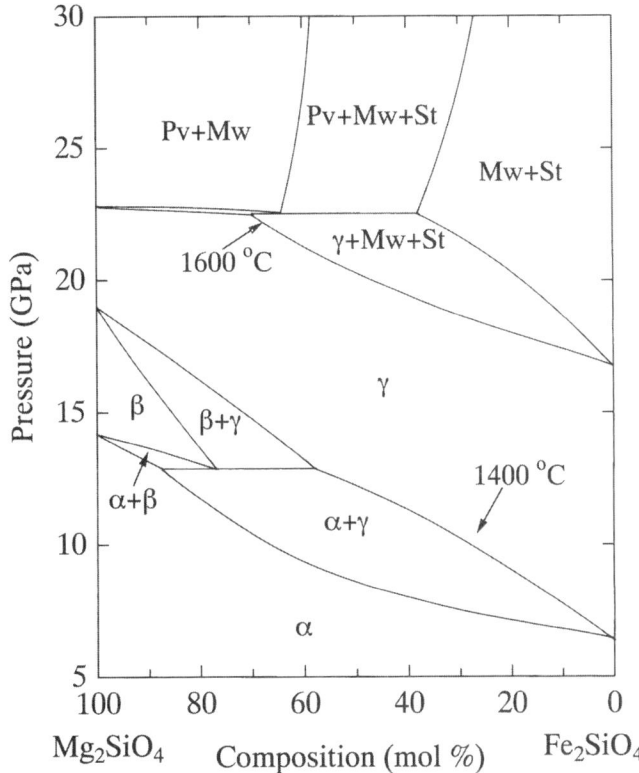

Figure 1. High-pressure phase relations in the system Mg_2SiO_4-Fe_2SiO_4. Olivine(α)–wadsleyite(β)–ringwoodite(γ) transition boundaries at 1400 °C were calculated using thermodynamic data from Akaogi et al. (1989), together with phase equilibrium data from Katsura and Ito (1989). The postspinel transition boundaries at 1600 °C were calculated using thermodynamic data from Akaogi et al. (1998a), with phase equilibrium data from Ito and Takahashi (1989). Pv—perovskite, Mw—magnesiowüstite, St—stishovite.

transition of mantle olivine composition yields a depth interval of more than 10 km, which is larger than those from seismic observations. Therefore, possible effects on the transition width have been experimentally and theoretically examined. It was found that wadsleyite and ringwoodite can contain water up to ~3 wt% as hydroxyl groups in the structures (Inoue et al., 1995; Kohlstedt et al., 1996). Presence of water makes the α-β transition interval wider than that in dry mantle due to OH in wadsleyite (Wood, 1995; Smyth and Frost, 2002). On the contrary, Mg-Fe partitioning between α, β, and garnet (Irifune and Isshiki, 1998; Frost, 2003a) makes the α-β transition interval narrower, and a nonlinear gradient in elastic properties across the α-β transition (Stixrude, 1997) allows the seismic discontinuity to be sharper compared with the transition interval simply estimated from the α-β transition loop for the fixed composition. The thickness of the discontinuity estimated seismologically may be biased to a narrower one than the true transition interval (Helffrich and Wood, 1996). Based on thermodynamically calculated α-β transition boundaries, Frost (2003b) indicated that these effects, particularly the Mg-Fe partitioning between α, β, and garnet, make the transition width as small as 6 km and that the width of the 410 km discontinuity can be explained by the olivine-wadsleyite transition in dry pyrolite.

Compared with the 410 km and 660 km discontinuities, seismological studies indicate that the 520 km discontinuity is small and has a rather broad velocity gradient of a 20–50 km range (Revenaugh and Jordan, 1991). It has also been suggested that the 520 km discontinuity may not be global but intermittent (Deuss and Woodhouse, 2001). The phase relations in the system Mg_2SiO_4-Fe_2SiO_4 indicate that the β-γ transition would occur at pressures corresponding to the 520 km discontinuity. Initial studies on elasticity of β- and γ-Mg_2SiO_4 have indicated that elastic properties of the two phases are very similar at ambient conditions, suggesting that the β-γ transition would not produce the seismic discontinuity (Sawamoto et al., 1984; Weidner et al., 1984). However, recent experimental measurements and theoretical studies on dependences of pressure, temperature, and Mg/Fe ratio on the elastic properties of β- and γ-$(Mg,Fe)_2SiO_4$ reveal that velocity jumps and impedance contrasts of the β-γ transition for mantle olivine can produce a visible discontinuity (Sinogeikin et al., 2003). By Mg-Fe partitioning among the coexisting β and γ phases and majoritic garnet, the β-γ transition width becomes narrower than that for the single β phase of the fixed Mg/Fe ratio, and the width is consistent with the seismic observation (Sinogeikin et al., 2003; Frost, 2003b). A seismological study by Deuss and Woodhouse (2001) on splitting of the 520 km discontinuity in a number of regions may suggest that the discontinuity corresponds to not only the β-γ transition but also exsolution of $CaSiO_3$ perovskite from majorite, as discussed later.

The postspinel transition of Mg-rich $(Mg,Fe)_2SiO_4$ spinel to $(Mg,Fe)SiO_3$ perovskite and $(Mg,Fe)O$ magnesiowüstite is accepted to be responsible for the 660 km discontinuity (Ito and Takahashi, 1989). Figure 1 shows equilibrium phase relations for the postspinel transition in the system Mg_2SiO_4-Fe_2SiO_4 at

(1989) and Akaogi et al. (1989). However, extrapolated pressures to 1400–1600 °C are almost the same as those measured by Katsura and Ito (1989) and by Akaogi et al. (1989). Therefore, a temperature profile estimated by Akaogi et al. (1989) could still be used as an average geotherm in the transition zone, in which temperatures at 410 and 660 km are ~1400 and 1600 °C, respectively.

The pressure interval of the olivine-wadsleyite transition is of special interest because seismic studies (e.g., Benz and Vidale, 1993) reveal that the thickness of the 410 km discontinuity varies considerably in different regions and that, in some areas, the discontinuity width is less than only 6 km, which is narrower than that expected from the previous α-β phase relation. Therefore, precise determination of the olivine-wadsleyite transition width and exploration of factors affecting the transition width may provide important constraints on temperature and chemical composition of the mantle near the 410 km depth. The olivine-wadsleyite transition interval was carefully examined by quench experiments (Fei and Bertka, 1999) and by in situ X-ray observation (Katsura et al., 2004). Both of the studies indicated that the α-β

1600 °C calculated by measured thermodynamic data by Akaogi et al. (1998a) in combination with phase relations by Ito and Takahashi (1989). The thermodynamic study (Akaogi et al., 1998a) confirmed the results by Ito and Takahashi (1989) that the postspinel transition boundary in the system has a negative dP/dT slope and that the pressure interval of the transition is <0.2 GPa, which is consistent with the very sharp seismic discontinuity at 660 km depth.

Using in situ X-ray multianvil experiments based on Anderson et al.'s (1989) gold pressure scale, Irifune et al. (1998) showed that the postspinel transition of Mg_2SiO_4 occurred at pressures 2 GPa lower than that of the 660 km depth, suggesting that the 660 km discontinuity could not be attributed to the postspinel transition and that mantle composition might be significantly different than currently thought. After this result was reported, several experimental studies were made to examine the accuracy of postspinel phase relations as well as of the pressure scales. In laser-heated diamond anvil cell experiments using platinum and ruby pressure scales, respectively, Shim et al. (2001) and Chudinovskikh and Boehler (2001) confirmed that the postspinel transition of Mg_2SiO_4 occurred at a pressure corresponding to the 660 km depth. Fei et al. (2004) demonstrated that the postspinel transition pressure of Mg_2SiO_4 based on the MgO pressure scale by Speziale et al. (2001) is consistent with the 660 km depth, but that the pressure using Anderson et al.'s (1989) gold scale is ~2 GPa lower than the 660 km depth. Although accuracies of the various pressure scales should be evaluated in more detail in the future, these experimental studies support the idea that the postspinel transition is responsible for the 660 km discontinuity, in contrast to Irifune et al.'s (1998) conclusion.

The effect of the negative slope of the postspinel transformation boundary on mantle convection has been extensively discussed. It has been established using studies of mantle convection simulation that a slope of about –3 MPa/K, determined by high-pressure experiments (Ito and Takahashi, 1989) and thermodynamic calculations (Ito et al., 1990; Akaogi and Ito, 1993), would work as a partial resistance to the mantle convection but would not impede the whole mantle convection (Christensen, 1995). This conclusion is consistent with the results from seismic tomography studies on behaviors of subducting slabs in the transition zone and the lower mantle (Fukao et al., 2001). However, less negative slopes for the postspinel transition have been reported recently by in situ X-ray multianvil experiments (Katsura et al., 2003; Fei et al., 2004). If this is the case, the effect of partial resistance to mantle convection would be less important. Because the experimental determination of the postspinel transition slope depends on accuracy in equations of state of the pressure standard materials, further studies would be necessary to refine both of pressure scales and the boundary slope.

The natural occurrence of $(Mg,Fe)SiO_3$ ilmenite and perovskite was reported for the first time in shocked meteorites by Tomioka and Fujino (1997) and Sharp et al. (1997). Both of the ilmenite- and perovskite-structured $(Mg,Fe)SiO_3$ phases have not yet been found in mantle-derived minerals and rocks, though some minerals with pyroxene compositions were interpreted as former perovskite by several investigators (Harte et al., 1999; Collerson et al., 2000).

It should be noted that the previous discussions on the α-β-γ transitions and the postspinel transition in the mantle were based on equilibrium phase relations determined by experiments and thermodynamic calculations. In the recent decade, kinetics and mechanisms of the transitions have been experimentally investigated, because they are particularly important in subduction zones due to the lower temperature than that in the surrounding mantle. Rubie and Ross (1994) concluded that metastable $(Mg_{0.9},Fe_{0.1})_2SiO_4$ olivine survives to greater depth than 550 km in cold slabs, and it transforms directly to ringwoodite under control of growth kinetics. Kerschhofer et al. (2000) studied the kinetics and mechanism of the intracrystalline olivine-ringwoodite transition in subducting slabs. Kubo et al. (2002) examined the kinetics and mechanism of the postspinel transition, suggesting that an overpressure of ~1 GPa is needed for the postspinel transition at ~700 °C, which is a possible lowest temperature at the 660 km depth. These results have been used to estimate buoyancy effects on subducting slabs.

PHASE TRANSITIONS OF PYROXENES AND GARNET

With increasing pressure, Ca-poor pyroxene (orthopyroxene) and Ca-rich pyroxene (clinopyroxene) are gradually dissolved into pyrope-rich garnet to form majorite, which is the most abundant mineral next to wadsleyite or ringwoodite in the transition zone. Majorite is expressed as a garnet-structured solid solution (s.s.) between end members of $M_4Si_4O_{12}$ and $M_3Al_2Si_3O_{12}$ (M = Mg, Fe, Ca). Since Moore and Gurney (1985) found majorite crystals as inclusions in kimberlitic diamonds, several occurrences of majorite from the mantle have been reported (e.g., Stachel, 2001). Based on high-pressure phase relations, it was interpreted that the majorites were derived from the transition zone.

Phase transformations of garnet end members and majorite have been examined in the past decade. At pressures of 25–27 GPa, $Mg_3Al_2Si_3O_{12}$ pyrope dissociates to $MgSiO_3$-rich perovskite s.s. + Al_2O_3-rich corundum s.s. at 1400–1800 °C, while pyrope first transforms to ilmenite and subsequently dissociates into perovskite s.s. + corundum s.s. below ~1200 °C (Irifune et al., 1996; Kubo and Akaogi, 2000; Akaogi et al., 2002). $Fe_3Al_2Si_3O_{12}$ almandine dissociates into a mixture of FeO wüstite, Al_2O_3 corundum, and stishovite at ~21 GPa (Conrad, 1998; Akaogi et al., 1998b). High-pressure phase relations in the system $Mg_4Si_4O_{12}$-$Mg_3Al_2Si_3O_{12}$ have been studied in detail because the system may serve as a simple model for transitions of majorite in the mantle (Kubo and Akaogi, 2000; Hirose et al., 2001). Results of the two studies are generally consistent when differences in adopted pressure scales and in experimental temperatures are considered. Figure 2A shows phase relations of the system at 1600 °C by Kubo and Akaogi (2000), which indicate complicated transition behaviors. In a $Mg_4Si_4O_{12}$-rich composition, an assemblage of γ-Mg_2SiO_4 + stishovite + majorite changes to ilmenite s.s. + majorite, further to ilmenite s.s., and finally to perovskite s.s., with increasing pressure. On the other hand, majorite with a $Mg_3Al_2Si_3O_{12}$-rich composition transforms to perovskite s.s. + majorite, and subsequently to perovskite s.s. + corundum s.s. When pyrolite mantle is adopted, majorite composition in the transition zone can be approximated as ~60 mol% $Mg_4Si_4O_{12}$–40 mol% $Mg_3Al_2Si_3O_{12}$. Therefore, it is expected that majorite transforms to perovskite s.s., intervening the field of majorite + perovskite s.s. at ~23–27 GPa along the normal mantle geotherm where temperature at 600 km depth is assumed to be 1600 °C with a temperature gradient of 0.35 °C/km. The majorite-perovskite transition may correspond to a steep seismic velocity gradient at depths of 660–720 km (Kubo and Akaogi, 2000).

Compared with the transitions along the normal mantle geotherm, phase transitions of majorite along a lower-temperature profile, such as that in subduction zones, are considerably different. Figure 2B shows phase relations (Akaogi et al., 2002) in the system $Mg_4Si_4O_{12}$-$Mg_3Al_2Si_3O_{12}$ at 1000 °C, which may be a representative temperature in subducting slabs in the transition zone. A complete solid solution of ilmenite is formed in the system $Mg_4Si_4O_{12}$-$Mg_3Al_2Si_3O_{12}$ at 20–27 GPa at 1000 °C, in contrast to ilmenite solid solution compositionally limited only to ~10 mol% $Mg_3Al_2Si_3O_{12}$ at 1600 °C in Figure 2A. Figures 2A and 2B indicate that the stability field of ilmenite s.s. expands toward $Mg_3Al_2Si_3O_{12}$ with decreasing temperature. Figure 3 shows mineral proportions in pyrolite at 570–770 km depth along the low-temperature profile where temperature is assumed to be 600 degrees lower than the normal mantle geotherm (Akaogi et al., 2002). The mineral proportions in Figure 3 were estimated from experimental results on postspinel transitions and majorite-ilmenite-perovskite transitions described here previously. As shown in the figure, formation of ilmenite may act as a buoyancy force for subducting slabs. Also, stability of ilmenite in a subducting slab would produce two steep gradients of seismic velocities associated with the majorite-ilmenite and ilmenite-perovskite transitions located, respectively, just above and below the abrupt velocity jump due to the postspinel transition.

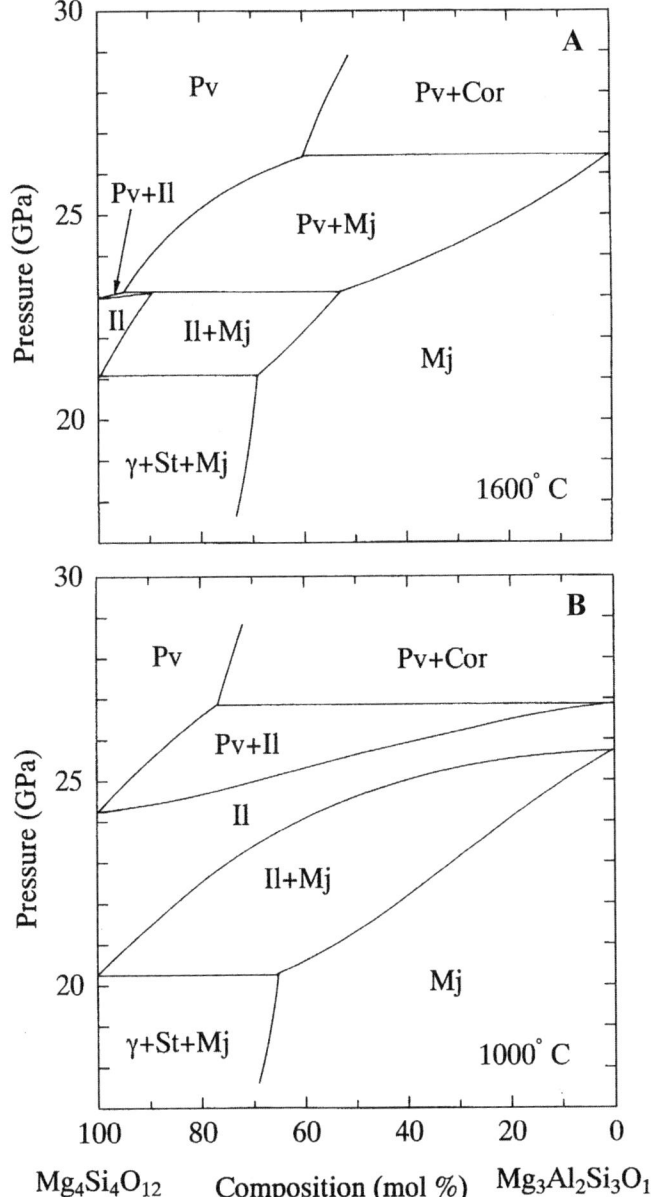

Figure 2. Phase relations in the system $Mg_4Si_4O_{12}$-$Mg_3Al_2Si_3O_{12}$ at (A) 1600 °C (modified from Kubo and Akaogi, 2000), and at (B) 1000 °C (modified from Akaogi et al., 2002). Mj—majorite, Il—ilmenite, Pv—perovskite, γ—ringwoodite, St—stishovite, Cor—corundum.

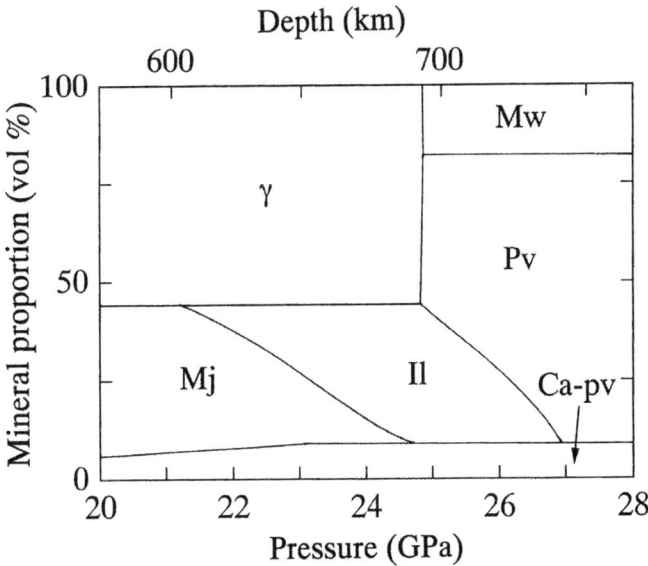

Figure 3. Mineral proportions in pyrolite at 570–770 km depth along the lower-temperature profile (modified from Akaogi et al., 2002). The lower-temperature profile is 600 degrees lower than the normal geotherm (see text). Mj—majorite, γ—ringwoodite, Il—ilmenite, Pv—Mg-perovskite, Ca-pv—Ca-perovskite, Mw—magnesiowüstite.

CaMgSi$_2$O$_6$ diopside is a major component of Ca-rich pyroxene of the upper mantle. Although several phase transition studies on diopside were reported, the results were controversial (Canil, 1994; Oguri et al., 1997). Very recently, detailed phase relations were reported for a wide pressure-temperature (P-T) range by Akaogi et al. (2004), the results of which are shown in Figure 4. At 17–18 GPa at temperatures below 1400 °C, diopside dissociates into CaSiO$_3$ perovskite and a MgSiO$_3$ component. The latter transforms to β-Mg$_2$SiO$_4$ + stishovite, γ-Mg$_2$SiO$_4$ + stishovite, ilmenite and perovskite with increasing pressure. Above 1400 °C, diopside first dissociates to Mg-rich (Mg,Ca)SiO$_3$ majorite and CaSiO$_3$ perovskite at ~18 GPa.

In pyrolite mantle, calcium is incorporated into Ca-rich pyroxene and garnet in the upper mantle. The Ca-rich pyroxene is dissolved into garnet at a depth range near the upper part of the transition zone. At pressures around 20 GPa, the Ca component is separated as CaSiO$_3$ perovskite from majorite. Although the phase relation on the exsolution of CaSiO$_3$ perovskite has not yet been examined in detail, dissociation pressures of CaMgSi$_2$O$_6$ diopside to a mixture of Mg-rich (Mg,Ca)SiO$_3$ majorite and CaSiO$_3$ perovskite shown in Figure 4 may represent the lower bound for the exsolution pressure of CaSiO$_3$ perovskite from mantle majorite.

Phase transition studies on CaSiO$_3$ indicate that CaSiO$_3$ wollastonite transforms to CaSiO$_3$ walstromite at ~3 GPa, and dissociates at ~9 GPa into an assemblage of Ca$_2$SiO$_4$ larnite and CaSi$_2$O$_5$ titanite, both of which further change to CaSiO$_3$ perovskite at ~14 GPa (Essene, 1974; Huang and Wyllie, 1975; Gasparik et al., 1994; Akaogi et al., 2004). Figure 5 shows the phase transitions in CaSiO$_3$ newly determined by Akaogi et al. (2004), in which the formation pressure of CaSiO$_3$ perovskite is ~3 GPa higher than that calculated by Gasparik et al. (1994). The natural occurrence of high-pressure Ca-silicate inclusions in diamonds in Kankan, Guinea, was reported by Joswig et al. (1999). The inclusions were CaSiO$_3$ walstromite and an assemblage of Ca$_2$SiO$_4$ larnite and CaSi$_2$O$_5$ titanite. Figure 5 indicates that these Ca-silicates were equilibrated at depths of the stability fields of walstromite and of larnite + titanite, although these high-pressure Ca-silicates might have been transformed from CaSiO$_3$ perovskite, which was stable in the deeper part of the mantle (Joswig et al., 1999).

PHASE TRANSITIONS IN BULK ROCKS

In recent years, phase transitions in pyrolite, peridotites, and basalt have been examined under the P-T conditions of upper part of the lower mantle (Hirose et al., 1999; Wood, 2000; Ono et al., 2001; Hirose, 2002; Nishiyama and Yagi, 2003). In the studies using multianvil apparatus, mineral proportions were estimated using mineral compositions analyzed by electron microprobe analyzers and mass balance calculations. Figure 6A indicates the mineral proportions in pyrolite as a function of pressure or depth down to ~1100 km depth along the normal mantle geotherm. In phase equilibrium studies on pyrolite as well as peridotite of very similar composition to pyrolite, Wood (2000) and Hirose (2002) indicated that at ~1600 °C, phase transitions of constituent minerals occur at pressures similar to those observed in simple systems described in the previous sections. The postspinel transition occurs at ~23 GPa in a very narrow pressure interval, consistent with a sharp 660 km discontinuity. When the postspinel transition occurs, (Mg,Fe)-perovskite with low Al$_2$O$_3$ content is first produced. At 23–26 GPa, the perovskite and majorite are enriched

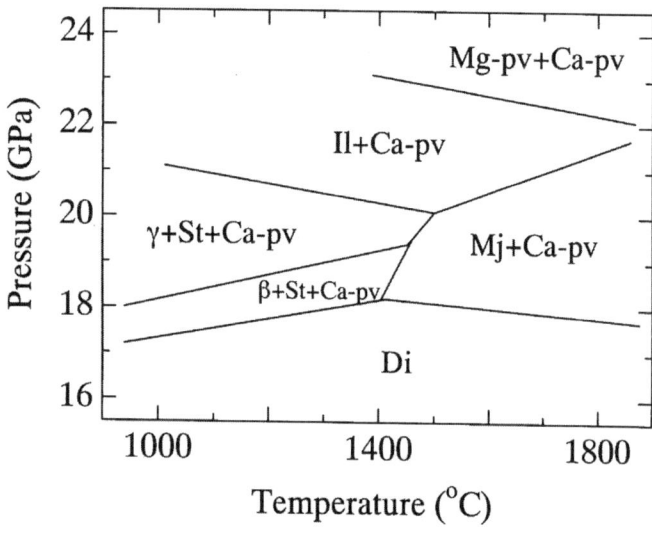

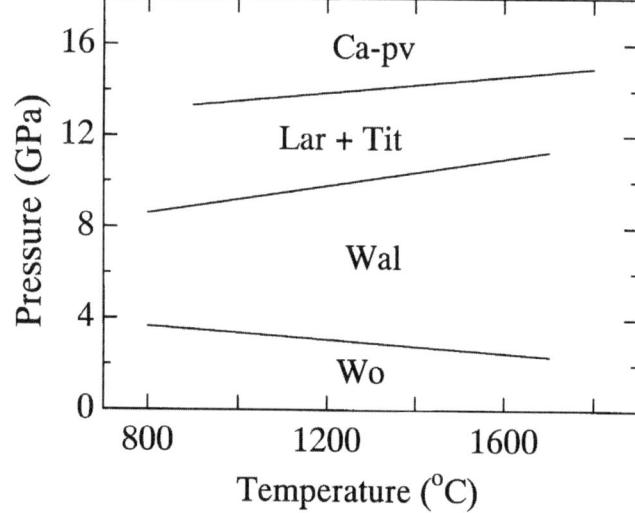

Figure 4. Phase relations in CaMgSi$_2$O$_6$ (modified from Akaogi et al., 2004). Mj—majorite, Di—diopside, β—wadsleyite, γ—ringwoodite, St—stishovite, Ca-pv—Ca-perovskite, Il—ilmenite, Mg-pv—Mg-perovskite.

Figure 5. Phase relations in CaSiO$_3$. Data sources are Essene (1974), Huang and Wyllie (1975), Gasparik et al. (1994), and Akaogi et al. (2004). Wo—CaSiO$_3$ wollastonite, Wal—CaSiO$_3$ walstromite, Lar—Ca$_2$SiO$_4$ larnite, Tit—CaSi$_2$O$_5$ titanite, Ca-pv—CaSiO$_3$ perovskite.

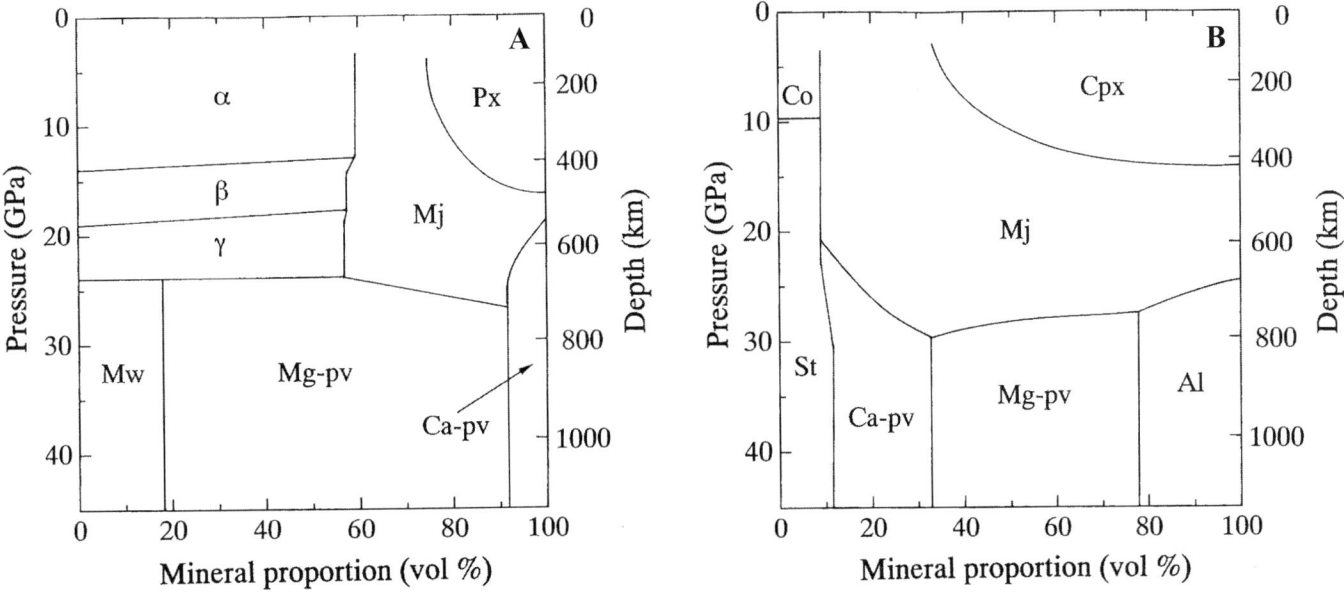

Figure 6. Mineralogical constitutions along the normal geotherm for (A) pyrolite and (B) mid-ocean-ridge basalt (MORB). Data sources are Irifune and Ringwood (1993), Hirose et al. (1999), Ono et al. (2001), and Aoki and Takahashi (2004). α—olivine, β—wadsleyite, γ—ringwoodite, Px—Ca-poor and Ca-rich pyroxenes, Mg-Pv—Mg-perovskite, Ca-pv—Ca-perovskite, Mw—magnesiowüstite, Co—coesite, St—stishovite, Mj—majorite, Cpx—Ca-rich clinopyroxene, Px—two pyroxenes, Cpx—clinopyroxene, Al—aluminous phase(s).

in Al_2O_3 with increasing pressure, and majorite disappears at ~26 GPa. The majorite-perovskite transition has a positive Clapeyron slope, in contrast to the postspinel transition, and it may produce a high gradient of seismic velocities at 660–720 km depth. $CaSiO_3$-rich perovskite is exsolved from majorite in the pressure range of ~20–25 GPa. It has been suggested that seismic observations on splitting of the 520 km discontinuity may be caused by two phase transitions: onset of $CaSiO_3$ perovskite exsolution at ~20 GPa and the β-γ transition of olivine (Deuss and Woodhouse, 2001).

It should be noted that the mineral proportions depend on temperature profile. Hirose (2002) and Nishiyama and Yagi (2003) observed that at 1800–2200 °C and 22–24 GPa the proportion of majorite in pyrolite increases with increasing temperature at the expense of ringwoodite, indicating that a reaction ringwoodite → majorite + magnesiowüstite occurs under those P-T conditions and that majorite becomes a dominant postspinel phase. At 30 GPa, (Mg,Fe,Al)-perovskite, $CaSiO_3$-rich perovskite, and magnesiowüstite coexist, and the mineral proportions do not change significantly with temperature. These experimental results suggest that in ascending high-temperature plumes, the perovskite → majorite transition would occur near the 660 km depth, which would have the effect of assisting plume upwelling due to its positive Clapeyron slope and low density of majorite (Hirose, 2002; Nishiyama and Yagi, 2003).

In oceanic lithosphere, mid-ocean-ridge basalt (MORB) and harzburgite overlie depleted pyrolite. The MORB layer has higher contents of SiO_2, Al_2O_3, CaO, FeO, and Na_2O than pyrolite. Therefore, the subducted MORB produces chemical heterogeneity in the deep mantle. High-pressure phase transition experiments of basalts under the P-T conditions of the lower mantle have been conducted in recent years. Figure 6B shows changes in mineral proportions in MORB with pressure or depth (Irifune and Ringwood, 1993; Hirose et al., 1999; Ono et al., 2001; Aoki and Takahashi, 2004). In transition zone conditions, Ca-rich pyroxene is dissolved in garnet, resulting in dominant majorite with small amounts of stishovite and $CaSiO_3$ perovskite, the latter of which begins to exsolve from majorite at ~20 GPa. The assemblage changes to (Mg,Fe,Al)-perovskite, $CaSiO_3$ perovskite, and aluminous phase at 25–28 GPa. The aluminous phase will be discussed in the later section. When an oceanic lithosphere is subducted into the mantle, a density difference appears between MORB (oceanic crust) and pyrolite due to the difference in mineralogy. Density calculations indicate that the assemblage of majorite, stishovite, and $CaSiO_3$ perovskite of MORB is denser than the mineral assemblage of surrounding pyrolite mantle at depths shallower than 660 km, and that the mineral assemblage of MORB is denser than pyrolite at depths deeper than ~720 km. However, at the depth range of 660–720 km, the assemblage of MORB is less dense than that of pyrolite. These density differences suggest that when oceanic lithospheres are subducted into the mantle, oceanic crust may be gravitationally trapped at 660–720 km, but would sink into deeper mantle if the slabs accumulate to sufficient thickness at the top of the lower mantle (Hirose et al., 1999).

PEROVSKITE IN THE LOWER MANTLE

As shown in Figure 6A, Mg-rich perovskite is the most abundant mineral in the pyrolitic lower mantle. Therefore, stability and physical properties of Mg-rich perovskite have been studied extensively in recent years. Phase relations and elastic properties of perovskite s.s. in the simple system $MgSiO_3$-$FeSiO_3$ were examined by a number of investigators (Hemley and Cohen, 1992, for review). The solubility of the $FeSiO_3$ component in perovskite increases with increasing pressure and temperature (Mao et al., 1997). In addition to $FeSiO_3$, Al_2O_3 is an important component in the perovskite solid solution in the lower mantle, as described in the previous sections. The incorporation of Al^{3+} allows perovskite to have different features from that of Fe^{2+}. There are two competing substitution mechanisms of Al^{3+} in $MgSiO_3$ perovskite. One is stoichiometric substitution:

$$Mg^{2+}_A + Si^{4+}_B = Al^{3+}_A + Al^{3+}_B, \qquad (1)$$

where subscripts A and B denote the eightfold A site and the octahedral B site, respectively, in perovskite structure. The other mechanism is nonstoichiometric substitution, leaving a vacancy in the oxygen site:

$$2Si^{4+}_B + O^{2-}_O = 2Al^{3+}_B + V_O, \qquad (2)$$

where subscript O denotes oxygen site, and V_O represents the oxygen site vacancy. Andrault et al. (1998) and Stebbins et al. (2001) showed that substitution mechanism 1 is dominant in perovskite solid solutions of stoichiometric compositions on the join $MgSiO_3$-Al_2O_3. However, Navrotsky et al. (2003) demonstrated that the nonstoichiometric substitution 2 prevails in perovskite solid solutions synthesized from starting materials of the $MgSiO_3$-$MgAlO_{2.5}$ system, indicating presence of defects in oxygen sites of the perovskite solid solutions. They demonstrated that the nonstoichiometric perovskites have lattice parameters and enthalpies different from those of corresponding stoichiometric perovskites in the system $MgSiO_3$-Al_2O_3. A ^{27}Al nuclear magnetic resonance (NMR) study by Stebbins et al. (2003) suggested that the defects are randomly distributed over oxygen sites of the perovskite structure. The defects in perovskite may have significant effects on thermodynamic and transport properties that should be clarified in more detail.

McCammon (1997) found by Mössbauer spectroscopy that ~50% of Fe was ferric in $(Mg,Fe)SiO_3$ perovskite containing 3.3 mol% of Al_2O_3. Lauterbach et al. (2000) observed a nonlinear increase of ferric iron content with increasing Al^{3+} content independently on oxygen fugacity. These results suggest that about half of total Fe in perovskite may be ferric in the lower mantle.

The dissolution of Al^{3+} and Fe^{3+} in $MgSiO_3$ perovskite could have considerable effects on physical and chemical properties of perovskite in the lower mantle. Zhang and Weidner (1999) reported a bulk modulus of Al-bearing $MgSiO_3$ perovskite that was a significantly smaller value than that of pure $MgSiO_3$ perovskite. The result implies that the lower-mantle composition so far estimated using elastic properties of Al-free $(Mg,Fe)SiO_3$ perovskite might be incorrect. Some subsequent studies indicated similar small bulk moduli for aluminous $MgSiO_3$ perovskites, while other studies revealed even higher values than that of pure $MgSiO_3$ perovskite (Andrault et al., 2001; Daniel et al., 2001; Kubo et al., 2000; Yagi et al., 2004). The controversy might arise from compression experiments using aluminous perovskites with different defect chemistry. Further study will be necessary to clarify the controversy by experiments on perovskite samples fully characterized on the defects.

Another interesting issue on defects in perovskite is possible incorporation of OH in the structure. Navrotsky (1999) proposed a possible mechanism for dissolution of hydrogen by protonation of oxygen vacancies of the defect perovskite:

$$V_O + O^{2-}_O + H_2O = 2OH^-_O, \qquad (3)$$

where the subscript O represents the oxygen site of perovskite structure. This indicates that OH^- could be incorporated in perovskite through the vacant sites. Using secondary ion mass spectroscopy and infrared spectroscopy, Murakami et al. (2002) reported that a significant amount of hydrogen was present in Mg-rich perovskite, Ca-perovskite, and magnesiowüstite in run products of natural peridotite composition. In contrast, Bolfan-Casanova et al. (2003) detected very low hydrogen content in aluminous (Mg,Fe)-perovskite by infrared spectroscopy. Therefore, further study would be needed to examine whether incorporation of OH^- is possible in lower-mantle perovskite.

Several hydrous silicates stable in *P-T* conditions of the transition zone and lower mantle have been examined in recent years. Because detailed discussion on the stability of the hydrous silicates is out of the scope of this review, we only refer to some recent papers (Litasov and Ohtani, 2003; Ohtani et al., 2004).

ALUMINOUS PHASES AND SILICA PHASES STABLE IN LOWER-MANTLE CONDITIONS

Mg-rich perovskite is the important host phase of aluminum in the lower mantle. However, as shown in Figure 6B, other aluminous phase(s) (~20% in volume) coexist with Mg-rich perovskite, $CaSiO_3$ perovskite, and stishovite in subducted MORB in lower-mantle conditions (Irifune and Ringwood, 1993; Kesson et al., 1994; Hirose et al., 1999; Ono et al., 2001). Irifune and Ringwood (1993) suggested that the aluminous phase with calcium ferrite structure was stable above 25 GPa, and Kesson et al. (1994) reported stability of calcium ferrite phase at 45–100 GPa. In a study of the $MgAl_2O_4$-$CaAl_2O_4$ system, Akaogi et al. (1999) found a new hexagonal aluminous phase with $2MgAl_2O_4 \cdot CaAl_2O_4$ composition, and suggested that the aluminous phase observed by Irifune and Ringwood (1993) was not of calcium ferrite structure but of the same as the hexagonal aluminous phase. The structure of $2MgAl_2O_4 \cdot CaAl_2O_4$ hexagonal aluminous phase was analyzed by Miura et al. (2000), and it is shown in Figure 7A in comparison with $MgAl_2O_4$ calcium ferrite of orthorhombic symmetry in Figure 7B (Kojitani et al., 2007). Both of the structures are based on double-octahedral chains running parallel to the *c*-axis with tunnels surrounded by the chains. The hexagonal phase has three kinds of cation sites: Ca ions occupy half of ninefold coordination sites, the other half of which are vacant, while Mg and Al occupy two distinct octahedral sites. In calcium ferrite structure, Mg and Al are in eightfold sites and octahedral sites, respectively.

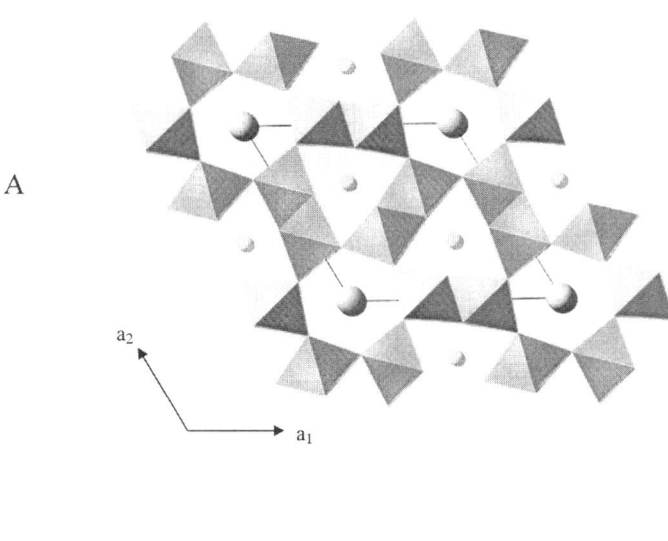

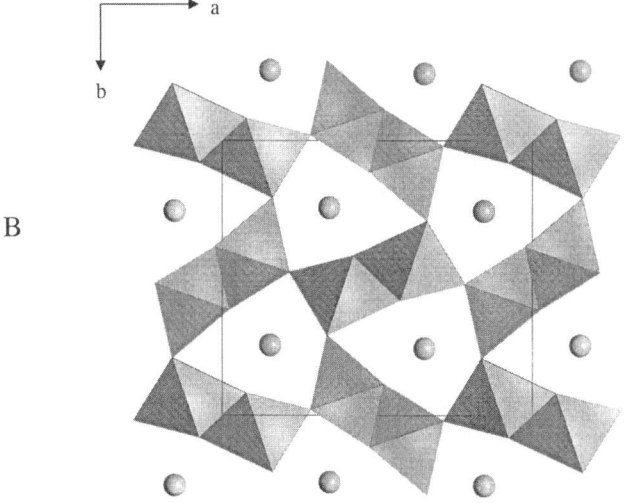

Figure 7. Structures of (A) $Mg_2CaAl_6O_{12}$ hexagonal aluminous phase (modified from Miura et al., 2000) and (B) $MgAl_2O_4$ calcium ferrite phase (Kojitani et al., 2007). In both of the structures, c-axis is perpendicular to the plane. In the structure of hexagonal aluminous phase, larger circles represent Ca, smaller circles represent Mg, and octahedra represent AlO_6. In the structure of calcium ferrite, circles show Mg, and octahedra show AlO_6. In both of the structures, double chains of octahedra run parallel to the *c*-axis.

Miyajima et al. (1999) reported that NAL (new aluminous) phase previously found in high-pressure assemblages of natural garnets has the structure identical to that of the hexagonal phase of $2MgAl_2O_4 \cdot CaAl_2O_4$ composition. We observed that the hexagonal phases with compositions of $MgAl_2O_4 \cdot NaAlSiO_4$ and $2MgAl_2O_4 \cdot KAlSiO_4$ are stable above ~15 GPa (Kobayashi et al., 2000; Iwabuchi et al., 2001), which suggests that the hexagonal phase could be a host of both of alkali and alkali-earth elements in the deep mantle. Miyajima et al. (2001) showed that, in run products of MORB synthesized at ~30 GPa, the hexagonal phase contained substantial amounts of Na and K, while the calcium ferrite phase preferentially incorporated Na. Guignot and Andrault (2004) reported that both the hexagonal aluminous phase and calcium ferrite phase are stable up to ~60 GPa, suggesting stability of the two phases in the deep lower mantle.

Subduction of sediments overlying oceanic crust and part of the continental crust has been discussed in geochemical point of view. High-pressure phase transitions of these silicic materials were examined by Irifune et al. (1994). They found that $KAlSi_3O_8$-rich hollandite (~30 vol%) and a Ca,Al,Si-rich phase (CAS) (~10 vol%) coexist with stishovite and garnet at pressures of the transition zone due to high concentrations of Si, K, Al, and Ca in materials of continental crust and sediments. Yagi et al. (1994) and Urakawa et al. (1994) reported phase relations in $KAlSi_3O_8$ in which hollandite is stable above 9 GPa. The hollandite structure accommodates relatively large cations such as K and Na in eightfold coordination sites. Further, Tutti et al. (2001) reported that $KAlSi_3O_8$ hollandite was synthesized in the experiments at pressures up to 95 GPa. However, a very recent study has indicated that a second-order transition of hollandite to hollandite-II occurs in $KAlSi_3O_8$ at around 25–30 GPa (Nishiyama et al., 2005). The experimental results reveal that hollandite-II of $KAlSi_3O_8$ would be stable in the wide pressure conditions of the lower mantle, while $KAlSi_3O_8$ hollandite is stable in the depth range shallower than the uppermost lower mantle. The CAS phase with the ideal composition of $CaAl_4Si_2O_{11}$ has the barium ferrite structure that contains octahedral (Al,Si) sites with some 5-fold (Si, Al) sites (Gautron et al., 1999). Although the stability field of CAS phase has not yet been tightly constrained, it would be stable in the transition zone pressure below ~1500 °C (Gautron et al., 1996).

Among the aluminous phases discussed here, hollandite and CAS phases have been discovered in shocked meteorites. $NaAlSi_3O_8$-rich hollandite was found in L6 chondrites (Gillet et al., 2000; Tomioka et al., 2000). $KAlSi_3O_8$-rich hollandite was found in a shocked Martian meteorite (Langenhorst and Poirier, 2000). Occurrence of the CAS phase in the system $CaAl_2Si_2O_{11}$-$NaAl_3Si_3O_{11}$ in a shocked Martian meteorite was reported by Beck et al. (2004).

High-pressure polymorphs of SiO_2 have been paid special attention as indicators of high pressure for rocks derived from the deep mantle and for meteorites and terrestrial rocks that have undergone shock events. Quartz transforms to coesite at ~3 GPa, which further transforms to stishovite at ~9 GPa. Precise determinations of the quartz-coesite and coesite-stishovite transition boundaries have been made because the two transitions have been widely used as pressure standards at high temperature (Bohlen and Boettcher, 1982; Zhang et al., 1996). The phase transition diagram in SiO_2 is shown in Figure 8. Both coesite and stishovite were first found in nature in the shocked terrestrial sandstones of Meteor Crater, Arizona (Chao et al., 1960, 1962). Also, coesite was found in high-pressure metamorphic rocks as well as in diamonds in some eclogites (Chopin, 1984; Meyer, 1987), and stishovite was found in several shocked meteorites (Langenhorst and Poirier, 2000).

Post-stishovite transitions have been studied from a viewpoint of a possible candidate of constituent minerals of the deep mantle. Diamond anvil experiments have indicated that a displacive transformation of stishovite to a $CaCl_2$-type structure occurs at around 50 GPa (Tsuchida and Yagi, 1989; Kingma et al., 1995; Ono et al., 2002). Phase transition of the $CaCl_2$-type phase to α-PbO_2–type phase was found at ~70–120 GPa by laser-heated diamond anvil experiments (Dubrovinsky et al., 1997; Murakami et al., 2003). The two transitions in SiO_2 are shown in Figure 8. The α-PbO_2–type SiO_2 was found in a shocked meteorite (El Goresy et al., 2000). The $CaCl_2$-type and α-PbO_2-structured phases may be present in the lower mantle, if basalt and continental crust materials are subducted into the deep lower mantle. However, the post-stishovite phases would not be stable in pyrolite mantle, because Mg-rich perovskite is stable up to ~120 GPa, at which point a new post-perovskite phase with $CaIrO_3$-type structure becomes stable, rather than dissociation into constituent oxides (Murakami et al., 2004; Oganov and Ono, 2004).

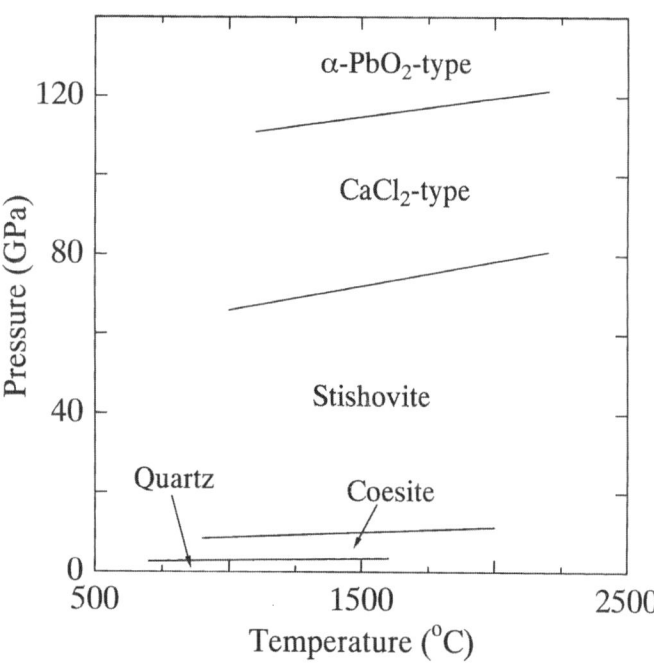

Figure 8. Phase transitions in SiO_2. Data sources are Bohlen and Boettcher (1982), Zhang et al. (1996), Ono et al. (2002), and Murakami et al. (2002). Stability fields of tridymite and cristobalite are not shown.

SOME PERSPECTIVES

In recent years, significant progress has been achieved in mantle mineralogy, as described herein. Phase transitions of mantle minerals have been clarified in detail for pyrolite, basalt, and continental crust materials under P-T conditions down to the deep lower mantle. Effects of the mineral phase transitions on the mantle dynamics have been extensively discussed. Also, several high-pressure silicate minerals stable under deep mantle conditions have been discovered in natural shocked meteorites. However, several important questions on mantle mineralogy still remain to be solved.

Concerning high-pressure techniques, pressure determination above ~20 GPa at high temperature is not accurate enough in current experimental techniques. Accurate determination of pressure is indispensable in order to establish phase relations that can be applied to more quantitatively clarify mineralogy and the thermal state of the transition zone and lower mantle, as well as to more accurately evaluate the effects of phase transitions on mantle dynamics. Simultaneous X-ray diffraction and acoustic measurements were carried out to establish accurate pressure scale at room temperature (Zha et al., 2000). It is important to extend the simultaneous measurements to high temperature for accurate determination of P-V-T relationships of appropriate pressure standard materials. Because the multianvil apparatus can be used to compress a sample of much larger volume than that of diamond anvil cell, technical innovation in the multianvil experiments with sintered diamond anvils is important to generate pressures corresponding to the core-mantle boundary and the core.

Possible presence of water in lower-mantle minerals, particularly in perovskite, should be examined in more detail. In relation to the incorporation of water, effects of trivalent cations (Al^{3+} and Fe^{3+}) on structure and properties of perovskite should also be studied. Although a possible spin transition of Fe^{2+} and Fe^{3+} from high spin state to low spin state was proposed more than three decades ago (Gaffney and Anderson, 1973), experimental evidences of the transition at 70–120 GPa have been reported very recently (Badro et al., 2004). Because Fe^{2+} and Fe^{3+} of low spin state in perovskite and magnesiowüstite would behave in different manners from those of high spin state, this high-low spin transition may affect seismic structures, dynamics, and chemical differentiation in the deep lower mantle. This should be examined in more detail. As shown here briefly, the post-perovskite transition of Mg-rich perovskite has been observed at P-T conditions near the lowermost mantle. Extensive investigations on the post-perovskite transition are currently being undertaken to clarify the nature of the D″ layer.

ACKNOWLEDGMENTS

The author thanks H. Kojitani for discussion and drawing Figure 7. Constructive comments by two anonymous reviewers were useful to improve the manuscript. This work was supported in part by Grant-in-aid of the Scientific Research (A) (No. 15204049) of the Japan Society for the Promotion of Science to M. Akaogi.

REFERENCES CITED

Agee, C.B., 1999, Phase transformations and seismic structure in the upper mantle and transition zone, *in* Hemley, R.J., ed., Ultrahigh-Pressure Mineralogy: Reviews in Mineralogy, v. 37, p. 165–203.

Akaogi, M., and Ito, E., 1993, Refinement of enthalpy measurement of $MgSiO_3$ perovskite and negative pressure-temperature slopes for perovskite-forming reactions: Geophysical Research Letters, v. 20, p. 1839–1842.

Akaogi, M., Ito, E., and Navrotsky, A., 1989, The olivine-modified spinel-spinel transitions in the system Mg_2SiO_4-Fe_2SiO_4: Calorimetric measurements, thermochemical calculation, and geophysical application: Journal of Geophysical Research, v. 94, p. 15,671–15,685.

Akaogi, M., Kojitani, H., Matsuzaka, K., Suzuki, T., and Ito, E., 1998a, Postspinel transformations in the system Mg_2SiO_4-Fe_2SiO_4: Element partitioning, calorimetry, and thermodynamic calculation, *in* Manghnani, M.H., and Yagi, T., eds., Properties of Earth and Planetary Materials at High Pressure and Temperature: Washington, D.C., American Geophysical Union Geophysical Monograph, v. 101, p. 373–384.

Akaogi, M., Ohmura, N., and Suzuki, T., 1998b, High pressure dissociation of $Fe_3Al_2Si_3O_{12}$ garnet: Phase boundary determined by phase equilibrium experiments and calorimetry: Physics of the Earth and Planetary Interiors, v. 106, p. 103–113, doi: 10.1016/S0031-9201(97)00084-8.

Akaogi, M., Hamada, Y., Suzuki, T., Kobayashi, M., and Okada, M., 1999, High pressure transitions in the system $MgAl_2O_4$-$CaAl_2O_4$: A new hexagonal aluminous phase with implication for the lower mantle: Physics of the Earth and Planetary Interiors, v. 115, p. 67–77, doi: 10.1016/S0031-9201(99)00076-X.

Akaogi, M., Tanaka, A., and Ito, E., 2002, Garnet-ilmenite-perovskite transitions in the system $Mg_4Si_4O_{12}$-$Mg_3Al_2Si_3O_{12}$ at high pressures and high temperatures: Phase equilibria, calorimetry and implications for mantle structure: Physics of the Earth and Planetary Interiors, v. 132, p. 303–324, doi: 10.1016/S0031-9201(02)00075-4.

Akaogi, M., Yano, M., Tejima, Y., Iijima, M., and Kojitani, H., 2004, High-pressure transitions of diopside and wollastonite: Phase equilibria and thermochemistry of $CaMgSi_2O_6$, $CaSiO_3$ and $CaSi_2O_5$-$CaTiSiO_5$ system: Physics of the Earth and Planetary Interiors, v. 143–144, p. 145–156, doi: 10.1016/j.pepi.2003.08.008.

Anderson, O.L., Isaak, D.G., and Yamamoto, S., 1989, Anharmonicity and the equation of state for gold: Journal of Applied Physics, v. 65, p. 1534–1543, doi: 10.1063/1.342969.

Andrault, D., Neuville, D.R., Flank, A.M., and Wang, Y., 1998, Cation sites in Al-rich $MgSiO_3$ perovskites: The American Mineralogist, v. 83, p. 1045–1053.

Andrault, D., Bolfan-Casanova, N., and Guignot, N., 2001, Equation of state of lower mantle (Al,Fe)-$MgSiO_3$ perovskite: Earth and Planetary Science Letters, v. 193, p. 501–508, doi: 10.1016/S0012-821X(01)00506-4.

Aoki, I., and Takahashi, E., 2004, Density of MORB eclogite in the upper mantle: Physics of the Earth and Planetary Interiors, v. 143–144, p. 129–143, doi: 10.1016/j.pepi.2003.10.007.

Badro, J., Rueff, J.P., Vanko, G., Monaco, G., Fiquet, G., and Guyot, F., 2004, Electronic transitions in perovskite: Possible nonconvecting layers in the lower mantle: Science, v. 305, p. 383–386, doi: 10.1126/science.1098840.

Beck, P., Gillet, P., Gautron, L., Daniel, I., and El Goresy, A., 2004, A new natural high-pressure (Na,Ca)-hexaluminosilicate [$(Ca_xNa_{1-x})Al_{3+x}Si_{3-x}O_{11}$] in shocked Martian meteorites: Earth and Planetary Science Letters, v. 219, p. 1–12, doi: 10.1016/S0012-821X(03)00695-2.

Benz, H.M., and Vidale, J.E., 1993, Sharpness of upper-mantle discontinuities determined from high-frequency reflections: Nature, v. 365, p. 147–150, doi: 10.1038/365147a0.

Bohlen, S.R., and Boettcher, A.L., 1982, The quartz-coesite transformation: A precise determination and the effects of other components: Journal of Geophysical Research, v. 87, p. 7073–7078.

Bolfan-Casanova, N., Keppler, H., and Rubie, D.C., 2003, Water partitioning at 660 km depth and evidence for very low water solubility in magnesium silicate perovskite: Geophysical Research Letters, v. 30, doi: 10.1029/2003GL017182.

Canil, D., 1994, Stability of clinopyroxene at pressure-temperature conditions of the transition zone: Physics of the Earth and Planetary Interiors, v. 86, p. 25–34, doi: 10.1016/0031-9201(94)05059-7.

Chao, E.C.T., Shoemaker, E.M., and Madsen, B.M., 1960, First natural occurrence of coesite: Science, v. 132, p. 220–222.

Chao, E.C.T., Fahey, J.J., Littler, J., and Milton, D.J., 1962, Stishovite: A new mineral from Meteor Crater, Arizona: Journal of Geophysical Research, v. 67, p. 419–421.

Chopin, C., 1984, Coesite and pure pyrope in high-grade blueschists of the western Alps: A first record and some consequences: Contributions to Mineralogy and Petrology, v. 86, p. 107–118, doi: 10.1007/BF00381838.

Christensen, U., 1995, Effects of phase transitions of mantle convection: Annual Review of Earth and Planetary Sciences, v. 23, p. 65–87, doi: 10.1146/annurev.ea.23.050195.000433.

Chudinovskikh, L., and Boehler, R., 2001, High-pressure polymorphs of olivine and the 660-km seismic discontinuity: Nature, v. 411, p. 574–577, doi: 10.1038/35079060.

Collerson, K.D., Hapugoda, S., Kamber, B.S., and Williams, Q., 2000, Rocks from the mantle transition zone: Majorite-bearing xenoliths from Malaita, southwest Pacific: Science, v. 288, p. 1215–1223, doi: 10.1126/science.288.5469.1215.

Conrad, P.G., 1998, The stability of almandine at high pressures and temperatures, *in* Manghnani, M.H., and Yagi, T., eds., Properties of Earth and Planetary Materials at High Pressure and Temperature: Washington, D.C., American Geophysical Union Geophysical Monograph, v. 101, p. 393–399.

Daniel, I., Cardon, H., Fiquet, G., Guyot, F., and Mezouar, M., 2001, Equation of state of Al-bearing perovskite to lower mantle pressure conditions: Geophysical Research Letters, v. 28, p. 3789–3792, doi: 10.1029/2001GL013011.

Deuss, A., and Woodhouse, J., 2001, Seismic observations of splitting of the mid-transition zone discontinuity in Earth's mantle: Science, v. 294, p. 354–357, doi: 10.1126/science.1063524.

Dubrovinsky, L.S., Saxena, S.K., Lazor, P., Ahuja, R., Eriksson, O., Wills, J.M., and Johansson, B., 1997, Experimental and theoretical identification of a new high-pressure phase of silica: Nature, v. 388, p. 362–365, doi: 10.1038/41066.

El Goresy, A., Dubrovinsky, L., Sharp, T.G., Saxena, S.K., and Chen, M., 2000, A monoclinic post-stishovite polymorph of silica in Shergotty meteorite: Science, v. 288, p. 1632–1634, doi: 10.1126/science.288.5471.1632.

Essene, E., 1974, High-pressure transformations in $CaSiO_3$: Contributions to Mineralogy and Petrology, v. 45, p. 247–250, doi: 10.1007/BF00383442.

Fei, Y., and Bertka, C.M., 1999, Phase transitions in the Earth's mantle and mantle mineralogy, *in* Fei, Y., Bertka, C.M., and Mysen, B., eds., Mantle Petrology: Field Observations and High Pressure Experimentation: A Tribute to Francis R. (Joe) Boyd: Houston, Geochemical Society Special Publication 6, p. 189–207.

Fei, Y., Orman, J.V., Li, J., Westrenen, W.V., Sanloup, C., Minarik, W., Hirose, K., Komabayashi, T., Walter, M., and Funakoshi, K., 2004, Experimentally determined postspinel transformation boundary in Mg_2SiO_4 using MgO as an internal pressure standard and its geophysical implications: Journal of Geophysical Research, v. 109, doi: 10.1029/2003JB002562.

Frost, D.J., 2003a, Fe^{2+}-Mg partitioning between garnet, magnesiowüstite, and $(Mg,Fe)_2SiO_4$ phases of the transition zone: The American Mineralogist, v. 88, p. 387–397.

Frost, D.J., 2003b, The structure and sharpness of $(Mg,Fe)_2SiO_4$ phase transformations in the transition zone: Earth and Planetary Science Letters, v. 216, p. 313–328, doi: 10.1016/S0012-821X(03)00533-8.

Fukao, Y., Widiyantoro, S., and Obayashi, M., 2001, Stagnant slabs in the upper and lower mantle transition region: Reviews in Geophysics, v. 39, p. 291–323, doi: 10.1029/1999RG000068.

Gaffney, E.S., and Anderson, D.L., 1973, Effect of low-spin Fe^{2+} on the composition of the lower mantle: Journal of Geophysical Research, v. 78, p. 7005–7014.

Gasparik, T., Wolf, K., and Smith, C.M., 1994, Experimental determination of phase relations in the $CaSiO_3$ system from 8 to 15 GPa: The American Mineralogist, v. 79, p. 1219–1222.

Gautron, L., Kesson, S.E., and Hibberson, W.O., 1996, Phase relations for $CaAl_2Si_2O_8$ (anorthite composition) in the system CaO-Al_2O_3-SiO_2 at 14 GPa: Physics of the Earth and Planetary Interiors, v. 97, p. 71–81, doi: 10.1016/0031-9201(96)03161-5.

Gautron, L., Angel, R.J., and Miletich, R., 1999, Structural characterization of the high-pressure phase of $CaAl_4Si_2O_{11}$: Physics and Chemistry of Minerals, v. 27, p. 47–51, doi: 10.1007/s002690050239.

Gillet, P., Chen, M., Dubrovinsky, L., and El Goresy, A., 2000, Natural $NaAlSi_3O_8$-hollandite in the shocked Sixiangkou meteorite: Science, v. 287, p. 1633–1636, doi: 10.1126/science.287.5458.1633.

Guignot, N., and Andrault, D., 2004, Equations of state of Na-K-Al host phases and implications for MORB density in the lower mantle: Physics of the Earth and Planetary Interiors, v. 143–144, p. 107–128, doi: 10.1016/j.pepi.2003.09.014.

Harte, B., Harris, J.W., Hutchison, M.T., Watt, G.R., and Wilding, M.C., 1999, Lower mantle mineral associations in diamonds from Sao Luiz, Brazil, in Fei, Y., Bertka, C.M., and Mysen, B.O., eds., Mantle Petrology: Field Observations and High Pressure Experimentation: A Tribute to Francis R. (Joe) Boyd: Houston, Geochemical Society Special Publication 6, p. 125–153.

Helffrich, G.R., and Wood, B.J., 1996, 410 km discontinuity sharpness and the form of the olivine α-β phase diagram: Resolution of apparent seismic contradictions: Geophysical Journal International, v. 126, p. F7–F12.

Hemley, R.J., and Cohen, R.E., 1992, Silicate perovskite: Annual Review of Earth and Planetary Sciences, v. 20, p. 553–600, doi: 10.1146/annurev.ea.20.050192.003005.

Hirose, K., 2002, Phase transitions in pyrolitic mantle around 670-km depth: Implications for upwelling of plumes from the lower mantle: Journal of Geophysical Research, v. 107, doi: 10.1029/2001JB000597.

Hirose, K., Fei, Y., Ma, Y., and Mao, H.K., 1999, The fate of subducted basaltic crust in the Earth's lower mantle: Nature, v. 397, p. 53–56, doi: 10.1038/16225.

Hirose, K., Fei, Y., Ono, S., Yagi, T., and Funakoshi, K., 2001, In situ measurements of the phase transition boundary in $Mg_3Al_2Si_3O_{12}$: Implications for the nature of the seismic discontinuities in the Earth's mantle: Earth and Planetary Science Letters, v. 184, p. 567–573, doi: 10.1016/S0012-821X(00)00354-X.

Huang, W.L., and Wyllie, P.J., 1975, Melting and subsolidus phase relationships for $CaSiO_3$ to 35 kilobars pressure: The American Mineralogist, v. 60, p. 213–217.

Inoue, T., Yurimoto, H., and Kudoh, Y., 1995, Hydrous modified spinel, $Mg_{1.75}SiH_{0.5}O_4$: A new water reservoir in the mantle transition zone: Geophysical Research Letters, v. 22, p. 117–120, doi: 10.1029/94GL02965.

Irifune, T., and Isshiki, M., 1998, Iron partitioning in a pyrolite mantle and the nature of the 410-km seismic discontinuity: Nature, v. 392, p. 702–705, doi: 10.1038/33663.

Irifune, T., and Ringwood, A.E., 1993, Phase transformations in subducted oceanic crust and buoyancy relationships at depths of 600–800 km in the mantle: Earth and Planetary Science Letters, v. 117, p. 101–110, doi: 10.1016/0012-821X(93)90120-X.

Irifune, T., Ringwood, A.E., and Hibberson, W.O., 1994, Subduction of continental crust and terrigenous and pelagic sediments: An experimental study: Earth and Planetary Science Letters, v. 126, p. 351–368, doi: 10.1016/0012-821X(94)90117-1.

Irifune, T., Koizumi, T., and Ando, J., 1996, An experimental study on the garnet-perovskite transformation in the system $MgSiO_3$-$Mg_3Al_2Si_3O_{12}$: Physics of the Earth and Planetary Interiors, v. 96, p. 147–157, doi: 10.1016/0031-9201(96)03147-0.

Irifune, T., Nishiyama, N., Kuroda, K., Inoue, T., Isshiki, M., Utsumi, W., Funakoshi, K., Urakawa, S., Uchida, T., Katsura, T., and Ohtaka, O., 1998, The postspinel phase boundary in Mg_2SiO_4 determined by in situ X-ray diffraction: Science, v. 279, p. 1698–1700, doi: 10.1126/science.279.5357.1698.

Ito, E., and Takahashi, E., 1989, Postspinel transformations in the system Mg_2SiO_4-Fe_2SiO_4 and some geophysical implications: Journal of Geophysical Research, v. 94, p. 10,637–10,646.

Ito, E., Akaogi, M., Topor, L., and Navrotsky, A., 1990, Negative pressure-temperature slopes for reactions forming $MgSiO_3$ perovskite from calorimetry: Science, v. 249, p. 1275–1278.

Iwabuchi, T., Akaogi, M., and Suzuki, T., 2001, Phase transition in the system $KAlSiO_4$-$MgAl_2O_4$ at high pressure [abs.]: Kyoto, Annual Meeting of the Japan Society High Press Scientific Technology, p. 146.

Joswig, W., Stachel, T., Harris, J.W., Baur, W.H., and Brey, G.P., 1999, New Ca-silicate inclusions in diamonds—Tracers from the lower mantle: Earth and Planetary Science Letters, v. 173, p. 1–6, doi: 10.1016/S0012-821X(99)00210-1.

Katsura, T., and Ito, E., 1989, The system Mg_2SiO_4-Fe_2SiO_4 at high pressures and temperatures: Precise determination of stabilities of olivine, modified spinel and spinel: Journal of Geophysical Research, v. 94, p. 15,663–15,670.

Katsura, T., Yamada, H., Shinmei, T., Kubo, A., Ono, S., Kanzaki, M., Yoneda, A., Walter, M.J., Ito, E., Urakawa, S., Funakoshi, K., and Utsumi, W., 2003, Post-spinel transition in Mg_2SiO_4 determined by high P-T in situ X-ray diffractometry: Physics of the Earth and Planetary Interiors, v. 136, p. 11–24, doi: 10.1016/S0031-9201(03)00019-0.

Katsura, T., Yamada, H., Nishikawa, O., Song, M., Kubo, A., Shinmei, T., Yokoshi, S., Aizawa, Y., Yoshino, T., Walter, M.J., and Ito, E., 2004, Olivine-wadsleyite transition in the system Mg_2SiO_4-Fe_2SiO_4: Journal of Geophysical Research, v. 109, doi: 10.1029/2003JB002438.

Kerschhofer, L., Rubie, D.C., Sharp, T.G., McConnell, J.D.C., and Dupas-Bruzek, C., 2000, Kinetics of intracrystalline olivine-ringwoodite transformation: Physics of the Earth and Planetary Interiors, v. 121, p. 59–76, doi: 10.1016/S0031-9201(00)00160-6.

Kesson, S.E., Fitz Gerald, J.D., and Shelley, J.M.G., 1994, Mineral chemistry and density of subducted basaltic crust at lower-mantle pressures: Nature, v. 372, p. 767–769, doi: 10.1038/372767a0.

Kingma, K.J., Cohen, R.E., Hemley, R.J., and Mao, H.K., 1995, Transformation of stishovite to a denser phase at lower-mantle pressure: Nature, v. 374, p. 243–245, doi: 10.1038/374243a0.

Kobayashi, M., Akaogi, M., and Suzuki, T., 2000, High pressure phase relations in the system $NaAlSiO_4$-$MgAl_2O_4$ [abs.]: Kyoto, 41st Annual Meeting of the Japan Society of High Press Scientific Technology, p. 197.

Kohlstedt, D.L., Keppler, H., and Rubie, D.C., 1996, Solubility of water in the α, β and γ phases of $(Mg,Fe)_2SiO_4$: Contributions to Mineralogy and Petrology, v. 123, p. 345–357, doi: 10.1007/s004100050161.

Kojitani, H., Hisatomi, R., and Akaogi, M., 2007, High-pressure phase relations and crystal chemistry of calcium ferrite-type solid solutions in the system $MgAl_2O_4$-Mg_2SiO_4: The American Mineralogist (in press).

Kubo, A., and Akaogi, M., 2000, Post-garnet transitions in the system $Mg_4Si_4O_{12}$-$Mg_3Al_2Si_3O_{12}$ up to 28 GPa: Phase relations of garnet, ilmenite and perovskite: Physics of the Earth and Planetary Interiors, v. 121, p. 85–102, doi: 10.1016/S0031-9201(00)00162-X.

Kubo, A., Yagi, T., Ono, S., and Akaogi, M., 2000, Compressibility of $Mg_{0.9}Al_{0.2}Si_{0.9}O_3$ perovskite: Proceedings of the Japan Academy, v. 76B, p. 103–107.

Kubo, A., Ito, E., Katsura, T., Shinmei, T., Yamada, H., Nishikawa, O., Song, M., and Funakoshi, K., 2003, In situ X-ray observation of iron using Kawai-type apparatus equipped with sintered diamond: Absence of β phase up to 44 GPa and 2100 K: Geophysical Research Letters, v. 30, doi: 10.1029/2002GL016394.

Kubo, T., Ohtani, E., Kato, T., Urakawa, S., Suzuki, A., Kanbe, Y., Funakoshi, K., Utsumi, W., Kikegawa, T., and Fujino, K., 2002, Mechanisms and kinetics of the post-spinel transformation in Mg_2SiO_4: Physics of the Earth and Planetary Interiors, v. 129, p. 153–171, doi: 10.1016/S0031-9201(01)00270-9.

Langenhorst, F., and Poirier, J.P., 2000, 'Eclogitic' minerals in a shocked basaltic meteorite: Earth and Planetary Science Letters, v. 176, p. 259–265, doi: 10.1016/S0012-821X(00)00028-5.

Lauterbach, S., McCammon, C.A., Aken, P.V., Langenhorst, F., and Seifert, F., 2000, Mossbauer and ELNES spectroscopy of $(Mg,Fe)(Si,Al)O_3$ perovskite: A highly oxidized component of the lower mantle: Contributions to Mineralogy and Petrology, v. 138, p. 17–26, doi: 10.1007/PL00007658.

Litasov, K., and Ohtani, E., 2003, Stability of various hydrous phases in CMAS pyrolite-H_2O system up to 25 GPa: Physics and Chemistry of Minerals, v. 30, p. 147–156, doi: 10.1007/s00269-003-0301-y.

Mao, H.K., Shen, G., and Hemley, R.J., 1997, Multivariable dependence of Fe-Mg partitioning in the lower mantle: Science, v. 278, p. 2098–2100, doi: 10.1126/science.278.5346.2098.

McCammon, C.A., 1997, Perovskite as a possible sink for ferric iron in the lower mantle: Nature, v. 387, p. 694–696, doi: 10.1038/42685.

Meyer, H.O.A., 1987, Inclusions in diamond, in Nixon, P.H., ed., Mantle Xenoliths: Chichester, Wiley, p. 501–522.

Miura, H., Hamada, Y., Suzuki, T., Akaogi, M., Miyajima, N., and Fujino, K., 2000, Crystal structure of $CaMg_2Al_6O_{12}$, a new Al-rich high pressure form: The American Mineralogist, v. 85, p. 1799–1803.

Miyajima, N., Fujino, K., Funamori, N., Kondo, T., and Yagi, T., 1999, Garnet-perovskite transformation under conditions of the Earth's lower mantle: An analytical transmission electron microscopy study: Physics of the Earth and Planetary Interiors, v. 116, p. 117–131, doi: 10.1016/S0031-9201(99)00127-2.

Miyajima, N., Yagi, T., Hirose, K., Kondo, T., Fujino, K., and Miura, H., 2001, Potential host phase of aluminum and potassium in the Earth's lower mantle: The American Mineralogist, v. 86, p. 740–746.

Moore, R.O., and Gurney, J.J., 1985, Pyroxene solid solution in garnets included in diamond: Nature, v. 318, p. 553–555, doi: 10.1038/318553a0.

Morishima, H., Kato, T., Suto, M., Ohtani, E., Urakawa, S., Utsumi, W., Shimomura, O., and Kikegawa, T., 1994, The phase boundary between α- and β-Mg_2SiO_4 determined by in situ X-ray observation: Science, v. 265, p. 1202–1203.

Murakami, M., Hirose, K., Yurimoto, H., Nakashima, S., and Takafuji, N., 2002, Water in Earth's lower mantle: Science, v. 295, p. 1885–1887, doi: 10.1126/science.1065998.

Murakami, M., Hirose, K., Ono, S., and Ohishi, Y., 2003, Stability of $CaCl_2$-type and α-PbO_2-type SiO_2 determined by in situ X-ray observations: Geophysical Research Letters, v. 30, doi: 10.1029/2002GL016722.

Murakami, M., Hirose, K., Kawamura, K., Sata, N., and Ohishi, Y., 2004, Post-perovskite phase transition in $MgSiO_3$: Science, v. 304, p. 855–858, doi: 10.1126/science.1095932.

Navrotsky, A., 1999, A lesson from ceramics: Science, v. 284, p. 1788–1799, doi: 10.1126/science.284.5421.1788.

Navrotsky, A., Schoenitz, M., Kojitani, H., Xu, H., Zhang, J., Weidner, D.J., and Jeanloz, R., 2003, Aluminum in magnesium silicate perovskite: Formation, structure, and energetics of magnesium-rich defect solid solutions: Journal of Geophysical Research, v. 108, p. 2330, doi: 10.1029/2002JB002055.

Nishiyama, N., and Yagi, T., 2003, Phase relation and mineral chemistry in pyrolite to 2200 °C under the lower mantle pressures and implications for dynamics on mantle plumes: Journal of Geophysical Research, v. 108, p. 2255, doi: 10.1029/2002JB002216.

Nishiyama, N., Rapp, R.P., Irifune, T., Sanehira, T., Yamazaki, D., and Funakoshi, K., 2005, Stability and P-V-T equation of state of $KAlSi_3O_8$-hollandite determined by in situ X-ray observations and implications for dynamics of subducted continental crust: Physics and Chemistry of Minerals, v. 32, p. 627–637, doi: 10.1007/S00269-005-0037-y.

Oganov, A.R., and Ono, S., 2004, Theoretical and experimental evidence for a post-perovskite phase of $MgSiO_3$ in Earth's D″ layer: Nature, v. 430, p. 445–448, doi: 10.1038/nature02701.

Oguri, K., Funamori, N., Sakai, F., Kondo, T., Uchida, T., and Yagi, T., 1997, High-pressure and high-temperature phase relations in diopside $CaMgSi_2O_6$: Physics of the Earth and Planetary Interiors, v. 104, p. 363–370, doi: 10.1016/S0031-9201(97)00029-0.

Ohtani, E., Litasov, K., Hosoya, T., Kubo, T., and Kondo, T., 2004, Water transport into the deep mantle and formation of a hydrous transition zone: Physics of the Earth and Planetary Interiors, v. 143–144, p. 255–269, doi: 10.1016/j.pepi.2003.09.015.

Ono, S., Ito, E., and Katsura, T., 2001, Mineralogy of subducted basaltic crust (MORB) from 25 and 37 GPa, and chemical heterogeneity of the lower mantle: Earth and Planetary Science Letters, v. 190, p. 57–63, doi: 10.1016/S0012-821X(01)00375-2.

Ono, S., Hirose, K., Murakami, M., and Isshiki, M., 2002, Post-stishovite phase boundary in SiO_2 determined by in situ X-ray determination: Earth and Planetary Science Letters, v. 197, p. 187–192, doi: 10.1016/S0012-821X(02)00479-X.

Revenaugh, J., and Jordan, T.H., 1991, Mantle layering from ScS reverberations; 2. The transition zone: Journal of Geophysical Research, v. 96, p. 19,763–19,780.

Ringwood, A.E., 1975, Composition and Petrology of the Earth's Mantle: New York, Wiley, 618 p.

Rubie, D.C., and Ross, C.R., II, 1994, Kinetics of the olivine-spinel transformation in subducting lithosphere: Experimental constraints and implications for deep slab processes: Physics of the Earth and Planetary Interiors, v. 86, p. 223–241, doi: 10.1016/0031-9201(94)05070-8.

Sawamoto, H., Weidner, D.J., Sasaki, S., and Kumazawa, M., 1984, Single-crystal elastic properties of the modified spinel (beta) phase of magnesium silicate: Science, v. 224, p. 749–751.

Sharp, T.G., Lingemann, M., Dupas, C., and Stöffler, D., 1997, Natural occurrence of $MgSiO_3$-ilmenite and evidence for $MgSiO_3$-perovskite in a shocked L chondrite: Science, v. 277, p. 352–355, doi: 10.1126/science.277.5324.352.

Shim, S.H., Duffy, T.S., and Shen, G., 2001, The post-spinel transformation in Mg_2SiO_4 and its relation to the 660-km seismic discontinuity: Nature, v. 411, p. 571–574, doi: 10.1038/35079053.

Sinogeikin, S.V., Bass, J.D., and Katsura, T., 2003, Single-crystal elasticity of ringwoodite to high pressures and high temperatures: Implications for 520 km seismic discontinuity: Physics of the Earth and Planetary Interiors, v. 136, p. 41–66, doi: 10.1016/S0031-9201(03)00022-0.

Smyth, J.R., and Frost, D.J., 2002, The effect of water on the 410-km discontinuity: An experimental study: Geophysical Research Letters, v. 29, doi: 10.1029/2001GL014418.

Speziale, S., Zha, C.S., Duffy, T.S., Hemley, R.J., and Mao, H.K., 2001, Quasi-hydrostatic compression of magnesium oxide to 52 GPa: Implications for the pressure-volume-temperature equation of state: Journal of Geophysical Research, v. 106, p. 515–528, doi: 10.1029/2000JB900318.

Stachel, T., 2001, Diamonds from the asthenosphere and the transition zone: European Journal of Mineralogy, v. 13, p. 883–892, doi: 10.1127/0935-1221/2001/0013/0883.

Stebbins, J.F., Kroeker, S., and Andrault, D., 2001, The mechanism of solution of aluminum oxide in $MgSiO_3$ perovskite: Geophysical Research Letters, v. 28, p. 615–618, doi: 10.1029/2000GL012279.

Stebbins, J.F., Kojitani, H., Akaogi, M., and Navrotsky, A., 2003, Aluminum substitution in $MgSiO_3$ perovskite: Multiple mechanisms by ^{27}Al NMR: The American Mineralogist, v. 88, p. 1161–1164.

Stixrude, L., 1997, Structure and sharpness of phase transitions and mantle discontinuities: Journal of Geophysical Research, v. 102, p. 14,835–14,852.

Suzuki, A., Ohtani, E., Morishima, H., Kubo, T., Kanbe, Y., Kondo, T., Okada, T., Terasaki, H., Kato, T., and Kikegawa, T., 2000, In situ determination of the phase boundary between wadsleyite and ringwoodite in Mg_2SiO_4: Geophysical Research Letters, v. 27, p. 803–806, doi: 10.1029/1999GL008425.

Tomioka, N., and Fujino, K., 1997, Natural $(Mg,Fe)SiO_3$-ilmenite and perovskite in the Tenham meteorite: Science, v. 277, p. 1084–1086, doi: 10.1126/science.277.5329.1084.

Tomioka, N., Mori, H., and Fujino, K., 2000, Shock-induced transition of $NaAlSi_3O_8$ feldspar into a hollandite structure in a L6 chondrite: Geophysical Research Letters, v. 27, p. 3997–4000, doi: 10.1029/2000GL008513.

Tsuchida, Y., and Yagi, T., 1989, A new, post-stishovite high-pressure polymorph: Nature, v. 340, p. 217–220, doi: 10.1038/340217a0.

Tutti, F., Dubrovinsky, L.S., Saxena, S.K., and Carlson, S., 2001, Stability of $KAlSi_3O_8$ hollandite-type structure in the Earth's lower mantle: Geophysical Research Letters, v. 28, p. 2735–2738, doi: 10.1029/2000GL012786.

Urakawa, S., Kondo, T., Igawa, N., Shimomura, O., and Ohno, H., 1994, Synchrotron radiation study on the high-pressure and high-temperature phase relations of $KAlSi_3O_8$: Physics and Chemistry of Minerals, v. 21, p. 387–391, doi: 10.1007/BF00203296.

Weidner, D.J., Sawamoto, H., Sasaki, S., and Kumazawa, M., 1984, Single-crystal elastic properties of the spinel phase of Mg_2SiO_4: Journal of Geophysical Research, v. 89, p. 7852–7860.

Wood, B.J., 1995, The effect of H_2O on the 410-kilometer seismic discontinuity: Science, v. 268, p. 74–76.

Wood, B.J., 2000, Phase transformations and partitioning relations in peridotite under lower mantle conditions: Earth and Planetary Science Letters, v. 174, p. 341–354, doi: 10.1016/S0012-821X(99)00273-3.

Yagi, A., Suzuki, T., and Akaogi, M., 1994, High pressure transitions in the system $KAlSi_3O_8$-$NaAlSi_3O_8$: Physics and Chemistry of Minerals, v. 21, p. 12–17, doi: 10.1007/BF00205210.

Yagi, T., Okabe, K., Nishiyama, N., Kubo, A., and Kikegawa, T., 2004, Complicated effects of aluminum on the compressibility of silicate perovskite: Physics of the Earth and Planetary Interiors, v. 143–144, p. 81–91, doi: 10.1016/j.pepi.2003.07.020.

Zha, C., Mao, H.K., and Hemley, R.J., 2000, Elasticity of MgO and a primary pressure scale to 55 GPa: Proceedings of the National Academy of Sciences of the United States of America, v. 97, p. 13,494–13,499, doi: 10.1073/pnas.240466697.

Zhang, J., and Weidner, D.J., 1999, Thermal equation of state of aluminum-enriched silicate perovskite: Science, v. 284, p. 782–784, doi: 10.1126/science.284.5415.782.

Zhang, J., Li, B., Utsumi, W., and Liebermann, R.C., 1996, In situ X-ray observations of the coesite-stishovite transition: Reversed phase boundary and kinetics: Physics and Chemistry of Minerals, v. 23, p. 1–10, doi: 10.1007/BF00202987.

Manuscript Accepted by the Society 14 August 2006

Properties of lower-mantle Al-(Mg,Fe)SiO₃ perovskite

D. Andrault[†]

Laboratoire Magmas et Volcans, Université Blaise Pascal, Clermont-Ferrand, France

ABSTRACT

The properties of the main lower-mantle phase appear to be more complex than expected. The common procedure of using the properties of the simplified MgSiO₃ (and [Mg,Fe]SiO₃) composition for direct analogy to the Al-bearing (Mg,Fe)SiO₃ lower-mantle perovskite can lead to significant misinterpretations. The presence of Al and Fe affects the equation of state, the defect population, the ability of this phase to insert minor and trace elements, and the transport properties, etc. Some difficulties remain for the quantitative determination of these effects because of two main reasons: many experimental techniques are ineffective because silicate perovskite is metastable at ambient conditions, and the crystal chemistry of Al-(Mg,Fe)SiO₃ perovskite is complex and can evolve with pressure, temperature, and chemical composition. This paper reviews the recent progress made in the determination of its properties and presents additional new results from our group. The original data concern the pressure-volume-temperature (P-V-T) equation of state of Al-(Mg,Fe)SiO₃ perovskite, the change of oxidation state (dismutation) of Fe^{2+} into a mixture of Fe^{3+} and Fe^0, which drives the lower-mantle oxygen fugacity to the Fe/(Mg,Fe)O buffer, and the stability of the (Mg,Fe)SiO₃ perovskite to the highest pressure and temperature conditions.

Keywords: lower mantle, silicate perovskite, crystal chemistry, elastic properties.

INTRODUCTION

A high diversity of minerals exists at Earth's surface because of an almost infinite variety of chemical compositions. Each mineral structure is well adapted with respect to the electronic configuration and ionic size of the different elements present in its chemical composition. It yields minimal formation energy ΔH_f, which is the dominant parameter in Gibbs free energy $G(P,T)$ at the smallest pressures and temperatures. With increasing depth, the mechanical (PdV) and thermal (TdS) terms of $G(P,T)$ compete with ΔH_f, which give rise to polymorphism. At shallow depth, the structural changes can remain minor, with breakdown of some symmetry elements (such as the transition between orthopyroxenes and clinopyroxenes), or changes of polyhedral stacking (such as the olivine to spinel transition), etc. These changes do not affect severely the local atomic configurations, but rather the three-dimensional structure of the minerals, which becomes progressively more compact. At greater depth, and especially below the 660 km discontinuity, the achievement of a maximal packing efficiency becomes a predominant requirement for $G(P,T)$ minimization. It yields drastic changes in local structure, such as the formation of SiO₆ octahedra in Al-bearing (Mg,Fe)SiO₃ perovskite (hereafter named Al-[Mg,Fe]SiO₃ perovskite) (Liu, 1974). This modification forces Si to a major change in its elec-

[†]E-mail: d.andrault@opgc.univ-bpclermont.fr.

tronic orbital hybridization from sp^3 (found in SiO$_2$ tetrahedron) to sp^{3}d^2. This type of coordination change occurs for almost all chemical elements between Earth's surface and the 660 km discontinuity. These modifications favor the PdV term of the lattice energy, but, in return, yield a higher ΔH_f because of the non-natural atomic configurations involved in achieving a maximal compactness.

The number of compact structures is very limited, which results in a drastic simplification of the mineralogy with increasing depth. It is already visible in the mantle transition zone, between 410 and 660 km depth, where most of the chemical elements enter in very few phases—the major ones are (Ca,Fe,Mg)$_3$Al$_2$Si$_3$O$_{12}$ garnets and (Mg,Fe)$_2$SiO$_4$ wadsleyite and ringwoodite spinels. Other phases, such as SiO$_2$ stishovite, KAlSi$_3$O$_8$ hollandite, etc., are expected in minor amounts or in material with specific chemical compositions. At the 660 km discontinuity, or at slightly higher depth for Al-rich materials, ringwoodite-spinel and majoritic-garnet transform into the mixture of Al-(Mg,Fe)SiO$_3$ and CaSiO$_3$ perovskites and (Mg,Fe)O ferropericlase (a phase also called magnesiowüstite), the three major phases of the lower mantle. In this reservoir, the amount of other phases could be very limited, or even negligible, because it appears that all the different elements of a pyrolitic composition can be included in one of these three phases (Andrault, 2003; Irifune, 1994; Nishiyama and Yagi, 2003). The Al-(Mg,Fe)SiO$_3$ perovskite phase is expected to represent more than 80% of the mass of the lower mantle. Due to the relative size of the lower mantle compared to other interior regions, it is likely the most abundant phase in our planet.

Due to its great importance, the Al-(Mg,Fe)SiO$_3$ perovskite phase has been extensively studied using all kinds of experimental and theoretical techniques. Major bulk properties are relatively well constrained experimentally, like the high-pressure MgSiO$_3$-Al$_2$O$_3$ phase diagram (Akaogi and Ito, 1999; Hirose et al., 2001; Irifune et al., 1996; Kubo and Akaogi, 2000), the P-V-T equation of state of (Mg,Fe)SiO$_3$ composition (Fiquet et al., 1998; Mao et al., 1991; Shim and Duffy, 2000; Utsumi et al., 1995; Wang et al., 1994), etc. Other important properties, such as transport properties, which are essential for modeling lower-mantle dynamics, remain known with limited precision, first because this phase is metastable and becomes amorphous easily at room pressure, and also because it can only be obtained in limited volume (~1 mm^3) after its synthesis at high pressure. Transport properties, for example, still have large uncertainties even when they are so important to the modeling of the deep Earth's properties. The picture is actually complicated by a complex crystal chemistry correlated with a large compliance of Al-(Mg,Fe)SiO$_3$ perovskite with respect to its chemical composition. Minor elements and structural defects could largely modify the bulk properties and, for this reason, the MgSiO$_3$-perovskite end member may not be an ideal analogue for modeling the lower mantle. The actual controversy about the effect of Al on the (Mg,Fe)SiO$_3$ perovskite bulk modulus illustrates this kind of difficulty. Fortunately, experimental techniques have rapidly evolved in the past few years, and a number of studies have been devoted to Al-(Mg,Fe)SiO$_3$ perovskite. Therefore, our knowledge of lower-mantle properties has recently improved.

CRYSTAL CHEMISTRY OF Al-(Mg,Fe)SiO$_3$ PEROVSKITE

Compactness of Orthorhombic Perovskite

The Al-(Mg,Fe)SiO$_3$ perovskite phase adopts the orthorhombic Pbnm space group symmetry for a large range of pressure, temperature, and composition (Fig. 1, perovskite structure). A main feature of this phase is that the Si and Mg (and also Fe) cations are too small to prevent from strong O-O repulsion in the oxygen sublattice. Therefore, this phase is metastable at ambient pressure, with easy amorphization upon various treatments. Magnesium silicate perovskite becomes stable at pressures above ~24 GPa, when the oxygen sublattice has already undergone severe compression, and it yields an atomic packing with optimal density. The CaSiO$_3$ composition provides the best model to discuss the compactness of the perovskite structure, thanks to its cubic symmetry. If we consider that Ca^{2+} and O^{2-} have about the same ionic size (ionic radii of 1.48 Å and 1.26 Å, respectively;

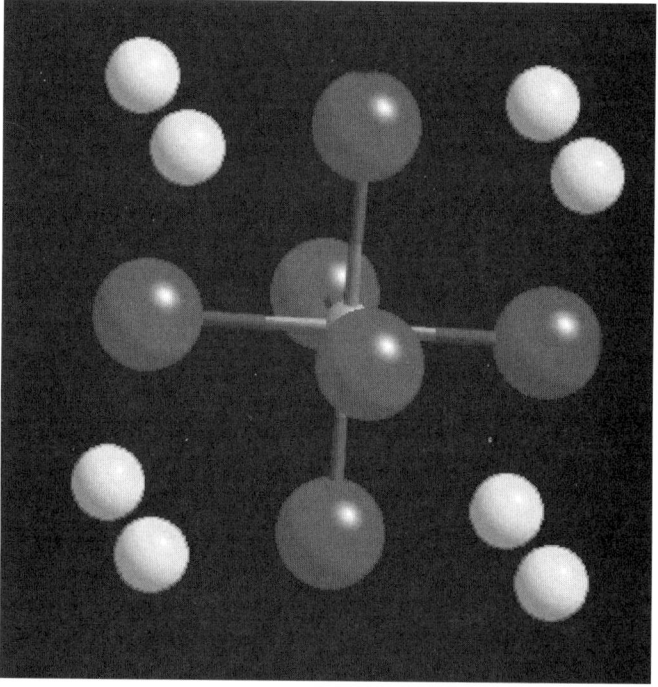

Figure 1. The orthorhombic structure of Al-(Mg,Fe)SiO$_3$ perovskite. Si cations are located in the smallest octahedra (smaller light atoms at center), while Mg and Fe share the distorted dodecahedral site (light). O are shown in dark. Al can be inserted into both polyhedra (via coupled substitution) or in the octahedral site only, and, in this case, O vacancies are required to balance the charge defect related to the insertion of this trivalent cation.

all atomic radii were extracted from Shannon and Prewitt, 1969), the cubic perovskite lattice can be described as a face-centered cubic (fcc) sublattice composed of Ca and O, with Si^{4+} cations inserted in 1/4 of the octahedral sites. Note that Si^{4+} cations are small compared to the size made available in O6-forming octahedral, and their insertion in the structure has limited effect on the structure compactness. For the $MgSiO_3$ and Al-$(Mg,Fe)SiO_3$ perovskite phases, the atomic packing is slightly different because the limited size of Mg^{2+} cations (ionic radius of ~0.86 Å) requires smaller Mg-O bonds compared to that of Ca-O, which induces the tilting of the SiO_6 octahedra into the unit cell lattice. Note that the distortion is relatively important and comparable to that found in the $GdFeO_3$ perovskite phase. Due to the distortion, the Mg site is no longer a perfect dodecahedron, but rather presents four shorter bonds in the 1.99–2.11 Å range followed by four additional ones at 2.28–2.29 and 2.43–2.45 Å (Farges et al., 1995; Parise et al., 1990). Nevertheless, the highest compactness of the perovskite lattice is mostly preserved along the symmetry reduction from cubic to orthorhombic Pbnm.

Mechanism of Al-Insertion in the $MgSiO_3$ Lattice

Although the first silicate perovskite sample was synthesized in a diamond anvil cell using $Mg_3Al_2Si_3O_{12}$ pyrope as starting material (Liu, 1974), the Al site occupancy remains uncertain. The first Al effect is to increase the transition pressure, from 23 to 24 GPa for pure $MgSiO_3$ to ~26.5 GPa for the pyrope composition (Akaogi and Ito, 1999; Hirose et al., 2001; Irifune et al., 1996; Katsura et al., 2003; Kubo and Akaogi, 2000) (Fig. 2, $MgSiO_3$-Al_2O_3 phase diagram). Thus, there is no doubt that the Al-$MgSiO_3$ perovskite phase can store the ~4 at% Al_2O_3 expected for a pyrolitic-type composition (Ringwood, 1979). The maximum Al-solubility in the perovskite remains controversial. Synthetic and natural pyrope compounds have been reported to fully transform into Al-$(Mg,Fe)SiO_3$ perovskite phase (Irifune et al., 1992; Ito et al., 1998; Kesson et al., 1995; Liu, 1974); however, careful analyses performed with the very high spatial resolution available in analytical transmission electron microscopes reveal the existence of additional phases, such as the so-called "new aluminous phase" (NAL) (Miyajima et al., 1999; Oguri et al., 2000). Note that the decompression of Al-saturated perovskite samples can yield formation of a $LiNbO_3$-type phase during decompression (Funamori et al., 1997). The grossular $Ca_3Al_2Si_3O_{12}$ has also been reported to transform into a perovskite phase at 30.2 GPa (Yusa et al., 1995).

Two substitution mechanisms are possible for the insertion of Al into the perovskite structure. In the first one, two Al^{3+} cations substitute for a pair of Mg^{2+} (or Ca^{2+}) and Si^{4+} to maintain the electroneutrality. The insertion of Al^{3+} (ionic radius of 0.675 Å) in both of the perovskite sites yields to the presence of a bigger cation in the octahedra, and a smaller cation in the dodecahedra, compared to Si^{4+} and Mg^{2+}, respectively. Both effects are expected to enhance the orthorhombic distortion. A comparable type of coupled substitution occurs along the pyrope-majorite joint, where 2 Al^{3+} exchange with a pair of Mg and Si in the octahedral site. The possibility of Al substitution on both perovskite sites is largely supported from first principle (ab initio) calculations, which show that pure corundum could undergo a phase transformation to a perovskite structure at very high pressures (Thomson et al., 1996). In a second substitution mechanism, Al enters in only one of the two perovskite sites, more likely the octahedral site, as observed in aluminous $XAlO_3$ perovskites (X^{3+} = Sc, Y, Gd, etc.). In this case, O vacancies should form to maintain the electroneutrality (Al^{3+} substitutes for Si^{4+} in $MgSiO_3$). Note that inserting 4 wt% Al_2O_3 into the silicate perovskite via this mechanism would result in the formation of ~2% of O vacancy, which represents a large number of structural defects. A simple equilibrium links the two substitution mechanisms (Υ represents O vacancies):

$$(Mg_{1-x}Al_x)(Al_xSi_{1-x})O_3 + 2x\ MgO \leftrightarrow (Mg_{1+x})(Al_{2x}Si_{1-x})O_{3+2x}\Upsilon_x.$$

Various techniques have been used to identify the type of Al substitution that prevails in different silicate perovskite samples.

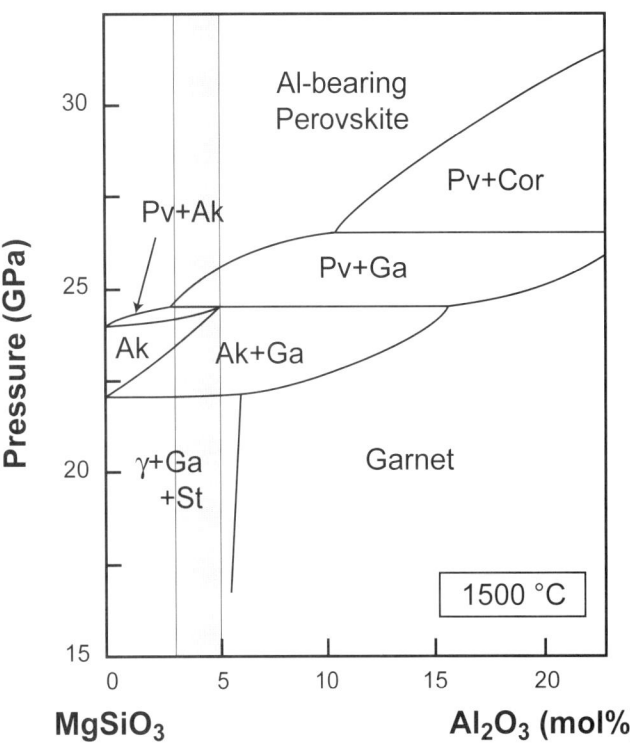

Figure 2. Phase relations in $MgSiO_3$-Al_2O_3 system at 15–32 GPa and 1500 °C (derived from Irifune et al., 1996). Ak, Cor, perovskite, Ga, γ, and St stand for Al-$MgSiO_3$ akimotoite, corundum, Al-$MgSiO_3$ perovskite, majoritic garnet, Mg_2SiO_4 ringwoodite, and stishovite. The shaded zone corresponds to Al_2O_3 content typically expected for pyrolitic lower mantle. At higher temperatures, the stability fields of akimotoite and the mixture of ringwoodite, stishovite and majorite reduce and disappear and give way to that of majoritic garnet (Hirose et al., 2001).

X-ray absorption analysis (XAFS) performed at the Mg, Al, and Si K-edges in Al-MgSiO$_3$ perovskite suggests a large majority of coupled substitution, without, however, quantitative assessment of the respective contents (Andrault et al., 1998b). More recently, ^{27}Al and ^{29}Si local structures were investigated by nuclear magnetic resonance (NMR). The signature of the two Al sites can be easily differentiated, because the isotropic chemical shifts are different (Stebbins et al., 2001). The ratio of site occupancy appears close to 1, a conclusion similar to that of the previous XAFS study. However, a second NMR study performed on samples with a different Mg/Si ratio showed evidence for a higher amount of Al substituted in the octahedral site, with the possible presence of significant amounts of O vacancies (Stebbins et al., 2003) (Fig. 3, NMR). Various ab initio calculations show that the energies involved in the two mechanisms of Al substitution are comparable, and thus the two types of defects are likely to coexist in the same perovskite phase. More precisely, the coupled substitution mechanism could be favored at higher pressures, because the presence of O vacancies works against an ideal compactness (Brodholt, 2000; Yamamoto et al., 2003). On the other hand, the O-vacancy mechanism could be favored at higher temperatures. Both experimental and theoretical work gives evidence for a higher formation entropy for the O-vacancy mechanism compared to the coupled substitution (Akber-Knutson and Bukowinski, 2004; Navrotsky et al., 2003).

Affinity between Al and Fe^{3+}

Several studies have addressed the effect of Al on Fe^{3+} content in Al-(Mg,Fe)SiO$_3$ perovskite. It happens to be dramatic, with a definite increase of the Fe^{3+} content for values up to 50% (McCammon, 1997). This effect induces a severe modification of the Fe-partitioning coefficient between the perovskite and the ferropericlase phases, resulting in higher Fe affinity for the Al-(Mg,Fe)SiO$_3$ perovskite structure (Wood and Rubie, 1996). An interesting behavior is that the formation of Fe^{3+} cations seems to be, in a large part, disconnected from the usual considerations of oxygen fugacity (fO_2) (Lauterbach et al., 2000). Thus, the strong atomic coupling between these two (Al^{3+},Fe^{3+}) cations relies on intrinsic properties of the silicate perovskite. An explanation may arise from the simplest arguments based on atomic radii, because Fe^{3+} (ionic radius of 0.785 Å or 0.69 Å for high-spin or low-spin electronic configuration, respectively) appears significantly bigger than Al^{3+}. Indeed, the substitution of Al^{3+} in both perovskite sites may be difficult due to its relatively small size compared to Mg^{2+} (and a fortiori to Fe^{2+}), and the occurrence of Fe^{3+} in the dodecahedral site could help resolve this problem. The crystal chemistry remains complicated, however, because Al^{3+} and Fe^{3+} contents are found to be different from each other in all available samples. In fact, different substitution mechanisms are possible for the insertion of the (Al^{3+},Fe^{3+}) cations in the MgSiO$_3$ perovskite structure. (Al-Al) and (Fe-Al) can be inserted via coupled substitution, and Al (and maybe Fe^{3+}) can also be inserted into the octahedral site solely with formation of O vacancies, using the same basic rules as those described previously for the Al-MgSiO$_3$ perovskite phase. Note that octahedral Fe^{3+} is expected to remain limited in number, as suggested from ab initio calculations (Richmond and Brodholt, 1998). From the Mössbauer analyses of various Al-(Mg,Fe)SiO$_3$ perovskite samples synthesized in a multianvil press, it has been shown that less than 25% of the (Al,Fe^{3+}) substitution occurs via the O-vacancy mechanism (Lauterbach et al., 2000) (Fig. 4, [Al,Fe^{3+}] in Al-[Mg,Fe]SiO$_3$ perovskite).

Minor Elements in the Al-(Mg,Fe)SiO$_3$ Perovskite Structure

Finally, since the crystal chemistry of the Al-(Mg,Fe)SiO$_3$ perovskite seems rather flexible to various types of structural defects, one can wonder to what extent the structure can integrate

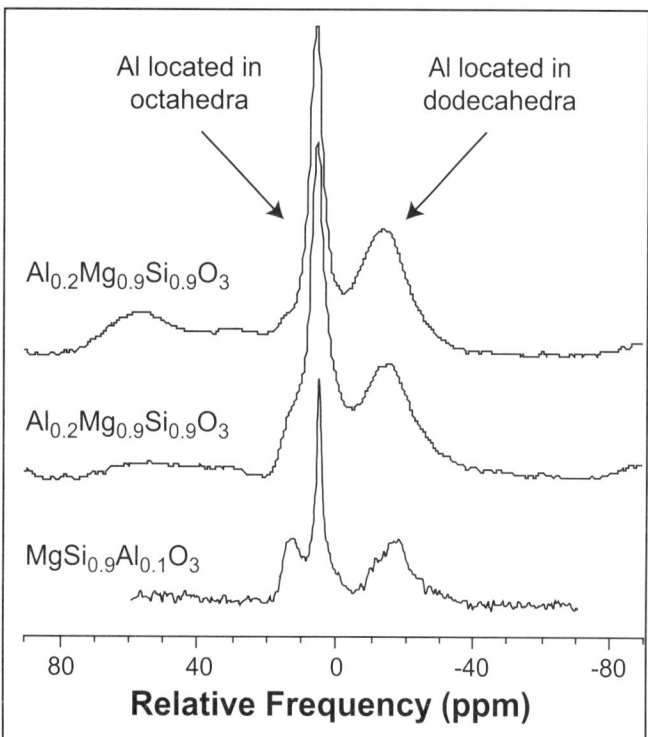

Figure 3. ^{27}Al nuclear magnetic resonance spectra of Al$_{0.2}$Mg$_{0.9}$Si$_{0.9}$O$_3$ (two different samples) and MgAl$_{0.1}$Si$_{0.9}$O$_3$ perovskites, collected at 18.8 T (derived from Stebbins et al., 2001, 2003). The narrow peak with an isotropic chemical shift at ~5.8 ppm is the signature of Al in a symmetrical, six-coordinated site. A second, larger, broad peak at ~−15 ppm results from a range in local structure (bond distances, angles, etc.) probably linked to cation disorder in the dodecahedral site. For the Al$_{0.2}$Mg$_{0.9}$Si$_{0.9}$O$_3$ samples, the simulation suggests similar cation occupancy for the two perovskite sites. For the MgAl$_{0.1}$Si$_{0.9}$O$_3$ sample, another peak found at ~15 ppm could correspond to Al in five-fold coordination. In this case, the peak ratio would suggest a higher Al content in the octahedral site and the presence of a significant amount of O vacancies.

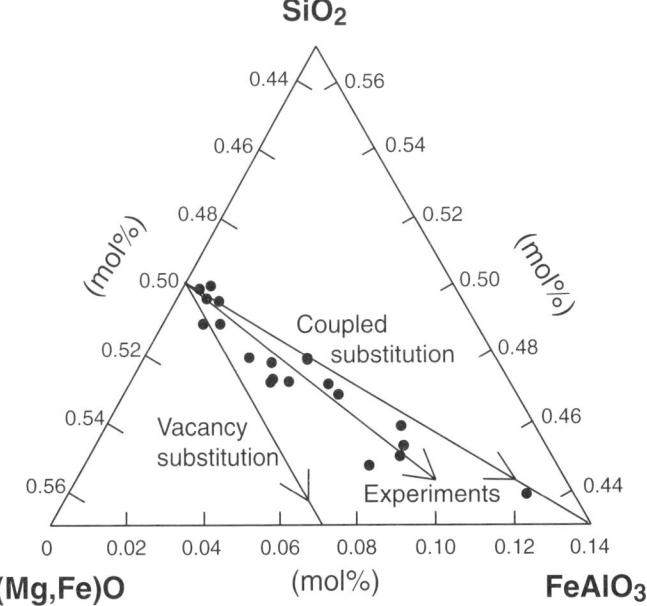

Figure 4. Ternary diagram between SiO_2, (Mg,Fe)O (divalent cations), and $(Al,Fe^{3+})_2O_3$ (trivalent cations) (derived from Lauterbach et al., 2000), showing the chemical paths expected for the insertion of trivalent cations into the $(Mg,Fe)SiO_3$-based structure using the two relevant substitution mechanisms. For the coupled substitution, the $(Mg,Fe)O/SiO_2$ ratio remains constant. On the contrary, this ratio increases for the O-vacancy mechanism, when Al predominantly replaces Si in the octahedral site. The chemical analysis, using electron microprobe and Mössbauer spectroscopy, of various samples of Al-$(Mg,Fe)SiO_3$ perovskite suggests that ~85% of the total amount of [Al,Fe] is inserted into the structure via coupled substitution.

It is important to establish the degree of compatibility of the different elements in one of the main lower-mantle phases, because all elements located at the grain boundaries are likely to be wiped out at any possible occasion, especially if the material undergoes partial melting. According to the high number of chemical analyses reported today for silicate perovskite synthesized from various starting material (see Table 1, various Al-$(Mg,Fe)SiO_3$ perovskite compositions), it seems likely that indeed many cations can find a suitable substitution mechanism. In a previous study, we used a specific technique to test the substitution mechanism for some chosen elements (Andrault, 2003). Using an $MgSiO_3$-based starting material with, for example, excess Na_2O and Al_2O_3, we could measure Na contents up to ~1% in the Al-$MgSiO_3$ perovskite, which is slightly above the pyrolitic Na content. We also found that different trivalent cations can be inserted in large amounts in the Al-$MgSiO_3$ perovskite lattice. Note that the $CaSiO_3$-perovskite phase is likely to be a preferential host for bigger cations, such as K (Corgne et al., 2003). It was also reported that (Mg,Fe)O ferropericlase could host elements such as Na (Kesson et al., 1998; Nishiyama and Yagi, 2003).

In any case, Al appears to be an important element that facilitates the substitutions and helps to resolve electroneutrality. Another dominant element, water (or the H-defect), could also play a comparable role. The presence of a significant amount of water in Al-$(Mg,Fe)SiO_3$ perovskite remains, however, uncertain and controversial. Previous works have suggested large amounts of water in Al-$MgSiO_3$ perovskite phases (Litasov et al., 2003; Murakami et al., 2002), but the most careful work performed on large perovskite single crystals does not show the presence of a significant amount of structural OH species (Bolfan-Casanova et al., 2003). Also, it has been shown from a careful infrared analysis that the water eventually present in some of the perovskite grains is not in the perovskite structure itself, but instead it is inserted into inclusions of a different mineral, such as the ringwoodite spinel (Fig. 5, H_2O in Al-$(Mg,Fe)SiO_3$ perovskite). This latter phase is likely to disappear at lower-mantle conditions, and therefore the water content in the Al-$(Mg,Fe)SiO_3$ perovskite structure

any other element, with variable size and electronic properties, found to be in minor or trace abundances in geological materials. In fact, the three major lower-mantle phases (Al-$[Mg,Fe]SiO_3$ perovskite, $CaSiO_3$ perovskite, and [Mg,Fe]O ferropericlase) appear to have the ability to store all the elements present in the pyrolitic-type composition, as no other phase was observed in sample charges recovered from experiments performed at lower-mantle pressure and temperature conditions (Irifune, 1994). Nevertheless, from these experiments it is difficult to tell if each element was really inserted into the lattice of one of these structures, or if some of them inhabited the grain boundaries. The grain boundary can indeed become a best refuge when a given element is present in amounts too small to form a separate phase. For higher amounts of "incompatible" elements, in mid-ocean-ridge basalt (MORB)–type material for example, other phases like the new aluminous– or the calcium-ferrite–type phases are observed in the high-pressure mineralogical assemblage (Guignot and Andrault, 2004; Hirose et al., 1999; Irifune and Ringwood, 1993; Kesson et al., 1994; Ono et al., 2001). Note that for this type of material, the Al-$(Mg,Fe)SiO_3$ perovskite phase adopts a radically different composition with up to 20% Fe and 10% Al per formula unit (Hirose et al., 1999; Ono et al., 2001).

TABLE 1. CHEMICAL COMPOSITION (MOL%) OF Al-$(Mg,Fe)SiO_3$ PEROVSKITE PHASES SYNTHESIZED FROM A WIDE VARIETY OF STARTING MATERIALS

Reference	A	B	C	D	E	F
P (GPa)	70	135	23.5	27	30	24
SiO_2	51.13	49.12	48.06	38.04	35.91	44.31
TiO_2	0.13	0.33	0.19	2.74	3.66	0.09
Al_2O_3	2.96	2.61	1.89	9.68	9.21	9.58
Cr_2O_3	0.18	0.17	0.16	–	–	–
FeO	3.53	3.34	4.47	19.58	18.09	17.4
MgO	41.59	43.48	44.34	27.07	32.55	28.51
CaO	0.476	0.94	0.76	1.8	0.27	0.13
Na_2O	0.0	0.0	0.14	0.86	0.3	–

Note: A–B: Kesson et al. (1998), and C: Wood (2000) are for pyrolitic-type materials, with an Al$(Mg,Fe)SiO_3$ perovskite phase found in contact with magnesiowüstite (A, B) and majoritic garnet (C). D: Hirose et al. (1999), and E: Ono et al. (2001) are for Al-$(Mg,Fe)SiO_3$ perovskite phases synthesized from mid-ocean-ridge basalt (MORB)–type materials. F is for Al-$(Mg,Fe)SiO_3$ perovskite phase synthesized from sediment-type material (Irifune et al., 1994).

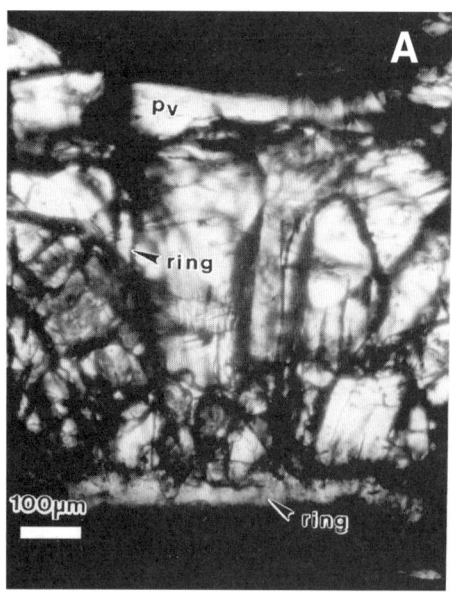
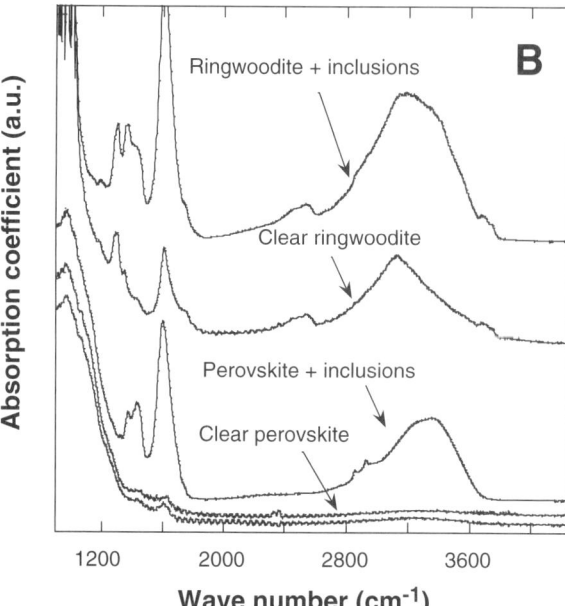

Figure 5. Infrared absorption coefficients of $(Mg,Fe)_2SiO_4$ ringwoodite-spinel and of $(Mg,Fe)SiO_3$ perovskite (derived from Bolfan-Casanova et al., 2003). (A) Optical photomicrograph showing $(Mg,Fe)SiO_3$ perovskite (white, pv) coexisting with ringwoodite ("ring") in a sample synthesized at 24 GPa and 1400 °C. (B) The Fourier transform infrared (FTIR) spectra were recorded in the OH-bending region for the different sample zones. The spectra recorded in regions of clear perovskite appear very flat, therefore pointing to very low water content. In some regions, the perovskite spectra appear largely contaminated by very small ringwoodite inclusions.

could be the most limited. Note that more water could be inserted into the (Mg,Fe)O phase (Bolfan-Casanova et al., 2002). Still, it is very likely that the total water content in the lower mantle is significantly lower than that expected in the transition zone (Bolfan-Casanova, 2005).

EQUATION OF STATE OF Al-(Mg,Fe)SiO$_3$ PEROVSKITE

While several experimental studies have been devoted to determination of bulk modulus and its pressure and temperature derivatives for various compositions of Al-(Mg,Fe)SiO$_3$ perovskite phases, only a few studies have addressed the shear properties at room (or moderate) pressure and temperature. The result is that our knowledge of the shear modulus and its derivatives, and, a fortiori, of the whole set of elastic constants, remains in the restricted field of ab initio calculations. Therefore, modeling of lower-mantle properties and dynamics, which requires a complete data set of bulk and shear moduli in an extended pressure and temperature range, can be done after compiling the most precise information of different experimental and theoretical reports (Table 2, elastic parameters).

Equation of State of (Mg,Fe)SiO$_3$ Perovskite

The PV-300K equation of state of (Mg,Fe)SiO$_3$ perovskite has been investigated in a large pressure range up to the core-mantle boundary conditions (Mao et al., 1991). The role of Fe is mainly to modify the room pressure density through a slight increase of the room pressure volume and a more severe change of the molar mass. Its effect on bulk properties appears negligible. Therefore, in subsequent studies devoted to the P-V-T equation of state, the MgSiO$_3$ perovskite phase has been used as an analogue to describe the lower-mantle Al-(Mg,Fe)SiO$_3$ perovskite elastic

TABLE 2. SET OF (Mg,Fe)SiO$_3$ THERMOELASTIC PARAMETERS

Ref.	A	B	C
ρ (g/cm^3)	4.1	4.11	4.1
$\Delta\rho$	1.03	1.07	1.09
K_0 (GPa)	264 (KS)	258 (KT)	250 (KT)
ΔK_0	20	–	–
$(dK/dP)T$	3.97	4.1	4
$(dK/dT)P$ (GPa/K)	–0.011	–0.029	–0.021
G_0 (GPa)	175	177	175
ΔG_0	–40	–	–40
$(dG/dP)T$	1.8	1.4	1.8
$(dG/dT)P$ (GPa/K)	–0.029	–0.024	–0.026
γ_0	1.31	1.33	–
Q	1.0	–	–
a_1 (10^5 K^{-1})	1.19	2.7	2.46
a_2 (10^8 K^{-2})	1.2	–0.165	–

Note: "Δ" indicates corrections for the Fe content. Given the iron content, X_{Fe}, the corrected value for a parameter M is $M_{(Mg,Fe)SiO_3} = (M_{MgSiO_3} + \Delta M^*X_{Fe})$. Thermal expansion at 1 bar is $\alpha(T) = a_1 + a_2T$. A, B, and C are data sets compiled from a wide range of experimental and ab initio calculation studies by Deschamps and Trampert (2004), Samuel et al. (2005), and Mattern et al. (2005), respectively.

properties. Diverse experimental techniques have been used for the determination of the $MgSiO_3$ perovskite P-V-T equation of state, including in situ X-ray diffraction in a laser-heated diamond anvil cell (Fiquet et al., 1998, 2000; Shim and Duffy, 2000) and in a multianvil press (Funamori and Yagi, 1993; Wang et al., 1994), and also ab initio calculations (Cohen, 1987; Karki et al., 2001; Matsui, 2000; Oganov et al., 2001), etc. (Fig. 6, P-V-T of $MgSiO_3$). Recently, theoretical calculations have predicted a significant effect of Fe content on shear modulus, with a decrease of up to 8% when adding 25 mol% of $FeSiO_3$ to $MgSiO_3$ perovskite (Kiefer et al., 2002).

Al-(Mg,Fe)SiO_3 Perovskite Equation of State: State of the Art

The equation of state of lower-mantle Al-$MgSiO_3$ perovskite has remained controversial in the past few years. Different studies have proposed that the effect of Al is to lower the perovskite bulk modulus (K_0), possibly down to values of 230 GPa (Daniel et al., 2001; Kubo et al., 2000; Yagi et al., 2004; Zhang and Weidner, 1999). However, these results are in strong disagreement with other studies that show an almost insignificant Al effect (Andrault et al., 2001; Daniel et al., 2004; Jackson et al., 2004; Walter et al., 2004; Yagi et al., 2004). Since both experimental trends have been reproduced in various research groups, it is likely that the controversy is at least in part due to intrinsic properties of Al-$MgSiO_3$ perovskite rather than to any particular experimental problems. Such controversy most probably results from the coexistence of two Al-substitution mechanisms in the Al-(Mg,Fe)SiO_3 perovskite lattice. Higher pressures could strongly favor the coupled substitution (Brodholt, 2000; Yamamoto et al., 2003), while higher temperatures could favor the O-vacancy mechanism (Akber-Knutson and Bukowinski, 2004; Navrotsky et al., 2003). Still, if this reasoning appears simple from a theoretical point of view, it remains based on semiquantitative statements, and it remains unclear how the changes in Al-substitution mechanism affect the perovskite elastic parameters.

Effect of Synthesis Conditions on the Al-(Mg,Fe)SiO_3 Perovskite Equation of State

In a recent study, we determined new compression curves for Al-$MgSiO_3$ perovskite samples synthesized at variable conditions of pressure and temperature (our unpubl. data). We used laser-heated diamond anvil cells (LH-DAC) and angle-dispersive X-ray diffraction at the ID30 beam-line of the ESRF (European Synchrotron Radiation Facility), using a similar procedure as in our previous work (Andrault et al., 2001). As starting materials, we used glasses of different compositions with and without excess of MgO or SiO_2. We found a minor effect of the compositions on the general interpretations. More details are provided elsewhere.

In a first set of three experiments, we synthesized the silicate perovskite at nominal pressures of 26, 35, and 47 GPa for temperatures below 2000 K. We then investigated the compression curves to 37, 46, and 56 GPa, respectively. The compression curves appeared to be significantly different from each other. The syntheses performed at increasing synthesis pressure yielded an increasing bulk modulus for Al-(Mg,Fe)SiO_3 perovskite, with a variation of more than 10% (Fig. 7A, equation of state, P effects). Also, the synthesis performed at 47 GPa yielded a bulk modulus of ~270 GPa, which was significantly higher than the K_0 value of 255–260 GPa accepted for the Al-free $MgSiO_3$ perovskite compound. In Figure 7, the room pressure volumes, V_0, may appear ambiguous, because (1) the effect of Al on the perovskite on V_0 is probably different for the two Al-substitution mechanisms (Navrotsky et al., 2003), and (2) the perovskite grains synthesized at a nominal pressure of 26 GPa may have contained slightly less Al than in the two other runs at 35 and 47 GPa, due to the particular shape of $MgSiO_3$-Al_2O_3 at moderate pressures.

We performed a second set of experiments using variable synthesis temperatures at a given pressure. Again, it resulted in a significant variation of the Al-$MgSiO_3$ perovskite compression curves (Fig. 7B, equation of state, T effect). The sample heated at modest temperatures of ~1800 K presented perovskite volumes in good agreement with those obtained for the highest synthesis pressure (Fig. 7A). In contrast, significantly higher

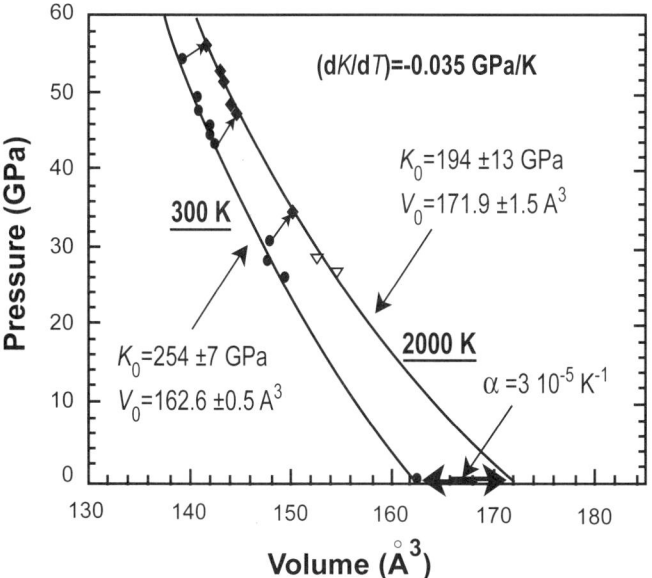

Figure 6. Pressure-volume (P-V) diagram showing $MgSiO_3$ perovskite unit-cell compression along two isotherms at 300 K and 2000 K (derived from Fiquet et al., 1998). Lines represent the best fit to the data, using isothermal second-order Birch-Murnaghan equation of state ($K'_0 = 4$).

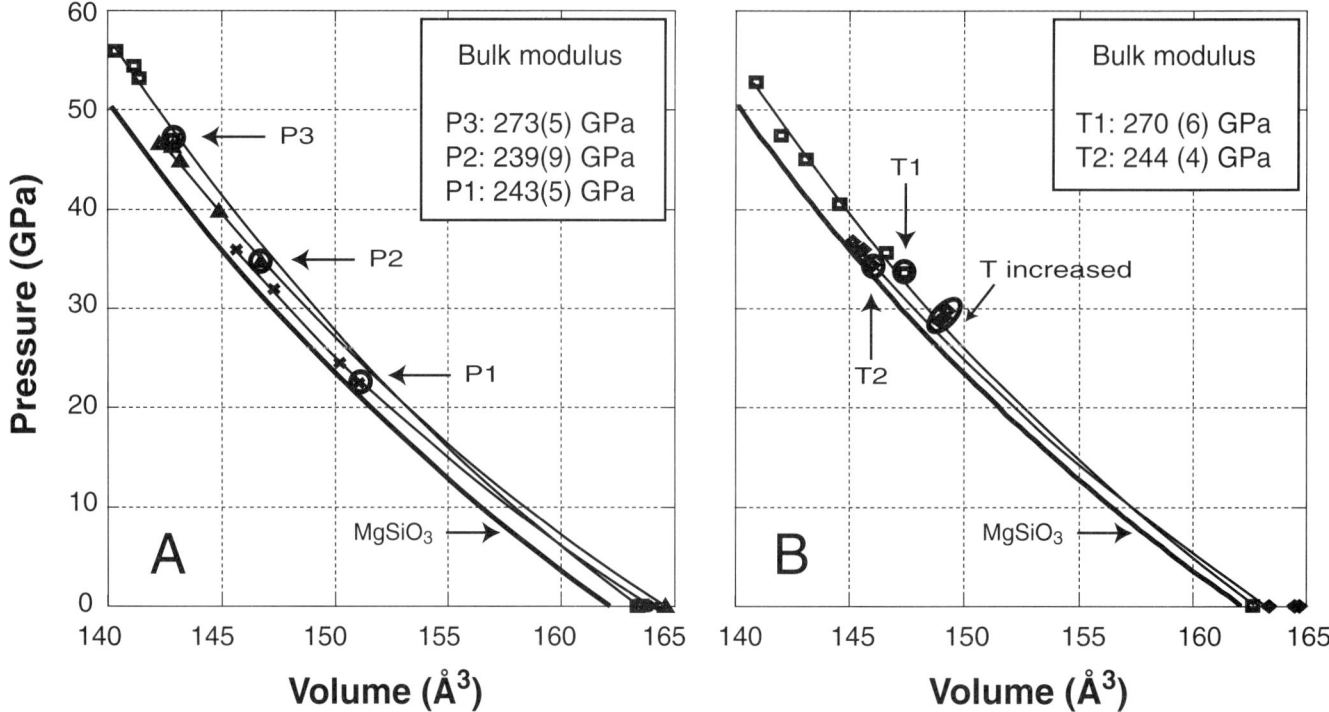

Figure 7. Volumes and compression curves obtained for Al-MgSiO$_3$ perovskite samples synthesized (A) at variable pressures of P1 = 26 GPa, P2 = 35 GPa, and P3 = 47 GPa. Higher bulk modulus is found for higher synthesis pressure; and (B) at variable temperatures. At a same pressure of ~34.5 GPa, the perovskite volume appears significantly higher after heating at 1800 K (T1) compared with heating to 2300 K (T2). Also, heating a sample loaded at ~30 GPa at increasing temperature yields progressive decrease of its unit-cell volume (derived from our unpubl. data). This effect is compatible with more Al substitution via the O-vacancy mechanism with increasing temperature (derived from our unpubl. data).

compressibility was found for another run performed at much higher temperatures of ~2300 K. A more detailed analysis of the compression curves is provided elsewhere.

In any case, we clearly confirmed that the Al-MgSiO$_3$ perovskite equation of state is significantly affected by the synthesis pressure and temperature conditions. According to the crystal chemistry of Al-MgSiO$_3$ perovskite, it is likely that the lower bulk modulus is correlated with a higher amount of Al substitution via of O vacancies. Following this principle, our results appear to be nicely compatible with an O-vacancy mechanism favored at lower pressures and higher temperatures, a trend perfectly compatible with previous experimental and theoretical reports (Akber-Knutson and Bukowinski, 2004; Brodholt, 2000; Navrotsky et al., 2003; Yamamoto et al., 2003). According to these new results, the controversy on the Al-MgSiO$_3$ perovskite elastic properties is readily explained. The different groups analyzed Al-MgSiO$_3$ perovskite samples with different types of defects population because of the variable pressure and temperature conditions available in LH-DAC and multianvil apparatus. Therefore, different bulk moduli resulted. It is worth noting that a pronounced effect of Al on the shear properties was observed for samples for which the bulk modulus K_0 was not much affected by Al (Jackson et al., 2004).

P-V-T Equation of State Relevant to Lower-Mantle Perovskite Phases

It is thus necessary to estimate which defect population dominates in the lower-mantle Al-(Mg,Fe)SiO$_3$ perovskite phase before we can define a relevant equation of state for it. There are different reasons to believe that the amount of O vacancies is much limited and consequently, that its characteristic bulk modulus is close to that of Al-free silicate perovskite. Concerning the deepest part of the lower mantle, our experimental results (our unpubl. data) and theoretical calculations (Brodholt, 2000; Yamamoto et al., 2003) agree for a limited amount of O vacancies that prevent ideal compactness. Then, the presence of Fe in the lower mantle, for Fe/(Mg + Fe) ratio close to 10%–12%, is expected to strongly favor (Al,Fe^{3+}) coupled substitution in the Al-(Mg,Fe)SiO$_3$ perovskite (Lauterbach et al., 2000). Finally, from a pure experimental point of view, it appears that the few reports devoted to the compression behavior of mantle-type Al-(Mg,Fe)SiO$_3$ perovskite are not controversial. Instead, they are in good agreement with a minor Al effect, and possibly a slight increase of the perovskite bulk modulus compared to the MgSiO$_3$ composition (Andrault et al., 2001; Ono et al., 2004) (Fig. 8, equation of state of mantle relevant Al-[Mg,Fe]SiO$_3$ perovskite phases). In our recent work, we also investigated the compression behavior

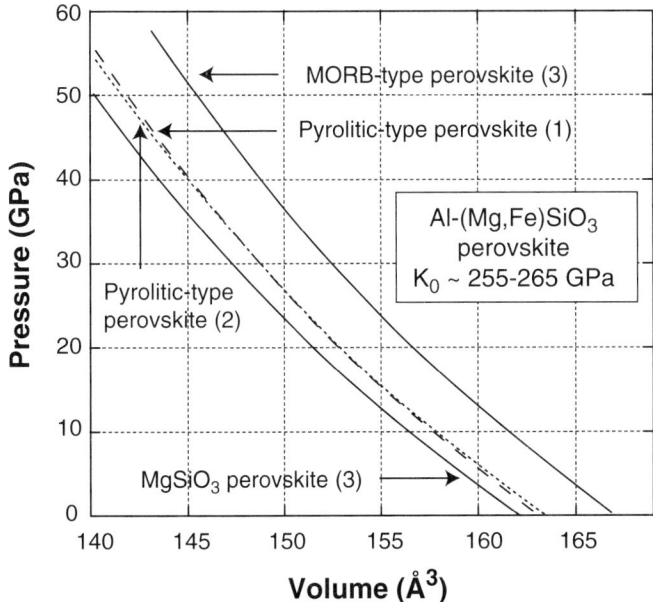

Figure 8. Pressure-volume (P-V) compression curves for perovskite phases with $MgSiO_3$ and Al-(Mg,Fe)SiO_3 of pyrolitic- and mid-ocean-ridge basalt (MORB)–type compositions. Experimental data are from 1(dashed)—Andrault et al. (2001), 2(dotted)—Ono et al. (2004), and 3(our unpubl. data). All phases agree well with K_0 of ~265 GPa (with K' fixed to 4), a value slightly higher than that of $MgSiO_3$ perovskite at ~257(2) GPa (our unpubl. data). The room pressure volume of MORB-type perovskite is much affected by the high Fe content.

of a perovskite phase relevant to MORB material and a refined bulk modulus of ~265 GPa. Therefore, the effect of Al on the Al-(Mg,Fe)SiO_3 perovskite bulk modulus remains quite limited for any geophysically relevant perovskite phase.

Questions may remain about the role of O vacancies in the uppermost part of the lowest lower mantle, when the pressure is insufficient to force the coupled Al (and maybe [Al,Fe^{3+}]) substitution. But in any case, it is clear that extrapolation to the highest lower-mantle pressures of the perovskite equation of state with a relatively low bulk modulus is largely irrelevant, because the defect population is expected to re-equilibrate as a function of depth, favoring coupled Al substitution and a higher bulk modulus in the deep mantle. Note that the effect on the shear properties may be more pronounced than for the bulk modulus K_0 (Li et al., 2005).

REFINING THE LOWER-MANTLE COMPOSITION USING THE ELASTIC PARAMETERS

Different geochemical models propose ambiguous lower-mantle compositions. This uncertainty results from the complex history of Earth's formation. First, the source material from which our planet was generated remains controversial, as it could be of composition similar to carbonous (CI) or enstatite (EH) chondrites, for example (Allègre et al., 1995; Javoy, 1995). Then, after Earth accretion, important events, including core formation and possible giant impacts with late veneers, could have significantly affected the mantle composition. Because the composition of the first ~200 km of the upper mantle is relatively well constrained from field measurements and direct observations, the uncertainty in global mantle composition is mostly transmitted to the deep mantle. Its Si/(Mg + Fe + Ca) ratio, in particular, could vary as much as from ~0.7 to ~1 according to pyrolitic- or EH-chondrite–derived models. This ratio is particularly critical because it defines the relative amounts of perovskite and ferropericlase phases, and in a more extreme case, the eventual presence of SiO_2 stishovite. Note that the uncertainty in mantle composition predominantly affects the amount of secondary phases. On the other hand, the presence of the Al-(Mg,Fe)SiO_3 perovskite phase can hardly be avoided for any geological material in the lower mantle.

Another procedure that infers the chemical composition of the lower mantle is the modeling of lower-mantle speeds of sound (Vp, Vs, and Vφ,) and density (ρ) profiles that are provided by seismology (Dziewonski and Anderson, 1981). For this, we need to use the elastic parameters of all potential lower-mantle phases (as an example, we report the complete data set for the [Mg,Fe]SiO_3 perovskite phase in Table 2). Another dominant parameter is the temperature profile in the lower mantle, which significantly affects the elastic properties of lower-mantle minerals. This profile remains unfortunately poorly constrained, with temperature uncertainties of up to more than 500 K for the deep lower mantle (Brown and Shankland, 1981; Bunge et al., 2001; Da Silva et al., 2000). Therefore, the temperature profile must also be refined together with the other main parameters; the Mg/Si and Fe/(Mg + Fe) ratio, the Ca and Al contents, etc. In these types of calculations, the effect of Al on the perovskite equation of state is usually neglected.

When modeling ρ and Vφ (or K) profiles only, the model yields a typical Fe/(Mg + Fe) ratio of ~12%, the presence of ~10% of ferropericlase, and a reasonable adiabatic temperature profile (Bina and Silver, 1990; Fiquet et al., 1998; Jackson, 1998; Sinelnikov et al., 1998; Wang et al., 1994; Yagi and Funamori, 1996). It should be noted, however, that different chemical models can adequately reproduce the Vφ and ρ profiles (Mattern et al., 2005) (Fig. 9, modeling the lower mantle). A specific temperature profile is calculated for each possible composition. Considering the Mg/Si ratio only, a higher-temperature profile is required for SiO_2-enriched materials. The reason is that higher SiO_2 content makes the material more stiff, an effect that can only be counterbalanced by higher temperatures, in order to stick to the seismological data. In other words, the seismological data could be compatible with a lower mantle made of pure (Mg,Fe,Ca)SiO_3 perovskite, if the temperature were relatively high. On the other hand, the Fe content can easily be adjusted to reproduce the lower-mantle densities, because its effect on the elastic properties remains limited.

The most recent calculations have taken advantage of the shear properties to refine the three seismological Vp, Vs, and ρ lower-mantle profiles and their anomalies (Deschamps and

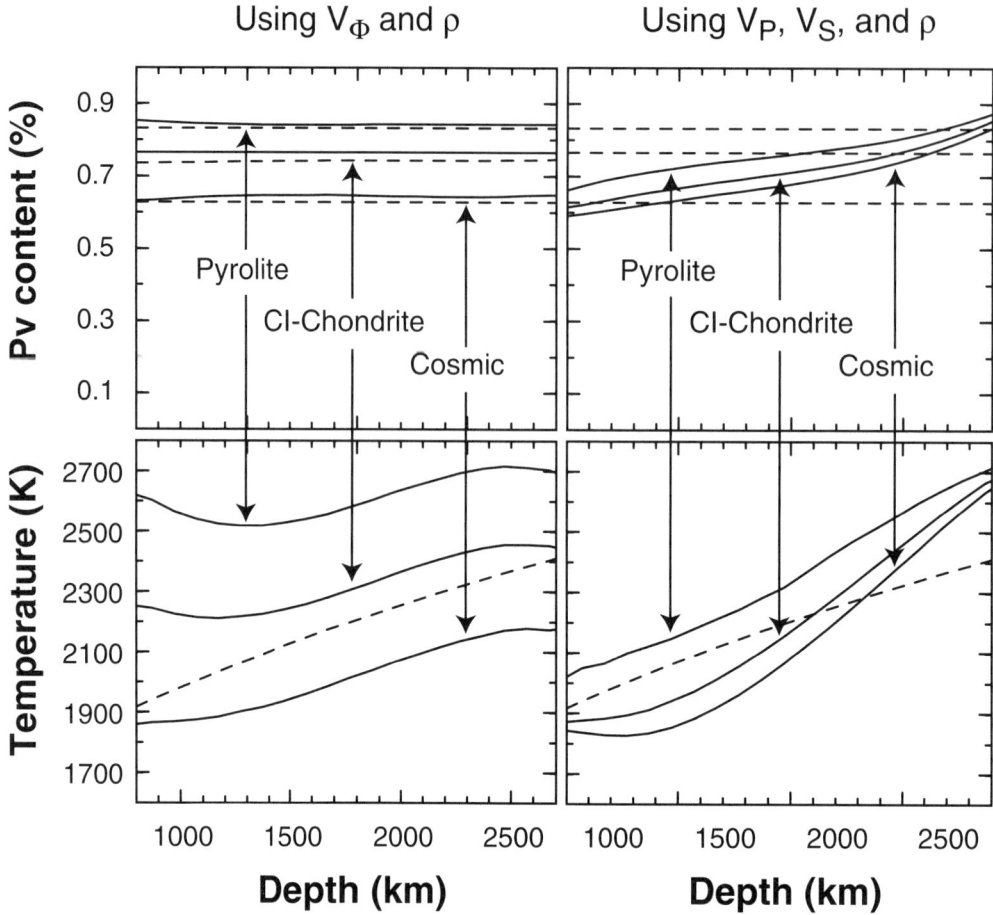

Figure 9. Models of lower-mantle Al-(Mg,Fe)SiO$_3$ perovskite (Pv) contents (upper frames) and temperature profiles (lower frames) reproducing the lower-mantle seismic profiles (PREM) of density ρ and bulk sound velocity V_Φ (left frames), or ρ and compressional V_P and shear V_S velocities (right frames) (derived from Mattern, 2005). A priori models (dashed lines) consist of pyrolite, CI-chondritic, or cosmic compositions, and adiabatic geotherm (Brown and Shankland, 1981). After numerical convergence, the refined compositional and thermal profiles (solid lines) are reported for each a priori model. The left panels shows that the numerical convergence is obtained for different mantle compositions, each of which correlate with a given temperature profile. Therefore, reproducing (ρ,V_Φ) seismic profiles yields insufficient constraints compared to the too many lower-mantle parameters to be refined. A much better numerical convergence is obtained for the compositional and thermal profiles when the calculation reproduces the three (ρ,V_P,V_S) seismic profiles.

Trampert, 2004; Mattern et al., 2005; Samuel et al., 2006). The method appears to potentially better constrain the chemical composition and the temperature profile. Best convergences are achieved with Fe and Si enrichments in the lowermost part of the lower mantle. This chemical stratification would help to explain specific seismological observations, such as the anticorrelation between bulk sound and shear-wave velocity anomalies parameters observed in this mantle region (Samuel et al., 2006). Also, the refined average geotherm could be slightly superadiabatic (e.g., Brown and Shankland, 1981), with temperature ranging from ~1900 K at the 660 km discontinuity to ~2700 K at the core-mantle boundary (Mattern et al., 2005). These conclusions should to be taken with some care, however, because the shear properties of the different lower-mantle phases that are essential for this numerical convergence are not yet known with sufficiently high accuracy.

PARTIAL Fe VALENCE CHANGE AND OXYGEN FUGACITY IN THE LOWER MANTLE

Coexistence of Fe^{3+} and Fe^0 in Sample Charges

The presence of Al induces the formation of Fe^{3+} in lower-mantle silicate perovskite. The way to produce large amounts of ferric iron from a starting material initially poor in Fe^{3+} remains unclear. For large-volume press experiments, oxygen could be diffusing from the experimental assemblage to the sample, or it could be the result of reduction of another material present in the sample vicinity. In diamond anvil cell experiments, the situation is different because there is often no material to provide oxygen to the sample. In a certain sense, this situation may be closer to that found in Earth's mantle. Indeed, for a pyrolitic-type material located in the lower mantle, the surrounding material is limited to Al-(Mg,Fe)SiO$_3$ perovskite itself, CaSiO$_3$ perovskite, and the (Mg,Fe)O ferropericlase. Some particular behavior is to be expected if the apparently unavoidable formation of some Fe^{3+} should occur. A few years ago, Mössbauer analyses evidenced the coexistence of an Al-(Mg,Fe)SiO$_3$ perovskite with ~5% ferric iron in a sample charge for which the oxygen fugacity (fO_2) was buffered to the Fe/FeO equilibrium (McCammon et al., 1992). Thus, significant amount of Fe^{3+} (in the perovskite phase) and Fe^0 (from the fO_2 buffer) were found to co-exist in the same sample. Similar results have been reported in other diamond anvil cell (Fujino et al., 1998) and multianvil press (Frost et al., 2004; Oguri et al., 2000; Wood and Rubie, 1996) experiments. These observations evidence that the power to oxidize some Fe^{2+} in Fe^{3+} (to be inserted in the Al-[Mg,Fe]SiO$_3$ perovskite lattice) is stronger than the power to oxidize Fe^0 into Fe^{2+} (to be inserted in the ferropericlase phase). In other words, the partial dismutation of Fe^{2+} into Fe^{3+} (in Al-[Mg,Fe]SiO$_3$ perovskite) and Fe^0 appears to occur at lower-mantle conditions.

Fe^{2+} Dismutation in the Lower Mantle

A few years ago (Andrault, 1999), we performed laser-heating experiments in a diamond anvil cell using starting materials made of synthetic (Mg$_{0.84}$,Fe$_{0.16}$)SiO$_3$ enstatite mixed, or not, with 8% Al$_2$O$_3$. Starting materials were loaded in Re-gaskets under Ar flux to prevent the capture of atmospheric oxygen into the high-pressure cavities. Pressures of ~45 GPa were observed from an external pressure gauge, previously calibrated using the ruby fluorescence technique. A defocused YAG laser was scanned all over the samples for several minutes, heating the whole sample volume to similar temperature. The temperature was evaluated to be ~2000 K from optical measurements. The samples were quenched to ambient conditions, extracted from the Re-gaskets, and mounted between copper grids. We carried out ion-thinning for microstructure observations and microanalyses into an analytical transmission electron microscope (ATEM, JEOL 200 SX coupled with a TRACOR X-ray fluorescence analysis).

We present microphotographs for the two types of samples, with and without Al$_2$O$_3$-additions (Fig. 10, Al-[Mg,Fe]SiO$_3$ perovskite microphotographs). In both cases, the main phase is the main lower-mantle silicate perovskite phase, the structure of which eventually amorphizes under the electron beam. The grain shape is polygonal, indicating that the samples have reached local equilibrium. We recognize the (Mg,Fe)SiO$_3$ perovskite twin structure already reported previously (Wang et al., 1990). More interestingly, we see evidence of the presence of small, dark, and round-shaped grains. These grains do not amorphize under the electron beam, and their diffraction pattern is compatible with the body-centered cubic (bcc) structure of iron. This observation is confirmed by X-ray fluorescence analysis, which shows evidence of a pure Fe composition for the small dark grains (Fig. 11, Al-[Mg,Fe]SiO$_3$ perovskite chemical analyses). The chemical analyses also show that the silicate phase has integrated the Al$_2$O$_3$, an effect confirmed by the absence of corundum in the rest of the sample. Due to the insertion of Al into the perovskite lattice and the presence of Fe^0, our observations are compatible with the following reaction:

$$(Mg_{1-x},Fe_x)SiO_3 + y[Al]_2O_3 \rightarrow (Mg_{1-x},Fe_{x-3z})[Fe_{2z},Al_{2y}]Si_{1-3z}O_3 + 3y - 2z + 3zSiO_2 + zFe, \quad (1)$$

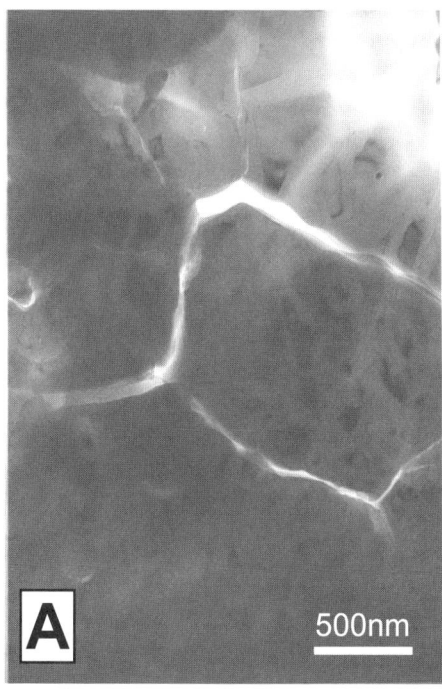 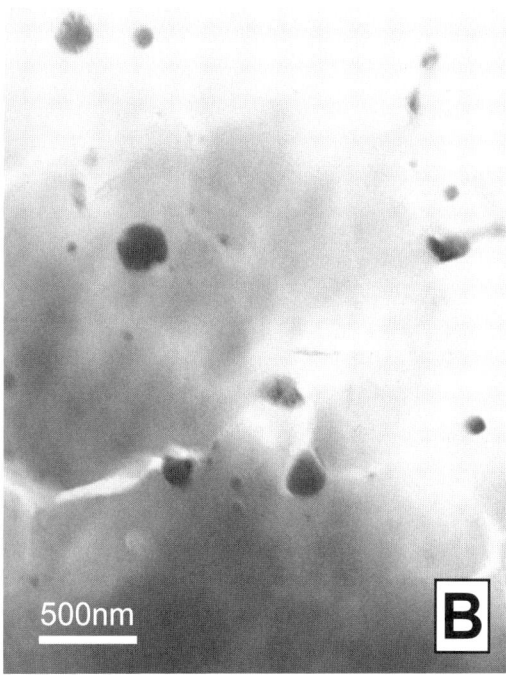

Figure 10. Microphotographs of samples recovered after experiment at ~45 GPa in a laser-heated diamond anvil cell. Starting materials were composed of (A) synthetic (Mg$_{0.84}$,Fe$_{0.16}$)SiO$_3$ enstatite and (B) mixture of the same enstatite material and corundum. Al-(Mg,Fe)SiO$_3$ perovskite matrixes show well-formed polygonal grains. In B, the small dark grains are nearly pure iron (see Fig. 11).

where parentheses and brackets correspond to divalent and trivalent cations, respectively. The obtained silicate perovskite phase adopts the $(Mg_a,Fe_b)[Fe_c,Al_d]Si_eO_3$ stoichiometry, with typical values of $a = 0.854$, $b = 0.04$, $c = 0.04$, $d = 0.065$, and $e = 0.968$, as suggested from multianvil work (McCammon, 1997). In agreement with Equation 1, the presence of Fe grains confirms the Fe^{2+} dismutation into Fe^{3+} and Fe^0. The SiO_2 particles are not easy to find in the sample because they show imaging contrast similar to the silicate perovskite. Equation 1 also proposes that the perovskite itself can be partially reduced to provide some oxygen for the formation of Fe^{3+}. In pyrolitic-type lower mantle, the ferropericlase phase is expected to be present, and the Fe-dismutation reaction becomes:

$$a(Mg,Fe)SiO_3 + b(Mg,Fe)O + y[Al]_2O_3 \rightarrow \qquad (2)$$
$$a'(Mg,Fe)[Fe,Al]SiO_3 + b'(Mg,Fe)O + zFe.$$

In this case $a' > a$, $b' < b$, and no stishovite phase is expected. According to this reaction, the partial dismutation of Fe^{2+} only requires slight modification of the perovskite/ferropericlase ratio in the lower-mantle mineral assemblage. This equilibrium also explains how the Fe-partitioning coefficient between the two phases is affected by the presence of Al (Wood and Rubie, 1996). Another main consequence is the coexistence of $(Mg,Fe)O$ and Fe^0 in Earth's lower mantle. This implies that the lower-mantle oxygen fugacity is buffered by $Fe/(Mg,Fe)O$ equilibrium.

PHASE TRANSFORMATION IN PBNM SILICATE PEROVSKITE

Four kinds of structural modifications have been proposed for $MgSiO_3$ and $Al-(Mg,Fe)SiO_3$ perovskite at the high pressures of the lowermost mantle. They concern subtle symmetry changes, electronic transition, modification of octahedral stacking, and decomposition into the mixture of SiO_2 and MgO. Some of these transformations remain controversial or have already been demonstrated to be unlikely in Earth's mantle. One should note that investigation of these extreme pressure and temperature conditions requires the use of very tiny samples loaded in laser-heated diamond anvil cells, and erroneous conclusions can arise from the nonequilibrium transformations that are produced when the deviatoric stresses and the temperature gradients are not correctly controlled.

Decomposition into an Oxide Mixture

The decomposition of $MgSiO_3$ has been proposed by two different groups in the 70–80 GPa region (Meade et al., 1995; Saxena et al., 1996). In both cases, the temperature gradients were not optimized and, in one study, the laser spot was maintained for several minutes focused on a single sample location. Several questions have been raised about the reproducibility and relevance of this transformation for the lower mantle. Indeed, other groups established the stability of the main silicate perovskite phase at even higher pressures for $MgSiO_3$ (Gong et al., 2004; Mao et

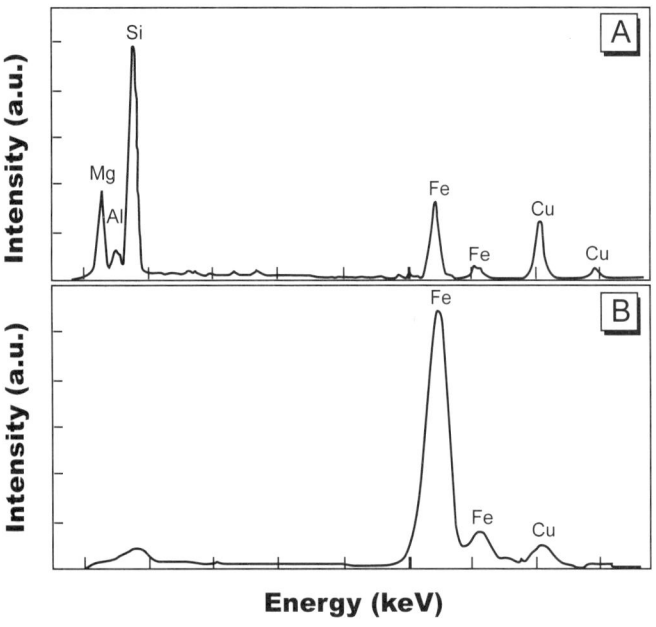

Figure 11. X-ray fluorescence microanalyses of minerals shown in Figure 10B. (A) The insertion of Al into $Al-(Mg,Fe)SiO_3$ perovskite was possible thanks to the extensive laser heating at high temperature. (B) Analysis of the small and dark grains shows that they are nearly pure iron. Some Cu was found due to pollution from the grids.

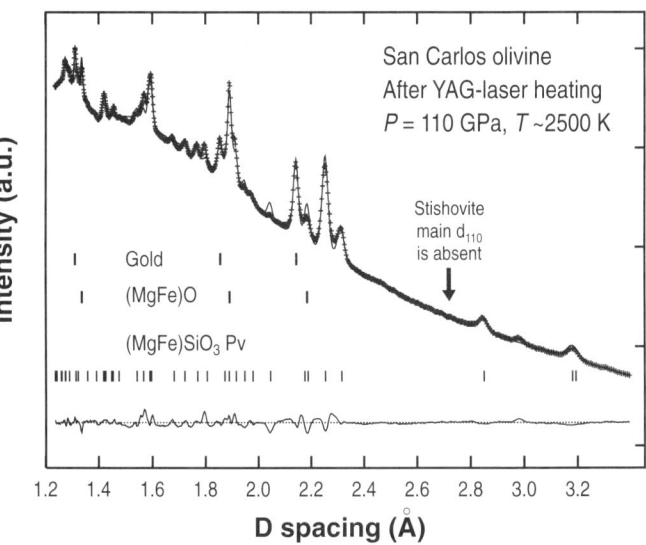

Figure 12. Mixture of $(Mg,Fe)SiO_3$ perovskite (Pv) and $(Mg,Fe)O$ ferropericlase (plus gold pressure marker) first synthesized from a San Carlos olivine at ~40 GPa using laser-heating, and then laser-annealed for several minutes at ~2500 K after recompression to 110 GPa. The X-ray diffraction pattern is reported together with a Rietveld fit for the phase mixture. The absence of stishovite evidences the absence of the magnesian silicate perovskite decomposition in this pressure-temperature range.

al., 1997; Serghiou et al., 1998) and Al-(Mg,Fe)SiO$_3$ compositions (Andrault, 2001; Kesson et al., 1995) (Fig. 12, stability of [Mg,Fe]SiO$_3$ perovskite). In the most significant work performed up to 100 GPa and 3000 K, the starting material made of a mixture of (Mg,Fe)O and SiO$_2$ was observed to recombine into the (Mg,Fe)SiO$_3$ perovskite phase, which definitively rules out silicate perovskite decomposition in these *P-T* conditions (Serghiou et al., 1998). It is probable that the decomposition was observed due to the Soret effect, which favors the motion of some atomic species relative to other ones in large temperature gradients, an effect already evidenced in laser-heated diamond anvil cells for (Mg,Fe)SiO$_3$ perovskite (Andrault and Fiquet, 2001; Campbell et al., 1992).

Subtle Changes of Symmetry

A subtle change of the perovskite structure has been reported at ~83 GPa and 1700 K (Shim et al., 2001). A new diffraction line was observed at 2.57–2.62 Å, and pressure evolution appeared compatible with the perovskite lattice compression behavior. In previous work performed on a mixture of MgSiO$_3$ perovskite and platinum at 85.6 GPa pressure, we also observed a new peak at this position, and a few others with variable intensities (unpublished work done in collaboration with G. Fiquet, D. Häusermann, and M. Kunz at the ID30 beamline of the ESRF) (Fig. 13, monoclinic [Mg,Fe]SiO$_3$ perovskite). This sample had been laser heated using a YAG laser and Pt absorber to more than 2000 K for several minutes, and, after the quench to 300 K, it remained loaded at this pressure without further treatment for more than 48 h. The diffraction pattern just after the quench showed the normal Pbnm perovskite features, but additional Bragg lines appeared after the waiting period. Also, after renewed laser heating to high temperatures, a set of new Bragg lines was found to evolve with temperature, with fewer lines than in the diffraction pattern presented in this paper.

For the diffraction pattern presented in Figure 13, the new lines observed, for example, at ~4.49, ~4.22, and ~2.60 Å cannot be explained by the normal Pbnm symmetry. Instead, they can be explained by a monoclinic cell with $a = 4.473$ Å, $b = 12.816$ Å, $c = 4.7649$ Å, and $\beta = 87.5°$. There are inconsistencies, however, between experimental and theoretical pattern, which may indicate a more complex symmetry. This monoclinic lattice would correspond to the doubling of the Pbnm unit cell along the *c*-axis. It is difficult at this point to refine a space group for this new perovskite lattice, however, because the set of Bragg lines evolves with the temperature conditions. Also, even if the diffraction peaks do not appear broad, we cannot be certain that our sample did not encountered deviatoric stresses that generated this subtle symmetry change. However, the occurrence of a low-symmetry MgSiO$_3$ perovskite lattice at low temperatures makes sense from a theoretical point of view. A similar transition is observed for various other perovskite compounds along the sequence of symmetry transitions (Glazer, 1972). The presence of a stability field at high pressures would be a strong argument for a higher dodecahedral compressibility compared to the octahedral site. This feature is largely compatible with previous structural refinements performed at low pressures (Kudoh et al., 1987), and it has been subject of discussion elsewhere (Andrault and Poirier, 1991; Ross et al., 2004). The continuous increase with pressure of the lattice distortion appears logical in the framework of the phase transition that yields to the post-perovskite CaIrO$_3$–phase at slightly higher pressure. The relevance for the lower-mantle properties in these subtle structural changes remains uncertain, however, because no similar structural modification has ever been reported for any Al-(Mg,Fe)SiO$_3$ perovskite, and because of the high lower-mantle temperatures.

Spin Transition

Another kind of transition has been reported for the spin state of Fe in the (Mg,Fe)SiO$_3$ perovskite, with two possible steps at 70 and 120 GPa. They are related to the partial and full electron pairing in Fe, respectively (Badro et al., 2004). Similar spin transition has also been reported for Al-bearing (Mg,Fe)SiO$_3$ perovskite at ~100 GPa by another group (Li et al., 2004). The main consequence of this spin transition is a significant volume change for the Fe cation. At room pressure, for example, the Fe^{2+} radius is reported to change from 0.92 Å to 0.75 Å along the high-spin to low-spin transition (Shannon and Prewitt, 1969). For the silicate perovskite, the density change related to this transition could be up to ~2%. This effect can yield important modifications of the Fe-partitioning coefficient between the silicate perovskite and

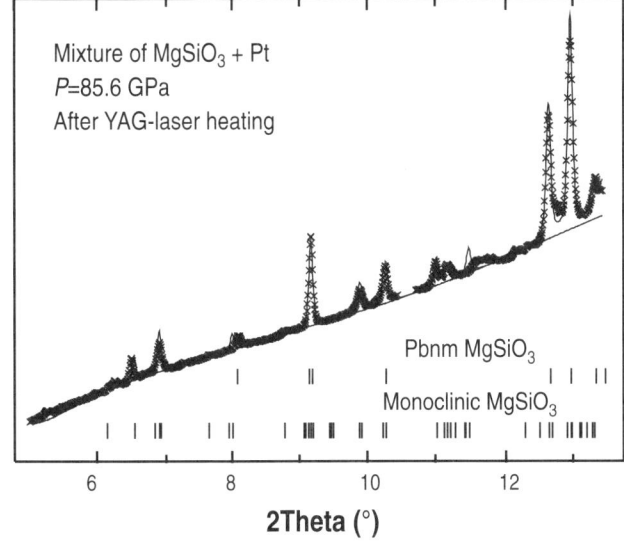

Figure 13. Tentative interpretation of the subtle symmetry change observed in MgSiO$_3$ perovskite in some high-pressure experiments. Here, the starting material was mixed with Pt powder and laser-annealed at high temperature in the diamond anvil cell for a nominal pressure of ~85.6 GPa. The classical Pbnm symmetry does not correctly explain the several new diffraction peaks observed with variable intensities. Instead, a new monoclinic lattice for the MgSiO$_3$ perovskite explains the new features relatively well.

ferropericlase phases. Note that iron also undergoes spin change in (Mg,Fe)O ferropericlase, possibly at shallower depth than for the perovskite phase (Badro et al., 2003). At the high temperature that prevails in the lower mantle, it is expected that the two high-spin and low-spin electron configurations coexist over a large pressure range, possibly extended over all the lower mantle, which would make no visible discontinuity from a seismological point of view (Sturhahn et al., 2005).

Post-Perovskite Structure

The final kind of transformation observed recently is a major change of the SiO_6 octahedral stacking that produces a post-perovskite structure. The proposed structure is that of $CaIrO_3$, which adopts the orthorhombic Cmcm space group. In this phase, some octahedra share edges, instead of corners as in perovskite (Murakami et al., 2004; Oganov and Ono, 2004). The density change is estimated to be more than 1%. The post-perovskite phase was first observed for the $MgSiO_3$ composition at ~120 GPa (Murakami et al., 2004). It seems that the kinetic effects are important, as several minutes of laser heating are required to complete the phase transformation. In a very recent work performed at the ID27 beamline of the ESRF, we also reproduced this phase transformation at ~110 GPa and 2500 K from both $MgSiO_3$ glass and $MgSiO_3$ perovskite starting material (Guignot et al., 2007). Another study proposed that Fe facilitates the phase transformation, as the post-perovskite phase has been synthesized at ~100 GPa for the $(Mg,Fe)SiO_3$ composition (Mao et al., 2004). Also, the presence of some Al does not seem to prevent the post-perovskite transformation, since a pyrolitic-type post-perovskite phase has been synthesized between 92 and 124 GPa (Murakami et al., 2005).

The possible stability of the Cmcm phase has been confirmed by ab initio calculations (Oganov and Ono, 2004; Tsuchiya et al., 2004). The elastic properties of this new phase, estimated from ab initio calculations, yield bulk moduli lower than those for silicate perovskite, which argues for significant modifications of the seismic properties at the transition pressure (Iitaka et al., 2004; Stackhouse et al., 2005). The occurrence of a post-perovskite phase is quite compatible with previous work based on seismic-wave analysis, which pointed out the ubiquitous presence of a phase transition at the base of the lower mantle (Sidorin et al., 1999). These findings contrast significantly, however, with the information previously provided by seismology. Indeed, many reports disagreed with a clear seismic discontinuity at a given depth, and instead suggested a complex lowermost mantle, with a large variability of the different seismic parameters (Kennett, 1998; Masters et al., 2000; Saltzer et al., 2001).

MELTING OF Al-(Mg,Fe)SiO₃ PEROVSKITE

Melting of the Perovskite Phase

The melting curve of $(Mg,Fe)SiO_3$ perovskite has been controversial in the past (Heinz et al., 1994), but recent experimental and theoretical studies tend to converge to a general agreement (Fig. 14, melting curve). Careful experiments performed in the multianvil press at 22–25 GPa report perovskite-phase melting of ~2773 K with a slope of 30 K/GPa (Ito and Katsura, 1992). In the pressure range from 25 to 60 GPa, two experimental studies performed with advanced laser-heated diamond anvil cell (LH-DAC) techniques agree well with each other. The Clapeyron slope appears relatively steep, with melting temperatures of ~3270 K

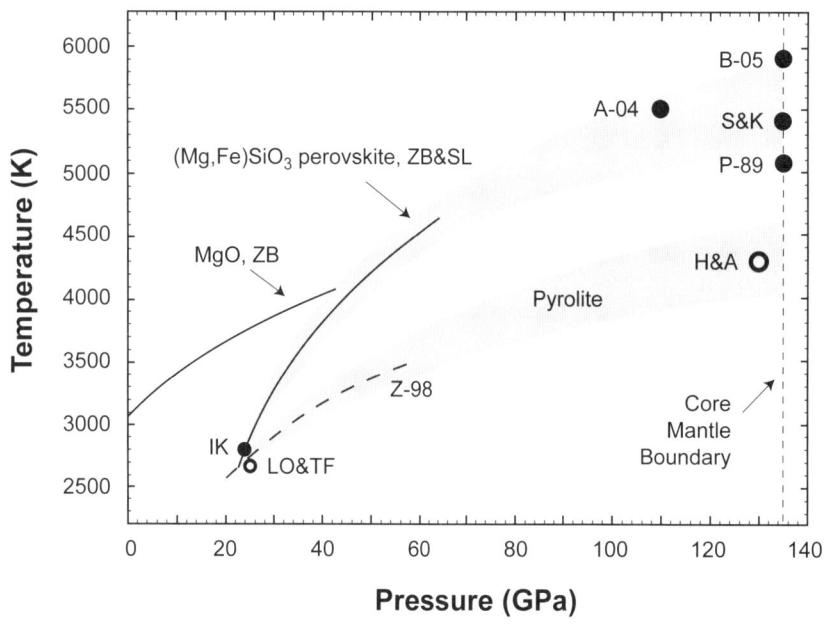

Figure 14. Melting curve of Al-(Mg,Fe)SiO₃ perovskite (solid circles and lines) and solidus of pyrolitic-relevant materials (open circles and dashed lines). References: IK (Ito and Katsura, 1992), ZB&SL (Zerr and Boehler, 1994; Shen and Lazor, 1995), A-04 (Akins et al., 2004), P-89 (Poirier, 1989), S&K (Stixrude and Karki, 2005), B-05 (Belonoshko et al., 2005), LO&TF (Litasov and Ohtani, 2002; Tronnes and Frost, 2002), ZB (Zerr and Boehler, 1994), Z-98 (Zerr et al., 1998), H&A (Holland and Ahrens, 1997).

and 4500 K at 30 GPa and 60 GPa, respectively (Shen and Lazor, 1995; Zerr, 1993). At even higher pressures, recent shock-wave experiments have been successfully applied to $MgSiO_3$-glass starting material to provoke melting of the perovskite phase at ~5500 K at a pressure of 117 GPa (Akins et al., 2004). This value plots slightly lower than extrapolations made from the LH-DAC experiments, but, in fact, all data appear to be in relatively good agreement with each other according to the poorly constrained curvature of the melting curve below 70 GPa. Note that the effect of Fe in the $(Mg,Fe)SiO_3$ solid solution seems to remain minor, in view of similar results obtained for Fe-free and Fe-bearing perovskite phases in two diamond anvil cells studies (Shen and Lazor, 1995; Zerr, 1993). In contrast, it has been suggested that the presence of Al_2O_3 in the Al-$(Mg,Fe)SiO_3$ phase could lower the melting point by ~300 K (Wang and Simmons, 1973).

From a theoretical point of view, two kinds of approaches have been developed to refine the high-pressure melting curve of $MgSiO_3$. The Lindemann law applied to a series of perovskite compounds yields a $MgSiO_3$ melting point at 5070 ± 625 K for the pressure at the core-mantle boundary (Poirier, 1989). On the other hand, recent ab initio calculations have suggested a melting point at the core-mantle boundary of 5400 ± 600 K (Stixrude and Karki, 2005) or ~5900 K (Belonoshko et al., 2005).

Lower-Mantle Melting of the Pyrolitic Composition

The importance of determining the melting curve of Al-$(Mg,Fe)SiO_3$ perovskite is clear; however, the lower-mantle melting properties are more directly connected to the pyrolite solidus and liquidus curves. At 25 GPa pressure, different large-volume press experiments report comparable solidus and liquidus temperatures at ~2550 and ~2700 K, respectively (Litasov and Ohtani, 2002; Tronnes and Frost, 2002). Above 31 GPa, the liquidus phase has been observed to change from ferropericlase to Al-$(Mg,Fe)SiO_3$ perovskite, which becomes the most refractory mineral in the lower mantle (Ito et al., 2004). This observation is nicely compatible with the crossing of the melting curves of each pure $(Mg,Fe)SiO_3$ and MgO minerals at ~40 GPa (Zerr, 1993; Zerr and Boehler, 1994). Accordingly, the shape of the liquidus curve is expected to correlate with the perovskite phase, while the solidus curve is expected to correlate with the melting curve of ferropericlase. The melting point of the third major lower-mantle phase, the $CaSiO_3$ perovskite phase, has been reported to occur at temperatures in-between (Ito et al., 2004). Additional LH-DAC experiments on pyrolitic-type material have reported solidus temperatures of ~2500 K and ~3500 K at 25 and 60 GPa, respectively (Zerr et al., 1998). This solidus curve shows a similar slope, but temperatures ~750 K lower compared to the melting curve of ferropericlase. At the higher pressures, melting of a $(Mg,Fe)_2SiO_4$ olivine starting material was detected at 4300 ±270 K for a pressure of 130 GPa using the shock-wave technique (Holland and Ahrens, 1997). Olivine composition is significantly different than pyrolite, but the onset of melting probably occurs at similar temperatures for both starting materials, because of eutectic relations between ferropericlase and silicate perovskite phases. Indeed, the data point at 4300 K and 130 GPa plots in good agreement with the experimental curves determined at moderated pressures for pyrolite.

TRANSPORT PROPERTIES

Despite the fact that Al-$(Mg,Fe)SiO_3$ perovskite transport properties are of great importance for modeling the dynamics of Earth's interior, few studies have been devoted to that subject. The main experimental difficulties are the stability field at high pressure disabling the classical measurements on bulk samples, the fragility of this metastable phase at room pressure, which prevents an optimal characterization of the structural defects, using transmission electron microscopy for example, and the limited Al-$(Mg,Fe)SiO_3$ perovskite volume (less than 1 mm^3) obtained after each synthesis in a multianvil press. Still, important experimental results are now available concerning atomic diffusion, deformation behavior, and electrical conductivity. From a general point of view, the transport properties can be addressed using the activation enthalpy ΔH^* and the activation volume V^* for the motion of characteristic structural defects (Poirier and Liebermann, 1984; Yamazaki and Karato, 2001). The defect can be punctual in the case of atomic diffusion and electrical conductivity, or linear for creep deformation. It is important to address the pressure and temperature dependencies of ΔH^* and V^* when deriving diffusion and rheological laws valid for the various lower-mantle depths. Heat diffusion behaves differently because it concerns the propagation of phonons, and sometimes of photons, and we will not consider this point in this paper. The presence of water may also significantly affect the transport properties, but, as described already, it is possible that the hydrogen content remains quite limited in the main lower-mantle phase.

Atomic Diffusivity

Atomic diffusion controls different processes of great importance. First, it helps grain growth after major phase transformations like the $Mg_2SiO_4 \rightarrow MgSiO_3 + MgO$ disproportionation at the 660 km discontinuity. It dominates the ability of the lower-mantle phases to re-equilibrate with each other. The various partitioning coefficients can be affected by changes of pressure and temperature, which implies diffusion of various elements between the main lower-mantle phases in order to retrieve the thermodynamic equilibrium. Also, the lower-mantle main material can undergo chemical interactions with materials from another provenance (descending slabs, ascending plumes, material from the D" region, etc.), and the longevity of these chemical heterogeneities is controlled by atomic diffusivity. Atomic diffusion is largely controlled by defect population, which can be rather complex in the Al-$(Mg,Fe)SiO_3$ perovskite phase. For Al-free $(Mg,Fe)SiO_3$ perovskite compounds, synthesized from decomposition of $(Mg,Fe)_2SiO_4$ olivine, for example, the major defects appear to be the Fe^{3+} cations, possibly located in the octahedral site and associated with O vacancies (Hirsch and Shankland, 1991). The case of point-defects in Al-bearing perovskite compounds is largely described in another section of this paper, where the two possible substitution mechanisms involving Al-coupled substitution on both perovskite sites and formation of O vacancies are described.

From an experimental point of view, atomic diffusivity is relatively well constrained for the pressure and temperature conditions of the uppermost lower mantle, because the measurement can be done from the chemical analysis of quenched samples (see Bejina et al., 2003, for a review). For silicate perovskite, atomic diffusion has been investigated for Si, Al, O, and (Mg,Fe). For Si, it has been reported that the activation energy is significantly smaller than in most of the other silicate phases, in which Si is usually found to be the slowest diffusive element. The reason is probably because of weaker Si-O bonds in sixfold compared to fourfold coordination sites (Yamazaki et al., 2000). However, the activation volume, V^*, which has been found to be negative in many other silicates (Bejina et al., 1999), remains undetermined for the Al-$(Mg,Fe)SiO_3$ perovskite. The Mg-Fe interdiffusion in the $(Mg,Fe)SiO_3$ perovskite appears to be of the same order of magnitude as the Si self-diffusion, which corresponds to Mg-Fe diffusion that is orders of magnitude slower than in other mantle minerals. On the other hand, oxygen diffusivity has been estimated to be relatively high in different experimental studies based on electrical conductivity measurements of Na-doped $MgSiO_3$ perovskite (Dobson, 2003; Xu and McCammon, 2002), in agreement with analogical experiments based on $CaTiO_3$ perovskite properties (Gautason and Muehlenbachs, 1993), and molecular dynamics calculations (Miyamoto, 1988). It has even been proposed that O^{2-} diffusion in silicate perovskite could dominate the electrical conductivity at the top of the lower mantle. Finally, Al diffusion was estimated to be very slow, ~1–2 orders of magnitude slower in Al-$(Mg,Fe)SiO_3$ perovskite than in majoritic garnet (Miyajima et al., 2001) (Fig. 15, Al diffusion in Al-[Mg,Fe]SiO_3 perovskite). Note that the analyses were often performed for simplified silicate perovskite compositions. In natural compounds, the eventual increased amount of structural defects and chemical disorder (provided by various types of substitutions) may affect the diffusion coefficients. Still, the diffusion coefficients appear to be very small, except perhaps for oxygen, and therefore we can expect that the lower-mantle chemical heterogeneities can hardly be resolved with geological time.

Electrical Conductivity

Concerning electrical conductivity, the latest reports agree that for a polaron-type mechanism with an activation energy of ~0.6–0.9 eV, the room temperature conductivity, σ_0, is between 10 and 200 S/m (Katsura et al., 1998; Xu et al., 1998). The activation energy has been controversial in the past, because an activation energy of 0.3–0.4 eV had been suggested from diamond anvil cell work (Poirier et al., 1996). However, in these experiments performed with relatively large temperature gradients, the observed ΔH^* could have been due to the presence of (Mg,Fe)O ferropericlase aggregates at the border of the laser-heated bands (Campbell et al., 1992). Note that the Al-$(Mg,Fe)SiO_3$ perovskite electrical conductivity is found to be rather small at room temperature, which makes the measurement difficult. The Al_2O_3 content appears to be a dominant parameter, with Al-bearing perovskite conductivity ~3.5 times greater than for Al-free perovskite (Xu et al., 1998). This trend appears rather natural considering the usual dominant effect of Fe^{3+} content on the electrical conductivity together with the strong correlation between Al and Fe^{3+} atoms in Al-$(Mg,Fe)SiO_3$ perovskite. It is also possible that the magnesium silicate perovskite electrical conductivity could be dependent

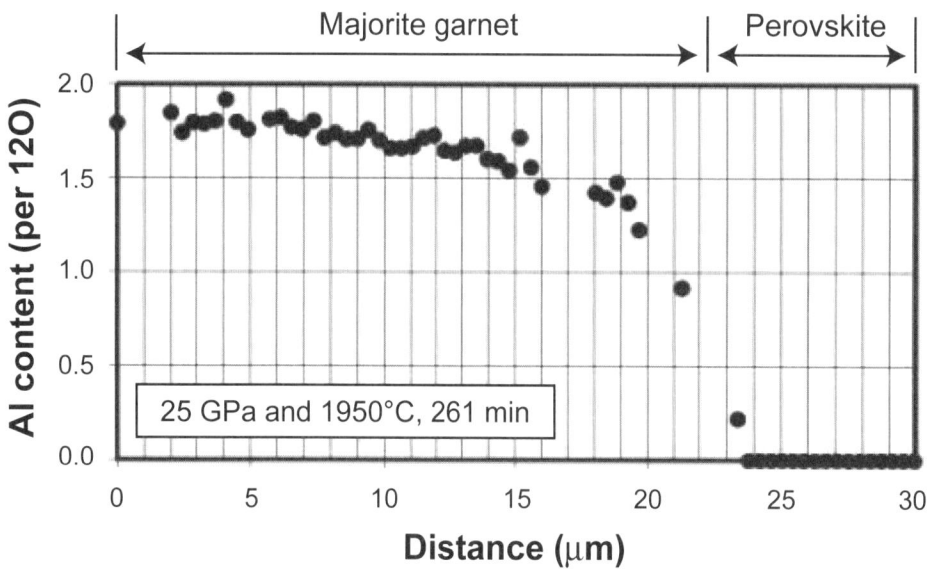

Figure 15. Aluminum concentration profile obtained at the surface boundary between majoritic garnet and $MgSiO_3$ perovskite after diffusion experiments performed at 25 GPa and 1950 °C for 261 min (derived from Miyajima et al., 2001). Al contents were measured using transmission electron microscope equipped with energy-dispersive spectrometers. Measured diffusion profile in perovskite is less than 1 µm in length, while profile in majoritic garnet is 5–20 µm long under the same pressure-temperature conditions. Thus, diffusion is 1–2 orders of magnitude faster in garnet than in perovskite.

on oxygen fugacity. However, thanks to its relatively high activation energy, it has been shown that typical pyrolitic material can fit the geophysical estimates of lower-mantle electrical conductivity without involving extremely high fO_2 values for the lower mantle (Katsura et al., 1998).

Creep and Viscosity

Twins appear to be the most abundant defects from observations of the (Mg,Fe)SiO$_3$-perovskite microstructure ((Wang et al., 1990; see perovskite grains microstructure in Fig. 10A), but the deformation mechanism is usually controlled by dislocations. Several studies have tentatively addressed the deformation slip of mantle perovskite using analogous perovskite compounds (Beauchesne and Poirier, 1989; Li et al., 1996), but it appears that perovskite phases do not constitute an analogue series for plastic deformation. More recently, deformation of MgSiO$_3$ perovskite was performed at lower-mantle pressure and temperature conditions; creep deformation was confirmed, and the major slip systems were determined (Cordier et al., 2004). In another work, it was shown that the uniaxial stress supported at room temperature by (Mg,Fe)SiO$_3$ perovskite aggregates increases continuously with pressure from ~2.6 GPa at 2 GPa to ~10.9 GPa at 32 GPa, an effect that evidences structural hardening with increasing pressure (Merkel et al., 2003). However, no quantitative measurements of mantle perovskite viscosity are presently available for relevant lower-mantle pressures and temperatures.

From a more theoretical point of view, the ΔH^* and V^* of the dislocation motion can be evaluated by two methods. In the first one, so-called homologous temperature scaling, the ΔH^* value is estimated from its usual good correlation with the melting temperature. This correlation suggests that melting occurs for a critical dislocation density. Using this approach, the pressure dependency of ΔH^*, and therefore the activation volume ($V^* = \partial H /\partial P)_S$, where S is entropy, can be determined from the Al-(Mg,Fe)SiO$_3$ perovskite melting curve at high pressures (Fig. 14). A second method consists in evaluating ΔH^* and V^* from an elastic model based on the strain energy of the crystal. In this case, the pressure and temperature dependencies of V^* are estimated using the P-V-T equation of state of perovskite. The two types of calculations yield similar conclusions. The mantle appears almost isoviscous when an adiabatic temperature profile is considered, except in the first 660–1000 km, where viscosity could be relatively higher (Poirier and Liebermann, 1984). One should note that the ferropericlase grains are expected to be much less viscous than the main perovskite phase, which can produce complex microstructure with highly strained ferropericlase grains (Yamazaki and Karato, 2001). It has also been proposed that the deformation behavior of the perovskite phase could be different just below the 660 km discontinuity, with a much softer material, because the grain size remains small after the phase transformation from ringwoodite-spinel (or majoritic-garnet) to the mixture of perovskite and ferropericlase (Solomatov et al., 2002).

GEOPHYSICAL IMPLICATIONS

In this paper, it is shown that MgSiO$_3$ perovskite does not provide a satisfactory analogue model for the properties of Earth's main lower-mantle phase. Direct implications arise from the specific Al-(Mg,Fe)SiO$_3$ perovskite properties.

1. Despite active controversy on the P-V-T equation of state, it is proposed that the (Al,Fe)-content could have negligible (or minor) effect on the elastic properties, at least on bulk moduli, while the effect on shear properties could be more pronounced. For bulk modulus, this statement is based on the observation that a major proportion of the (Al,Fe^{3+}) trivalent cations is inserted into the MgSiO$_3$ perovskite structure via the coupled substitution mechanism. This mechanism affects the silicate perovskite elastic properties less than the Al insertion via formation of O vacancies. It cannot be ruled out, however, that a significant amount of O vacancies may be found in Al-(Mg,Fe)SiO$_3$ perovskite, especially at the top of the lower mantle where pressure remains moderate. Therefore the bulk modulus could be smaller in this region. However, extrapolation of this reduced bulk modulus to the elevated lower-mantle depths is certainly irrelevant. Using data sets compatible with these principles, the most recent models, based on the inversion of seismic profiles, suggest a mantle with a gradual enrichment in Si and Fe in the lowermost part of the lower mantle. It should be noted that these calculations do not include yet the recently reported phase transition to the post perovskite phase, which could significantly modify these conclusions.

2. The presence of Al enhances the ability of the perovskite structure to accept various minor and trace elements. Al can be inserted in both octahedral and dodecahedral sites, and therefore it can easily charge balance the insertion of any additional element, such as Fe^{3+} (by Al insertion in Si^{4+}-based octahedra) or Na$^+$ (by Al insertion in Mg^{2+}-dodecahedra). Some substitution mechanisms maybe more complex, but it is accepted that all the elements found in the pyrolitic composition can be hosted into the phase mixture made of Al-(Mg,Fe)SiO$_3$ perovskite, ferropericlase, and CaSiO$_3$ perovskite. For compositions enriched in incompatible elements, like in MORBs, other phases are expected because the insertion abilities of the main lower-mantle phases are exceeded.

3. The strong atomic correlation between Al and Fe^{3+} in the Al-(Mg,Fe)SiO$_3$ perovskite structure makes this material a strong O consumer. It induces low oxygen fugacity in the lower mantle. This effect is expected to enhance the Fe^{3+} content below the 660 km discontinuity, in conjunction with the reduction of some other material present in the vicinity of the Al-(Mg,Fe)SiO$_3$ perovskite phase. This material may be a small proportion of the perovskite phase itself, or, more likely, the ferropericlase phase. The reduction implies the occurrence of small Fe drops that

are experimentally observed in typical lower-mantle conditions. Therefore, the lower mantle fO_2 is expected to be buffered by Fe/(Mg,Fe)O equilibrium. It is possible that the Fe dismutation plays a dominant role in late core growth, estimated to concern ~15% of its actual mass (Allègre et al., 1982). Some authors have suggested that some FeO could have diffused slowly from the silicate material to the outer core (Ito et al., 1995; Ringwood, 1977), but, it is also possible that some Fe^0 is extracted from the mantle. Indeed, the Fe^0 can be provided after the direct segregation of the small Fe drops at the core-mantle boundary, or in two steps, first with dissolution of some FeO in the liquid outer core, followed by O diffusion from the outer core back to the mantle through the core-mantle boundary. This additional O component would easily oxidize the small Fe particles eventually present in the lower mantle. Comparable speculations can be found elsewhere (e.g., Komiya, 2004).

4. The transport properties can certainly be largely affected by the complex crystal chemistry of Al-(Mg,Fe)SiO_3 perovskite. For electrical conductivity, it has been shown that the presence of Al increases conductivity by a factor of ~3.5. Experimental measurements of atomic diffusion do not yet cover the complete range of chemical composition representative of the lower mantle. Al seems to be the least diffusive element in silicate perovskite, probably because it is strongly involved in different coupled substitutions, and, therefore, its motion imposes simultaneous diffusion of other species (atoms and/or structural defects). Therefore, different compositions of Al-(Mg,Fe)SiO_3 perovskite phases may coexist in the lower mantle. For example, it is unlikely that the Al-free phase produced from the decomposition of olivine at 660 km re-equilibrates with the Al-bearing composition produced after the phase transformation of majoritic garnet at slightly greater depth. On the contrary, O appears to be the fastest diffusive element, possibly producing ionic conductivity at moderate lower-mantle depths.

5. Mantle perovskite appears stable in most of the lower mantle, except probably in the vicinity of the core-mantle boundary, where it may undergo a phase transition to a post-perovskite structure. All recent experimental investigations and ab initio calculations agree on this matter and also on the $CaIrO_3$ structure for this post-perovskite phase. The consequences of this transition have not yet been fully investigated, but they are likely to be important, especially for our vision of the lowermost part of the lower mantle. Additional work may be required for variable chemical compositions and temperature ranges. Also, the high-temperature elastic properties of this new phase have been experimentally investigated only recently (Guignot et al., 2007). Therefore, it is still difficult to tell to what extent the chemical and seismological models will converge to eventually propose a distinct lower-mantle reservoir at the highest lower-mantle depths. Also, other minor Al-(Mg,Fe)SiO_3 perovskite transformations such as subtle symmetry changes and electronic transitions may have some important effects on mantle properties at shallower depths, but both remain to be confirmed for the elevated temperatures relevant to the lower mantle.

ACKNOWLEDGMENTS

I warmly thank E. Ohtani for his invitation to participate in this volume, N. Bolfan-Casanova for her active participation and great help with the different topics covered in this paper, J.D. Bass, L. Dubrovinsky, K. Hirose, and E. Ohtani for constructive corrections and comments about the manuscript, and A. Bouhifd, G. Fiquet, A.M. Flank, N. Guignot, D. Häusermann, M. Kunz, M. Mezouar, H. Samuel, J.F. Stebbins, and D. Rubie for fruitful collaborations. I sincerely apologize for the inevitable lack of acknowledgments of some previous works. This work was supported by Institute National des Sciences d'Univers (INSU-CNRS), Institut de Minéralogie et de Physique des Milieux Condensés (IMPMC), Institut de Physique du Globe de Paris (IPGP), and ESRF.

REFERENCES CITED

Akaogi, M., and Ito, E., 1999, Calorimetric study on majorite-perovskite transition in the system $Mg_4Si_4O_{12}$–$Mg_3Al_2Si_3O_{12}$: Transition boundary with positive pressure-temperature slopes: Earth and Planetary Science Letters, v. 114, p. 129–140.

Akber-Knutson, S., and Bukowinski, M.S., 2004, The energetics of aluminum solubility into $MgSiO_3$ perovskite at lower mantle conditions: Earth and Planetary Science Letters, v. 220, p. 317–330, doi: 10.1016/S0012-821X(04)00065-2.

Akins, J.A., Luo, S.N., Asimov, P.D., and Ahrens, T.J., 2004, Shock-induced melting of $MgSiO_3$ perovskite and implications for melts in Earth's lowermost mantle: Geophysical Research Letters, v. 31, L14612, doi: 10.1029/2004GL020237.

Allègre, J.A., Dupré, B., and Brévart, O., 1982, Chemical aspects of the formation of the core: Philosophical Transaction of the Royal Society of London, v. A306, p. 49–59.

Allègre, J.A., Poirier, J.P., Humler, E., and Hofmann, A.W., 1995, The chemical composition of the Earth: Earth and Planetary Science Letters, v. 134, p. 515–526, doi: 10.1016/0012-821X(95)00123-T.

Andrault, D., 1999, A possible Redox Equilibrium for the Earth's Lower Mantle: Boston, American Geophysical Union Spring Meeting.

Andrault, D., 2001, Evaluation of (Mg,Fe) partitioning between perovskite and magnesiowüstite up to 120 GPa: Journal of Geophysical Research, v. 106, p. 2079–2087, doi: 10.1029/2000JB900362.

Andrault, D., 2003, Cationic substitution in $MgSiO_3$ perovskite: Physics and Chemistry of Minerals, v. 4200, p. 1–12.

Andrault, D., and Fiquet, G., 2001, Synchrotron radiation and laser-heating in a diamond anvil cell: The Review of Scientific Instruments, v. 72, no. 2, p. 1283–1288, doi: 10.1063/1.1343866.

Andrault, D., and Poirier, J.P., 1991, Evolution of the distortion of perovskites under pressure: An EXAFS study of $BaZrO_3$, $SrZrO_3$, and $CaGeO_3$: Physics and Chemistry of Minerals, v. 18, p. 91–105, doi: 10.1007/BF00216602.

Andrault, D., Fiquet, G., Itié, J.P., Richet, P., Gillet, P., Haüsermann, D., and Hanfland, M., 1998a, Thermal pressure in a laser-heated diamond-anvil cell: An X-ray diffraction study: European Journal of Mineralogy, v. 10, p. 931–940.

Andrault, D., Neuville, D., Flank, A.M., and Wang, Y., 1998b, Cation coordination sites in Al-$MgSiO_3$ perovskite: The American Mineralogist, v. 83, p. 1045–1053.

Andrault, D., Bolfan-Casanova, N., and Guignot, N., 2001, Equation of state of the lower mantle Al-(Mg,Fe)SiO_3 perovskite: Earth and Planetary Science Letters, v. 193, p. 501–508, doi: 10.1016/S0012-821X(01)00506-4.

Badro, J., Fiquet, G., Guyot, F., Rueff, J.-P., Struzhkin, V.V., Vankó, G., and Monaco, G., 2003, Iron partitioning in Earth's mantle: Toward a deep lower mantle discontinuity: Science, v. 300, p. 789–791, doi: 10.1126/science.1081311.

Badro, J., Rueff, J.-P., Vankó, G., Monaco, G., Fiquet, G., and Guyot, F., 2004, Electronic transitions in perovskite: Possible nonconvecting layers in the lower mantle: Science, v. 305, p. 383–386, doi: 10.1126/science.1098840.

Beauchesne, S., and Poirier, J.P., 1989, Creep of barium titanate perovskite: A contribution to a systematic approach to the viscosity of the lower mantle: Physics of the Earth and Planetary Interiors, v. 55, p. 187–199, doi: 10.1016/0031-9201(89)90242-2.

Bejina, F., Jaoul, O., and Liebermann, R.C., 1999, Activation volume of Si diffusion in San Carlos olivine: Implications for upper mantle rheology: Journal of Geophysical Research, v. 104, p. 25,529–25,542, doi: 10.1029/1999JB900270.

Bejina, F., Jaoul, O., and Liebermann, R.C., 2003, Diffusion in minerals at high pressure: A review: Physics of the Earth and Planetary Interiors, v. 139, p. 3–20, doi: 10.1016/S0031-9201(03)00140-7.

Belonoshko, A.B., Skorodumova, N.V., Rosengren, A., Ahuja, R., Johansson, B., Burakovsky, L., and Preston, D.L., 2005, High pressure melting of $MgSiO_3$: Physical Review Letters, v. 94, doi: 10.1103/PhysRevLett.94.195701.

Bina, C.R., and Silver, P.G., 1990, Constraints on the lower mantle composition and temperature from density and bulk sound velocity profiles: Geophysical Research Letters, v. 17, p. 1153–1156.

Bolfan-Casanova, N., 2005, Water in the Earth's mantle: Mineralogical Magazine, v. 69, p. 227–255.

Bolfan-Casanova, N., Mackwell, S., Keppler, H., McCammon, C.A., and Rubie, D.C., 2002, Pressure dependence of H solubility in magnesiowüstite up to 25 GPa: Implications for the storage of water in the Earth's lower mantle: Geophysical Research Letters, v. 29, no. 10, p. 1449, doi: 10.1029/2001GL014457.

Bolfan-Casanova, N., Keppler, H., and Rubie, D.C., 2003, Water partitioning at 660 km depth and evidence for very low water solubility in magnesium silicate perovskite: Geophysical Research Letters, v. 30, doi: 10.1029/2003GL017182.

Brodholt, J.P., 2000, Pressure-induced changes in the compression mechanism of aluminous perovskite in the Earth's mantle: Nature, v. 407, p. 620–622, doi: 10.1038/35036565.

Brown, J.M., and Shankland, T., 1981, Thermodynamic parameters in the Earth as determined from seismic profiles: Geophysical Journal of the Royal Astronomical Society, v. 66, p. 579–596.

Bunge, H.P., Ricard, Y., and Matas, J., 2001, Non-adiabaticity in mantle convection: Geophysical Research Letters, v. 28, p. 879–882, doi: 10.1029/2000GL011864.

Campbell, A.J., Heinz, D.L., and Davis, A.M., 1992, Material transport in laser-heated diamond anvil cell melting experiments: Geophysical Research Letters, v. 19, no. 10, p. 1061–1064.

Cohen, R.E., 1987, Elasticity and equation of state of $MgSiO_3$ perovskite: Geophysical Research Letters, v. 14, p. 1053–1056.

Cordier, P., Ungar, T., Zsoldos, L., and Tichy, G., 2004, Dislocation creep in $MgSiO_3$ perovskite at conditions of the Earth's uppermost lower mantle: Nature, v. 428, p. 837–840, doi: 10.1038/nature02472.

Corgne, A., Allan, N.L., and Wood, B.J., 2003, Atomistic simulations of trace element incorporation into the large site of $MgSiO_3$ and $CaSiO_3$ perovskites: Physics of the Earth and Planetary Interiors, v. 139, no. 1–2, p. 113–127, doi: 10.1016/S0031-9201(03)00148-1.

Daniel, I., Cardon, H., Fiquet, G., Guyot, F., and Mezouar, M., 2001, Equation of state of Al-bearing perovskite to lower mantle pressure conditions: Geophysical Research Letters, v. 28, p. 3789–3792, doi: 10.1029/2001GL013011.

Daniel, I., Bass, J.D., Fiquet, G., Cardon, H., Zhang, J., and Hanfland, M., 2004, Effect of aluminium on the compressibility of silicate perovskite: Geophysical Research Letters, v. 31, p. L15608, doi: 10.1029/2004GL020213.

Da Silva, C., Wentzcovitch, R.M., Patel, A., Price, G.D., and Karato, S., 2000, The composition and geotherm of the lower mantle: Constraints from the elasticity of silicate perovskite: Physics of the Earth and Planetary Interiors, v. 118, p. 103–109, doi: 10.1016/S0031-9201(99)00133-8.

Deschamps, F., and Trampert, J., 2004, Towards a lower mantle reference temperature and composition: Earth and Planetary Science Letters, v. 222, p. 161–175, doi: 10.1016/j.epsl.2004.02.024.

Dobson, D.P., 2003, Oxygen ionic conduction in $MgSiO_3$ perovskite: Physics of the Earth and Planetary Interiors, v. 139, p. 55–64, doi: 10.1016/S0031-9201(03)00144-4.

Dziewonski, A., and Anderson, D., 1981, Preliminary reference Earth model: Physics of the Earth and Planetary Interiors, v. 25, p. 297–356, doi: 10.1016/0031-9201(81)90046-7.

Farges, F., Guyot, F., Andrault, D., and Wang, Y., 1995, Local structure around Fe in $Mg_{0.9}Fe_{0.1}SiO_3$: An X-ray absorption spectroscopy study at Fe-K edge: European Journal of Mineralogy, v. 6, p. 303–312.

Fiquet, G., Andrault, D., Dewaele, A., Charpin, T., Kunz, M., and Haüsermann, D., 1998, *P-V-T* equation of state of $MgSiO_3$ perovskite: Physics of the Earth and Planetary Interiors, v. 105, p. 21–31, doi: 10.1016/S0031-9201(97)00077-0.

Fiquet, G., Dewaele, A., Andrault, D., Kunz, M., and Le Bihan, T., 2000, Thermoelastic properties and crystal structure of $MgSiO_3$ perovskite at lower mantle pressure and temperature conditions: Geophysical Research Letters, v. 27, p. 21–24, doi: 10.1029/1999GL008397.

Frost, D.J., Liebske, C., Langenhorst, F., McCammon, C.A., Trønnes, R., and Rubie, D.C., 2004, Experimental evidence for the existence of iron-rich metal in the Earth's lower mantle: Nature, v. 428, p. 409–412, doi: 10.1038/nature02413.

Fujino, K., Miyajima, N., Yagi, T., Kondo, T., and Funamori, N., 1998, Analytical electron microscopy of the garnet-perovskite transformation in a laser heated diamond anvil cell, in Manghnani, M.H., and Yagi, T., eds., Properties of Earth and planetary materials at high pressure and temperature: Washington, D.C., American Geophysical Union, p. 409–417.

Funamori, N., and Yagi, T., 1993, High pressure and high temperature in situ X-ray observation of $MgSiO_3$ perovskite under lower mantle conditions: Geophysical Research Letters, v. 20, no. 5, p. 387–390.

Funamori, N., Yagi, T., Miyajima, N., and Fujino, K., 1997, Transformation of garnet from orthorhombic perovskite to $LiNbO_3$ phase on release of pressure: Science, v. 275, p. 513–515, doi: 10.1126/science.275.5299.513.

Gautason, B., and Muehlenbachs, K., 1993, Oxygen diffusion in perovskite: Implications for electrical conductivity in the lower mantle: Science, v. 260, p. 518–521.

Glazer, A.M., 1972, The classification of tilted octahedra in perovskites: Acta Crystallographia, v. B28, p. 3384–3392.

Gong, Z.Z., Fei, Y., Dai, F., Zhang, L., and Jing, F.Q., 2004, Equation of state and phase stability of mantle perovskite up to 140 GPa shock pressure and its geophysical implications: Geophysical Research Letters, v. 31, p. L04614, doi: 10.1029/2003GL019132.

Guignot, N., and Andrault, D., 2004, Equations of state of Na-K-Al host phases in the lower mantle and implications for MORB density in the lower mantle: Physics of the Earth and Planetary Interiors, v. 134–144, p. 107–128.

Guignot, N., Andrault, D., Morard, G., Bolfan-Casanova, N., and Mezouar, M., 2007, Thermoelastic properties of post-perovskite phase $MgSiO_3$ determined experimentally at core-mantle boundary P-T conditions: Earth and Planetary Science Letters, v. 256, no.1–2, p. 162–168, doi: 10.1016/j.epsl.2007.01.025.

Heinz, D.L., Knittle, E., Sweeney, J.S., Williams, Q., and Jeanloz, R., 1994, Technical comments: High-pressure melting of $(Mg,Fe)SiO_3$ perovskite: Science, v. 264, p. 279–281.

Hirose, K., Fei, Y., Ma, Y., and Mao, H.K., 1999, The fate of subducted basaltic crust in the Earth's lower mantle: Nature, v. 397, p. 53–56, doi: 10.1038/16225.

Hirose, K., Fei, Y., Ono, S., Yagi, T., and Funakoshi, K., 2001, In situ measurements of the phase transition boundary in $Mg_3Al_2Si_3O_{12}$: Implications for the nature of the seismic discontinuities in the Earth's mantle: Earth and Planetary Science Letters, v. 189, no. 3–4, p. 177–188, doi: 10.1016/S0012-821X(01)00359-4.

Hirsch, L.M., and Shankland, T., 1991, Point defects in $(Mg,Fe)SiO_3$ perovskite: Geophysical Research Letters, v. 18, no. 7, p. 1305–1308.

Holland, K.G., and Ahrens, T.J., 1997, Melting of $(Mg,Fe)_2SiO_4$ at the core-mantle boundary of the Earth: Science, v. 275, p. 1623–1625, doi: 10.1126/science.275.5306.1623.

Iitaka, T., Hirose, K., Kawamura, K., and Murakami, M., 2004, The elasticity of the $MgSiO_3$ post-perovskite phase in the Earth's lowermost mantle: Nature, v. 430, p. 442–445, doi: 10.1038/nature02702.

Irifune, T., 1994, Absence of an aluminous phase in the upper part of the Earth's lower mantle: Nature, v. 370, p. 131–133, doi: 10.1038/370131a0.

Irifune, T., and Ringwood, A.E., 1993, Phase transformations in subducted oceanic crust and buoyancy relationships at depth of 600–800 km in the mantle: Earth and Planetary Science Letters, v. 117, p. 101–110, doi: 10.1016/0012-821X(93)90120-X.

Irifune, T., Adachi, Y., Fujino, K., Ohtani, E., Yoneda, A., and Sawamoto, H., 1992, A performance test for WC anvils for multianvil apparatus and phase transformation in some aluminous minerals up to 28 GPa, *in* Syono,

Y., and Manghnani, M.H., eds., High-Pressure Research: Application to Earth and Planetary Sciences: Tokyo, Terra Scientific Publications Company, p. 43–50.

Irifune, T., Ringwood, A.E., and Hibberson, W.O., 1994, Subduction of continental crust and terrigenous and pelagic sediments: An experimental study: Earth and Planetary Science Letters, v. 126, p. 351–368, doi: 10.1016/0012-821X(94)90117-1.

Irifune, T., Koizumi, T., and Ando, J.I., 1996, An experimental study of the garnet-perovskite transformation in the system $MgSiO_3$-$Mg_3Al_2Si_3O_{12}$: Physics of the Earth and Planetary Interiors, v. 96, p. 147–157, doi: 10.1016/0031-9201(96)03147-0.

Ito, E., and Katsura, T., 1992, Melting of ferromagnesian silicates under lower mantle conditions, in Syono, Y., and Manghnani, M.H., eds., High-Pressure Research: Application to Earth and Planetary Sciences: Tokyo, Terra Scientific Publication Company, p. 315–322.

Ito, E., Morooka, K., Ujike, O., and Katsura, T., 1995, Reactions between molten iron and silicate melts at high pressure: Implications for the chemical evolution of the Earth's core: Journal of Geophysical Research, v. 100, no. B4, p. 5901–5910, doi: 10.1029/94JB02645.

Ito, E., Kubo, A., Katsura, T., Akaogi, M., and Fujita, T., 1998, High-pressure transformation of pyrope ($Mg_3Al_2Si_3O_{12}$) in a sintered diamond cubic anvil assembly: Geophysical Research Letters, v. 25, no. 6, p. 821–824, doi: 10.1029/98GL00519.

Ito, E., Kubo, A., Katsura, T., and Walter, M.J., 2004, Melting experiments of mantle materials under lower mantle conditions with implications for magma ocean differentiation: Physics of the Earth and Planetary Interiors, v. 143–144, p. 397–406, doi: 10.1016/j.pepi.2003.09.016.

Jackson, I., 1998, Elasticity, composition and temperature of the Earth's mantle: A reappraisal: Geophysical Journal International, v. 134, p. 291–311, doi: 10.1046/j.1365-246x.1998.00560.x.

Jackson, J.M., Zhang, J., and Bass, J.D., 2004, Sound velocities and elasticity of aluminous $MgSiO_3$ perovskite: Implications for aluminum heterogeneity in Earth's lower mantle: Geophysical Research Letters, v. 31, no. 10, p. L10614.1–L10614.4.

Javoy, M., 1995, The integral enstatite chondrite model for the Earth: Earth and Planetary Science Letters, v. 22, p. 2219–2222.

Karki, B.B., Stixrude, L., and Wentzcovitch, R.M., 2001, High-pressure elastic properties of major materials of Earth's mantle from first principles: Review in Geophysics, v. 39, no. 4, p. 507–534, doi: 10.1029/2000RG000088.

Katsura, T., Sato, K., and Ito, E., 1998, Electrical conductivity of silicate perovskite at lower mantle conditions: Nature, v. 395, p. 493–495, doi: 10.1038/26736.

Katsura, T., Yamada, H., Shinmei, T., Kubo, A., Ono, S., Kanzaki, M., Yoneda, A., Walter, M.J., Ito, E., Urakawa, S., Funakoshi, K., and Utsumi, W., 2003, Post-spinel transition in Mg_2SiO_4 determined by high P-T in situ X-ray diffractometry: Physics of the Earth and Planetary Interiors, v. 136, p. 11–24, doi: 10.1016/S0031-9201(03)00019-0.

Kavner, A., and Duffy, T.S., 2001, Pressure-volume-temperature paths in the laser-heated diamond anvil cell: Journal of Applied Physics, v. 89, no. 3, p. 1907–1914, doi: 10.1063/1.1335827.

Kennett, B., 1998, On the density distribution within the Earth: Geophysical Journal International, v. 132, p. 374–382, doi: 10.1046/j.1365-246x.1998.00451.x.

Kesson, S.E., Fitz Gerald, J.D., and Shelley, J.M., 1994, Mineral chemistry and density of subducted basaltic crust at lower mantle pressures: Nature, v. 372, p. 767–769, doi: 10.1038/372767a0.

Kesson, S.E., Fitz Gerald, J.D., Shelley, J.M., and Withers, R.L., 1995, Phase relations, structure and crystal chemistry of some aluminous silicate perovskites: Earth and Planetary Science Letters, v. 134, p. 187–201, doi: 10.1016/0012-821X(95)00112-P.

Kesson, S.E., Fitz Gerald, J.D., and Shelley, J.M., 1998, Mineralogy and dynamics of a pyrolite lower mantle: Nature, v. 393, p. 252–255, doi: 10.1038/30466.

Kiefer, B., Stixrude, L., and Wentzcovitch, R.M., 2002, Elasticity of (Mg,Fe)SiO_3-perovskite at high pressure: Geophysical Research Letters, v. 11, p. 34–38.

Komiya, T., 2004, Material circulation model including chemical differentiation within the mantle and secular variation of temperature and composition of the mantle: Physics of the Earth and Planetary Interiors, v. 146, no. 1–2, p. 333–367, doi: 10.1016/j.pepi.2003.03.001.

Kubo, A., and Akaogi, M., 2000, Post-garnet transitions in the system $Mg_4Si_4O_{12}$-$Mg_3Al_2Si_3O_{12}$ up to 28 GPa: Phase relations of garnet, ilmenite and perovskite: Physics of the Earth and Planetary Interiors, v. 121, p. 85–102, doi: 10.1016/S0031-9201(00)00162-X.

Kubo, A., Yagi, T., Ono, S., and Akaogi, M., 2000, Compressibility of $Mg_{0.9}Al_{0.2}Si_{0.9}O_3$ perovskite: Proceedings of the Japan Academy, v. 16, p. 103–107.

Kudoh, Y., Ito, E., and Takeda, H., 1987, Effect of pressure on the crystal structure of perovskite-type $MgSiO_3$: Physics and Chemistry of Minerals, v. 14, p. 350–354, doi: 10.1007/BF00309809.

Lauterbach, S., McCammon, C.A., Van Aken, P., Langenhorst, F., and Seifert, F., 2000, Mössbauer and ELNES spectroscopy of (Mg,Fe)(Si,Al)O_3 perovskite: A highly oxidized component of the lower mantle: Contributions to Mineralogy and Petrology, v. 138, p. 17–26, doi: 10.1007/PL00007658.

Li, J., Struzhkin, V.V., Mao, H.-K., Shu, J., Hemley, R.J., Fei, Y., Mysen, B., Dera, P., Prakapenka, V., and Shen, G., 2004, Electronic spin state of iron in lower mantle perovskite: Proceedings of the National Academy of Sciences of the United States of America, v. 101, p. 14,027–14,030, doi: 10.1073/pnas.0405804101.

Li, L., Brodholt, J.P., Stackhouse, S., Weidner, D.J., Alfredsson, M., and Price, G.D., 2005, Elasticity of (Mg,Fe)(Si,Al)O_3 perovskite at high pressure: Earth and Planetary Science Letters, v. 240, no. 2, p. 529–536, doi: 10.1016/j.epsl.2005.09.030.

Li, P., Karato, S., and Wang, Z., 1996, High-temperature creep in fine-grained polycrystalline $CaTiO_3$, an analogue material of (Mg,Fe)SiO_3 perovskite: Physics of the Earth and Planetary Interiors, v. 95, p. 19–36, doi: 10.1016/0031-9201(95)03107-3.

Litasov, K., and Ohtani, E., 2002, Phase relations and melt compositions in CMAS–pyrolite–H_2O system up to 25 GPa: Physics of the Earth and Planetary Interiors, v. 134, no. 1–2, p. 105–127, doi: 10.1016/S0031-9201(02)00152-8.

Litasov, K., Ohtani, E., Langenhorst, F., Yurimoto, H., Kubo, T., and Kondo, T., 2003, Water solubility in Mg-perovskites and water storage capacity in the lower mantle: Earth and Planetary Science Letters, v. 211, p. 189–203, doi: 10.1016/S0012-821X(03)00200-0.

Liu, L.G., 1974, Silicate perovskite from phase transformation of pyrope-garnet at high-pressure and temperature: Geophysical Research Letters, v. 1, p. 277–288.

Mao, H.K., Hemley, R.J., Fei, Y., Shu, J.F., Chen, L.C., Jephcoat, A.P., Wu, Y., and Bassett, W.A., 1991, Effect of pressure, temperature and composition on the lattice parameters and density of three (Fe,Mg)SiO_3 perovskites up to 30 GPa: Journal of Geophysical Research, v. 96, no. B5, p. 8069–8079.

Mao, H.K., Shen, G., and Hemley, R.J., 1997, Multivariable dependence of Fe-Mg partitioning in the lower mantle: Science, v. 278, p. 2098–2100, doi: 10.1126/science.278.5346.2098.

Mao, W.L., Shen, G., Prakapenka, V.B., Meng, Y., Campbell, A.L., Heinz, D.L., Shu, J., Hemley, R.J., and Mao, H.K., 2004, Ferromagnesian postperovskite silicates in the D" layer of the Earth: Proceedings of the National Academy of Sciences of the United States of America, v. 101, no. 45, p. 15,867–15,869, doi: 10.1073/pnas.0407135101.

Masters, G., Laske, G., Bolton, H., and Dziewonski, A., 2000, The relative behavior of shear velocity, bulk sound speed, and compressional velocity in the mantle: Implications for chemical and thermal structure, in Karato, S., Forte, A.M., Liebermann, R.C., Masters, G., and Stixrude, L., eds., Earth's Deep Interior: Mineral Physics and Tomography from the Atomic to the Global Scale: American Geophysical Union Geophysical Monograph 117, p. 63–87.

Matsui, M., 2000, Molecular dynamics simulation of $MgSiO_3$ perovskite and the 660-km discontinuity: Physics of the Earth and Planetary Interiors, v. 121, p. 77–84, doi: 10.1016/S0031-9201(00)00161-8.

Mattern, E., 2005, Composition et température dans le manteau profond: Interprétations minéralogiques des observations sismologiques [Ph.D. thesis]: Lyon, France, Ecole Normale Supérieure, 231 p.

Mattern, E., Matas, J., Ricard, Y., and Bass, J.D., 2005, Lower mantle composition and temperature from mineral physics and thermodynamic modeling: Geophysical Journal International, v. 160, p. 973–990.

McCammon, C.A., 1997, Perovskite as a possible sink for ferric iron in the lower mantle: Nature, v. 387, p. 694–696, doi: 10.1038/42685.

McCammon, C.A., Rubie, D.C., Ross, C.R., II, Seifert, F., and O'Neill, H.S.C., 1992, Mössbauer spectroscopy of $^{57}Fe_{0.05}Mg_{0.95}SiO_3$ perovskite at 80 and 298 K: The American Mineralogist, v. 77, p. 894–897.

Meade, C., Mao, H.K., and Hu, J., 1995, High-temperature phase-transition and dissociation of (Mg,Fe)SiO_3 perovskite at lower mantle pressures: Science, v. 268, p. 1743–1745.

Merkel, S., Wenk, H.R., Badro, J., Montagnac, G., Gillet, P., Mao, H.K., and Hemley, R.J., 2003, Deformation of $(Mg_{0.9},Fe_{0.1})SiO_3$ perovskite aggregates up to 32 GPa: Earth and Planetary Science Letters, v. 209, p. 351–360, doi: 10.1016/S0012-821X(03)00098-0.

Miyajima, N., Fujino, K., Funamori, N., Kondo, T., and Yagi, T., 1999, Garnet-perovskite transformation under conditions of the Earth's lower mantle: An analytical transmission electron microscopy study: Physics of the Earth and Planetary Interiors, v. 116, p. 117–131, doi: 10.1016/S0031-9201(99)00127-2.

Miyajima, N., Langenhorst, F., Frost, D.J., and Rubie, D.C., 2001, Aluminium diffusion in silicate perovskite and majoritic garnet: Berichte Mineralogischen Gesellschaft, v. 13, p. 122.

Miyamoto, M., 1988, Ion migration on $MgSiO_3$ perovskite and olivine by molecular dynamics calculations: Physics and Chemistry of Minerals, v. 15, p. 601–604, doi: 10.1007/BF00311032.

Murakami, M., Hirose, K., Yurimoto, H., Nakashima, S., and Takafuji, N., 2002, Water in the Earth's lower mantle: Science, v. 295, p. 1885–1887, doi: 10.1126/science.1065998.

Murakami, M., Hirose, K., Kawamura, K., Sata, N., and Ohishi, Y., 2004, Post-perovskite phase transition in $MgSiO_3$: Science, v. 304, p. 855–858, doi: 10.1126/science.1095932.

Murakami, M., Hirose, K., Sata, N., and Ohishi, Y., 2005, Post-perovskite phase transition and mineral chemistry in the pyrolitic lowermost mantle: Geophysical Research Letters, v. 32, p. L03304, doi: 10.1029/2004GL021956.

Navrotsky, A., Schoenitz, M., Kojitani, H., Xu, H., Zhang, J., Weidner, D.J., and Jeanloz, R., 2003, Aluminum in magnesium silicate perovskite: Formation, structure, and energetics of magnesium-rich defect solid solutions: Journal of Geophysical Research, v. 108, no. B7, p. 2330, doi: 10.1029/2002JB002055.

Nishiyama, N., and Yagi, T., 2003, Phase relations and mineral chemistry in pyrolite to 2200°C under the lower mantle pressures and implications for dynamics of mantle plumes: Journal of Geophysical Research, v. 108, doi: 10.1029/2002JB002216.

Oganov, A.R., and Ono, S., 2004, Theoretical and experimental evidence for a post-perovskite phase of $MgSiO_3$ in the Earth's D" layer: Nature, v. 430, p. 445–448, doi: 10.1038/nature02701.

Oganov, A.R., Brodholt, J.P., and Price, G.D., 2001, Ab-initio elasticity and thermal equation of state of $MgSiO_3$: Earth and Planetary Science Letters, v. 184, p. 555–560, doi: 10.1016/S0012-821X(00)00363-0.

Oguri, K., Funamori, N., Uchida, T., Miyajima, N., Yagi, T., and Fujino, K., 2000, Post-garnet transition in natural pyrope: A multi-anvil study based on in situ X-ray diffraction and transmission electron microscopy: Physics of the Earth and Planetary Interiors, v. 122, p. 175–186, doi: 10.1016/S0031-9201(00)00178-3.

Ono, S., Ito, E., and Katsura, T., 2001, Mineralogy of subducted basaltic crust (MORB) from 25 to 37 GPa, and chemical heterogeneity of the lower mantle: Earth and Planetary Science Letters, v. 190, p. 57–63, doi: 10.1016/S0012-821X(01)00375-2.

Ono, S., Kikegawa, T., and Iizuka, T., 2004, The equation of state of orthorhombic perovskite in a peridotitic mantle composition to 80 GPa: Implications for chemical composition of the lower mantle: Physics of the Earth and Planetary Interiors, v. 145, p. 9–17, doi: 10.1016/j.pepi.2004.01.005.

Parise, J.B., Wang, Y., Yeganeh-Haeri, A., Cox, D.E., and Fei, Y., 1990, Crystal structure and thermal expansion of $(Mg,Fe)SiO_3$ perovskite: Geophysical Research Letters, v. 17, p. 2089–2092.

Poirier, J.P., 1989, Lindemann law and the melting temperature of perovskites: Physics of the Earth and Planetary Interiors, v. 54, p. 364–369, doi: 10.1016/0031-9201(89)90253-7.

Poirier, J.P., and Liebermann, R.C., 1984, On the activation volume for creep and its variation with depth in the Earth's lower mantle: Physics of the Earth and Planetary Interiors, v. 35, p. 283–293, doi: 10.1016/0031-9201(84)90022-0.

Poirier, J.P., Goddat, A., and Peyronneau, J., 1996, Ferric iron dependence of the electrical conductivity of the Earth's lower mantle material: Philosophical Transaction of the Royal Society of London, v. 354, p. 1361–1369.

Richmond, N.C., and Brodholt, J.P., 1998, Calculated role of aluminum in incorporation of ferric iron into magnesium silicate perovskite: The American Mineralogist, v. 83, p. 947–951.

Ringwood, A.E., 1977, Composition of the core and implication for the origin of the Earth: Geochemical Journal, v. 11, p. 111–135.

Ringwood, A.E., 1979, Origin of the Earth and Moon: New York, Springer, 295 p.

Ross, N.L., Zhao, J., and Angel, R.J., 2004, High-pressure structural behavior of $GdAlO_3$ and $GdFeO_3$ perovskites: Journal of Solid State Chemistry, v. 177, no. 10, p. 3768–3775, doi: 10.1016/j.jssc.2004.07.002.

Saltzer, R., Hilst, V.D., and Karason, H., 2001, Comparing P and S wave heterogeneity in the mantle: Geophysical Research Letters, v. 28, p. 1335–1338, doi: 10.1029/2000GL012339.

Samuel, H., Farnetani, C.G., and Andrault, D., 2006, Chemically denser material in the lowermost mantle: Origin and induced seismic velocity anomalies, in Bass, J.D., van der Hilst, R.D., and Trampert, J., eds., Structure, Evolution, and Composition of the Earth's Mantle: American Geophysical Union Monograph, p. 101–116.

Saxena, S.K., Dubrovinsky, L.S., Lazor, P., Cerenius, Y., Haggkvist, P., Hanfland, M., and Hu, J., 1996, Stability of perovskite ($MgSiO_3$) in the Earth's mantle: Science, v. 274, p. 1357–1359, doi: 10.1126/science.274.5291.1357.

Serghiou, G., Zerr, A., and Boehler, R., 1998, $(Mg,Fe)SiO_3$-perovskite stability under lower mantle conditions: Science, v. 280, p. 2093–2095, doi: 10.1126/science.280.5372.2093.

Shannon, R.D., and Prewitt, C.T., 1969, Effective ionic radii and oxides and fluorides: Acta Crystallographia, v. B25, p. 925–946.

Shen, G., and Lazor, P., 1995, Measurement of melting temperatures of some minerals under lower mantle conditions: Journal of Geophysical Research, v. 100, no. B9, p. 17,699–17,713, doi: 10.1029/95JB01864.

Shim, S.H., and Duffy, T.S., 2000, Constraints on the P-V-T equation of state of $MgSiO_3$ perovskite: The American Mineralogist, v. 85, p. 354–363.

Shim, S.H., Duffy, T.S., and Shen, G., 2001, Stability and structure of $MgSiO_3$ perovskite to 2300-km depth conditions: Science, v. 293, p. 2437–2440, doi: 10.1126/science.1061235.

Sidorin, I., Gurnis, M., and Helmberger, D.V., 1999, Evidence for a ubiquitous seismic discontinuity at the base of the mantle: Science, v. 286, p. 1326–1331, doi: 10.1126/science.286.5443.1326.

Sinelnikov, Y.D., Chen, G., Neuville, D., Vaughan, M.T., and Liebermann, R.C., 1998, Ultrasonic shear wave velocity of MgSiO3 perovskite at 8 GPa and 800 K and lower mantle composition: Science, v. 281, p. 677–679, doi: 10.1126/science.281.5377.677.

Solomatov, V.S., El-Khozondar, R., and Tikare, V., 2002, Grain size in the lower mantle: Constraints from numerical modeling of grain growth in two phase systems: Physics of the Earth and Planetary Interiors, v. 129, p. 265–282, doi: 10.1016/S0031-9201(01)00295-3.

Stackhouse, S., Brodholt, J.P., Wookey, J., Kendall, J.M., and Price, G.D., 2005, The effect of temperature on the seismic anisotropy of the perovskite and post-perovskite polymorphs of $MgSiO_3$: Earth and Planetary Science Letters, v. 230, p. 1–10, doi: 10.1016/j.epsl.2004.11.021.

Stebbins, J.F., Kroeker, S., and Andrault, D., 2001, The mechanism of solution of aluminium oxide in mantle perovskite ($MgSiO_3$): Geophysical Research Letters, v. 28, p. 615–619, doi: 10.1029/2000GL012279.

Stebbins, J.F., Kojitani, H., Akaogi, M., and Navrotsky, A., 2003, Aluminum substitution in $MgSiO_3$ perovskite: Investigation of multiple mechanisms by ^{27}Al NMR: The American Mineralogist, v. 88, p. 1161–1164.

Stixrude, L., and Karki, B.B., 2005, Structure and freezing of $MgSiO_3$ liquid in the Earth's lower mantle: Science, v. 310, p. 297–299, doi: 10.1126/science.1116952.

Sturhahn, W., Jackson, J.M., and Lin, J.F., 2005, The spin state of iron in minerals of the Earth's lower mantle: Geophysical Research Letters, v. 32, p. L12307, doi: 10.1029/2005GL022802.

Thomson, K.T., Wentzcovitch, R.M., and Bukowinski, M.S., 1996, Polymorphs of alumina predicted by first principles: Putting pressure on the ruby pressure scale: Science, v. 274, p. 1880–1882, doi: 10.1126/science.274.5294.1880.

Tronnes, R.G., and Frost, D.J., 2002, Peridotite melting and mineral–melt partitioning of major and minor elements at 22–24.5 GPa: Earth and Planetary Science Letters, v. 197, no. 1–2, p. 117–131, doi: 10.1016/S0012-821X(02)00466-1.

Tsuchiya, T., Tsuchiya, J., Umemoto, K., and Wentzcovitch, R.M., 2004, Phase transition in $MgSiO_3$ perovskite in the Earth's lower mantle: Earth and Planetary Science Letters, v. 224, p. 241–248, doi: 10.1016/j.epsl.2004.05.017.

Utsumi, W., Funamori, N., Yagi, T., Ito, E., Kikegawa, T., and Shimomura, O., 1995, Thermal expansivity of $MgSiO_3$ perovskite under high pressures to 20 GPa: Geophysical Research Letters, v. 22, p. 1005–1008, doi: 10.1029/95GL00584.

Walter, M.J., Kubo, A., Yoshino, T., Brodholt, J., Koga, K.T., and Ohishi, Y., 2004, Phase relations and equation-of-state of aluminous Mg-silicate perovskite and implications for the Earth's lower mantle: Earth and

Planetary Science Letters, v. 222, p. 501–516, doi: 10.1016/j.epsl.2004.03.014.

Wang, H., and Simmons, G., 1973, Elasticity of some mantle crystal structures, 2. Rutile GeO_2: Journal of Geophysical Research, v. 78, p. 1262–1273.

Wang, Y., Guyot, F., Yeganeh-Haeri, A., and Liebermann, R.C., 1990, Twinning in $MgSiO_3$ perovskite: Science, v. 248, p. 468–471.

Wang, Y., Weidner, D.J., Liebermann, R.C., and Zhao, Y., 1994, *P-V-T* equation of state of $(Mg,Fe)SiO_3$ perovskite: Constraints on composition of the lower mantle: Physics of the Earth and Planetary Interiors, v. 83, p. 13–40, doi: 10.1016/0031-9201(94)90109-0.

Wood, B.J., 2000, Phase transformations and partitioning relations in peridotite under lower mantle conditions: Earth and Planetary Science Letters, v. 174, p. 341–354, doi: 10.1016/S0012-821X(99)00273-3.

Wood, B.J., and Rubie, D.C., 1996, The effect of alumina on phase transformations at the 660-kilometer discontinuity from Fe-Mg partitioning experiments. Science, v. 273, p. 1522–1524.

Xu, Y., and McCammon, C.A., 2002, Evidence for ionic conductivity in lower mantle $(Mg,Fe)SiO_3$ perovskite: Journal of Geophysical Research, v. 107, no. B10, doi: 10.1029/2001JB000677.

Xu, Y., McCammon, C.A., and Poe, B.T., 1998, The effect of alumina on the electrical conductivity of silicate perovskite: Science, v. 282, p. 922–924, doi: 10.1126/science.282.5390.922.

Yagi, T., and Funamori, N., 1996, Chemical composition of the lower mantle inferred from the equation of state of $MgSiO_3$ perovskite: Philosophical Transactions of the Royal Society of London, v. 354, p. 1371–1384.

Yagi, T., Okabe, K., Nishiyama, N., Kubo, A., and Kikegawa, T., 2004, Complicated effects of aluminum on the compressibility of silicate perovskite: Physics of the Earth and Planetary Interiors, v. 143–144, p. 81–91, doi: 10.1016/j.pepi.2003.07.020.

Yamamoto, T., Yuen, D.A., and Ebisuzaki, T., 2003, Substitution mechanism of Al ions in $MgSiO_3$ perovskite under high pressure conditions from first-principles calculations: Earth and Planetary Science Letters, v. 206, p. 617–625, doi: 10.1016/S0012-821X(02)01099-3.

Yamazaki, D., and Karato, S., 2001, Some mineral physics constraints on the rheology and geothermal structure of Earth's lower mantle: The American Mineralogist, v. 86, p. 385–391.

Yamazaki, D., Kato, T., Yurimoto, H., Ohtani, E., and Toriumi, M., 2000, Silicon self diffusion in $MgSiO_3$ perovskite at 25 GPa: Physics of the Earth and Planetary Interiors, v. 119, p. 299–309, doi: 10.1016/S0031-9201(00)00135-7.

Yusa, H., Yagi, T., and Shimobayashi, N., 1995, A new unquenchable high-pressure polymorph of $Ca_3Al_2Si_3O_{12}$: Physics of the Earth and Planetary Interiors, v. 92, p. 25–31, doi: 10.1016/0031-9201(95)03057-4.

Zerr, A., 1993, Melting of $(Mg,Fe)SiO_3$-perovskite to 625 kbar: Indication of a high melting temperature in the lower mantle: Science, v. 262, p. 553–555.

Zerr, A., and Boehler, R., 1994, Constraints on the melting temperature of the lower mantle from high pressure experiments on MgO and magnesiowüstite: Nature, v. 371, p. 506–508, doi: 10.1038/371506a0.

Zerr, A., Diegeler, A., and Boehler, R., 1998, Solidus of Earth's mantle: Science, v. 281, p. 243–246, doi: 10.1126/science.281.5374.243.

Zhang, J., and Weidner, D.J., 1999, Thermal equation of state of aluminum-enriched silicate perovskite: Science, v. 284, p. 782–784, doi: 10.1126/science.284.5415.782.

MANUSCRIPT ACCEPTED BY THE SOCIETY 14 AUGUST 2006

Discovery of post-perovskite phase transition and implications for the nature of the D″ layer of the mantle

Kei Hirose[†]
Katsuyuki Kawamura
Department of Earth and Planetary Sciences, Tokyo Institute of Technology, Meguro, Tokyo 152-8551, Japan

ABSTRACT

$MgSiO_3$ perovskite is a principal mineral in the upper part of the lower mantle, but its stability and possible phase transition at greater depths have long been uncertain. Recently, a new high-pressure $MgSiO_3$ polymorph called "post-perovskite" was discovered above 125 GPa and 2500 K on the basis of X-ray diffraction measurements in a laser-heated diamond anvil cell (LH-DAC). Crystal structure of post-perovskite was first determined to be orthorhombic (space group: *Cmcm*) by molecular dynamics (MD) calculations. The first-principles theoretical calculations also confirmed the stability of this new phase. These results suggest that $MgSiO_3$-rich post-perovskite is a predominant mineral below 2500–2700 km depth near the base of the mantle. The D″ layer has long been the most enigmatic region in Earth's interior. The post-perovskite phase can account for the large seismic anomalies observed in the D″ region, such as D″ discontinuity, polarization anisotropy, and anticorrelation between S-wave and bulk sound velocities. The long-term enigma may be explained with this newly discovered crystal.

Keywords: post-perovskite, perovskite, D″ layer, core-mantle boundary, phase transition.

INTRODUCTION

Since the seismic observations showed unexplained features in the lowermost mantle, a phase transition that could occur in this region has long been a subject of debate (e.g., Lay and Helmberger, 1983; Wysession et al., 1998). Sidorin et al. (1999) suggested a solid-solid phase transition with a positive Clapeyron slope near the base of the mantle in order to explain the topography of the D″ layer. On the other hand, experimental results on the stability of $MgSiO_3$ perovskite, a principal mineral in the lower mantle, have been controversial for more than a decade.

Recent developments in synchrotron X-ray diffraction measurements have enabled us to investigate phase equilibria at high pressures and high temperatures up to the conditions of the core-mantle boundary (e.g., Shen et al., 2001; Watanuki et al., 2001). Here, we briefly review the previous studies on the stability and phase transition of $MgSiO_3$ perovskite. We then report our results on the discovery of a post-perovskite phase transition based on X-ray diffraction measurements (Murakami et al., 2004; Shim et al., 2004). The crystal structure of post-perovskite was first determined by molecular dynamics (MD) simulations using X-ray diffraction data (Murakami et al., 2004). The details of crystal

[†]E-mail: kei@geo.titech.ac.jp.

structure determination are also described in this paper. The post-perovskite phase transition was later examined in (Mg,Fe)SiO$_3$ (Mao et al., 2004, 2005), Mg$_3$Al$_2$Si$_3$O$_{12}$ (Tateno et al., 2005), natural pyrolite (KLB-1 peridotite) (Murakami et al., 2005; Ono and Oganov, 2005), and mid-ocean-ridge basalt (MORB) compositions (Ono et al., 2005; Hirose et al., 2005a).

Theory has played a significant role in the determination of stability and elasticity of MgSiO$_3$ perovskite and post-perovskite (Tsuchiya et al., 2004a, 2004b; Iitaka et al., 2004; Oganov and Ono, 2004). The effects of Al$_2$O$_3$ and FeO have also been examined theoretically (Caracas and Cohen, 2005; Akber-Knutson et al., 2005; Mao et al., 2005). In addition, high-temperature calculations have been made on the stability (Tsuchiya et al., 2004a; Oganov and Ono, 2004) and elasticity of post-perovskite (Stackhouse et al., 2005; Wentzcovitch et al., 2006).

A phase transition of MgSiO$_3$-rich perovskite in the lowermost mantle has significant geophysical implications, especially for seismology and geodynamics. The lower mantle is a seismically "quiet" region except at its bottom. Large seismic anomalies are observed in the D″ layer, including the D″ discontinuity (Wysession et al., 1998), S-wave polarization anisotropy (Lay et al., 1998), anticorrelation between S-wave and bulk sound velocities (Masters et al., 2000), and an ultralow velocity zone (Garnero et al., 1998). Since the origin of these anomalies has always been uncertain, the D″ layer has long been the most enigmatic region inside the planet. This long-term enigma may be reconciled with newly discovered MgSiO$_3$ post-perovskite.

The post-perovskite phase transition is an exothermic reaction. Theory has predicted the Clapeyron slope of the post-perovskite phase transition boundary in MgSiO$_3$ to be +7.5 to +10 MPa/K (Tsuchiya et al., 2004a; Oganov and Ono, 2004). Experiments have also demonstrated large positive values of the Clapeyron slope (Hirose and Fujita, 2005; Ono and Oganov, 2005; Hirose et al., 2006). The exothermic reaction near the base of the mantle has large effects on the formation of plumes and heat transport to shallower depths (Nakagawa and Tackley, 2004; Matyska and Yuen, 2005).

BRIEF REVIEWS ON THE STABILITY OF PEROVSKITE

Knittle and Jeanloz (1987) reported a conservation of orthorhombic (Mg,Fe)SiO$_3$ perovskite (space group: Pbnm) up to 127 GPa. However, their pressure measurement included large uncertainty because it was not determined at high temperature. In addition, the value of 127 GPa was measured before heating, and the pressure was dropped to 112 GPa after temperature quenching.

Wolf and Bukowinski (1987) theoretically suggested that orthorhombic MgSiO$_3$ perovskite exhibits critical soft-mode behavior and undergoes second-order phase transitions to tetragonal and cubic structures with increasing temperature. Wang et al. (1990) made similar suggestions based on the observation of twin microstructures in MgSiO$_3$ perovskite. Meade et al. (1995) also reported the phase transition of orthorhombic MgSiO$_3$ perovskite to cubic phase on the basis of X-ray diffraction measurements. In contrast, theoretical calculations have confirmed the stability of the orthorhombic perovskite structure relative to cubic under all lower mantle conditions (Wentzcovitch et al., 1993; Stixrude and Cohen, 1993).

Meade et al. (1995) suggested that MgSiO$_3$ perovskite dissociates into mixed oxides above 70 GPa. However, the same group later argued that the dissociation may have been caused by a large thermal gradient in the laser-heated sample (Mao et al., 1997). Saxena et al. (1996, 1998) reported similar observations based on both laser-heating (LH) and resistance-heating diamond anvil cell (DAC) experiments.

Kesson et al. (1998) performed phase equilibria experiments on natural pyrolite composition and claimed stability of Al-bearing (Mg,Fe)SiO$_3$ perovskite to 130 GPa on the basis of transmission electron microprobe (TEM) observations on recovered samples. It is noted, however, that they measured pressure only at room temperature before heating. Pressure could have been significantly reduced due to stress release by heating. The conventional ruby-fluorescence measurement was used to estimate pressure, which is sometimes difficult to measure above ~70 GPa. It gives higher pressure than that of the sample, when the sample becomes very thin and ruby is compressed directly by the diamond anvils.

Recent synchrotron X-ray diffraction experiments by Shim et al. (2001) showed no evidence for dissociation, but they observed one additional peak in the diffraction patterns of MgSiO$_3$ perovskite. Since this peak cannot be explained by Pbnm perovskite, they suggested a subtle change in the perovskite structure above 83 GPa. This additional peak, however, may be assigned to PtC that was formed by a chemical reaction between Pt and diamond (anvil) (Oganov and Ono, 2004).

DISCOVERY OF POST-PEROVSKITE PHASE TRANSITION

The phase transition of MgSiO$_3$ perovskite was discovered by a drastic change in the X-ray diffraction pattern. The angle-dispersive X-ray diffraction measurements were conducted at BL10XU of SPring-8, using high brilliance synchrotron X-rays from the in-vacuum type undulator. The diffraction patterns were collected in situ at high pressure-temperature (*P-T*) in LH-DAC. Recent technical developments in the generation of high *P-T* conditions that are stably for long durations (approx. more than 4 h) and the ability to focus X-ray beams on a small sample in DAC have enabled us to search for a new phase in deep lower-mantle conditions.

In 2002, we first recognized the phase transition of MgSiO$_3$ perovskite in a natural pyrolite mantle composition. Gel starting material and NaCl pressure medium were used in these experiments (Murakami et al., 2005). The sample was coated by a thin film of gold that was used for a pressure standard (Tsuchiya, 2003). Heating was generated with a focused multimode continuous-wave Nd-YAG (yttrium-aluminum-garnet) laser from both sides. It is well known that pyrolitic mantle consists predominantly of Al-bearing (Mg,Fe)SiO$_3$ perovskite together with minor amounts of (Mg,Fe)O magnesiowüstite and CaSiO$_3$ perovskite in the lower-mantle conditions. However, a X-ray diffraction pattern taken at 115 GPa and 2550 K did not include peaks from MgSiO$_3$-rich perovskite. Instead, we observed thirteen unknown peaks in the diffraction pattern. The interpretation of these unknown peaks was not made at that time.

We then switched the sample to pure MgSiO$_3$ and performed X-ray diffraction experiments at a similar P-T range (Murakami et al., 2004). Experiments were conducted using MgSiO$_3$ gel starting material mixed with platinum powder (Jamieson et al., 1982). The pure MgSiO$_3$ gel (unmixed with platinum) was used also as a thermal insulator. We usually used amorphous gel starting material in order to avoid kinetic hindering of phase transitions. Moreover, the sample volume largely decreased when a high-pressure phase was synthesized from gel, which significantly reduced the deviatoric stress in the sample.

We conducted two separate sets of experiments on pure MgSiO$_3$. In the first run, perovskite was first synthesized by heating at 105–114 GPa and 2250–2300 K for 11 min. All peaks were indexed by Pbnm perovskite and platinum (Fig. 1A). We further compressed this sample to 127 GPa at room temperature and reheated to 2500–2600 K at 127–134 GPa. Thirteen new diffraction peaks appeared within 9 min. These new peaks grew and peaks from perovskite became weak with further heating for a total of 70 min (Fig. 1B). Two-dimensional (2-D) diffraction images showed circular Debye rings for these new peaks. Similarly, in the second set of experiments, perovskite was synthesized from the amorphous starting material by heating to 1700–1970 K for 15 min at 69–73 GPa. We subsequently compressed this sample to 122 GPa at room temperature. With heating to 2200–2300 K at 128–129 GPa, intermittently for a total of 120 min by opening/closing the laser shutter several times at fixed press load, the new peaks were again observed in the diffraction patterns within 10 min of heating and were the same as those in the first run. This sample was then decompressed to 97 GPa at room temperature. The new peaks were still recognizable after decompression, although they had broadened. After heating to 2000–2200 K for 10 min at 89–101 GPa, the new peaks disappeared, and the diffraction pattern changed back to that consisting only of Pbnm perovskite and platinum. This indicated that observed change in the diffraction pattern was reversible, and therefore a chemical reaction between sample and gasket material was unlikely to have occurred.

These new peaks did not correspond to the possible dissociation products of MgO or high-pressure polymorphs of SiO$_2$ (Fig. 1). They indicated that MgSiO$_3$ perovskite did not dissociate, but transformed to a new high-pressure form above 125 GPa and 2500 K (Fig. 2).

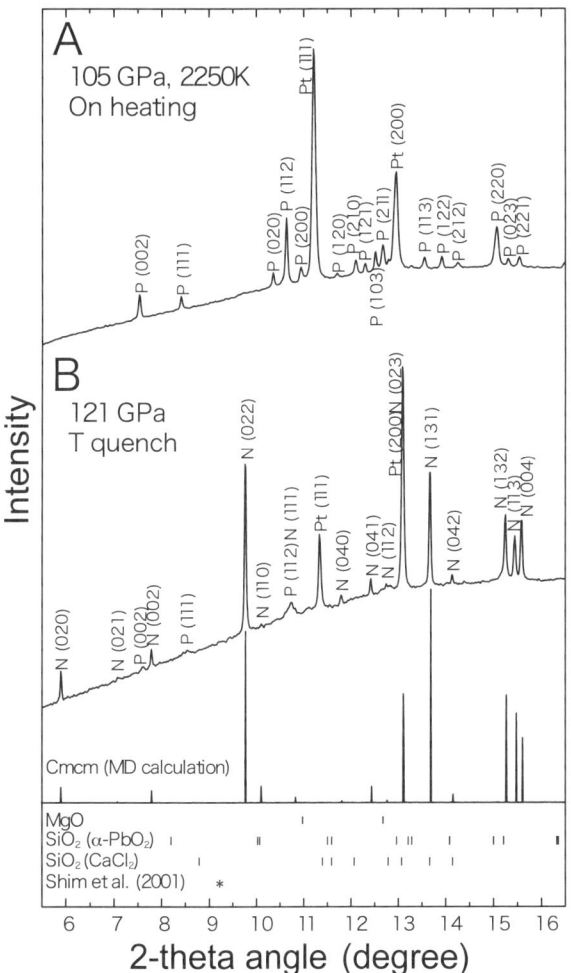

Figure 1. X-ray diffraction patterns of MgSiO$_3$ at (A) 105 GPa and 2250 K and (B) at 121 GPa and 300 K from Murakami et al. (2004). P—Pbnm perovskite; Pt—platinum; N—new phase (post-perovskite), MD—molecular dynamics calculations. The calculated peak positions of MgO and both α-PbO$_2$–type and CaCl$_2$-type SiO$_2$ are shown by small ticks. An extra peak position of MgSiO$_3$ perovskite reported by Shim et al. (2001) is also indicated by a star. The calculated powder X-ray diffraction pattern of post-perovskite was corrected for Lorentz, polarization, and multiplicity factors for comparison with the observed pattern.

DETERMINATION OF CRYSTAL STRUCTURE

The 13 diffraction peaks of the new MgSiO$_3$ polymorph (called post-perovskite) can be best indexed by an orthorhombic cell with lattice parameters of $a = 4.021$ Å, $b = 4.912$ Å, and $c = 6.093$ Å. This unit-cell cannot be reconciled with the perovskite-type structures. The unit-cell volume is smaller by ~1% from that of perovskite if $Z = 4$. Since the volume change was not large, it was expected that post-perovskite would have sixfold Si and eightfold Mg coordination, the same as perovskite.

We determined the crystal structure of post-perovskite, which possesses these lattice parameters, on the basis of structural relaxation from random structures by MD calculations (Murakami et al., 2004). The appropriate number of atoms (8 Mg + 8 Si + 24 O; $Z = 8$ for a double unit-cell because of the small a

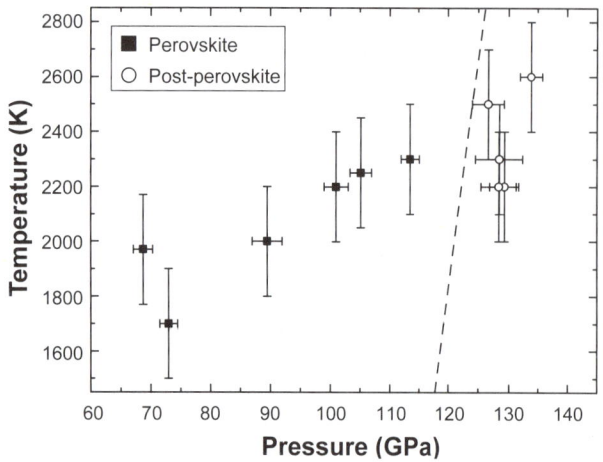

Figure 2. Phase diagram of MgSiO$_3$ from Murakami et al. (2004). Solid squares and open circles indicate the stabilities of perovskite and post-perovskite, respectively. Dashed line shows the phase transition boundary proposed by Sidorin et al. (1999) to explain the topography of the D″ discontinuity by solid-solid phase transition.

parameter) were positioned randomly and detached from each other in the MD cell with the experimentally observed dimensions. The MD calculations were performed with the full Ewald method for electrostatic interactions, the velocity Verlet algorithm with a 2 fs time step for integration of equations of motions of atoms, and (NVT) and (NTP) ensembles. The MD cell dimensions were fixed in the calculations. The interatomic potential model was prepared to reproduce the perovskite-type MgSiO$_3$ structure well at 300 K and 109 GPa.

The initial temperature of the system was set to be 5000 K. It was decreased stepwise by 1 K (2 fs/step) to 0 K with structural relaxation at each step by MD calculations (Fig. 3). At the end of the simulations, we obtained the structure in which O atoms are regularly arranged and all 16 cations (Si and Mg ions) are six- or eightfold coordinated. This procedure was repeated independently 10 times. Similar structures were found in 5 of 10 calculations (Fig. 4). We then artificially exchanged the positions of some Si and Mg atoms for all Si and Mg to be 6- and 8-coordinated, respectively. The structural relaxation was again performed on this arrangement. The resulting structure was stable for a reasonably long duration in MD calculations at 130 GPa and 3000 K.

The structure factors were calculated for this structure in *P1* space group using all atoms in the MD cell in order to determine the unit-cell and space group. The systematic extinction of structure factors $F(hkl)$ in reciprocal space indicated that the *a* and *b* parameters should be double and half of those originally estimated, respectively. Moreover, crystal structure was found to be *C*-face centered with a glide plane along the *c*-axis perpendicular to *a*-axis. After exchanging *a*- and *b*-axes (with glide place along *c*-axis perpendicular to *b*-axis), only Cmc2$_1$ (no. 36) or Cmcm (no. 63) were possible space groups. Since the presence of a center of symmetry was suggested from the observation of atomic arrangements, we concluded that Cmcm was the plausible space group. Atomic coordinates of 20 atoms were easily determined, because (1) 4 Mg atoms and 4 Si atoms should be in special posi-

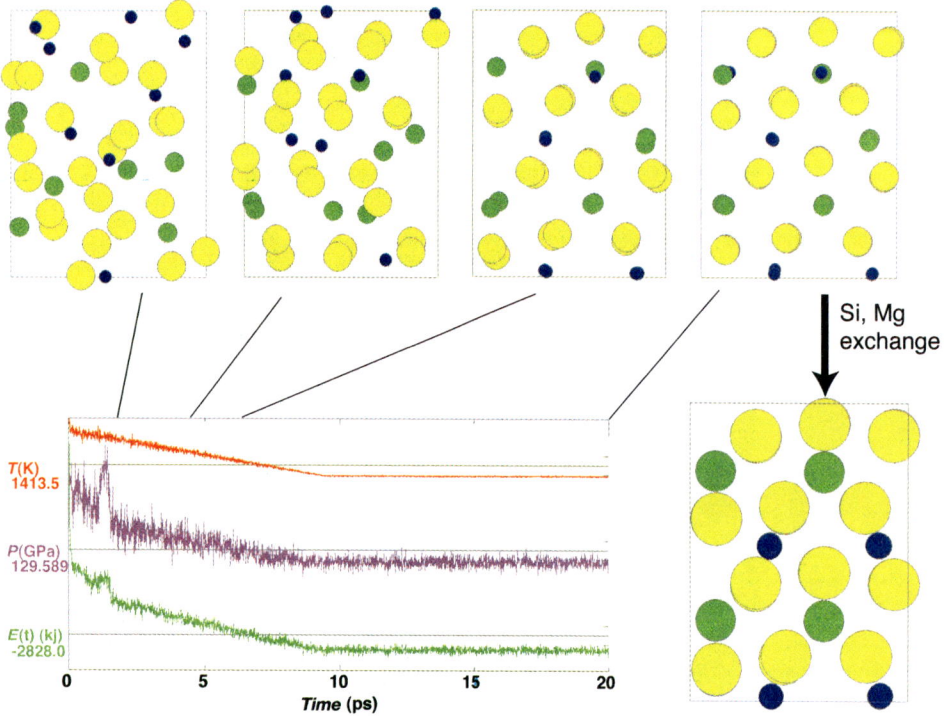

Figure 3. An example of molecular dynamics (MD)–aided structural relaxation from random structures, showing changes in structure as well as temperature (*T*), pressure (*P*), and total energy (*E*) with time. Dark blue, green, and yellow balls indicate Si, Mg, and O atoms, respectively. At the end of the simulation, six Si and two Mg atoms were six-coordinated and two Si and six Mg were eight-coordinated. We then exchanged two Si and Mg atoms for all Si and Mg to be six- and eight-coordinated, respectively.

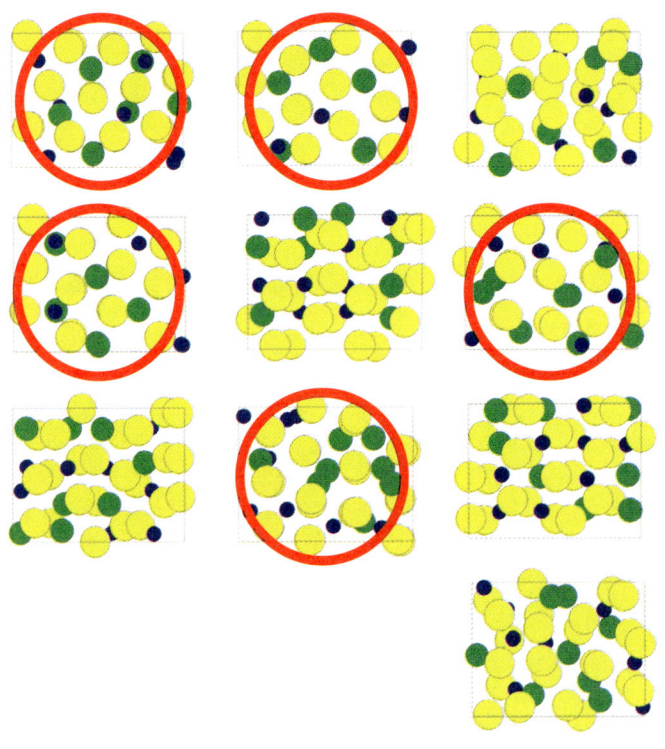

Figure 4. Results of structural relaxation from random structures by molecular dynamics (MD) simulations. Balls are same as Figure 3. O atoms are regularly arranged, and all of sixteen cations are six- or eight-coordinated in 5 of 10 calculations (shown by circles). Crosses indicate different structures.

tions and (2) 4 and 8 O atoms are respectively in the equivalent positions (it is noted that there must be 16 equivalent positions if they are in general positions). The X-ray diffraction pattern calculated from these crystal data is shown in Figure 1. The calculated powder X-ray diffraction pattern reproduces not only peak positions but also intensities of all of the observed new peaks. Post-perovskite is denser than perovskite and coexisting in the diffraction pattern by 1.0%–1.2% at ~120 GPa and 300 K.

Crystal structure of $MgSiO_3$ post-perovskite is shown in Figure 5. It has sixfold Si and eightfold Mg coordination, and the SiO_6 octahedra share edges to make an octahedral chain-like rutile-type structure. These chains run along the *a*-axis and are interconnected to each other by apical oxygen atoms in the direction of the *c*-axis to form edge and apex shared octahedral sheets. The octahedral sheets are stacked along the *b*-axis with interlayered Mg^{2+} ions. The MD calculations suggest that the *b*-axis is more compressible than the *a*- and *c*-axes.

We further performed first-principles calculations in order to refine the crystal structure and to investigate the stability of $MgSiO_3$ post-perovskite (Iitaka et al., 2004). The calculations were carried out at zero temperature with the initial models of perovskite and post-perovskite phases provided by the MD calcu-

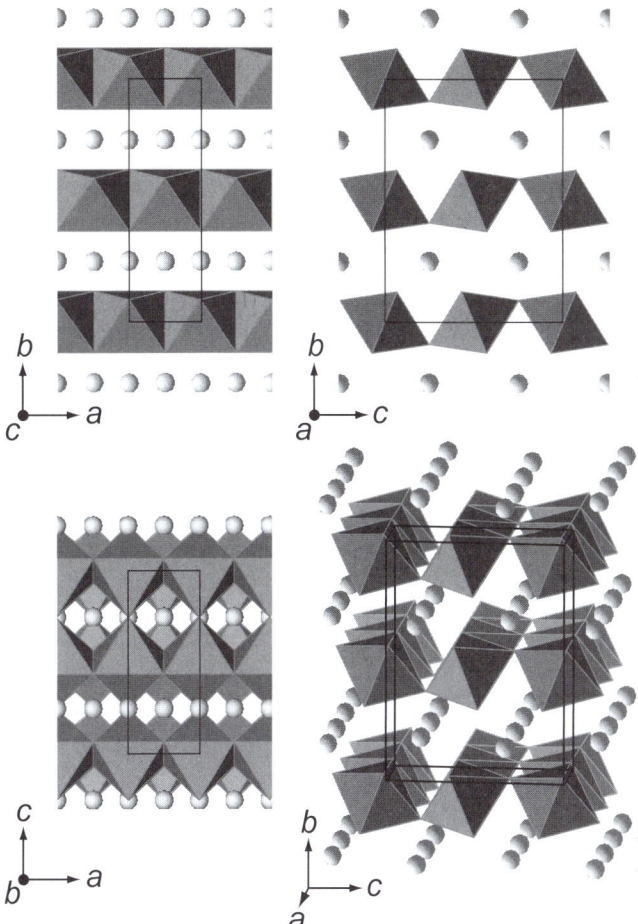

Figure 5. Crystal structure of post-perovskite projected along (001), (100), and (010) directions, and a stereoscopic view showing the layer stacking structure (Murakami et al., 2004). Coordination polyhedra of oxygen atoms around Si atoms are shown as octahedra, and the Mg^{2+} ions are shown as balls. Bold line indicates the unit cell.

lations. Next, the lattice constants and the positions of the atoms were optimized within the given symmetry to minimize enthalpy. The resulting crystal data for both perovskite and post-perovskite are presented in Table 1 (Iitaka et al., 2004).

This procedure was repeated by changing the external pressure from 80 GPa to 130 GPa in steps of 10 GPa. Results show that the post-perovskite becomes stable relative to perovskite above 98 GPa at $T = 0$ K (Fig. 6). Tsuchiya et al. (2004a) and Oganov and Ono (2004) reported similar results at 0 K. They also investigated the phase transition boundary at high temperatures using ab initio simulations. Their results showed that the post-perovskite phase transition boundary has a positive Clapeyron slope of 7.5–10 MPa/K.

ISOSTRUCTURAL COMPOUNDS

After we determined the crystal structure, a number of compounds isostructural with $MgSiO_3$ post-perovskite were found in

TABLE 1. CRYSTAL DATA OF POST-PEROVSKITE AND PEROVSKITE PHASES						
	Post-perovskite (120 GPa, 0 K)			Perovskite (120 GPa, 0 K)		
Crystal system	Orthorhombic			Orthorhombic		
Space group	Cmcm (no. 63)			Pbnm (no. 62)		
Cell parameters						
a (Å)	2.455			4.289		
b (Å)	8.051			4.557		
c (Å)	6.099			6.264		
Z	4			4		
V (Å^3)	120.55			122.43		
Atomic coordinates	x	y	z	x	y	z
Mg	0.000	0.253	0.250	0.477	0.925	0.250
Si	0.000	0.000	0.000	0.000	0.000	0.000
O1	0.000	0.928	0.250	0.884	0.033	0.250
O2	0.000	0.636	0.442	0.817	0.307	0.557
Interatomic distances (Å)						
Si-O	1.632(×2)	1.681(×4)		1.645(×2)	1.649(×2)	1.657(×2)
Average		[1.665]				[1.650]
Mg-O	1.867(×2)	1.943(×4)	2.084(×2)	1.815 2.042(×2)	1.848(×2) 2.185(×2)	1.885
Average		[1.959]				[1.981]
Shortest Si-Si		2.455			3.129	
Shortest Si-Mg		2.547			2.494	
Shortest Mg-Mg		2.455			2.997	

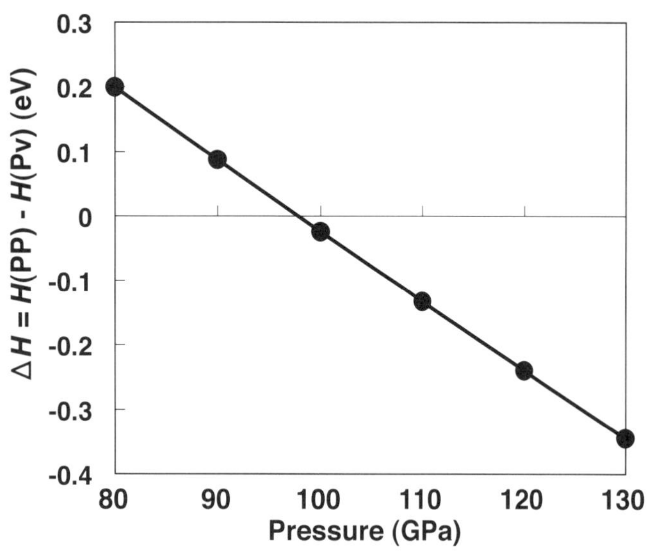

Figure 6. The enthalpy difference between the perovskite phase (Pv) and post-perovskite (PP) phase as a function of pressure at $T = 0$ K (Iitaka et al., 2004). Post-perovskite is favored over perovskite at pressures above 98 GPa.

stacking structure, same as MgSiO$_3$ post-perovskite, but differ in that (1) alkali ions are eight-coordinated and form a layer between the octahedral sheets and (2) divalent cations occupy the tetrahedral site attached to the octahedral sheet.

Only CaIrO$_3$ is isostructural as oxides. According to Shannon (1976), the ionic radius of Ir^{4+} is 0.625 Å, which is larger than 0.605 Å of Ti^{4+} and smaller than 0.690 Å of Sn^{4+} and 0.720 Å of Zr^{4+}. All of CaTiO$_3$, CaSnO$_3$, and CaZrO$_3$ have perovskite structure at ambient pressure, but only CaIrO$_3$ adopts different structure. These are not yet well understood.

It is well known that a wide range of chemical compositions with A^{II}B^{IV}X$_3$-type (Roman numbers indicate valence state) compounds crystallize as perovskite structure. How about post-perovskite structure? The Mg^{2+} site in post-perovskite is remarkably smaller than that in perovskite (Iitaka et al., 2004). It should be therefore difficult for large cations, such as Fe^{2+}, to replace Mg^{2+} in post-perovskite. However, if the ionic radii of transition metal elements are significantly reduced by the change in the spin state (e.g., Badro et al., 2003), substitution may be possible. High-pressure experimental studies have shown so far that MgSiO$_3$, MgGeO$_3$, MnGeO$_3$, and CaIrO$_3$ transform from perovskite to post-perovskite phase among A^{II}B^{IV}X$_3$-type compounds (Hirose et al., 2005b; Hirose and Fujita, 2005; Tateno et al., 2006).

Ijjaali et al. (2004) reported that both ThMnSe$_3$ and UMnSe$_3$ have needle-like crystal habit. When we compare U^{IV}Mn^{II}Se$^{II}_3$ and Mg^{II}Si^{IV}O$^{II}_3$, Si^{IV}-O^{II} bonding is stronger than Mn^{II}-Se^{II} bonding, and Mg^{II}-O^{II} is weaker than U^{IV}-Se^{II} bonding (when coordination number is the same, as this case, bonding is stronger for pairs with higher valence state). These indicate that bonding within the octahedral sheet is stronger and bonding between the octahedral sheets is weaker in MgSiO$_3$ post-perovskite than in UMnSe$_3$. It is therefore expected that the MgSiO$_3$ post-perovskite is more anisotropic and has elongated platy crystal habit.

the literature. These include UI$_3$ (octahedral site is vacant) (Levy et al., 1975), UFeS$_3$ (Noel and Padiou, 1976), AgTaS$_3$ (Wada et al., 1992; Kim et al., 1997), ThMnSe$_3$ and UMnSe$_3$ (Ijjaali et al., 2004), and CaIrO$_3$ ($a = 3.145$, $b = 9.857$, and $c = 7.296$ Å) (McDaniel and Schneider, 1972). In addition, KCuZrQ$_3$ (Q = S, Se, Te) (Mansuetto et al., 1992), CsLnZnSe$_3$ (Mitchell et al., 2002), CsLnMSe$_3$ (M = Zn, Cd, Hg) (Mitchell et al., 2003), CsLnMnSe$_3$ and AYbZnQ$_3$ (A = Rb, Cs; Q = S, Se, Te) (Mitchell et al., 2004) have similar structures. They have octahedral sheet

IMPLICATIONS FOR SEISMIC ANOMALIES IN THE LOWERMOST MANTLE

MgSiO$_3$ post-perovskite is a predominant mineral in the lowermost mantle, coexisting with minor amounts of magnesiowüstite and CaSiO$_3$ perovskite (Murakami et al., 2005; Ono and Oganov, 2005). Phase transition from MgSiO$_3$-rich perovskite to post-perovskite can cause large seismic anomalies. The elasticity of MgSiO$_3$ perovskite and post-perovskite has been estimated by first-principles calculations (Fig. 7) (Iitaka et al., 2004; Tsuchiya et al., 2004b; Oganov and Ono, 2004). Their results are quite consistent with each other. Since the D″ region is a convection boundary, strong horizontal shear flow is expected. Therefore, the preferred orientation of post-perovskite needs to be considered in the interpretation of seismic anomalies. The lattice-preferred orientation (LPO) depends on the deformation mechanisms, particularly the slip systems. Since the post-perovskite phase has an unusually large elastic constant for shear along the layering plane, prediction of the slip system is not straightforward. On the basis of structural considerations, it has been suggested that the layering plane (010) is a dominant slip plane (Murakami et al., 2004; Iitaka et al., 2004; Oganov and Ono, 2004). However, recent theoretical predictions by Oganov et al. (2005) and plastic deformation experiments on analogue material by Merkel et al. (2006) have shown that deformation of post-perovskite occurs predominantly by slip on (110) or (100). This indicates that lattice planes (110) or (100) become aligned parallel to the horizontal shear flow in the D″ region.

D″ Discontinuity

The D″ seismic discontinuity is observed at 2600–2700 km depth (119–125 GPa), several hundreds of kilometers above the core-mantle boundary (Wysession et al., 1998). A seismic velocity increase at the D″ discontinuity is up to 3% for both S and P waves, although the discontinuity is not observed in many localities, especially for P-wave velocity. It has been thought that such velocity increase is caused by thermal or chemical anomalies.

Both experiments and theory have shown that the pressure of the post-perovskite phase transition in pure MgSiO$_3$ matches the depth of the D″ discontinuity (Murakami et al., 2004; Tsuchiya et al., 2004a; Ono and Oganov, 2005; Hirose et al., 2006). In addition, the post-perovskite phase transition in natural pyrolitic mantle composition occurs at a depth very similar to that in MgSiO$_3$ (Hirose et al., 2006). Compositional effects of Fe and Al on transition pressure have been investigated both experimentally and theoretically. Mao et al. (2004) argued that incorporation of Fe^{2+} significantly expands the stability of post-perovskite relative to perovskite. This was later supported by theory (Caracas and Cohen, 2005; Mao et al., 2005) and partitioning experiments in an (Mg,Fe)SiO$_3$-(Mg,Fe)O system (Kobayashi et al., 2005). In contrast, similar partitioning experiments in natural pyrolite composition by Murakami et al. (2005) have demonstrated that Fe predominantly partitions into magnesiowüstite rather than post-perovskite, which suggests that Fe shrinks the stability of post-perovskite. Such apparent discrepancies may be due to the difference in the valence state of Fe. Both valence state and spin state of Fe in natural systems need to be further investigated in order to understand the partitioning behavior of Fe. Both experiments and theory consistently have shown that Al stabilizes perovskite relative to post-perovskite (Tateno et al., 2005; Caracas and Cohen, 2005; Akber-Knutson et al., 2005).

The elasticity of MgSiO$_3$ perovskite and post-perovskite calculated at $T = 0$ K suggests that the velocity increase is 1.0% for S waves and −0.1% for P waves at the phase transition if both are isotropic aggregates (Iitaka et al., 2004). The S-wave velocity increase becomes larger at higher temperatures (Stackhouse et al., 2005), whereas temperature has little effect on the P-wave discontinuity. These calculations are generally consistent with the observations that the velocity discontinuity is much more observable for S waves than P waves. The magnitude of velocity increase observed, however, seems much greater than the calculations, especially for P waves. Murakami et al. (2005) pointed out that a larger seismic discontinuity is formed by the onset of strong preferred orientation of post-perovskite. It is still uncertain whether the observed large velocity jump compared to mineral physics predictions is a result of anisotropy, focusing by topography enhanced by lateral temperature variations, or the effect of chemical variations associated with slab fragments.

Polarization Anisotropy

Significant polarization anisotropy has also been observed within the D″ layer, especially under the circum-Pacific regions, where the horizontally polarized S-wave velocity (V_{SH}) is faster by 1%–3% than the vertically polarized S-wave velocity (V_{SV}) (e.g., Lay et al., 1998; Panning and Romanowicz, 2004). This polarization anisotropy is not reconciled with MgSiO$_3$ perovskite. Preferred orientation of perovskite causes opposite polarization

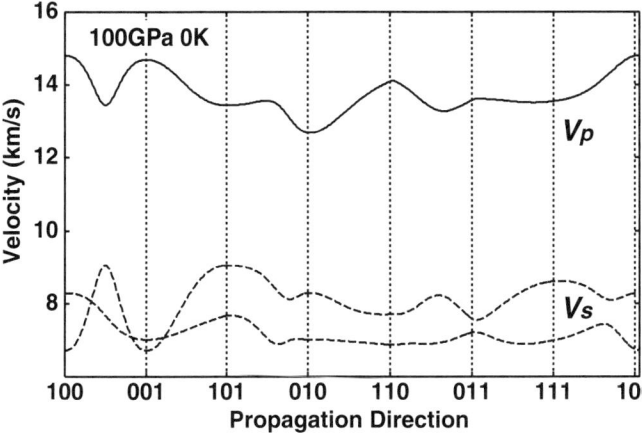

Figure 7. The variation of P- and S-wave velocities of MgSiO$_3$ post-perovskite at 100 GPa and 0 K as a function of propagation direction (Iitaka et al., 2004).

anisotropy ($V_{SV} > V_{SH}$) under horizontal shear flow (Wentzcovitch et al., 1998). Alternatively, Yamazaki and Karato (2002) suggested that anisotropy is caused by the preferred orientation of magnesiowüstite. However, it is noted that magnesiowüstite is a proportionally minor phase (<20 vol%) in the mantle.

The observed polarization anisotropy can be explained by the preferred orientation of post-perovskite. If we assume that the (110) or (100) as slip plane is aligned parallel to the horizontal shear flow in the D″ layer, the polarization anisotropy of S-waves is estimated as ~4% (V_{SH}/V_{SV} = 1.04) (Oganov et al., 2005; Merkel et al., 2006). This result is consistent with the observations, if we consider the effect of imperfect crystal alignments.

Both D″ discontinuity and polarization anisotropy are likely caused by the preferred orientation of post-perovskite. It explains why anisotropy is observed only below the D″ discontinuity. In fact, the D″ discontinuity and polarization anisotropy are observed in similar areas (Wysession et al., 1998; Panning and Romanowicz, 2004).

Anticorrelation between S-Wave and Bulk Sound Velocities

The large-scale seismic velocity models show the anticorrelation between S-wave and bulk sound velocities in the very deep portions of the mantle (e.g., Su and Dziewonski, 1997; Masters et al., 2000). Since such anticorrelation is not expected only by the temperature anomaly, Karato and Karki (2001) claimed significant chemical heterogeneities both in Fe and Ca contents in the lowermost mantle.

The post-perovskite phase transition is an exothermic reaction. Theory has predicted a positive Clapeyron slope of +7.5 to +10 MPa/K for the post-perovskite phase transition in pure $MgSiO_3$ (Tsuchiya et al., 2004a; Oganov and Ono, 2004). Experiments have demonstrated similar values of +5 to +11 MPa/K, depending on the pressure scale (Ono and Oganov, 2005; Hirose et al., 2006). It is expected that perovskite is dominant in hot regions, whereas post-perovskite is a major mineral in cold regions at a given depth near the phase transition boundary. The calculations show faster S-wave velocity and slower bulk sound velocity for post-perovskite than for perovskite at equivalent pressure. The negative correlation between S-wave and bulk sound velocities in global seismic tomography images can thus be attributed to the difference in perovskite/post-perovskite ratio in some depth range of the lowermost mantle. The difference in perovskite/post-perovskite ratio is due to the difference in depth of phase transition, which is caused by the temperature anomaly.

ACKNOWLEDGMENTS

O. Shimomura, H. Suematsu, and Y. Tatsumi are acknowledged for their support in the construction of the laser-heated diamond anvil cell (LH-DAC) system at BL10XU of SPring-8. Discussions on lower-mantle seismicity with S. Kaneshima, N. Kobayashi, and T. Okamoto were fruitful. Comments by D. Andrault, E. Ohtani, and two other anonymous reviewers improved the manuscript.

REFERENCES CITED

Akber-Knutson, S., Steinle-Neumann, G., and Asimow, P.D., 2005, Effect of Al on the sharpness of the $MgSiO_3$ perovskite to post-perovskite phase transition: Geophysical Research Letters, v. 32, p. L14303, doi: 10.1029/2005GL023192.

Badro, J., Fiquet, G., Guyot, F., Rueff, J.P., Struzhkin, V.V., Vanko, G., and Monaco, G., 2003, Iron partitioning in Earth's mantle: Toward a deep lower mantle discontinuity: Science, v. 300, p. 789–791, doi: 10.1126/science.1081311.

Caracas, R., and Cohen, R.E., 2005, Effect of chemistry on the stability and elasticity of the perovskite and post-perovskite phases in the $MgSiO_3$-$FeSiO_3$–Al_2O_3 system and implications for the lowermost mantle: Geophysical Research Letters, v. 32, p. L16310, doi: 10.1029/2005GL023164.

Garnero, E., Revenaugh, J., Williams, Q., Lay, T., and Kellogg, L., 1998, Ultralow velocity zone at the core-mantle boundary, in Gurnis, M., Wysession, M.E., Knittle, E., and Buffet, B.A., eds., The Core-Mantle Boundary Region: American Geophysical Union Geodynamics series, 28, p. 319–334.

Hirose, K., and Fujita, Y., 2005, Clapeyron slope of the post-perovskite phase transition boundary in $CaIrO_3$: Geophysical Research Letters, v. 32, p. L13313, doi: 10.1029/2005GL023219.

Hirose, K., Takafuji, N., Sata, N., and Ohishi, Y., 2005a, Phase transition and density of subducted MORB crust in the lower mantle: Earth and Planetary Science Letters, v. 237, p. 239–251, doi: 10.1016/j.epsl.2005.06.035.

Hirose, K., Kawamura, K., Ohishi, Y., Tateno, S., and Sata, N., 2005b, Stability and equation of state of $MgGeO_3$ post-perovskite phase: The American Mineralogist, v. 90, p. 262–265, doi: 10.2138/am.2005.1702.

Hirose, K., Sinmyo, R., Sata, N., and Ohishi, Y., 2006, Determination of post-perovskite phase transition boundary in $MgSiO_3$ using Au and MgO internal pressure standards: Geophysical Research Letters, v. 33, p. L01310, doi: 10.1029/2005GL024468.

Iitaka, T., Hirose, K., Kawamura, K., and Murakami, M., 2004, The elasticity of the $MgSiO_3$ post-perovskite phase in the Earth's lowermost mantle: Nature, v. 430, p. 442–445, doi: 10.1038/nature02702.

Ijjaali, I., Mitchell, K., Huang, F.Q., and Ibers, J.A., 2004, Syntheses and characterization of the actinide manganese selenides $ThMnSe_3$ and $UMnSe_3$: Journal of Solid State Chemistry, v. 177, p. 257–261, doi: 10.1016/j.jssc.2003.08.004.

Jamieson, J.C., Fritz, J.N., and Manghnani, M.H., 1982, Pressure measurement at high temperature in X-ray diffraction studies: Gold as a primary standard, in Akimoto, S., and Manghnani, M.H., eds., High-Pressure Research in Geophysics: Tokyo, Center for Academic Publications Japan, p. 27–48.

Karato, S., and Karki, B.B., 2001, Origin of lateral variation of seismic wave velocities and density in the deep mantle: Journal of Geophysical Research, v. 106, p. 21,771–21,784, doi: 10.1029/2001JB000214.

Kesson, S.E., Fitz Gerald, J.D., and Shelley, J.M., 1998, Mineralogy and dynamics of a pyrolite lower mantle: Nature, v. 393, p. 252–255, doi: 10.1038/30466.

Kim, C., Yun, H., Lee, Y., Shin, H., and Liou, K., 1997, Structure and electrical conductivity of $AgTaS_3$: Journal of Solid State Chemistry, v. 132, p. 389–393, doi: 10.1006/jssc.1997.7478.

Knittle, E., and Jeanloz, R., 1987, Synthesis and equation of state of $(Mg,Fe)SiO_3$ perovskite to over 100 GPa: Science, v. 235, p. 668–670.

Kobayashi, Y., Kondo, T., Ohtani, E., Hirao, N., Miyajima, N., Yagi, T., Nagase, T., and Kikegawa, T., 2005, Fe-Mg partitioning between $(Mg,Fe)SiO_3$ post-perovskite, perovskite, and magnesiowüstite in the Earth's lower mantle: Geophysical Research Letters, v. 32, p. L19301, doi: 10.1029/2005GL023257.

Lay, T., and Helmberger, D.V., 1983, A lower mantle S-wave triplication and the velocity structure of D″: Geophysical Journal of the Royal Astronomical Society, v. 75, p. 799–837.

Lay, T., Williams, Q., Garnero, E.J., Kellogg, L., and Wysession, M.F., 1998, Seismic wave anisotropy in the D″ region and its implications, in Gurnis, M., Wysession, M.E., Knittle, E., and Buffett, B.A., eds., The Core-Mantle Boundary Region: American Geophysical Union Geodynamics series 28, p. 299–318.

Levy, J.H., Taylor, J.C., and Wilson, P.W., 1975, The structure of uranium(III) triiodide by neutron diffraction: Acta Crystallographica, v. B31, p. 880–882.

Mansuetto, M., Keane, P.M., and Ibers, J.A., 1992, Synthesis, structure, and conductivity of the new group IV chalcogenides, $KCuZrQ_3$ (Q = S, Se, Te): Journal of Solid State Chemistry, v. 101, p. 257–264, doi: 10.1016/0022-4596(92)90182-U.

Mao, H.K., Shen, G., and Helmley, R.J., 1997, Multivariable dependence of Fe-Mg partitioning in the lower mantle: Science, v. 278, p. 2098–2100, doi: 10.1126/science.278.5346.2098.

Mao, W.L., Shen, G., Prakapenka, V.B., Meng, Y., Campbell, A.J., Heinz, D., Shu, J., Hemley, R.J., and Mao, H.K., 2004, Ferromagnesian post-perovskite silicates in the D″ layer of the Earth: Proceedings of the National Academy of Sciences of the United States of America, v. 101, p. 15,867–15,869, doi: 10.1073/pnas.0407135101.

Mao, W.L., Meng, Y., Shen, G., Prakapenka, V.B., Campbell, A.J., Heinz, D.L., Shu, J., Caracas, R., Cohen, R.E., Fei, Y., Hemley, R.J., and Mao, H.-K., 2005, Iron-rich silicates in the Earth's D″ layer: Proceedings of the National Academy of Sciences of the United States of America, v. 102, p. 9751–9753, doi: 10.1073/pnas.0503737102.

Masters, G., Laske, G., Bolton, H., and Dziewonski, A., 2000, The relative behavior of shear velocity, bulk sound velocity, and compressional velocity in the mantle: Implications for chemical and thermal structure, in Karato, S., Forte, A., Liebermann, R., Masters, G., and Stixrude, L., eds., Earth's Deep Interior: Washington, D.C., American Geophysical Union, p. 63–88.

Matyska, C., and Yuen, D.A., 2005, The importance of radiative heat transfer on superplumes in the lower mantle with the new post-perovskite phase change: Earth and Planetary Science Letters, v. 234, p. 71–81, doi: 10.1016/j.epsl.2004.10.040.

McDaniel, C.L., and Schneider, S.J., 1972, Phase relation in the $CaO-IrO_2$-Ir system in air: Journal of Solid State Chemistry, v. 4, p. 275–280, doi: 10.1016/0022-4596(72)90117-X.

Meade, C., Mao, H.K., and Hu, J., 1995, High-temperature phase transition and dissociation of $(Mg,Fe)SiO_3$ perovskite at lower mantle pressures: Science, v. 268, p. 1743–1745.

Merkel, S., Kubo, A., Miyagi, L., Speziale, S., Duffy, T.S., Mao, H.-K., and Wenk, H.-R., 2006, Plastic deformation of $MgGeO_3$ post-perovskite at lower mantle pressures: Science, v. 311, p. 644–646, doi: 10.1126/science.1121808.

Mitchell, K., Haynes, C.L., McFarland, A.D., Van Duyne, R.P., and Ibers, J.A., 2002, Tuning of optical band gaps: Syntheses, structures, magnetic properties, and optical properties of $CsLnZnSe_3$ (Ln = Sm, Tb, Dy, Ho, Er, Tm, Yb, and Y): Inorganic Chemistry, v. 41, p. 1199–1204, doi: 10.1021/ic011200u.

Mitchell, K., Huang, F.Q., McFarland, A.D., Haynes, C.L., Somers, R.C., Van Duyne, R.P., and Ibers, J.A., 2003, The $CsLnMSe_3$ semiconductors (Ln = rare-earth element, Y; M = Zn, Cd, Hg): Inorganic Chemistry, v. 42, p. 4109–4116, doi: 10.1021/ic020733f.

Mitchell, K., Huang, F.Q., Caspi, E.N., McFarland, A.D., Haynes, C.L., Somers, R.C., Jorgensen, J.D., Van Duyne, R.P., and Ibers, J.A., 2004, Syntheses, structure, and selected physical properties of $CsLnMnSe_3$ (Ln = Sm, Gd, Tb, Dy, Ho, Er, Tm, Yb, Y) and $AYbZnQ_3$ (A = Rb, Cs; Q = S, Se, Te): Inorganic Chemistry, v. 43, p. 1082–1089, doi: 10.1021/ic030232+.

Murakami, M., Hirose, K., Kawamura, K., Sata, N., and Ohishi, Y., 2004, Post-perovskite phase transition in $MgSiO_3$: Science, v. 304, p. 855–858, doi: 10.1126/science.1095932.

Murakami, M., Hirose, K., Sata, N., and Ohishi, Y., 2005, Post-perovskite phase transition and crystal chemistry in the pyrolitic lowermost mantle: Geophysical Research Letters, v. 32, p. L03304, doi: 10.1029/2004GL021956.

Nakagawa, T., and Tackley, P., 2004, Effects of a perovskite–post-perovskite phase change mantle boundary in compressible mantle: Geophysical Research Letters, v. 31, p. L16611, doi: 10.1029/2004GL020648.

Noel, H., and Padiou, J., 1976, Structure cristalline de $FeUS_3$: Acta Crystallographica, v. B32, p. 1593–1595.

Oganov, A.R., and Ono, S., 2004, Theoretical and experimental evidence for a post-perovskite phase of $MgSiO_3$ in Earth's D″ layer: Nature, v. 430, p. 445–448, doi: 10.1038/nature02701.

Oganov, A.R., Martonak, R., Laio, A., Raiteri, P., and Parrinello, M., 2005, Anisotropy of Earth's D″ layer and stacking faults in the $MgSiO_3$ post-perovskite phase: Nature, v. 438, p. 1142–1144, doi: 10.1038/nature04439.

Ono, S., and Oganov, A.R., 2005, In situ observations of phase transition between perovskite and $CaIrO_3$-type phase in $MgSiO_3$ and pyrolitic mantle composition: Earth and Planetary Science Letters, v. 236, p. 914–932, doi: 10.1016/j.epsl.2005.06.001.

Ono, S., Ohishi, Y., Isshiki, M., and Watanuki, T., 2005, In situ X-ray observations of phase assemblages in peridotite and basalt compositions at lower mantle conditions: Implications for density of subducted oceanic plate: Journal of Geophysical Research, v. 110, B02208, doi: 10.1029/2004JB003196.

Panning, M., and Romanowicz, B., 2004, Inferences on flow at the base of Earth's mantle based on seismic anisotropy: Science, v. 303, p. 351–353, doi: 10.1126/science.1091524.

Saxena, S.K., Dubrovinsky, L.S., Lazor, P., Cerenius, Y., Häggkvist, P., Hanfland, M., and Hu, J.Z., 1996, Stability of perovskite ($MgSiO_3$) in the Earth's mantle: Science, v. 274, p. 1357–1359, doi: 10.1126/science.274.5291.1357.

Saxena, S.K., Dubrovinsky, L.S., Lazor, P., and Hu, J.Z., 1998, In situ X-ray study of perovskite ($MgSiO_3$): Phase transition and dissociation at mantle conditions: European Journal of Mineralogy, v. 10, p. 1275–1281.

Shannon, R.D., 1976, Revised effective ionic radii and systematic studies of interatomic distances in halides and calcogenides: Acta Crystallographica, v. A32, p. 751–767.

Shen, G., Rivers, M.L., Wang, Y., and Sutton, S.R., 2001, Laser heated diamond cell system at the Advanced Photon Source for in situ X-ray measurements at high pressure and temperature: The Review of Scientific Instruments, v. 72, p. 1273–1282, doi: 10.1063/1.1343867.

Shim, S.H., Duffy, T.S., and Shen, G., 2001, Stability and structure of $MgSiO_3$ perovskite to 2300-kilometer depth in Earth's mantle: Science, v. 293, p. 2437–2440, doi: 10.1126/science.1061235.

Shim, S.H., Duffy, T.S., Jeanloz, R., and Shen, G., 2004, Stability and crystal structure of $MgSiO_3$ perovskite to the core-mantle boundary: Geophysical Research Letters, v. 31, p. L10603, doi: 10.1029/2004GL019639.

Sidorin, I., Gurnis, M., and Helmberger, D.V., 1999, Evidence for a ubiquitous seismic discontinuity at the base of the mantle: Science, v. 286, p. 1326–1329, doi: 10.1126/science.286.5443.1326.

Stackhouse, S., Brodholt, J.P., Wookey, J., Kendall, J.-M., and Price, G.D., 2005, The effect of temperature on the seismic anisotropy of the perovskite and post-perovskite polymorphs of $MgSiO_3$: Earth and Planetary Science Letters, v. 230, p. 1–10, doi: 10.1016/j.epsl.2004.11.021.

Stixrude, L., and Cohen, R.E., 1993, Stability of orthorhombic $MgSiO_3$-perovskite in the Earth's lower mantle: Nature, v. 364, p. 613–616, doi: 10.1038/364613a0.

Su, W.-J., and Dziewonski, A.M., 1997, Simultaneous inversion for 3-D variations in shear and bulk velocity in the mantle: Physics of the Earth and Planetary Interiors, v. 100, p. 135–156, doi: 10.1016/S0031-9201(96)03236-0.

Tateno, S., Hirose, K., Sata, N., and Ohishi, Y., 2005, Phase relations in $Mg_2Al_2Si_3O_{12}$ to 180 GPa: Effect of Al on post-perovskite phase transition: Geophysical Research Letters, v. 32, p. L15306, doi: 10.1029/2005GL023309.

Tateno, S., Hirose, K., Sata, N., and Ohishi, Y., 2006, High-pressure behavior of $MnGeO_3$ and $CdGeO_3$ perovskites and the post-perovskite phase transition: Physics and Chemistry of Minerals, v. 32, p. 721–725, doi: 10.1007/s00269-005-0049-7.

Tsuchiya, T., 2003, First-principles prediction of the P-V-T equation of state of gold and the 660-km discontinuity in Earth's mantle: Journal of Geophysical Research, v. 108, 2462, doi: 10.1029/2003JB002446.

Tsuchiya, T., Tsuchiya, J., Umemoto, K., and Wentzcovitch, R.M., 2004a, Phase transition in $MgSiO_3$ perovskite in the Earth's lower mantle: Earth and Planetary Science Letters, v. 224, p. 241–248, doi: 10.1016/j.epsl.2004.05.017.

Tsuchiya, T., Tsuchiya, J., Umemoto, K., and Wentzcovitch, R.M., 2004b, Elasticity of post-perovskite $MgSiO_3$: Geophysical Research Letters, v. 31, p. L14603, doi: 10.1029/2004GL020278.

Wada, H., Onoda, M., and Nozaki, H., 1992, Structure and properties of a new compound $AgTaS_3$: Journal of Solid State Chemistry, v. 97, p. 29–35, doi: 10.1016/0022-4596(92)90005-G.

Wang, Y., Guyot, F., Yegane-Haeri, A., and Lievermann, R.C., 1990, Twinning in $MgSiO_3$ perovskite: Science, v. 248, p. 468–471.

Watanuki, T., Shimomura, O., Yagi, T., Kondo, T., and Issiki, M., 2001, Construction of laser-heated diamond anvil cell system for in situ X-ray diffraction study at SPring-8: The Review of Scientific Instruments, v. 72, p. 1289–1292, doi: 10.1063/1.1343869.

Wentzcovitch, R.M., Martins, J.L., and Price, G.D., 1993, Ab initio molecular dynamics with variable cell shape: Application to $MgSiO_3$ perovskite:

Physical Review Letters, v. 70, p. 3947–3950, doi: 10.1103/PhysRevLett.70.3947.

Wentzcovitch, R.M., Karki, B.B., Karato, S., and Da Silva, C.R.S., 1998, High pressure elastic anisotropy of $MgSiO_3$ perovskite and geophysical implications: Earth and Planetary Science Letters, v. 164, p. 371–378, doi: 10.1016/S0012-821X(98)00230-1.

Wentzcovitch, R.M., Tsuchiya, T., and Tsuchiya, J., 2006, MgSiO3 post-perovskite at D″ conditions: Proceedings of the National Academic Sciences, p. 103, p. 543–546.

Wolf, G.H., and Bukowinski, S.T., 1987, Theoretical study of the structural properties and equations of state of $MgSiO_3$ and $CaSiO_3$ perovskites: Implications for lower mantle composition, *in* Manghnani, M.H., and Syono, Y., eds., High-Pressure Research in Mineral Physics: Part 2. Mineral Physics: Tokyo and Washington, D.C., Terra Publications and American Geophysical Union Geophysical Monograph 39, p. 313–331.

Wysession, M.E., Lay, T., Revenaugh, J., Williams, Q., Garnero, E., Jeanloz, R., and Kellog, L., 1998, The D″ discontinuity and its implications, *in* Gurnis, M., Wysession, M.E., Knittle, E., and Buffet, B.A., eds., The Core-Mantle Boundary Region: American Geophysical Union Geodynamics series, 28, p. 273–297.

Yamazaki, D., and Karato, S.-I., 2002, Fabric development in (Mg,Fe)O during large strain, shear deformation: Implications for seismic anisotropy in Earth's lower mantle: Physics of the Earth and Planetary Interiors, v. 131, p. 251–267, doi: 10.1016/S0031-9201(02)00037-7.

MANUSCRIPT ACCEPTED BY THE SOCIETY 14 AUGUST 2006

High-pressure phase transformations in the system FeO-MgO

I.Yu. Kantor[†]
A.P. Kantor
L.S. Dubrovinsky
C.A. McCammon
Bayerisches Geoinstitut, Universität Bayreuth, Bayreuth, 95440, Germany

ABSTRACT

Experimental studies of $Fe_{1-x}O$ wüstite and $(Mg_{0.8}Fe_{0.2})O$ ferropericlase were performed at room temperature and high pressure using a diamond anvil cell technique. Mössbauer spectroscopy indicated a magnetic ordering transition in FeO at 4.7 GPa, while no evidence of magnetic ordering was detected in ferropericlase up to at least 50 GPa. A structural cubic-to-rhombohedral distortion was observed in both materials using powder X-ray diffraction: at 12 GPa for FeO and at 35 GPa for ferropericlase. This suggests that the structural distortion is independent of magnetic ordering in the FeO-MgO system. A complete single-crystal elasticity tensor of FeO was determined up to 10 GPa, and only minor changes of elastic moduli were detected at the magnetic transition pressure, indicating relatively weak magnetoelastic coupling in FeO at high pressures. The existence of a rhombohedral distortion in ferropericlase with mantle composition at high pressures and the absence of magnetic ordering have important implications for the interpretation of seismological data with respect to lower-mantle inhomogeneity.

Keywords: wüstite, ferropericlase, high-pressure, phase transitions.

INTRODUCTION

The MgO-FeO system has attracted the interest of scientists for many decades. Periclase MgO, wüstite FeO, and their solid solutions are of fundamental importance in geoscience because (Mg,Fe)O is believed to be one of the main components of Earth's lower mantle (the second most abundant phase after [Mg,Fe]SiO$_3$ perovskite). MgO is a unique material compared to other oxides; periclase is stable in one structural type (NaCl-like structure) up to pressures of at least 227 GPa (Duffy et al., 1995). Its compressibility and elastic properties at high pressures are well known (Zha et al., 2000), and MgO is often used to calibrate internal pressure in synchrotron high-pressure experiments (Anderson et al., 1992). The structure and properties of MgO can also be calculated from first principles with high accuracy (Oganov et al., 2003). In contrast to periclase, FeO (wüstite) is one of the most enigmatic binary oxides (Mao et al., 1996) (in this paper, we will refer to wüstite as the stoichiometric composition FeO, even though a minor degree of nonstoichiometry is always present, even under reducing conditions). Iron is a polyvalent element, and FeO always contains some amount of trivalent iron, where the charge excess is compensated by iron vacancies. Its

---

[†]E-mail: innokenty.kantor@uni-bayreuth.de.

Kantor, I.Yu., Kantor, A.P., Dubrovinsky, L.S., and McCammon, C.A., 2007, High-pressure phase transformations in the system FeO-MgO, in Ohtani, E., ed., Advances in High-Pressure Mineralogy: Geological Society of America Special Paper 421, p. 47–55, doi: 10.1130/2007.2421(04). For permission to copy, contact editing@geosociety.org. ©2007 Geological Society of America. All rights reserved.

chemical formula therefore should be written as $Fe_{1-x}O$. The problems of defect formation in wüstite, defect clustering, and effect of defects on physical properties are mostly beyond the subject of this study.

One of the most important physical properties of FeO is magnetism: wüstite is an antiferromagnet with a Néel temperature (T_N) around 200 K at ambient pressure, where T_N varies with nonstoichiometry (Seehra and Srinivasan, 1984). At low temperatures FeO also undergoes a structural distortion from a cubic to a rhombohedrally distorted structure (Willis and Rooksby, 1953). The same transitions occur upon compression at room temperature (Shu et al., 1998; Yagi et al., 1985; Mao et al., 1996). Magnetic properties of FeO at high pressures are not so well known: it has been reported that magnetization appears between 0.3 and 8 GPa (Nasu, 1994), and disappears above 100 GPa due to high-spin to low-spin transition (Pasternak et al., 1997).

This type of phase transformation is probably of second order or weakly first order, and therefore no significant density discontinuity is expected at the transition pressure. Nevertheless, it could be important for geophysics, since elasticity anomalies are expected to occur in the vicinity of the transition, resulting in rapid elastic mode softening and decrease of shear-wave velocities (Sumino et al., 1980).

For decades, it was believed that this structural transition was driven by magnetic ordering and, hence, coincided with the magnetic transition. In this case, the implications for geophysics were considered minimal, because long-range magnetic ordering is unlikely to exist at conditions within Earth's mantle due to the high temperatures. For (Mg,Fe)O solid solutions, the rhombohedral distortion has so far only been reported in Fe-rich members (Shu et al., 1998), and it was suggested to be associated with magnetic ordering. The goal of the present study is to investigate relations between magnetic ordering, rhombohedral distortion, and elastic mode softening in FeO and in ferropericlase close to the predicted mantle composition $(Mg_{0.8}Fe_{0.2})O$.

EXPERIMENTAL DETAILS

Sample Preparation and Characterization

All samples used in this study were synthetic materials. The elasticity of FeO was measured on oriented single crystals, while Mössbauer spectroscopic and X-ray diffraction measurements were performed on polycrystalline samples. A single crystal of FeO was synthesized by the floating zone technique in an arc furnace (Berthon et al., 1979). The measured cell parameter was $a = 4.3068 \pm 0.0001$ Å, corresponding to the composition $Fe_{0.94}O$ (McCammon and Liu, 1984).

A polycrystalline sample of FeO was synthesized from pure iron in a gas-flow furnace. Iron powder was treated at 900 °C in a controlled CO/CO_2 gas flow with an oxygen fugacity of 10^{-16} atm. The measured cell parameter was 4.308(2) Å, corresponding to the composition $Fe_{0.95}O$. A ferropericlase sample with nominal $(Mg_{0.8}Fe_{0.2})O$ composition was obtained by mixing stoichiometric amounts of MgO and Fe_2O_3, heating overnight at 1200 °C in reducing conditions ($\log fO_2 = -17.4$) using a CO/CO_2 gas-flow furnace, and quenching into water. The amount of Fe^{3+} (expressed as $Fe^{3+}/\Sigma Fe$) in the ferropericlase sample was ~4.7% as determined by Mössbauer spectroscopy. All samples studied using Mössbauer spectroscopy were enriched (from 60% to 80%) with the isotope ^{57}Fe.

Diamond Anvil Cell (DAC) Technique

High-pressure in situ studies were performed using a diamond anvil cell (DAC) technique with modified Merrill-Basset DACs. In single-crystal experiments with FeO, the diamond culet size was 600 μm in diameter, and a stainless-steel gasket pre-indented to 100 μm thickness was used. Powder samples were studied in DACs with culet sizes from 250 to 350 μm, and gaskets were made of Re or Mo. The powder sample thicknesses were 30–60 μm. Experiments below 10 GPa (elasticity and some Mössbauer spectroscopic measurements of FeO) were performed using an ethanol:methanol:water mixture in the proportions 16:3:1, which provided true hydrostatic conditions in the sample chamber. The X-ray diffraction study of FeO up to 30 GPa was performed using a NaCl pressure-transmitting medium. Experiments with ferropericlase were performed without any pressure-transmitting medium, but stress in the samples was relaxed by scanning with a laser at low power (temperature below 1200 K, and the thermal emission was observed with a near-infrared sensitive camera). Pressure measurements were based on a ruby-fluorescence scale (Mao et al., 1986).

High-Pressure Mössbauer Spectroscopy

The ^{57}Fe Mössbauer spectra were recorded at room temperature in transmission mode on a constant-acceleration Mössbauer spectrometer using a nominal 370 MBq ^{57}Co high-specific-activity source in a 12 μm Rh matrix (point source). The velocity scale was calibrated relative to 25 μm α-Fe foil using the positions certified for (former) National Bureau of Standards standard reference material no. 1541; line widths of 0.36 mm/s for the outer lines of α-Fe were obtained at room temperature. Collection time for each spectrum varied from 12 to 36 h. Normally, pressure was measured before and after each Mössbauer spectrum collection, and the typical pressure drift was less than 1 GPa. NORMOS (Brand, 1997) and RECOIL (Lagarec, 2002) software packages were used for the treatment of experimental Mössbauer spectra.

X-Ray Diffraction

X-ray diffraction (XRD) measurements of FeO in a DAC were performed at beam-line ID 30 at the European Synchrotron Radiation Facility ESRF. Angle-dispersive diffraction was carried out using a constant 0.3738 Å wavelength. The MAR 345 image-plate system at 324 mm distance from the sample was used as a detector. Data processing was performed using the FIT2D software package (Hammersley, 2004). The size of the beam on the sample was ~10 µm × 9 µm full width at half maximum (FWHM). Pure NaCl powder was added to the wüstite sample in a 1:1 proportion to act as a pressure-transmitting medium and as an internal pressure calibration standard (Brown, 1999).

Gigahertz Ultrasonic Interferometry

Ultrasonic interferometry is a well-known method for the determination of elastic moduli of solids (bulk modulus, K, and shear modulus, G, as well as individual elastic constants, C_{ij}). Ultrasonic interferometry at high frequencies (gigahertz ultrasonic interferometry [GUI]; 0.6–2.1 GHz) enables traveltime measurements to be made with high accuracy on very small samples, because of the short wavelength in relation to the sample size. Single-crystal samples with a thickness of <100 µm allow the use of a DAC for the generation of high pressures.

High-frequency electronic pulses generate short packets of coherent sinusoidal waves, and a thin-film ZnO transducer produces the acoustic P waves from these packets. An acoustic signal travels into the Y-Al garnet (YAG) or Al_2O_3 buffer rod and further through one of the diamond anvils to inside the sample, and the signal is reflected from each interface and then counted. The measured traveltime of acoustic waves through the sample can be converted into sound-wave velocities (using the known sample thickness), and individual elastic constants can be calculated from the solutions of the Christoffel equations. S waves are generated from P waves using specially designed P-to-S conversion buffer rods (Jacobsen et al., 2002a, 2002b).

RESULTS

FeO (Wüstite)

Mössbauer Spectroscopy

The paramagnetic Mössbauer spectrum of nonstoichiometric $Fe_{1-x}O$ is a nonsymmetric doublet (Fig. 1). Different authors use models consisting of several (three or four) doublets and singlet positions, but no commonly accepted assignment of subspectral positions exists (Long and Grandjean, 1991, and references therein). The complex structure of the spectrum arises from the nonstoichiometric defect structure of wüstite.

The experimental Mössbauer spectra collected in the DAC were fitted to two Lorentzian doublets corresponding to Fe^{2+} and Fe^{3+} (Fig. 1). Above ~5 GPa, a new broad subspectral component appeared, and its relative area grew to 90% at 13 GPa. This new component can be described as a magnetic sextet with broad line width. The isomer shift of the new component was the same as that of the paramagnetic doublet within experimental uncertainty, and the hyperfine field (B_{hf}) increased with pressure (Fig. 2). The pressure dependence of B_{hf} did not follow a Brillouin curve, consistent with predictions from molecular field theory. Instead, all of the data can be described by an Ising model with the equation

$$\frac{B_{hf}(P)}{B_{hf}(\infty)} = \left[1 - \frac{P_N}{P}\right]^{2\beta}, \quad (1)$$

where $B_{hf}(P)$ is the hyperfine field at a given pressure P, $B_{hf}(\infty)$ is the maximum possible hyperfine field (at infinite pressure), P_N is the Néel transition pressure, and β is the critical exponent. A log-log plot of our experimental data indeed shows a linear trend, and the slope gives the value of 2β (Fig. 3). The resulting values of the fit parameters are $P_N = 4.7 \pm 0.2$ GPa; $B_{hf}(\infty) = 23.2 \pm 0.5$ T (teslas), and $\beta = 0.385 \pm 0.01$, which is close to the widely accepted value $\beta \approx 1/3$ predicted for the vanishing of the order parameter at the critical point in magnetic systems. The large uncertainty in B_{hf} in the vicinity of the magnetic ordering transition is due to the low amount of the magnetic subspectral component and its extremely large line width.

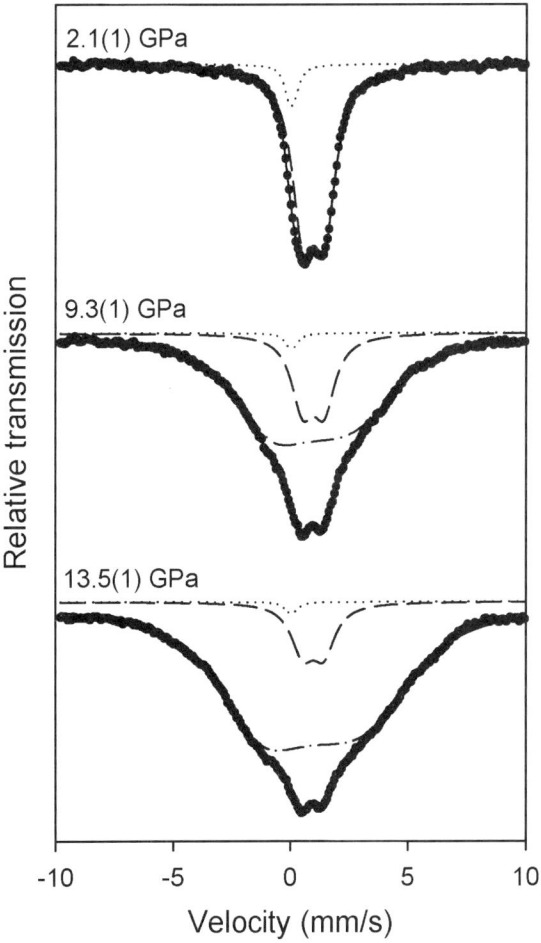

Figure 1. Room temperature Mössbauer spectra of polycrystalline FeO at various pressures. Circles show experimental spectra, solid line—total fit, dashed line—Fe^{2+} paramagnetic doublet, dotted line—Fe^{3+} doublet, dash-dotted line—magnetic sextet.

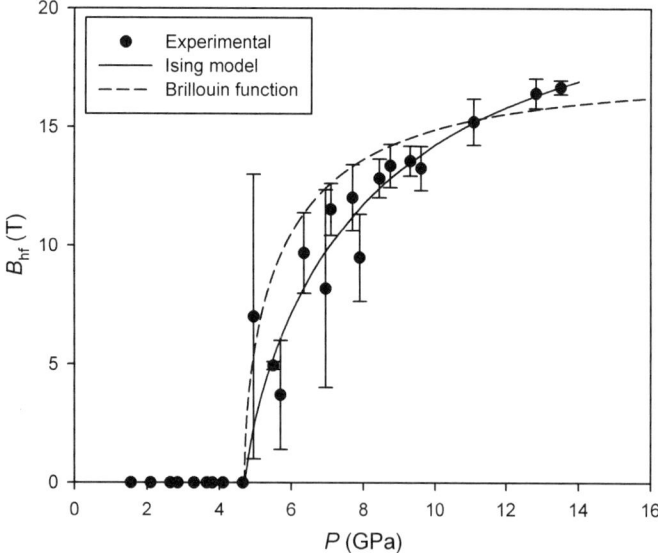

Figure 2. Pressure dependence of the hyperfine field corresponding to the magnetic Mössbauer subspectral component of FeO. Circles show observed values, solid line—fit to Ising model, dashed line—Brillouin function for S = 2.

used to calculate the density change and the change in sample length with pressure. The complete elastic tensor was obtained using the pure-mode solutions to the Christoffel equations (Brugger, 1965) and is plotted in Figure 4. The isotropic adiabatic bulk (K_S) and shear (G) moduli were calculated using the variational approach of Hashin and Strikman (Hashin and Strikman, 1962) and are also plotted in Figure 4.

A change in slope was observed for both C_{11} and C_{12} elastic constants at 4.7 ± 0.2 GPa (Fig. 4). We tested and confirmed the reversibility of the anomaly by measuring t_p<100> on both compression and decompression. The unusual behavior was also clear in the variation of the isotropic moduli. The elastic constant C_{44} decreased continuously with pressure. Quadratic polynomial fits to experimental elastic constants data were made, and polynomial coefficients are listed in Table 1.

X-Ray Diffraction

Powder X-ray diffraction measurements of FeO were performed in the DAC in order to investigate the cubic-to-rhombohedral distortion. This transition results in splitting of some of the diffraction lines (hereafter we use the subscripts "c" and "r" after the Miller indices of FeO to denote cubic and rhombohedral,

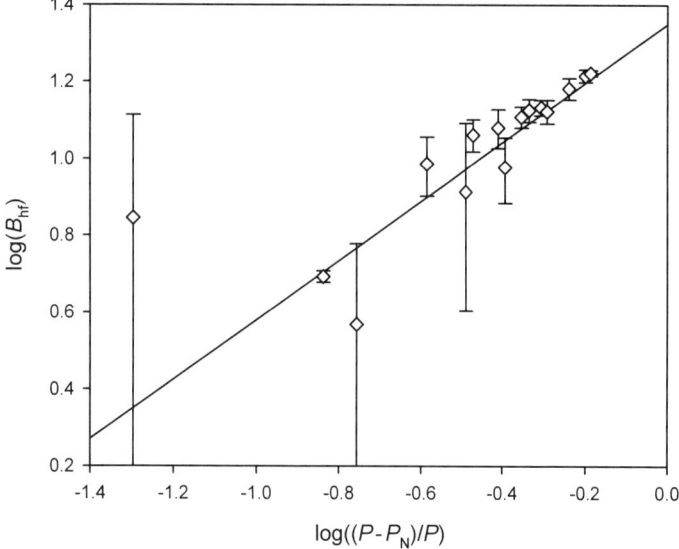

Figure 3. Effect of pressure on the hyperfine magnetic field of FeO shown on a log-log plot. The line represents a linear fit to the data.

Gigahertz Ultrasonic Interferometry

FeO single crystals of 60–70 μm diameter were oriented along <100> and <111> directions and polished to ~40 μm thickness. The measured single-crystal traveltime data t_p<100>, t_s<100>, and t_s<111> were converted to sound-wave velocities using the calculated sample thickness determined initially from bench-top velocity measurements (Jacobsen et al., 2002a). An isothermal equation of state for FeO (Jackson et al., 1990) was

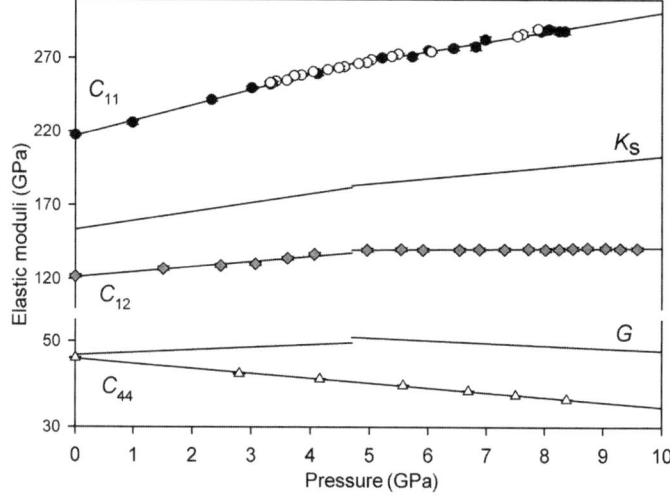

Figure 4. Single-crystal and bulk elastic moduli of FeO. Solid and open circles—compression and decompression paths for C_{11} elastic constant, diamonds—C_{12} elastic constant, triangles—C_{44} elastic constant. Lines show linear fits to the data.

TABLE 1. COEFFICIENTS OF THE LINEAR FITS TO THE ELASTIC MODULI OF FeO TO THE EQUATION A + B × P (GPa)

Modulus and pressure range	A	B
C_{11} P < 4.7 GPa	217 ± 1	10.5 ± 0.3
4.7 < P < 10 GPa	234.9 ± 0.5	6.6 ± 0.2
C_{12} P < 4.7 GPa	121 ± 1	3.5 ± 0.7
4.7 < P < 10 GPa	138.4 ± 0.5	0.27 ± 0.7
C_{44} P < 4.7 GPa	45.9 ± 0.7	-1.13 ± 0.03
K_S P < 4.7 GPa	153 ± 1	6.1 ± 0.4
4.7 < P < 10 GPa	165 ± 1	3.7 ± 0.3
G P < 4.7 GPa	46.8 ± 0.4	0.6 ± 0.1
4.7 < P < 10 GPa	49 ± 1	-0.22 ± 0.03

respectively). Specifically, the $(111)_c$ reflection splits into $(101)_r$ and $(003)_r$; $(220)_c$ splits into $(110)_r$ and $(104)_r$; and $(311)_c$ splits into $(021)_r$, $(113)_r$, and $(015)_r$. The most intense reflection, $(200)_c$, remains unsplit. In the vicinity of the structural transition, the degree of the distortion is small, and line splitting can be seen as line broadening. In the studied pressure range, only the $(111)_c$ and $(200)_c$ reflections of FeO did not overlap with any of the NaCl reflections. The rhombohedral distortion in FeO was thus extracted from the asymmetric broadening of the $(111)_c$ reflection. In the rhombohedral phase, the latter reflection splits into two, where $(101)_r$ is expected to be at lower angles and three times lower in intensity compared with the $(003)_r$ reflection. Fitting of the $(111)_c$ reflection as two peaks gives the degree of splitting (as a difference between the two peak positions), which provides a good criterion for detecting the rhombohedral transition (Fig. 5). The transition pressure was found to be 12 ± 1 GPa, in good agreement with previous reports on FeO (Yagi et al., 1985; Dubrovinsky et al., 2000b).

$(Mg_{0.8}Fe_{0.2})O$ (Ferropericlase)

Mössbauer Spectroscopy

Mössbauer spectra of $(Mg_{0.8}Fe_{0.2})O$ ferropericlase were collected up to 56 GPa. The spectra were consistent with single-phase (Mg,Fe)O, and no evidence for chemical inhomogeneity (e.g., FeO-rich regions) was found. A typical Mössbauer spectrum of the ferropericlase sample at 47 GPa is shown in Figure 6. We deconvoluted the Mössbauer spectra according to the model of two Lorentzian quadrupole doublets, one for Fe^{2+} and one for Fe^{3+} (Dobson et al., 1998), where the amount of Fe^{2+} and Fe^{3+} was determined based on the absorption area of the relevant dou-

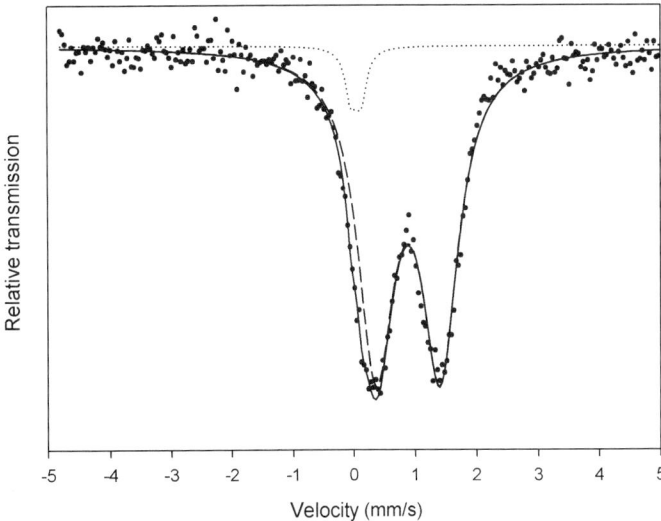

Figure 6. Mössbauer spectrum of $(Mg_{0.8}Fe_{0.2})O$ ferropericlase at 47 GPa. Circles—experimental data, solid line—total fit, dashed line—Fe^{2+} doublet, dotted line—Fe^{3+} doublet.

blet. Within the pressure range from 3 to 56 GPa, the Fe^{3+} amount remained almost constant and fluctuated around 4.7 ± 0.4%. The isomer shift of the main doublet depended linearly on pressure and had a slope of about −0.003 mm/s per GPa within the studied pressure range. We observe a strong non-monotonous pressure dependence of the quadrupole splitting: it increased up to pressures of ~36–37 GPa and decreased at higher pressures (Fig. 7). Quadrupole splitting is a sensitive indicator of the distortion of the local iron environment. For example, in $FeSiO_3$ clinoferrosilite, the distortion of the $M2(Fe^{2+})$ octahedron is coupled to a rapid decrease of quadrupole splitting (McCammon and Tennant, 1996). We infer that the changes in quadrupole splitting are also due to changes in the distortion of the local Fe^{2+} environment. No evidence of magnetic ordering or line broadening in Mössbauer spectra was observed up to the highest pressure reached in this study (56 GPa).

X-Ray Diffraction

X-Ray diffraction spectra of the ferropericlase sample fit perfectly to a cubic rock-salt structure up to 35 GPa. At pressures higher than 36 GPa, some of the diffraction peaks started to broaden. From the X-ray diffraction images, we clearly saw the splitting of the $(220)_c$ reflection, although in the integrated pattern, the line splitting was not pronounced. Figure 8 shows the pressure variation of the full width at half maximum (FWHM) of the $(220)_c$ and $(200)_c$ peaks. Rapid broadening of the $(220)_c$ reflection started at 35 GPa. This broadening could not have been due to stress development because the FWHM of the $(200)_c$ reflection remained almost constant. Besides, above 35 GPa splitting (or broadening) of $(111)_c$, $(311)_c$, and $(222)_c$ reflections was observed, in full agreement with the rhombohedral symmetry. From the

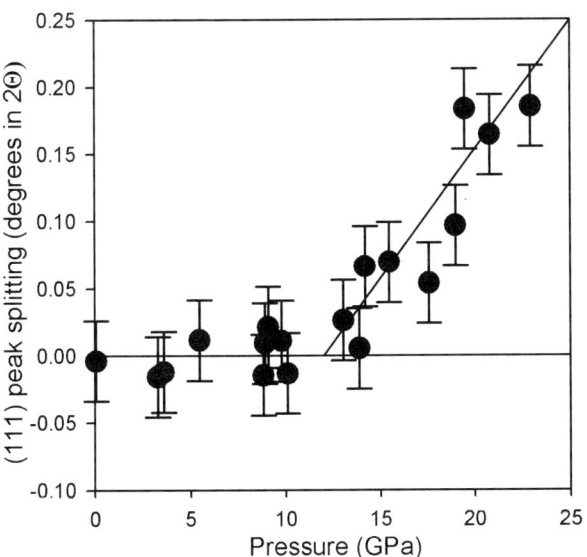

Figure 5. Splitting of the cubic (111) reflection measured in polycrystalline FeO as a function of pressure. The rhombohedral distortion starts at ~12 GPa. Lines are given as guides for the eye.

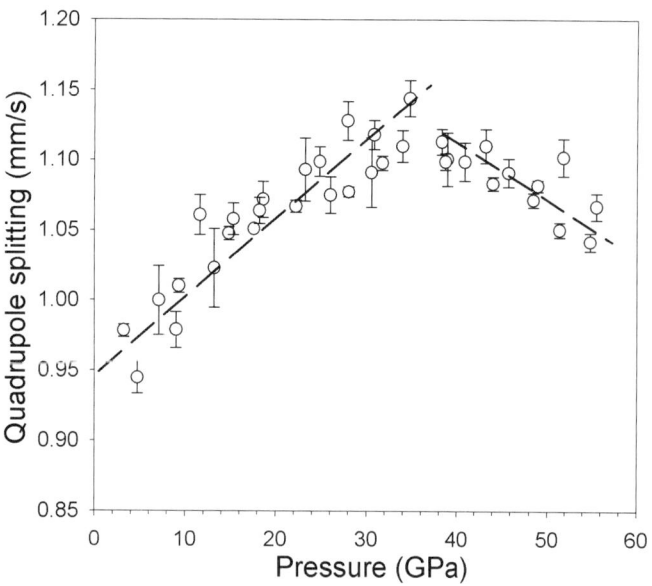

Figure 7. Variation in quadrupole splitting of the main doublet of $(Mg_{0.8}Fe_{0.2})O$ ferropericlase with pressure. Lines are given as guides for the eye.

XRD data, the transition pressure was determined to be 35 ± 1 GPa. Pressure-volume (P-V) data were fit to a second-order Birch-Murnaghan isothermal equation of state with values of $K_{300} = 158(5)$ GPa and $V_0 = 11.53(1)$ cm³/mol for the cubic structure and $K_{300} = 170(7)$ GPa and $V_0 = 11.21(1)$ cm³/mol for the rhombohedral structure (Fig. 9). Although these equation-of-state parameters imply volume discontinuity at the transition point, the quality of our data (and lack of points in the vicinity of phase transition) does not allow us to distinguish between the first- and second-order transformation.

DISCUSSION

Our combined Mössbauer spectroscopic, X-ray diffraction, and elasticity measurements of nonstoichiometric FeO imply that the cubic-to-rhombohedral transition does not coincide with the magnetic ordering (Néel) transition. At room temperature in the pressure range from ~5 to ~15 GPa, a cubic magnetically ordered phase of FeO exists. A commonly accepted type-II antiferromagnetic structure, usually assigned to FeO, is not consistent with cubic symmetry.

Random-phase Green's-function theory for AF-II face cubic centered (fcc) structures predicts that spin-phonon interactions result in changes of the interaxial angle (α) and cell dimensions (a) (Morosin, 1970; Rodbell and Owen, 1964). Namely,

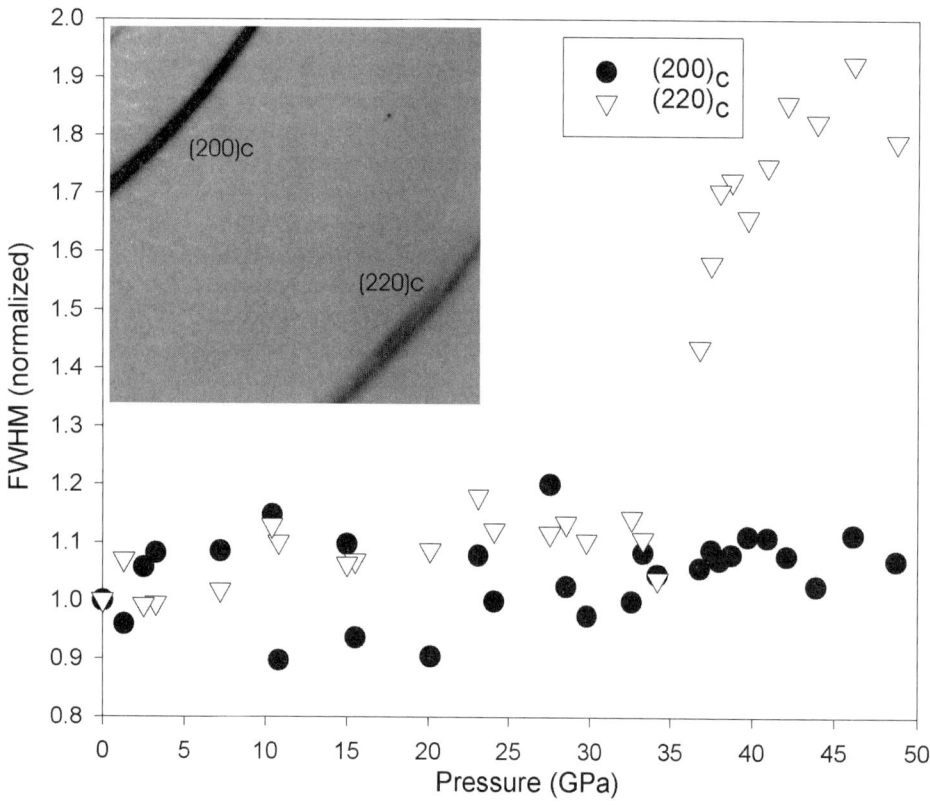

Figure 8. Normalized full width at half maximum (FWHM) for the cubic (200) and (220) reflections of $(Mg_{0.8}Fe_{0.2})O$ ferropericlase. The phase transformation starts at ~35 GPa. Inset: part of the image plate diffraction image at 47 ± 2 GPa, showing splitting of $(220)_c$ reflection.

$$\Delta\alpha = Nz_1 J_1 \varepsilon_1 [(S_i \times S_j)_{nnP} - (S_i \times S_j)_{nn^a}] / 24 C_{44}, \quad (2)$$

$$\delta a / a = N_{z2} J_2 \varepsilon_2 (S_i \times S_j)_{nnn^a} / 6(C_{11} + 2C_{12}), \quad (3)$$

where α is the rhombohedral setting interaxial angle ($\alpha = 60°$ for the cubic structure); N is the number of spins in the system; z_1 (= 12) and z_2 (= 6) are the number of nearest-neighbor (nn) and next-nearest-neighbor (nnn) metal ions; $<...>_{nnP}$, $<...>_{nn^a}$, and $<...>_{nnn^a}$ refer to the thermal averages over parallel nn, antiparallel nn, and antiparallel nnn, respectively; J_1 and J_2 are nn and nnn exchange interactions; $\varepsilon_1 = -r(\partial \ln J_1)/\partial r$ and $\varepsilon_2 = -r (\partial \ln J_2)/\partial r$; r is the interspin distance; and finally C_{11}, C_{12}, and C_{44} are the appropriate (cubic) elastic constants.

Equation 3 implies that in AF-II structures, nnn interactions are coupled with changes in C_{11}, C_{12}, and volume striction. It is reasonable to assume that a cubic antiferromagnetic structure of FeO exists (at room temperature and pressures between ~5 and ~17 GPa), and that in a cubic magnetic structure, the nnn interactions are the same as in the rhombohedral AF-II structure. Since C_{44} changes smoothly and continuously up to at least 9 GPa and $\Delta\alpha = 0$ in this pressure range, the nn spin distribution in the cubic magnetic phase of FeO is not equal to that in the classical AF-II structure.

In the classical rhombohedral (AF-IIr) structure, the parallel spins form sheets within (111) planes of the lattice, where each adjacent sheet is antiparallel. Magnetic interactions between sheets give rise to a slightly attractive or repulsive force resulting in rhombohedral distortion of the lattice below the Néel transition. In the AF-IIr structure, all the next-nearest neighbors are antiparallel, and only half of the nearest neighbors are antiferromagnetically coupled, while the other half are coupled ferromagnetically (Fig. 10A). Assuming nnn interactions (of FeO) to be the same as in the AF-IIr type, we propose a structure where the spins are distributed in the first coordination shell in a different way: six nn with antiparallel spins lie in the same close-packing layer as the central ion, while six nn with parallel spins lie in upper and lower hexagonal layers (Fig. 10B). This cubic antiferromagnetic structure (AF-IIc) is closely related to the classical AF-IIr type, but the symmetry remains cubic. Relative to the cubic fcc lattice, the proposed magnetic structure has doubled cell dimensions and a space group $Fd\bar{3}m$, while the crystallographic space group remains $Fm\bar{3}m$. The AF-IIc structure can be considered as an intermediate step toward the AF-IIr structure, since every hexagonal layer consists primarily of ions with parallel spins and only one quarter of ions have antiparallel spins.

Another important result is that the magnetic ordering transition in FeO is not exactly coupled with the structural distortion. This result is important for understanding the fundamental interactions that exist in this group of materials. First of all, the nature of structural distortions in transition metal mono-oxides at low temperatures and high pressures should be revised: if the structural transition is not driven by magnetization, some other kind of interatomic interaction or specific features of metal-oxygen chemical bonding should be considered. The second implication is that magnetoelastic coupling in this material group is perhaps not so strong as previously believed (Struzhkin et al., 2001), and a rapid mode softening is not connected with magnetic interactions but only with the structural phase transition.

According to the present study, the Néel temperature of FeO increases quite fast with the slope of ~20 K per GPa. Structural transition temperature increases with much smaller slope of ~6 K per GPa. This implies that increasing pressure and temperature would stabilize the proposed cubic antiferromagnetic phase of FeO.

In $(Mg_{0.8}Fe_{0.2})O$ ferropericlase, no magnetic ordering was observed up to 50 GPa, and at higher pressures, Fe transformed to the low-spin state in $(Mg_{0.83}Fe_{0.17})O$ ferropericlase (Badro et al., 2003); therefore, magnetism is not expected to exist in Earth's lower mantle, otherwise pure FeO would exist, for example, as a product of ferropericlase partial decomposition (Dubrovinsky et al., 2000a).

CONCLUSIONS

In summary, we have observed a magnetic transition in FeO at 4.7 ± 0.2 GPa by means of Mössbauer spectroscopy and a rhombohedral distortion at ~12 GPa by X-ray diffraction. The complete third-order elastic tensor (C_{ij}) of FeO to 9.6 GPa was determined using gigahertz ultrasonic interferometry in the DAC. We monitored with high precision a pressure-induced mode softening of the C_{44} elastic constant by 20% at 10 GPa, consistent with previous ultrasonic measurements to 3 GPa (Jackson et al., 1990). An unusual discontinuity in the pressure derivatives of C_{11} and C_{12} at 4.7 ± 0.2 GPa is consistent with the pressure at which magnetic ordering starts, as observed by high-pressure Mössbauer spectroscopy. The results indicate that an intermediate,

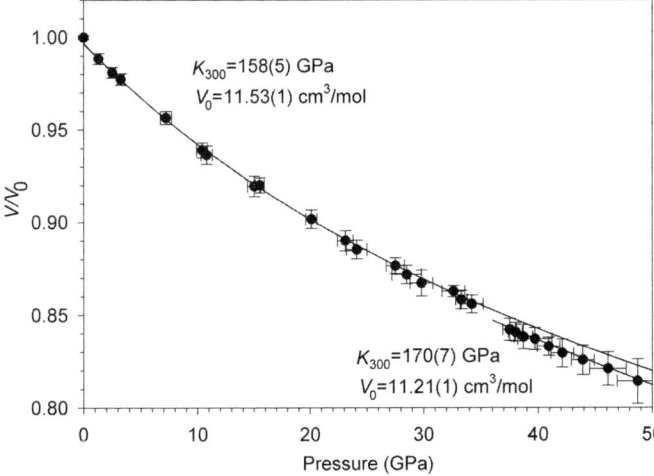

Figure 9. Compressibility of $(Mg_{0.8}Fe_{0.2})O$ ferropericlase. Circles are observed values, lines are fits to a second-order Birch-Murnaghan equation of state for cubic ($P < 35$ GPa) and rhombohedral ($P > 35$ GPa) phases.

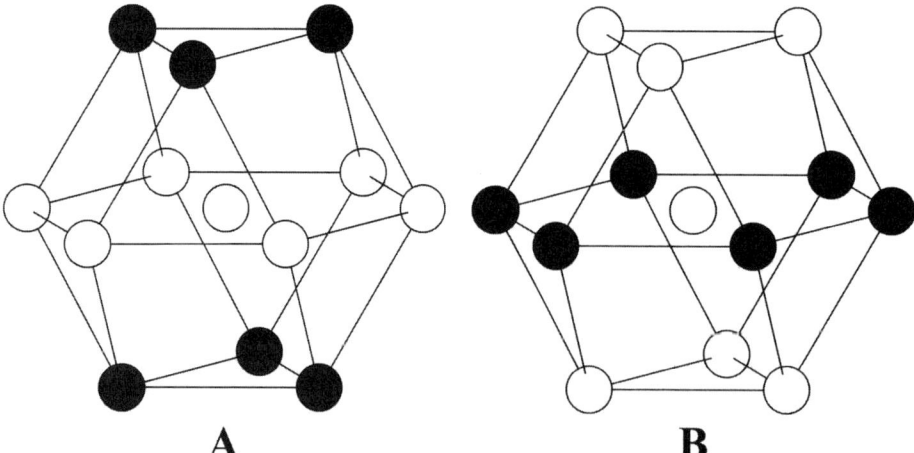

Figure 10. Proposed nearest-neighbor environment of the metal ion in the AF-IIr (A) and AF-IIc (B) structures. Ions with positive spins are represented by open circles, and ions with negative spins are shown by filled circles. The cubic <111> direction is vertical.

partially magnetic but still cubic, phase of FeO exists at room temperature in the pressure range from ~5 to ~17 GPa, and we have proposed a magnetic structure for this phase. In ferropericlase $(Fe_{0.2}Mg_{0.8})O$, the opposite behavior was observed: a structural distortion occurred at 35 ± 2 GPa with no evidence for magnetic ordering. These results indicate that magnetic ordering is decoupled from the structural distortion in FeO and (Mg,Fe)O.

ACKNOWLEDGMENTS

We would like to acknowledge S. Jacobsen, H. Reichmann, and W. Crichton for their assistance with performing the experiments. This study was financially supported by the Deutsche Forschungsgemeinschaft (DFG).

REFERENCES CITED

Anderson, O.L., Oda, H., and Isaak, D.G., 1992, A model for the computation of thermal expansivity at high T and high compression: MgO as an example: Geophysical Research Letters, v. 19, p. 1987–1990.

Badro, J., Fiquet, G., Guyot, F., Rueff, J.P., Struzhkin, V.V., Vankó, G., and Monaco, G., 2003, Iron partitioning in Earth's mantle: Toward a deep lower mantle discontinuity: Science, v. 300, p. 789–791, doi: 10.1126/science.1081311.

Berthon, J., Revcolevschi, A., Morikawa, H., and Touzelin, B., 1979, Growth of wüstite (Fe_{1-x}O) crystals of various stoichiometries: Journal of Crystal Growth, v. 47, p. 736–738, doi: 10.1016/0022-0248(79)90020-4.

Brand, R.A., 1997, Normos software package: http://www.wissel-instruments.de/produkte/software.html (last accessed February 2007).

Brown, J.M., 1999, The NaCl pressure standard: Journal of Applied Physics, v. 86, p. 5801–5808, doi: 10.1063/1.371596.

Brugger, K., 1965, Pure modes for elastic waves in crystals: Journal of Applied Physics, v. 36, p. 759–768, doi: 10.1063/1.1714215.

Dobson, D.P., Cohen, N.S., Pankhurst, Q.A., and Brodholdt, J.P., 1998, A convenient method for measuring ferric iron in magnesiowüstite (MgO-Fe_{1-x}O): The American Mineralogist, v. 83, p. 794–798.

Dubrovinsky, L.S., Dubrovinskaia, N.A., Saxena, S.K., Annersten, H., Hålenius, E., Harryson, H., Tutti, F., Rekhi, S., and Le Bihan, T., 2000a, Stability of ferropericlase in the lower mantle: Science, v. 289, p. 430–432, doi: 10.1126/science.289.5478.430.

Dubrovinsky, L.S., Dubrovinskaia, N., Saxena, S., and Le Bihan, T., 2000b, X-ray diffraction under non-hydrostatic conditions in experiments with diamond anvil cell: Wüstite (FeO) as an example: Materials Science and Engineering, v. A288, p. 187–190.

Duffy, T.S., Hemley, R.J., and Mao, H.K., 1995, Equation of state and shear strength at multimegabar pressures: Magnesium oxide to 227 GPa: Physical Review Letters, v. 74, p. 1371–1374, doi: 10.1103/PhysRevLett.74.1371.

Hammersley, A.H., 2004, Fit2D software: http://www.esrf.fr/computing/scientific/FIT2D/ (last accessed February 2007).

Hashin, Z., and Strikman, S., 1962, A variational approach to the theory of the elastic behaviour of polycrystals: Journal of the Mechanics and Physics of Solids, v. 10, p. 343–352, doi: 10.1016/0022-5096(62)90005-4.

Jackson, I., Khanna, S.K., Revcolevschi, A., and Berthon, J., 1990, Elasticity, shear mode softening and high-pressure polymorphism of wüstite Fe_{1-x}O: Journal of Geophysical Research, v. 95, p. 21,671–21,685.

Jacobsen, S.D., Reichmann, H.-J., Spetzler, H.A., Mackwell, S.J., Smyth, J.R., Angel, R.J., and McCammon, C.A., 2002a, Structure and elasticity of single-crystal (Mg,Fe)O and a new method of generating shear waves for gigahertz ultrasonic interferometry: Journal of Geophysical Research, v. 107, p. 2037–2051, doi: 10.1029/2001JB000490.

Jacobsen, S.D., Spetzler, H.A., Reichmann, H.-J., Smyth, J.R., Mackwell, S.J., Angel, R.J., and Bassett, W.A., 2002b, Gigahertz ultrasonic interferometry at high P and T: New tools for obtaining a thermodynamic equation of state: Journal of Physics Condensed Matter, v. 14, p. 11,525–11,530, doi: 10.1088/0953-8984/14/44/510.

Lagarec, K., 2002, Recoil software: http://www.isapps.ca/recoil (last accessed February 2007).

Long, G.J., and Grandjean, F., 1991, Mössbauer effect, magnetic and structural studies of wüstite Fe_{1-x}O, in Catlow, C.R.A., ed., Advances in Solid State Chemistry. Volume 2: London, JAI Press, p. 187–221.

Mao, H.-K., Xu, J., and Bell, P.M., 1986, Calibration of the ruby pressure gauge to 800 kbar under quasihydrostatic conditions: Journal of Geophysical Research, v. 91, p. 4673–4678.

Mao, H.-K., Shu, J., Fei, Y., Hu, J., and Hemley, R.J., 1996, The wüstite enigma: Physics of Earth and Planetary Interiors, v. 96, p. 135–145, doi: 10.1016/0031-9201(96)03146-9.

McCammon, C.A., and Liu, L.-G., 1984, The effects of pressure and temperature on non-stoichiometric wüstite, Fe_xO: The iron-rich phase boundary: Physics and Chemistry of Minerals, v. 10, p. 106–113, doi: 10.1007/BF00309644.

McCammon, C.A., and Tennant, C., 1996, High-pressure Mössbauer study of synthetic clinoferrite, $FeSiO_3$, in Mineral Spectroscopy: A Tribute to Roger G. Burn: Houston, American Geochemical Society Special Publication 5, p. 281–288.

Morosin, B., 1970, Exchange striction effects in MnO and MnS: Physical Review B: Condensed Matter and Materials Physics, v. 1, p. 236–243.

Nasu, S., 1994, High pressure Mössbauer spectroscopy using a diamond anvil cell: Hyperfine Interactions, v. 90, p. 59–75, doi: 10.1007/BF02069118.

Oganov, A.R., Gillan, M.J., and Price, G.D., 2003, Ab initio lattice dynamics and structural stability of MgO: The Journal of Chemical Physics, v. 118, p. 10,174–10,182, doi: 10.1063/1.1570394.

Pasternak, M.P., Taylor, R.D., Jeanloz, R., Li, H., Nguyen, J.H., and McCammon,

C.A., 1997, High pressure collapse of magnetism in $Fe_{0.94}O$: Mössbauer spectroscopy beyond 100 GPa: Physical Review Letters, v. 79, p. 5046–5049, doi: 10.1103/PhysRevLett.79.5046.

Rodbell, D.S., and Owen, J., 1964, Sublattice magnetization and lattice distortions in MnO and NiO: Journal of Applied Physics, v. 35, p. 1002–1003, doi: 10.1063/1.1713351.

Seehra, M.S., and Srinivasan, G., 1984, Magnetic studies of non-stoichiometric Fe_zO and evidence for magnetic defect clusters: Journal of Physics C: Condensed Matter, v. 17, p. 883–892.

Shu, J., Mao, H.K., Hu, J., Fei, Y., and Hemley, R.J., 1998, High-pressure phase transition in magnesiowüstite $(Fe_{1-x}Mg_x)O$: Eos (Transactions, American Geophysical Union), v. 79(17), Spring Meeting Supplement, abstract M21A-01.

Struzhkin, V.V., Mao, H.K., Hu, J., Schwoerer-Böhning, M., Shu, J., Hemley, R.J., Sturhahn, W., Hu, M.Y., and Alp, E.E., 2001, Nuclear inelastic X-ray scattering of FeO to 48 GPa: Physical Review Letters, v. 87, 255501 (4 pages).

Sumino, Y., Kumazawa, M., Nishizawa, O., and Pluschkell, W., 1980, The elastic constants of single crystal $Fe_{1-x}O$, MnO and CoO, and the elasticity of stoichiometric magnesiowüstite: Journal of Physics of the Earth, v. 28, p. 475–495.

Willis, B.T.M., and Rooksby, H.P., 1953, Change of structure of ferrous oxide at low temperature: Acta Crystallographica, v. 6, p. 827–831, doi: 10.1107/S0365110X53002441.

Yagi, T., Suzuki, T., and Akimoto, S., 1985, Static compression of wüstite $(Fe_{0.98}O)$ to 120 GPa: Journal of Geophysical Research, v. 90, p. 8784–8788.

Zha, C.-S., Mao, H.-K., and Hemley, R.J., 2000, Elasticity of MgO and a primary pressure scale to 55 GPa: Proceedings of the National Academy of Sciences of the United States of America, v. 97, p. 13,494–13,499, doi: 10.1073/pnas.240466697.

MANUSCRIPT ACCEPTED BY THE SOCIETY 14 AUGUST 2006

High-pressure mineral assemblages in shocked meteorites and shocked terrestrial rocks: Mechanisms of phase transformations and constraints to pressure and temperature histories

Philippe Gillet[†]
Laboratoire de Sciences de la Terre, Ecole Normale Supérieure de Lyon et Université Lyon I (UMR CNRS 5570), 46 Allée d'Italie, 69364 Lyon Cedex, France

Ahmed El Goresy
Bayerisches Geoinstitut, Universität Bayreuth, D-95440 Bayreuth, Germany

Pierre Beck
Laboratoire de Sciences de la Terre, Ecole Normale Supérieure de Lyon et Université Lyon I (UMR CNRS 5570), 46 Allée d'Italie, 69364 Lyon Cedex, France

Ming Chen
Guangzhou Institute of Geochemistry, Chinese Academy of Sciences, Wushan, Guangzhou 510640, China

ABSTRACT

After an overview of the most recent results of static high-pressure and high-temperature experiments, we present a review of the mineralogy of shocked meteorites. The high-pressure minerals in these rocks result either from solid-state reactions or from the crystallization of melts at high pressures. Comparisons of naturally shocked samples with samples processed in dynamic experiments must be made with extreme caution. The durations of the equilibrium shock pressure experienced by meteorites can vary over at least three orders of magnitude (10^{-2} s to 10 s), and they lie within the lower range of the duration of static experiments conducted in diamond anvil cells or multianvil apparatus. We emphasize that dynamic experiments up to 130 GPa have never produced any reconstructive solid-state phase transition or liquidus high-pressure minerals that offer a reliable calibration of the continuum of shock pressures and temperatures. The solid-state transformations observed in shocked meteorites are in many cases incomplete and provide only insights into the initial stages of high-pressure phase transitions, crystallization, and chemical interdiffusion. In contrast, the natural high-pressure species crystallized from silicate liquids at high pressures and temperatures provide more precise information on the pressures and temperatures reached during a shock event on the parental asteroid. The kinetics of phase

---

[†]E-mail: philippe.gillet@ens-lyon.fr.

transitions and diffusion of trace elements permit meaningful estimates of the pressure, temperature, and shock durations. We also present information on new dense minerals (C and TiO$_2$) in terrestrial shocked rocks in impact craters and discuss their relevance to a reliable estimate of pressure and temperature conditions.

Keywords: meteorites, shock, mineralogy, high pressure, melting, chondrites, SNC meteorites, diamond.

INTRODUCTION

The formation of planets in the solar system and their subsequent evolution was in great part governed by impacts at various scales. Recent impacts (younger than 5 million years) on the Moon and Mars are also responsible for the delivery of fragments of these planetary bodies to Earth (Gladman, 1997). Numerical simulations have been extensively used to model the accretion and growth of planetoids and planets from collisions among asteroids (Chambers and Wetherill, 1998; Inaba et al., 2001), as well as the delivery to the Earth of Martian and lunar meteorites (Head et al., 2002; Artemieva and Ivanov, 2004). It has been shown, and further confirmed by the cratering rate curve of the Moon and Mars, that the collisions and impacts were more numerous and involved larger bodies in the earliest history of the solar system. In the most recent history, the number and size of impacts have been less important (Hartmann and Neukum, 2001). These shock events are recorded on the surface of planets by impact craters of various sizes. The shock history of Earth's crust can be addressed in detail through the study of the dynamically deformed rocks and shock melts within the terrain of individual impact craters. However, such a study is only promising for craters excavated in recent impact events, in the last several hundred million years, e.g., Ries in Germany and Popigai in Russia, which have ejecta that have not been subjected to late metamorphic overprinting. Shocked meteorites, depending on their origin, provide an important source of information for understanding the dynamics of impacts at different stages of the evolution of the solar system. Shock events in the solar system are best recorded in L-meteorites, are less pronounced in H-meteorites, and are even less pronounced in carbonaceous chondrites, which are meteorites that are 4.5 billion years old and that often show features related to intensive shock deformation, which in some cases can be accurately dated (McConville et al., 1988; Bogard, 1995). Achondrites of the SNC clan, which are probably fragments of Mars, are also heavily shocked. They provide a unique opp ortunity for studying deformations and phase transitions that have resulted from recent impact events in the solar system, since they have been recently ejected from Mars. Impact metamorphism has thus been a key process in the solar system. In addition to being responsible for planetary evolution, impact metamorphism of planetary rocks also provides valuable information on the phase transformations that can occur in deeply subducted terrestrial rocks.

In this article, we review the mineralogy of shocked meteorites and terrestrial rocks and infer the pressure and temperature conditions that prevailed during shock events, as well as the time scales of the dynamically induced high-pressure and high-temperature regimes.

PETROLOGY AND MINERALOGY OF UNSHOCKED METEORITES

Chondrites are undifferentiated meteorites with gradual degrees of chemical equilibration among their different mineral constituents. The degree of equilibration can be scaled between 3 (primitive or unequilibrated) and 6 (highly equilibrated). A complete review of the mineralogy of chondrites can be found in Brearley and Jones (1998).

Chondrules are the most common component, especially in the least-equilibrated members (Zanda, 2004). They consist of ferromagnesian silicates (olivine, pyroxene), feldspars, and/or glass, with less abundant chromite and phosphates. They are of igneous origin. They were probably formed during flash-heating events in the early solar nebula. However, the exact mechanisms of their formation are still debated. The second component is the matrix; it consists of fine-grained mixture of silicates (mostly olivine, pyroxene, feldspar, oxides, metal, sulfides, phosphates, carbonates, sulfates, and organic components), representing material that accreted mechanically without melting in the solar nebula. A third class of components consists of refractory and mafic inclusions (e.g., calcium-aluminum–rich inclusions; Grossman, 1975) that are more abundant in carbonaceous than in ordinary chondrites. These unique objects emerged from high-temperature melting-evaporation-condensation processes in the earliest episodes of the solar nebula (Grossman et al., 2002). Chondritic matrices also contain traces of interstellar grains, such as diamond, silicon carbide, corundum, hibonite, and graphite (Nittler, 2003). They are usually more abundant in the less-equilibrated classes 3–4.

Achondrites are differentiated meteorites that are of planetary origin and that have experienced a wide range of recrystallization and differentiation within small (e.g., Vesta) or large bodies (e.g., Mars or the Moon). They include rocks that have experienced limited magmatic and metamorphic processes, like acapulcoites, to well-differentiated rocks, such as basalts, eucrites, or the Martian meteorites. Many of these meteorites display igneous or recrystallized textures. The mineralogy of achondrite meteorites is dominated by the following minerals: olivine, pyroxenes, feldspars, silica polymorphs, and accessory minerals of magmatic or hydrothermal origin.

HIGH-PRESSURE MINERALOGY AND PETROLOGY: LABORATORY EXPERIMENTS

The behavior at high pressures and temperatures of minerals and rocks is essential for understanding the shock history of planetary and meteoritic materials. Olivines $(Mg,Fe)_2SiO_4$, orthopyroxenes $(Mg,Fe)SiO_3$, clinopyroxenes $(CaMgSi_2O_6)$, garnets $([Mg,Fe,Ca]_3Al_2Si_3O_{12})$, the SiO_2 polymorphs, and feldspars are the dominant minerals in meteorites. Peridotite-related rocks and basalts are the major rocks forming differentiated planetary materials. In this section, we provide a brief review concerning the high-pressure phase equilibria in peridotitic, chondritic, and basaltic compositions. Other extensive information can be found in Fiquet (2001) and Fei and Bertka (1999). These articles also provide reviews and references for the structures and physical properties of these essential Earth-forming minerals. A special emphasis is given to the most recent results.

Experiments on Individual Minerals

The $(Mg,Fe)_2SiO_4$ System

The phase transformations of $(Mg,Fe)_2SiO_4$ olivine have been extensively studied. With increasing pressure (depth), olivine (α-$[Mg,Fe]_2SiO_4$) reconstructively transforms to wadsleyite (β-$[Mg,Fe]_2SiO_4$), wadsleyite transforms to ringwoodite (γ-$[Mg,Fe]_2SiO_4$), and finally ringwoodite dissociates to an assemblage of silicate-perovskite ($[Mg,Fe]SiO_3$-pv) with Si, in sixfold coordination, and magnesiowüstite ($[Mg,Fe]O$) (Fig. 1) (Wang et al., 1997). The iron content has a strong effect on the phase boundaries and stability fields (Fig. 1). Agee (1998) and Fei and Bertka (1999) have reviewed existing data and interpretations on this system. The kinetics of these transitions have also been established (Rubie and Ross, 1994; Kerschhofer et al., 1996, 1998, 2000; Mosenfelder et al., 2000; Kubo et al., 2002), and in situ determination of the phase boundaries by X-ray diffraction has been completed (e.g., Hirose et al., 2001; Katsura et al., 2003).

Multianvil experiments have revealed that the olivine-ringwoodite phase transition takes place by two different mechanisms, depending on grain size and the thermal stress, which thus dictate the ringwoodite nucleation mode and rate: (1) incoherent phase transition along grain boundaries leading to polycrystalline ringwoodite with triple junctions. This mechanism is confined to olivine grains <100 μm. (2) Coherent intracrystalline nucleation and growth of ringwoodite lamellae along stacking faults in the (100) planes of large single crystals of olivine followed by semicoherent nucleation of wadsleyite at the olivine-ringwoodite interface (Kerschhofer et al., 1996, 1998, 2000). The incoherent growth rate is one to three orders of magnitude faster than the intracrystalline growth rate (Kerschhofer et al., 2000).

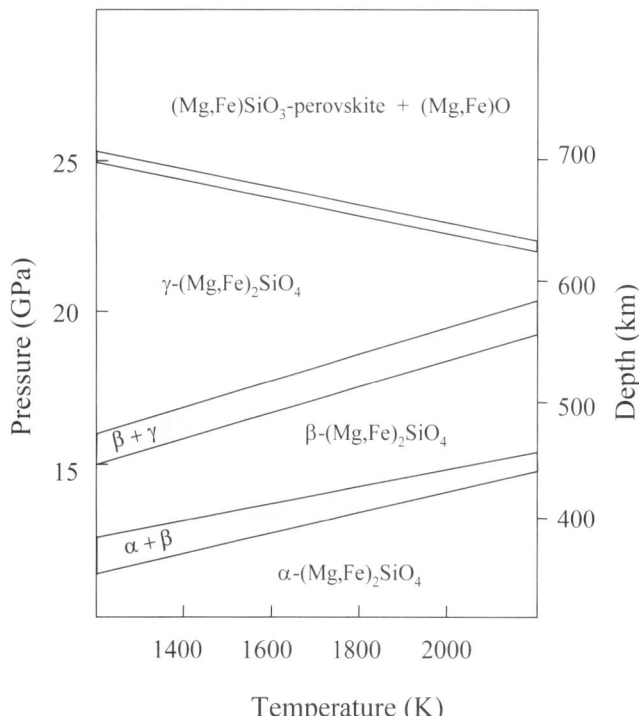

Figure 1. Phase diagram of $(Mg,Fe)_2SiO_4$ at high pressures and temperatures (after Katsura and Ito, 1989). The actual composition for these experiments was $(Mg_{0.9},Fe_{0.1})_2SiO_3$.

The $(Mg,Fe)SiO_3$ System

Enstatite, the Mg end member of orthopyroxenes, undergoes a series of polymorphic transitions to form denser phases with the following structures: clinoenstatite, ilmenite (akimotoite), non-cubic garnet (majorite), and perovskite (pv) (Akaogi et al., 1998; Chudinovskikh and Boehler, 2004) (Fig. 2). At specific pressure-temperature (P-T) conditions, the $MgSiO_3$ polymorphs dissociate into wadsleyite + SiO_2 (stishovite) or ringwoodite + SiO_2 (stishovite) (see Fig. 2). $MgSiO_3$-ilmenite, like $MgSiO_3$-pv, has Si in sixfold coordination. $MgSiO_3$-majorite contains Si in both four- and sixfold coordinations. For Fe-rich compositions, perovskite, ilmenite, and majorite structures are not stable, and mixtures of Fe_2SiO_4-spinel + stishovite are observed instead. $(Mg,Fe)SiO_3$-perovskite has been extensively studied with a special emphasis on the effect of Fe and Al substitutions on its physical properties (McCammon, 1997; Daniel et al., 2001; Yagi et al., 2004).

Very recently, in situ X-ray diffraction (XRD) measurements have shown that $MgSiO_3$ perovskite transforms to a new high-pressure post-perovskite form that has stacked SiO_6-octahedral sheet structure above 125 GPa and 2500 K (corresponding to 2700 km depth near the base of the mantle) with an increase in density of 1.0%–1.2% (Murakami et al., 2004).

The $CaSiO_3$-$CaMgSi_2O_6$ System

Earlier investigations have shown that diopside breaks down at high pressures into a mixture of $CaSiO_3$-perovskite and

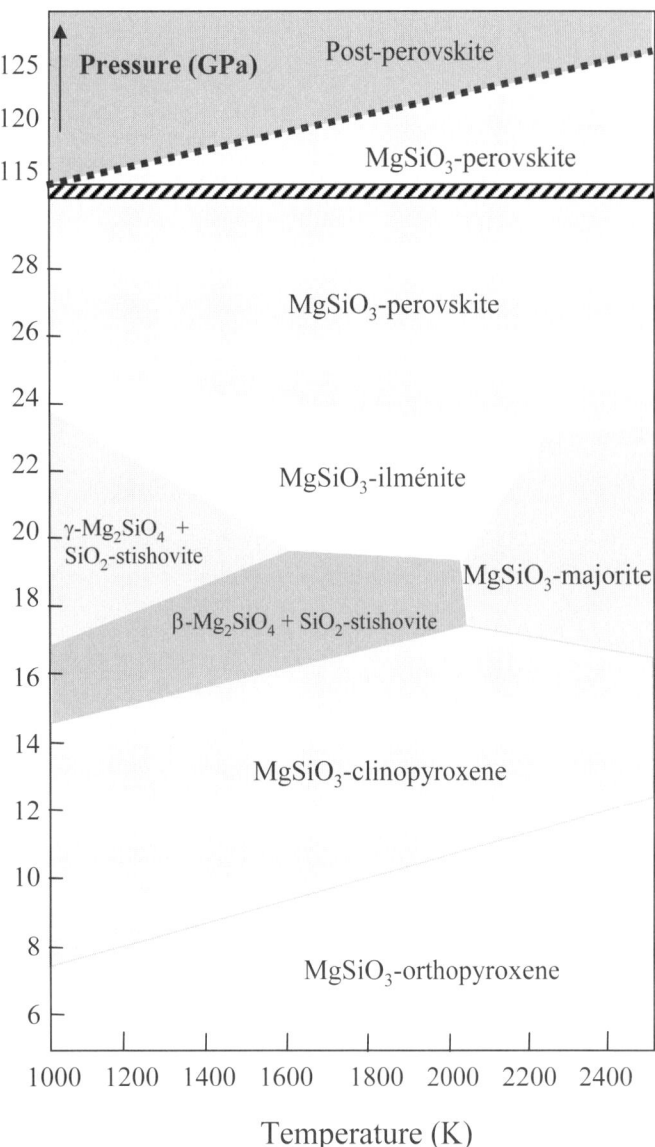

Figure 2. Phase diagram of $MgSiO_3$ after Matas (1999). The $MgSiO_3$-perovskite to post-perovskite phase boundary is from Murakami et al. (2004).

ron slope. At 1900 K and 15 GPa, larnite + titanite recombine to $CaSiO_3$-perovskite (Akaogi et al., 2004).

Phase transitions in Al-enriched $CaSiO_3$ and pure $CaSiO_3$ perovskites have been investigated at 24–75 GPa and 300–2250 K by in situ X-ray diffraction (Kurashina et al., 2004). It has been shown that Al-enriched $CaSiO_3$ transforms above 1840 K and 50 GPa from a $GdFeO_3$-type orthorhombic perovskite structure to a cubic structure. Similarly the end-member, $CaSiO_3$, is slightly distorted from cubic symmetry at low temperatures and transforms to a cubic structure with increasing temperature. The transition occurs at ~580 K and 52 GPa, at much lower temperature than the Al_2O_3-enriched composition at similar pressure.

The $(Mg,Fe)SiO_3$-$(Mg,Fe)_3Al_2Si_3O_{12}$ System

This chemical system exhibits quite complex phase relations depending on both the P-T conditions and the Al content of the starting material (Gasparik, 1992; Irifune et al., 1996; Fei and Hirose, 1999; Kubo and Akaogi, 2000; Akaogi et al., 2002). Depending on the P-T conditions, pyrope $Mg_3Al_2Si_3O_{12}$ transforms either to $Mg_3Al_2Si_3O_{12}$-ilmenite at 24 GPa and 1300 K, or at higher temperatures and pressure to $Mg_3Al_2Si_3O_{12}$-ilmenite or to an assemblage of Al-bearing $MgSiO_3$-pv + Al_2O_3 corundum. For intermediate compositions, the solubility of Al_2O_3 in the various phases is a function of both pressure and temperature conditions. Majoritic garnets are able to incorporate other elements, and most compositions can be described as a solid solution between normal garnets $M_3Al_2Si_3O_{12}$ and an alumina-free end-member $M_3(MSi)Si_3O_{12}$ (majorite component), where M = Ca, Fe, or Mg. Additional substitution mechanisms must be taken into account to explain for instance the incorporation of Na and Cr.

The SiO_2 System

The SiO_2 system and its very dense natural mineral species have been extensively studied at high pressure and temperature (El Goresy et al., 2004; Hemley et al., 1994). The phase relations indicate that the low-pressure polymorphs (quartz, cristobalite, tridymite) transform at high temperature and increasing pressure to coesite, stishovite, and a type-I $CaCl_2$-structured phase, respectively (Hemley et al., 1994; Kingma et al., 1995). A series of displacive phase transitions to species denser than stishovite has emerged from both first-principles calculations and diamond anvil cell (DAC) experiments. Theoretical calculations indicate that stishovite inverts displacively to the type-I $CaCl_2$-structured polymorph (space group Pnnm) at pressures above 48 GPa (Teter et al., 1998). At pressures of 70–85 GPa, the type-I $CaCl_2$-structured polymorph transforms to an α-PbO_2–like structure, which, like stishovite or the type-I $CaCl_2$ modifications, contains silicon in distorted octahedra, but with kinked chains of the SiO_6 octahedra (Dubrovinsky et al., 1997; Teter et al., 1998). High-pressure DAC experiments have revealed that stishovite undergoes several phase transitions to denser polymorphs above 48 GPa (Hemley et al., 1994; Kingma et al., 1995; Belonoshko et al., 1996; Dubrovinsky et al., 1997). In situ high-pressure DAC experiments at high temperature have shown that an α-PbO_2–like

(Ca,Mg)SiO_3-majorite or $MgSiO_3$-ilmenite (Gasparik, 1990). More recent studies (Akaogi et al., 2004) have established that at 18 GPa, diopside dissociates to $CaSiO_3$-rich perovskite + Mg-rich (Mg,Ca)SiO_3 tetragonal garnet above ~1700 K. The solubilities of $CaSiO_3$ in garnet and $MgSiO_3$ in perovskite increase with temperature. At 18 GPa, below ~1700 K, diopside dissociates to $CaSiO_3$-perovskite + β-Mg_2SiO_4 + stishovite. With increasing pressure, the (Mg,Si) phases coexisting with $CaSiO_3$-perovskite change to γ-Mg_2SiO_4 + stishovite, to ilmenite, and finally to Mg-perovskite.

Wollastonite ($CaSiO_3$) transforms to the walstromite structure, and further dissociates to Ca_2SiO_4 larnite + $CaSi_2O_5$ titanite. The latter transition occurs at 9–11 GPa with a positive Clapey-

structure (space group Pbcn or Pb2n) could stably exist above ~85 GPa (Dubrovinsky et al., 1997). This finding is supported by results of ab initio calculations (Teter et al., 1998). However, the calculated pressures for the inversion from the type-I $CaCl_2$ to the Pnc2 type vary from 98 GPa (Karki et al., 1997) to >120 GPa (Belonoshko et al., 1996).

Recently, DAC experiments have also been conducted on cristobalite (Dubrovinsky et al., 2001a; Dubrovinskaia et al., 2001). They resulted in an entirely different path of high-pressure phase transitions to phases denser than stishovite. In contrast to the theoretical predictions for equilibrium transformations starting with stishovite, cristobalite transforms directly to an α-PbO_2–type species (space group Pbcn or Pb2n) at $P \geq 40$ GPa (Dubrovinsky et al., 2001b; Dubrovinskaia et al., 2001). Several low-pressure polymorphs of silica are usually not pure SiO_2. It is emphasized here that the effect of minor components in silica on the stabilities of the dense phases and the path of the phase transitions have not been explored in any theoretical calculation or laboratory experiment so far. Taking these features into account should help to realistically evaluate the pressure and temperature conditions of formation of natural silica polymorphs denser than stishovite.

Murakami et al. (2003) reported that the type-I $CaCl_2$ silica undergoes further transition to the α-PbO_2–type structure above 121 GPa at 2400 K. Shim et al. (2001) predicted from their DAC experiments that type-I $CaCl_2$ silica polymorph is a stable phase at pressures equivalent to 2400 km depth. In addition, a variety of new SiO_2 polymorphs that have different degrees of silicon ordering within hexagonal close-packed oxygen in both octahedral and tetrahedral sites were recently synthesized at pressures between 30 and 100 GPa at high temperatures (Dubrovinsky et al., 2004).

(Ca,Na,K) Aluminum Silicates

High-pressure experiments performed on basaltic and crustal rocks have also produced various dense aluminum-bearing silicates, which may act as potential hosts for Al, alkali, and alkali-earth elements in the deep Earth. A K-, Na, and Ca-containing aluminosilicate has already been synthesized at 21.8 GPa and 1200 °C and characterized as a new hexagonal aluminous phase or NAL phase by Miura et al. (2000). Alkali feldspars in the system $KAlSi_3O_8$-$NaAlSi_3O_8$ and plagioclases in the system $NaAlSi_3O_8$-$CaAl_2Si_2O_8$ are abundant minerals in Earth's crust. Their behavior under moderate pressure and temperature conditions are well established. For instance, the albite to jadeite + quartz transition is a well-calibrated metamorphic reaction used to infer the P-T conditions of feldspar-rich blueschist and eclogitic rocks. The behavior of alkali feldspars at P-T conditions of Earth's mantle has been experimentally studied. It has been shown that $KAlSi_3O_8$ and $NaAlSi_3O_8$ transform at 1300 K into a denser phase with the hollandite structure at 12 GPa and 21–24 GPa, respectively (Ringwood et al., 1967; Liu, 1978a; Yagi et al., 1994), in which Al and Si occupy octahedral sites. It was the second silicate, after SiO_2-stishovite, to display sixfold coordination of Si. Several other silicates with the hollandite structure have also been synthesized (Madon et al., 1989). Experiments performed in the $KAlSi_3O_8$-$NaAlSi_3O_8$ system at pressures between 5 and 23 GPa and temperatures ranging from 1000 to 1500 K have provided more details on hollandite-structured feldspars solid-solutions in the system $KAlSi_3O_8$-$NaAlSi_3O_8$ (Yagi et al., 1994). $KAlSi_3O_8$-hollandite is stable at pressures in excess of 10 GPa (Tutti et al., 2000, 2001). In comparison, $NaAlSi_3O_8$-albite dissociates to jadeite + stishovite below 19 GPa. Both phases recombine then at 19 GPa to the hollandite-structured $NaAlSi_3O_8$, and the latter dissociates at 23 GPa to $NaAlSiO_4$ (calcium ferrite structure) + stishovite. $NaAlSi_3O_8$-hollandite seems to be unstable at such conditions, and it has been concluded that it should be stable at much higher temperatures. It has also been shown that the solubility of the $NaAlSi_3O_8$ component in the hollandite structure increases at 1300 K with increasing pressure and that the maximum solubility is reached at 40 mol% $NaAlSi_3O_8$ at 23 GPa. $KAlSi_3O_8$ (space group I4/m) inverts above 20 GPa to another dense polymorph (hollandite II) with the space group I2/m (Sueda et al., 2004; Ferroir et al., 2006). However, this phase is nonquenchable and inverts back to hollandite I.

Other Ca-bearing aluminosilicates have also been synthesized. The first synthesis of a calcium aluminum silicate phase (CAS phase, $CaAl_4Si_2O_{11}$) was reported by Irifune and Ringwood (1993). Single crystals of this phase were further synthesized and fully characterized by Gautron et al. (1996, 1997, 1999). It has also been reported to occur with other aluminous silicates in melting experiments on basalts (Miyajima et al., 2001; Hirose and Fei, 2002). This new phase may have a hexagonal Ba-ferrite structure with silicon in four- and sixfold coordination at room temperature and ambient pressure. Other K-, Na-, and Ca-aluminosilicates have been synthesized but are yet not fully characterized (Wang and Takahashi, 1999; Miyajima et al., 2001; Hirose and Fei, 2002).

Accessory Minerals: Rutile and Zircon

Rutile TiO_2 and $ZrSiO_4$ zircon are two important accessory minerals in mantle and continental rocks, and they host numerous trace elements, such as Nb, Ta, Hf, U, Th, and Zr. Structural phase transformations at high pressures might change their crystal chemistry and alter the fractionation and partitioning of these elements under deep Earth conditions.

The response of TiO_2 to high pressure has been a subject of many experimental investigations. Because rutile is isostructural with stishovite (SiO_2), its behavior at high pressures and temperatures has been thought to offer an analogy to explore the mechanisms and path of post-stishovite phase transitions at more convenient laboratory conditions. High-pressure experiments on TiO_2 have revealed the existence of several high-pressure polymorphs (McQueen et al., 1967; Linde and DeCarli, 1969; Liu, 1978b; Haines and Léger, 1997; Olson et al., 1999; Dubrovinsky et al., 2001b; Withers et al., 2003). Experiments up to total pressures in excess of 55 GPa have established the existence of four dense polymorphs: an orthorhombic α-PbO_2 phase (TiO_2 II;

space group Pbcn), stable below 14 GPa at 300 K, a monoclinic baddeleyite-structured phase (M I space group P2$_1$/c) with 7-coordinated Ti, stable above 14 GPa, an orthorhombic (O I space group Pbca), stable above 28 GPa, and an orthorhombic cotunnite (PbCl$_2$)-structured polymorph (O II space group Pnma), with 9-coordinated Ti, stable above 55 GPa. None of the polymorphs denser than the α-PbO$_2$ phase has been recovered in experiments upon decompression at ambient temperatures. All of them convert upon pressure release to the α-PbO$_2$ phase.

Rutile transforms by shock-wave compression at peak pressures of 20 GPa to the α-PbO$_2$ polymorph (Linde and DeCarli, 1969). The monoclinic baddeleyite-structured phase has never been obtained in any shock-loading experiment. Instead, the α-PbO$_2$ phase has been recovered from shock-loading experiments at $P > 20$ GPa (Linde and DeCarli, 1969). This result has been interpreted as suggestive of inversion of the baddeleyite-structured polymorph to the α-PbO$_2$ phase upon decompression. It has further been inferred that rutile, in analogy to MnF$_2$, transforms by shock compression in the pressure range of 72 GPa to a distorted fluorite- or fluorite-type structure (Kusaba et al., 1988). Theoretical studies have been conducted in order to explore the nature of the predicted cubic phase by examining both the CaF$_2$ and the FeS$_2$ model systems (Dewhurst and Lowther, 1996; Muscat et al., 2002; Dubrovinsky et al., 2001b). It has been demonstrated that the pyrite polymorph is generally more stable than the fluorite-type structure, although neither of the two structural types can be stabilized with respect to rutile at pressures below 60 GPa (Dubrovinsky et al., 2001b). However, heating at a temperature of 1900–2100 K at 48 GPa has revealed a cubic polymorph, the reflections of which could be interpreted in terms of a CaF$_2$-structured TiO$_2$ phase. Notwithstanding the numerous experimental results, a reliable P-T phase diagram up to 80 GPa has not been established so far.

High-pressure experiments on zircon have revealed that it transforms to a scheelite-type structure at 12 GPa at 900 °C (Reid and Ringwood, 1969). The high-pressure polymorph is 11% denser than zircon. Nonetheless, this phase transformation is not accompanied by any change in the Si- and Zr-coordination; Si atoms are in tetrahedral sites, and Zr atoms are in eight-coordinated (distorted cube) sites. It has been recently shown that ZrSiO$_4$ (scheelite structure) decomposes to ZrO$_2$ (cotunnite structure) and SiO$_2$ (stishovite) between 1700 and 2100 K in the pressure range 15–25 GPa (Tange and Takahashi, 2004). The high-pressure polymorph of zircon is expected to exist in Earth's upper mantle, and a search in kimberlites may be successful. The zircon-scheelite–type phase transition starts at dynamic loading conditions at 30 GPa and is complete around 53 GPa (Kusaba et al., 1985). This equilibrium shock pressure is at the upper pressure bound experienced in many shocked crystalline rocks in terrestrial meteorite craters. Nonetheless, no systematic search has been conducted so far. The first natural occurrence was recently reported from marine sediments in the upper continental slope off New Jersey (Glass and Liu, 2001). The occurrence is interpreted to be present in impact ejecta layers of the Chesapeake Bay. Natural postshock melting above 1676 °C usually leads to breakdown of zircon to baddeleyite + silica (El Goresy, 1964, 1965). This dissociation reaction is considered a diagnostic feature for recognition of natural dynamic events in Earth's crust.

Experiments on Peridotites, Pyrolite, and Mid-Ocean-Ridge Basalt (MORB)

Most of the previously described minerals are encountered when natural rock (peridotites, basalts, granitic rocks, and meteorites) or models (e.g., pyrolite or piclogite) are subjected to varying pressure and temperature conditions. The melting and subsolidus phase relations of natural peridotites have been determined up to lower-mantle conditions (26 GPa and 1900 K) (Takahashi and Ito, 1987; Herzberg and Zhang, 1996; Funamori et al., 2000). Similar experiments have been carried out on the synthetic compositions, pyrolite and piclogite. The pressure and temperature of mineralogical changes are quite similar in all these rocks. The main differences concern the relative abundances of the various high-pressure minerals. For instance, at pressure and temperature conditions of the mantle transition zone, the piclogite model gives a rock with a mineralogy dominated by majoritic garnet, while the pyrolite model leads to a rock dominated by wadsleyite (Bass and Anderson, 1984; Ringwood, 1996).

Differentiated rocks like basalts are chemically different from peridotites. They are, in particular, richer in Si, Al, Fe, K, and Na. At pressures <10 GPa, their mineralogy is dominated by garnet and clinopyroxene. With increasing pressure, this mineral assemblage transforms into the association stishovite + majorite, and at lower-mantle conditions, to an assemblage of stishovite + CaSiO$_3$-pv + MgSiO$_3$-pv + aluminosilicate phase (Hirose et al., 1999; Ono et al., 2001; Aoki and Takahashi, 2004). In fact, different aluminosilicates have been observed during subsolidus and melting experiments of basalts at high pressures, depending on details of the chemical composition of the starting material. They include: Al-Na rich phases that have a calcium ferrite–type structure (Hirose et al., 1999), different Al-Ca– and Al-K–rich phases (Wang and Takahashi, 1999; Miyajima et al., 2001; Hirose and Fei, 2002). Wang and Takahashi (1999) reported that K is mainly incorporated in three Al-K phases (I, II, and III) (4.9–14.2 wt% K$_2$O) at pressures of 15–25.5 GPa. Na is in turn mainly incorporated in garnet, magnesiowüstite, and the K phases. Majorite is able to incorporate up to 20 wt% of Al$_2$O$_3$, compared to ~8 wt% in peridotite compositions.

MINERALOGY OF SHOCKED METEORITES

Generalities on Shock Metamorphism and Shock Features in Rocks

The fundamental physics of shock-wave propagation have already been addressed in detail in numerous articles and textbooks. For this reason, this topic will be only briefly addressed here. For more details, see articles by Grady et al. (1975), Grady

(1980), Jeanloz (1980), Melosh (1989), Asay and Shahinpoor (1993), and Sharp and DeCarli (2006).

The compression wave created at the impact interface to the target propagates in the impacted matter and induces a high-pressure state followed by adiabatic decompression. Brittle solids such as MgO or SiO_2 seem to undergo significant or total loss of strength (Grady, 1980). There is also a significant difference between the response of single crystals and polycrystalline material (Grady, 1980). Shock properties of the meteoritic minerals vary over a wide range. Consequently, the initial peak shock pressure induced in the shock front may vary over an order of magnitude in a solid medium from mineral to mineral. This results in a heterogeneous shock temperature distribution. Pressure equilibration is then reached in a microsecond after the initial shock. The term "peak shock pressure" consequently denotes the averaged peak pressure rather than the nanosecond duration peaks localized in submillimeter-sized regions (Sharp and DeCarli, 2006). Hence, the pressure heterogeneities relax more rapidly than the temperature decay due to conduction to the cold neighboring matrix (e.g., Xie et al., 2006). At the adiabatic decompression, a release wave is produced. The time that elapses between the birth of the shock wave and the production of the release wave determines the duration of the shock pulse. The postshock temperature denotes the residual energy, i.e., the waste heat set free at the foot of the release adiabat. Knowing its magnitude, it is possible to roughly estimate the temperature increase during the adiabatic compression of a polymineralic sample, e.g., meteorite, to the shocked state P-V. The postshock temperature can be very high (up to 3000 K) and, depending on its magnitude and the cooling rate of the shocked target, can erase shock deformation features or induce the back transformation of high-pressure minerals formed at the maximum pressure. Lateral variation in the high-pressure liquidus assemblages in shock melt veins from their edges to the matrix to the interior of the veins can give clues about phase transformations and crystallization events at high-pressure, before or during decompression, and event conditions (Xie et al., 2006). Depending on the size of the impactor, the duration of a natural shock event ranges from milliseconds to seconds (Chen et al., 2004a; Ohtani et al., 2004; Beck et al., 2005). In a rock, interfaces between minerals with different elastic properties or the presence of an initial porosity strongly affects the propagation of the shock wave by wave reverberation and attenuation due to differences of shock impedance of the neighboring minerals. Thus, the pressure and temperature distribution in a shocked rock is heterogeneous, and the shock features are heterogeneously recorded.

Depending on the shock history and the initial texture and composition of the rock, the following modifications are observed in shocked rocks (Leroux, 2001):
- fracturing,
- plastic deformation,
- amorphization (pressure-induced crystal to amorphous phase transition),
- polymorphic phase transitions,
- high-pressure melting and crystallization of liquidus high-pressure phases, and
- metamorphic reactions.

In the following, we will only focus on the mineralogical and petrological transformations associated with a shock event on planetary materials (Fig. 3). In meteorites, these transformations occur mostly in very localized zones called shock melt veins and shock melt pockets (Fig. 4). The shock melt veins are easily observed on polished rock sections as black veins running throughout the sample. Their width ranges between micrometer and centimeter scale. The melt pockets are roughly spherical or shapeless objects typically from 0.5 to 2 mm in diameter; in some cases, they can reach a few centimeters.

Shock Mineralogy in Chondrites

Shock-wave propagation induces different changes in all the minerals in chondrites. The high-pressure polymorphs generally form within shock melt veins and at their margins. However, their formation can extend in the matrix as far as 300 μm from the vein edge (Chen et al., 2004a; El Goresy et al., 2005). The formation of high-pressure minerals can occur in different settings and involves various mechanisms within the shock melt veins. The shock melt veins often enclose subrounded fragments of the host chondrites, or individual matrix grains, which have been entrained in the melt of the shock melt veins and which have been transformed by a solid-state reaction to polycrystalline high-pressure phases. In this case, the chemical composition of the high-pressure phases is almost identical to that of the host minerals. The mineral constituents in the matrix of the shock melt veins usually result from the crystallization at high pressures of a melt of bulk chondritic composition and thus contain mixtures of high-pressure liquidus crystalline phases, rounded metal-sulfide droplets, and glass, representing quenched liquids (Chen et al., 1996). The entrained meteorite fragments and mineral grains in the meteorite melt are subjected to high pressures under hydrostatic conditions. It is therefore unrealistic to believe that pressure and temperature excursions, as suggested by Stöffler and co-workers, prevailed during shock melt vein formation (Stöffler et al., 1988, 1991). The chemistry of high-pressure liquidus minerals is significantly different from the host minerals in the less-shocked portion of the meteorite. The mineralogy of the liquidus phases should constrain the pressure-temperature conditions at which they crystallized from the meteorite melt (Agee et al., 1995; Chen et al., 1996; Gillet et al., 2000; Xie et al., 2006). In some cases, significant textural, mineralogical, and chemical differences can be observed between the walls and the central parts of the same shock melt vein and may vary from vein to another in a given meteorite (Xie et al., 2006). The crystallization of high-pressure liquidus minerals in melt veins indicates that the vein quenched at high pressures. The crystallization is driven by heat conduction to the surrounding cooler host rock and hence commences rapidly at the vein margins (Langenhorst and Poirier, 2000a; Xie et al., 2006). The transition from large

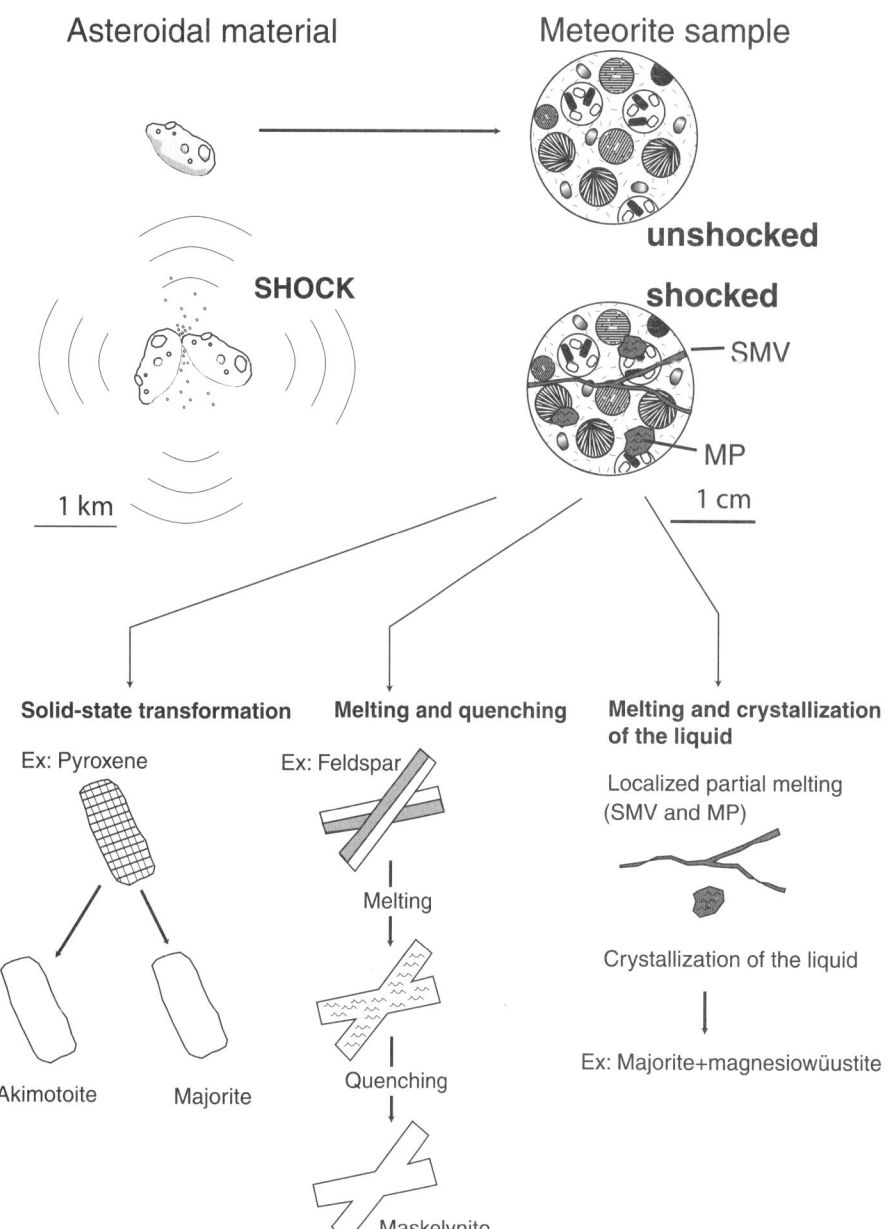

Figure 3. Scheme of the different high-pressure and high-temperature mineralogical transformations observed in shocked meteorites. Abbreviations: SMV—shock melt vein; MP—melt pockets (see Fig. 7). Shocks among asteroids or with planets generate meteorites that can fall on to Earth. A shock event is recorded in particular by the formation of shock melt veins and melt pockets, in which three major transformation types can be outlined. Solid-state transitions, such as the olivine-ringwoodite or the pyroxene-akimotoite transformations, are abundant; they occur in most cases in the shock melt veins and melt pockets, but intracrystalline ringwoodite can in some cases be encountered in the bulk rock (see Fig. 5). Melting and quenching of individual minerals can occur; one of the most common examples of this transformation is the formation of maskelynite at the expense of feldspars. Melt pockets and shock melt veins represent material melted during shock and crystallized at high pressure.

euhedral crystals of high-pressure minerals in the vein center to dendritic crystals of similar phases at the vein margins is indicative of crystallization at a given pressure accompanied by a temperature gradient to the cooler chondritic matrix. Most of these observations are in favor of cooling of the veins by thermal conduction rather than by adiabatic cooling related to pressure release. The survival of high-pressure metastable mineral assemblages requires a rapid quench before decompression below a critical temperature, below which the back transformation to low-pressure assemblages is kinetically inhibited. High-pressure phase transformations are also observed in minerals in the vicinity of the shock melt vein but for distances on the order of 300 μm from the nearest veins (El Goresy et al., 2005). This observation emphasizes the critical role of thermal stress for inducing the high-pressure transitions in the bulk meteorite. The mineralogy of the shock melt vein in chondrites is very probably an example of the accreting Earth in its early formational history during the crystallization of the magma ocean.

Ringwoodite was the first natural high-pressure polymorph of $(Mg,Fe)_2SiO_4$ olivine discovered in a natural sample, the Tenham chondrite (Binns, 1970). Wadsleyite was then reported in numerous shocked chondrites, either formed at the expense of olivine or replacing ringwoodite (Putnis and Price, 1979; Madon and Poirier, 1983; Price, 1983) (Fig. 5). In most cases, the chemical composition of the ringwoodite and wadsleyite crystals is very close to the olivine present in the unshocked part of the

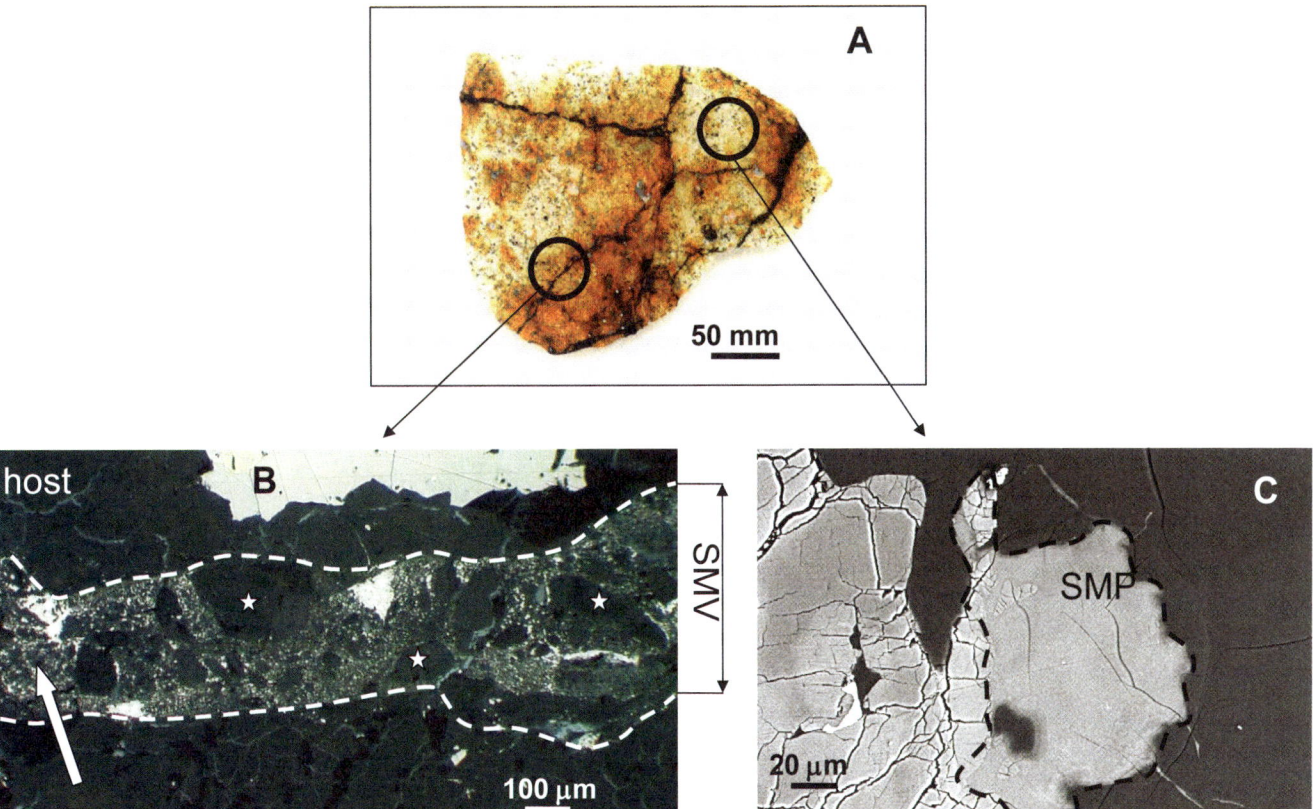

Figure 4. Photographs of shock melt veins (SMV) and shock melt pockets (SMP) in shocked meteorites. (A) Optical photograph in plane-polarized light of a thin section of a shocked chondrite. The black veins running throughout the samples are shock melt veins. (B) Scanning electron micrograph (SEM) of a shock melt vein in the Tenham chondrite. Stars indicate individual matrix grains that have been entrained in the melt of the shock melt veins and that have been transformed by a solid-state reaction to polycrystalline high-pressure phases. The arrow points to parts of the shock melt vein that resulted from the partial crystallization at high pressures of a melt and that contains mixtures of high-pressure liquidus crystalline minerals, rounded metal-sulfide droplets, and glass that resulted from quenching of silicate liquids. (C) SEM photograph of a melt pocket in an SNC meteorite.

chondrites, suggesting a formation by a solid-state mechanism from entrained olivine crystals in the melt of the shock melt vein. The transformation of olivine to one or other high-pressure polymorph may involve the formation of numerous defects (stacking faults and dislocations) followed by the nucleation and growth of the high-pressure polymorph. The solid-state olivine-ringwoodite phase transition is induced through different mechanisms: (1) incoherent grain-boundary nucleation and interface-controlled growth commencing at the olivine grain boundaries and advancing inward, resulting in a ringwoodite polycrystalline aggregate with triple junctions within the boundaries of the parental olivine crystal (Chen et al., 1996, 2004a; El Goresy et al., 2005)—this transition is confined to small olivine grains (<100 µm) within the shock melt veins; and (2) intracrystalline lamellar nucleation and growth of ringwoodite along distinct planes in large olivine grains (Ohtani et al., 2004; Chen et al., 2004a; Beck et al., 2005). Recently, Chen et al. (2004a) have shown that the intracrystalline olivine-to-ringwoodite transformation in shocked chondrites can proceed through different mechanisms. In one mechanism,

growth of ringwoodite intracrystalline lamellae can occur along (100) planes of olivine as observed in static multianvil experiments (Kerschhofer et al., 1996, 1998, 2000) or along {101} planes in the presence of high thermal stress (El Goresy et al., 2005). The latter intracrystalline ringwoodite-olivine growth is diffusion controlled and has not been produced experimentally so far. The experiments conducted by Kerschhofer et al. (1996, 1998, 2000) suggest that the intracrystalline phase transformation mechanism along the (100) planes is kinetically controlled by the ringwoodite nucleation and growth rates, which are mainly temperature dependent. Olivine grains with the (100) ringwoodite lamellae are confined to the meteoritic matrix in the 300 µm area in the vicinity of the shock melt veins (Chen et al., 2004a; El Goresy et al., 2005). In contrast, olivine with intracrystalline ringwoodite lamellae along {101} planes are encountered either inside the shock melt veins or in the 100 µm zone bordering the shock melt veins. The intracrystalline ringwoodite lamellae along (100) definitely nucleated and grew at lower temperatures than those oriented along {101}. This demonstrates that the

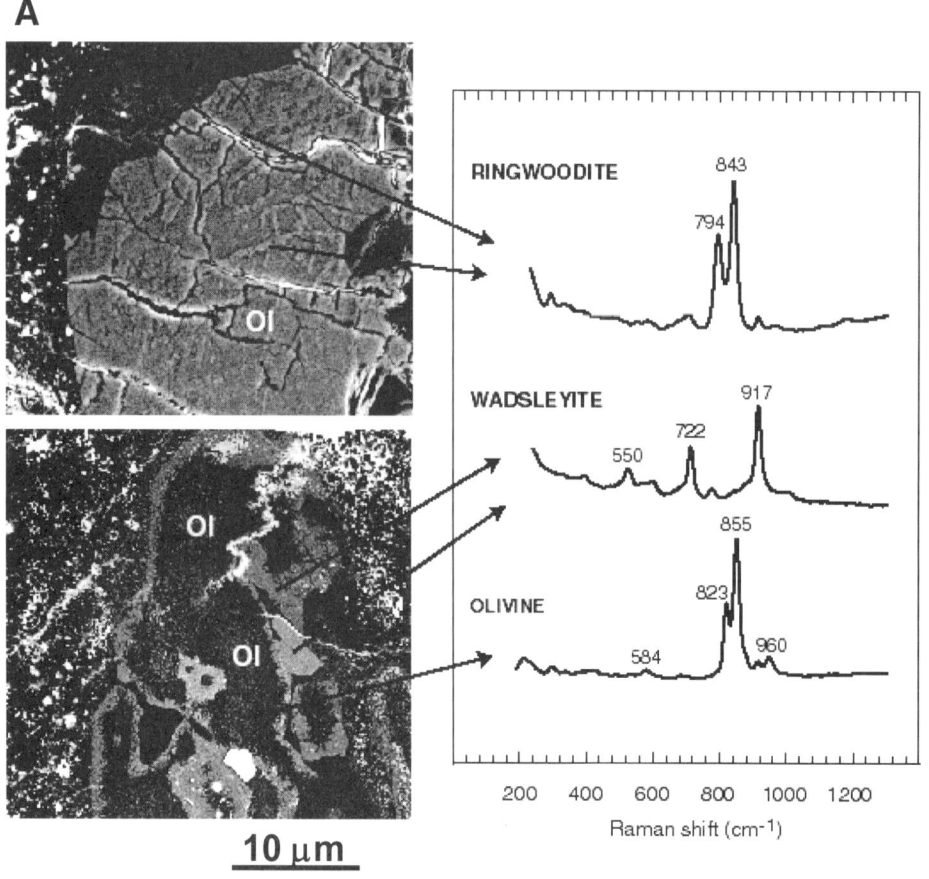
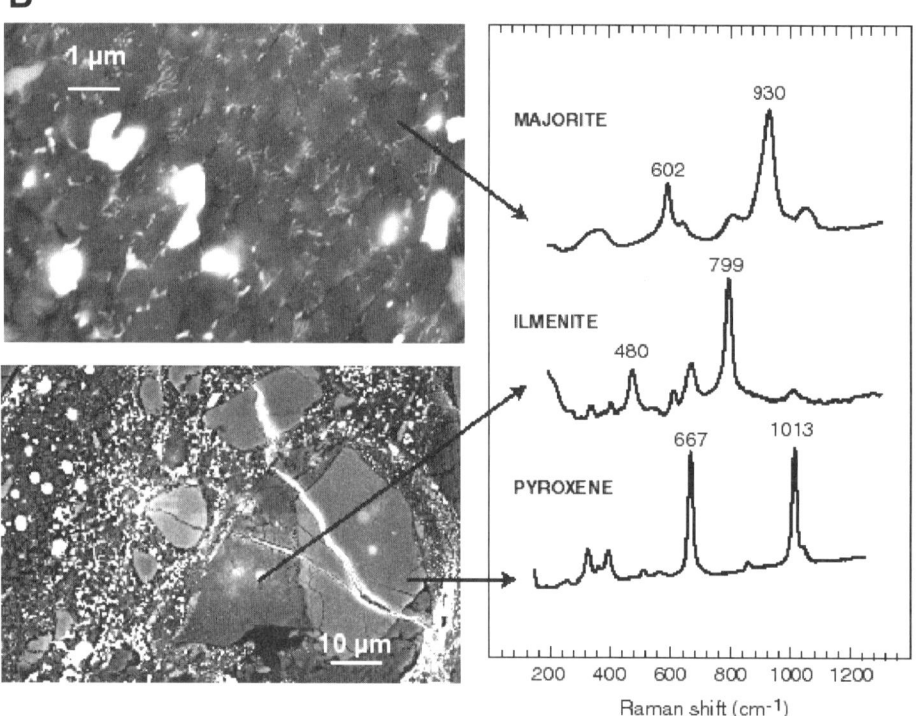

Figure 5. Scanning electron micrographs (SEM) of high-pressure minerals observed in shock melt veins of shocked meteorites. Raman spectra allow an unambiguous determination of the nature of these minerals. (A) $(Mg,Fe)_2SiO_4$ polymorphs olivine (Ol), ringwoodite, and wadsleyite in a shocked chondrite. The upper photograph shows an olivine crystal that was entrained in the shock melt vein and partially transformed into ringwoodite (bright lamellae). The lower photograph depicts a similar olivine grain partially transformed into wadsleyite at its rim. (B) $(Mg,Fe)SiO_3$ polymorphs pyroxene, akimotoite, and majorite in a shocked chondrite. The upper photograph displays a liquidus assemblage, consisting of majorite and magnesiowüstite in the Tenham chondrite. The lower photograph shows a pyroxene crystal partially transformed into the ilmenite-structured mineral akimotoite.

magnitude of thermal stress is a very important parameter controlling the type of intracrystalline olivine-ringwoodite phase transition (El Goresy et al., 2005). Shock melt veins in some L-6 chondrites, Umbarger and Tenham, contain liquidus ringwoodite intergrown with akimotoite (Xie and Sharp, 2004; MAPS, 39, 12, 2043–2054, 2004). In rare cases, such as the Tenham meteorite (Xie and Sharp, 2003), some ringwoodite crystals show enrichments in Al and Fe, indicating that they formed from the melt and not by a solid-state mechanism.

The association $FeSiO_4$-spinel + SiO_2-stishovite has been observed in a shocked chondrite (Xie and Sharp, 2002). This assemblage is very localized in shock melt veins and has been interpreted as the crystallization product of a FeO-SiO_2–rich melt. The potential precursor for such melts can be found in unshocked meteorites in the form of alteration zones.

Majorite has been found in two distinct settings corresponding to two formation mechanisms (Chen et al., 1996). In association with ringwoodite and wadseleyite (Fig. 5), coarse-grained majorite has a composition close to that of the low-Ca pyroxene outside the shock melt vein, which argues for a solid-state transformation of former orthopyroxenes entrained in the melt. In this case, the majorite crystals are full of dislocations due to imperfect growth (Voegelé et al., 2000). In the second setting, fine-grained idiomorphic majorite crystals are closely associated with silicate melt, Fe-Ni metal blebs, and magnesiowüstite that crystallized from the bulk chondritic shock-induced melt (Chen et al., 1996). The composition of the latter majorite is characterized by enrichment in Al, Ca, Cr, and Na, and the crystals are defect-free (Fig. 6). Xie et al. (2006), for example, reported a composition $Na_{0.03}Fe_{0.15}Mg_{0.77}Ca_{0.06}Al_{0.03}(Al_{0.06},SiO_{0.94})O_3$. This majorite occurrence has been interpreted, in accordance with Chen et al. (1996), by the crystallization at high pressure from a silicate melt having a composition close to that of the bulk composition of the meteorite. On the basis of high-pressure melting experiments on peridotite and chondrites, the Ca-rich majorite + magnesiowüstite assemblage crystallized between 2050 K and 2200 K and 20–24 GPa. Ca-rich majorite can also form by the solid-state transformation of former clinopyroxenes induced by heating of the melt (Malavergne et al., 2001; Xie and Sharp, 2003).

The $(Mg,Fe)SiO_3$ ilmenite, called akimotoite, is also a rather common phase unambiguously encountered in shock melt veins of chondrites (Fig. 7). It was simultaneously first observed by Sharp et al. (1997) and Tomioka and Fujino (1997, 1999). Well-crystallized akimotoite is now currently observed in heavily shocked chondrites (Fig. 5). Akimotoite is often associated with ringwoodite and a glassy phase of $(Mg,Fe)SiO_3$ composition, the latter of which may represent former $(Mg,Fe)SiO_3$ perovskite amorphized during the pressure decrease (Sharp et al., 1997). Tomioka and Fujino (1997) reported evidence for crystalline perovskite and akimotoite in the Tenham meteorite. According to these authors, both akimotoite and perovskite originated from pyroxene through a solid-state transformation. Sharp et al. (1997) encountered akimotoite in a different setting and texture strongly suggestive that the akimotoite-perovskite intergrowths were formed by crystallization of a high-pressure and high-temperature melt. Sharp et al. (1997) and Tomioka and Fujino (1997) may have indeed encountered two different petrographic settings that emerged from solid-state transformation and crystallization from the chondritic liquid, respectively. Both akimotoite and perovskite contain slightly higher amounts of Na than those

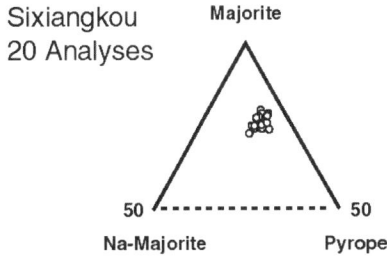

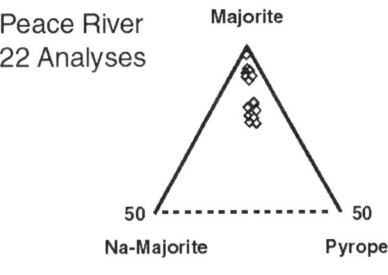

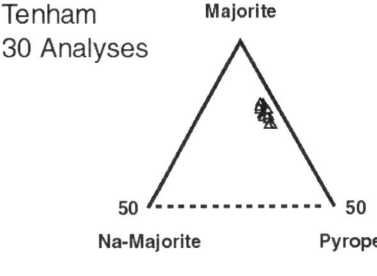

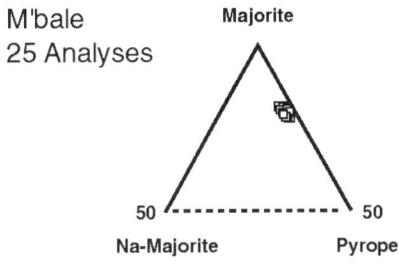

Figure 6. Chemical composition of majoritic garnets in the shock melt veins of four different chondritic meteorites: Sixiangkou, Peace River, Tenham, and M'bale. These diagrams show the various compositional ranges observed for these minerals. The compositions are plotted in the triangle majorite–Na-majorite–pyrope. The mean chemical compositions are given in terms of molar volume of five compositional end members. Mj—Mg-Fe majorite; Na-Mj—Na majorite; Ca-Mj—Ca majorite; Py—pyrope; Uv—uvarovite. Since the bulk compositions of the L-chondritic melts are similar, the compositional differences of the majorites in the four meteorites may have resulted from differences in the quenching temperatures.

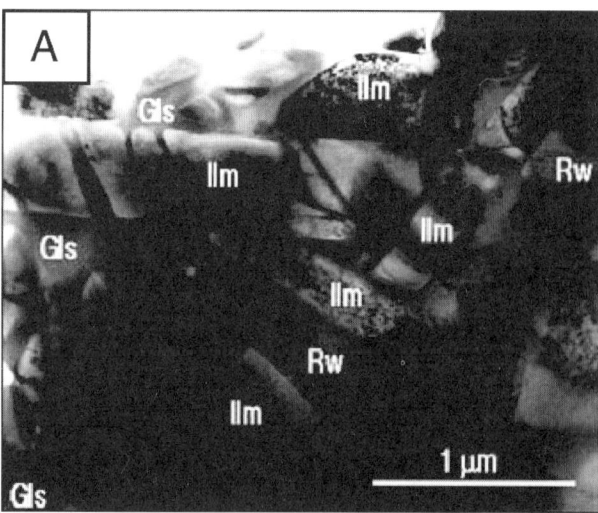
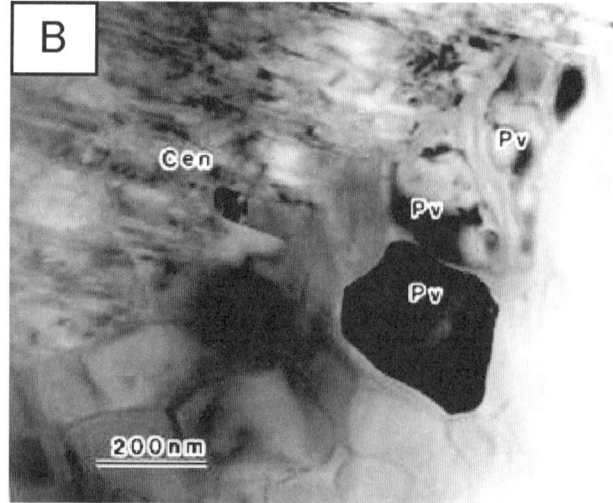

Figure 7. Transmission electron photographs of the (Mg,Fe)SiO$_3$ polymorphs, akimotoite and perovskite, in the Acfer 040 meteorite (after Sharp et al., 1997). (A) Assemblage of akimotoite (Ilm) and ringwoodite (Rw). (B) (Mg,Fe)SiO$_3$-perovskite crystals (Pv) in contact with (Mg,Fe)SiO$_3$-clinoenstatite (Cen). Amorphous material (Gls) is also present.

reported from experiments on peridotites compositions (Nishiyama and Yagi, 2003), but they lie within the uncertainties linked to the thermal emission microprobe (TEM) used for chemical characterization. Further chemical analyses must be undertaken to confirm these observations and to rule out possible analytical artifacts. The clear and unambiguous observation of natural (Mg,Fe)SiO$_3$-perovskite remains a challenge to be confirmed by a complete diffraction pattern of a naturally quenched grain. The new post-perovskite phase, which is stable above 100 GPa, should not be observed in S6 chondrites, since we have no indication at the moment of such shock conditions in meteorites.

More recently, compelling evidence for CaSiO$_3$-perovskite has been reported from the shock melt vein of a chondrite (Tomioka and Kimura, 2003). Fine-grained assemblages of majorite crystals and glassy grains rich in CaSiO$_3$ have been interpreted as the dissociation products of former diopside crystals. The CaSiO$_3$-rich glass is inferred to have crystallized in a perovskite structure vitrified during the pressure release. It is emphasized that majorite contains in that case significant amounts of Ca.

The shock melt veins of chondrites also contain enclaves of feldspathic composition at the wall of the shock melt vein or as fragments within the shock melt vein (Fig. 8). These zones, often called maskelynite, have been interpreted as diaplectic glasses formed by shock-induced solid-state amorphization or as quenched products from a high-temperature and high-pressure event before decompression from the shock event (Chen and El Goresy, 2000). Detailed investigations have shown that very often these glassy zones are in fact a mixture of glass and crystallites of quite similar compositions. The crystallites have a hollandite structure (Gillet et al., 2000), and the compositions lie between the NaAlSi$_3$O$_8$ and CaAl$_2$Si$_2$O$_8$ end members. This high-pressure polymorph of oligoclase is now named lingunite.

In addition to these major high-pressure phases of silicates, high-pressure polymorphs of accessory minerals are also encountered in the shock melt vein. This is the case for phosphate minerals. Whitlockite (Ca$_9$MgH[PO$_4$]$_7$) and apatite(Ca$_5$[PO$_4$]$_3$[F,OH,Cl]) are among the most important phosphate minerals found in chondrites (Brearley and Jones, 1998). The behavior of these minerals at high pressures is of great interest since they could act as important hosts for rare earth elements and other large lithophile elements at depth in Earth's mantle. Although experiments have revealed that apatite (Ca$_5$[PO$_4$]$_3$[F,OH,Cl]) would decompose to a phase assemblage of dense Ca$_3$(PO$_4$)$_2$ + Ca(F,OH,Cl)$_2$ at high pressure and temperature (Murayama et al., 1986), Chen et al. (1995) first found a high-pressure polymorph of phosphates, namely a polymorph of chlorapatite in the Sixiangkou meteorite. This high-pressure polymorph contains appreciable amounts of chlorine similar to chlorapatite in the host meteorite. The phase transformation from chlorapatite to a high-pressure polymorph took place without decomposition. The exact conditions of this phase transformation are unknown, since static experiments have not been conducted on chlorapatite so far. However, the stability of this high-pressure species probably lies below 23 GPa and 2000 °C as inferred from the coexisting assemblage majorite-pyrope solid-solution + magnesiowüstite (Chen et al., 1996). More recently, Xie et al. (2002) reported a high-pressure polymorph of merrillite with the structure of trigonal γ-Ca$_3$(PO$_4$)$_2$ in a shock melt vein of the Suizhou meteorite. This high-pressure polymorph, similar to the synthetic one (Murayama et al., 1986), occurs as polycrystalline aggregate up to 40 μm in size in the shock veins surrounded by the fine-grained matrix, and it has an identical composition with the merrillite in the host meteorite, and no chlorine was detected. This γ-Ca$_3$(PO$_4$)$_2$ is the product of the solid-state transformation

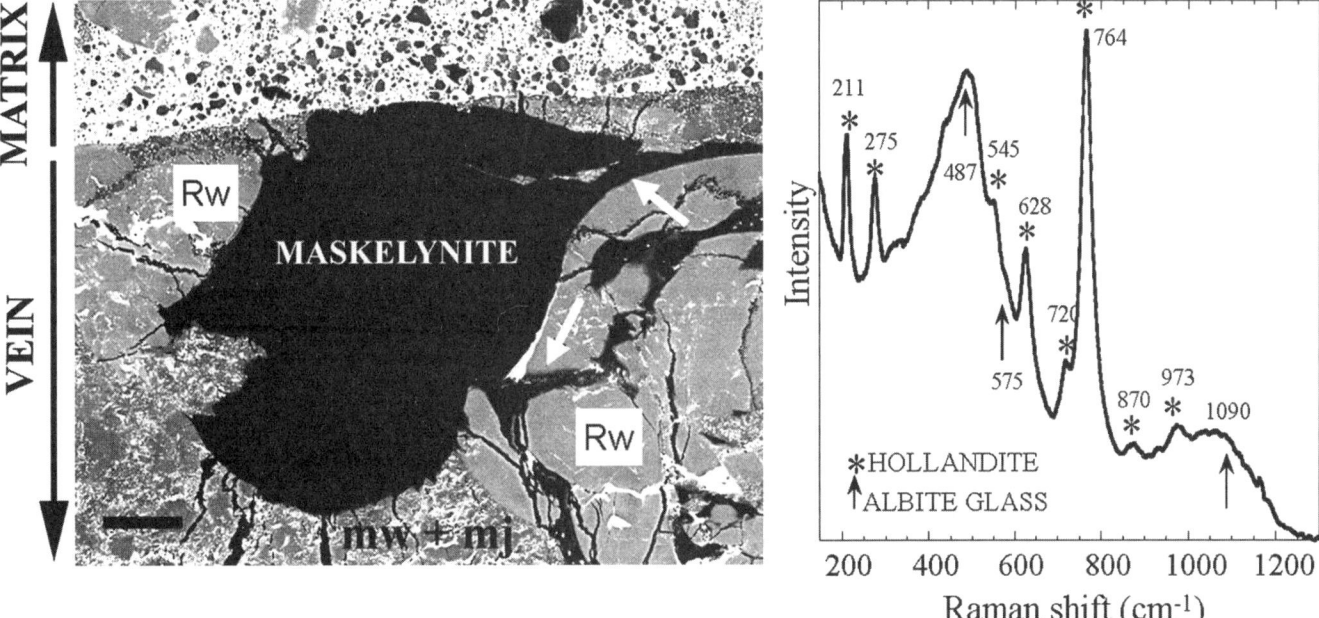

Figure 8. Hollandite in the Sixiangkou meteorite (after Gillet et al., 2000). Scanning electron micrographs (SEM) of a "maskelynite" obtained with a field emission SEM in backscattered electron (BSE) mode. Scale bar is 10 μm. Maskelynite is surrounded by ringwoodite crystals (Rw) and by majorite-pyrope$_{ss}$ + magnesiowüstite (mw + mj). The arrows indicate fractures filled with material of the same composition as "maskelynite." The Raman spectrum shows broad peaks (arrows) corresponding to NaAlSi$_3$O$_8$ glass, but the sharp peaks (*) belong to a mineral with a hollandite structure. The chemical composition is close to Ab$_{80}$An$_{12}$Or$_8$.

of merrillite crystals. Such phosphates should be stable throughout the upper mantle and transition zone.

Spinels of composition (Mg,Fe)(Al,Cr)$_2$O$_4$ are important accessory minerals in mantle peridotites and extraterrestrial rocks. Chen et al. (2003a) reported the first natural occurrence of two high-pressure and high-temperature new minerals with postspinel structures in the shock vein of the Suizhou chondrite. The original cubic chromite spinel (Fe,Mg,Mn)(Cr,Al,V,Fe)$_2$O$_4$ had been transformed into two dense postspinel polymorphs with orthorhombic structures, among them, the CaFe$_2$O$_4$-type and CaTi$_2$O$_4$-type structures, respectively. The CaFe$_2$O$_4$-type polymorph was distributed in the area outside the shock veins, whereas the CaTi$_2$O$_4$-type polymorph occurred within the shock veins and close to the shock veins. Some chromite grains consisted of three polymorphic zones, including the two postspinels and the unaltered chromite in a temperature gradient from the shock veins to the matrix, respectively. The occurrence of high-pressure polymorphs of chromite supports a solid-state mechanism for the transformation of chromite to the dense CaFe$_2$O$_4$-type and CaTi$_2$O$_4$-type polymorphs, respectively. There is no evidence for an intermediate decomposition process. Using high-pressure experiments, Chen et al. (2003b) established that such compositional chromite-spinel transforms to the CaFe$_2$O$_4$-type polymorph at 12.5 GPa and then to the CaTi$_2$O$_4$-type polymorph above 20 GPa. The formation of these two new minerals from initial FeCr$_2$O$_4$ spinel from the chondrite lies within the pressure range of the transition zone.

Shock Mineralogy in Achondrites

In what follows, we will review the effect of shock on achondrites with special emphasis on Martian (SNC) meteorites. The shock-induced melting and mineralogical changes provide unique natural analogues for studying the formation of high-pressure minerals in rocks that have chemical and mineralogical compositions close to that of terrestrial basaltic rocks. Since the shock durations are short, the high-pressure transformations are incomplete and thus provide only constraints on the initial stages of partial melting processes and chemical transport that took place in these rocks during the dynamic event on Mars.

The shergottites, nakhlites, chassignite, and ALH84001 form a class of igneous meteorites believed to have originated from Mars (e.g., Treiman et al., 2000; McSween, 2002). They are among the most-studied achondrites because, in the absence of returned Martian samples, they provide direct insights into the differentiation and magmatic history of the red planet. A few of these meteorites are old cumulate rocks, such as the nakhlites (ca. 1.3 Ga) and ALH84001 (ca. 4.5 Ga). The shergottites are younger, and they demonstrate that Mars was still active 170 m.y. ago. Several large shock events on the SNC parent body (Mars) induced a variety of deformational features and a series of high-pressure phase transitions now preserved in Martian meteorites. As in chondrites, the most striking shock features in SNC meteorites are shock melt veins and melt pockets in which a large number of high-pressure polymorphs were formed by shock-induced solid-state phase transformation (Langenhorst and Poirier, 2000b). Pervasive transformations in the bulk of the meteorite are also observed (Malavergne et al., 2001; Sharp et al., 1999).

Transformations in Bulk Rock: Maskelynite, Stishovite, and the Post-Stishovite Silica Polymorphs

Among the abundant characteristics of Martian meteorites is the occurrence of maskelynite, a glass of labradorite composition, between the ferromagnesian silicates. Chen and El Goresy (2000) have previously presented ample evidence that maskelynite in many SNC meteorites is a dense glass quenched from a shock-induced dense melt at high pressures and is not a diaplectic plagioclase glass. Specifically, the lack of any shock-induced fractures in the maskelynite, presence of maskelynite melt offshoot veins in shock-induced fractures in the neighboring pyroxene crystals and the presence of pervasive radiating fractures from the maskelynite surfaces in the neighboring silicates are clear evidence for melting and solidification at high pressure followed by relaxation of the glass after decompression (Chen and El Goresy, 2000).

The Shergotty meteorite contains two crystalline post-stishovite SiO_2 polymorphs: an orthorhombic (α-PbO_2–like phase; seifertite) and another dense monoclinic polymorph (baddeleyite structure), respectively (Sharp et al., 1999; El Goresy et al., 2000, 2004; Dera et al., 2002). Every silica grain containing any of the new post-stishovite silica polymorph consists of mosaics of domains (10–60 μm in size), each of which displays two orthogonal sets of crystalline and dense glass lamellae, which are each <250 nm wide and have different brightness in scanning electron microscopy (Sharp et al., 1999; El Goresy et al., 2000, 2004) (Fig. 9). Electron microprobe analyses with a defocused beam on the widest seifertite and the dense SiO_2-glass lamellae and on areas poor in lamellae showed almost pure SiO_2 with minor concentrations in Na_2O (0.40 wt%) and Al_2O_3 (1.14 wt%). The two post-stishovite polymorphs occur in individual multiphase grains coexisting with or enclosed in maskelynite and SiO_2 dense glass and have a morphology of prismatic grains or rhombic-shaped cross sections typical of orthorhombic β-tridymite or cristobalite morphology (El Goresy et al., 2004). The occurrence of multiple post-stishovite SiO_2 phases is consistent with the complexity of post-stishovite structures that have been investigated experimentally and theoretically (Belonoshko et al., 1996; Dubrovinsky et al., 1997; Hemley et al., 1994; Kingma et al., 1995; Teter et al., 1998). The silica phase transformations induced by the dynamic event on the Shergotty parent body produced multiple dense silica polymorphs in the same original tridymite or cristobalite grains. Some of these metastable post-stishovite polymorphs did not survive the postshock conditions. The well-oriented SiO_2 dense glass lamellae are very probably vitrified products of preexisting metastable post-stishovite silica that converted to dense glass during or after decompression. Post-stishovite silica grains enclosed in maskelynite display drag and twist-

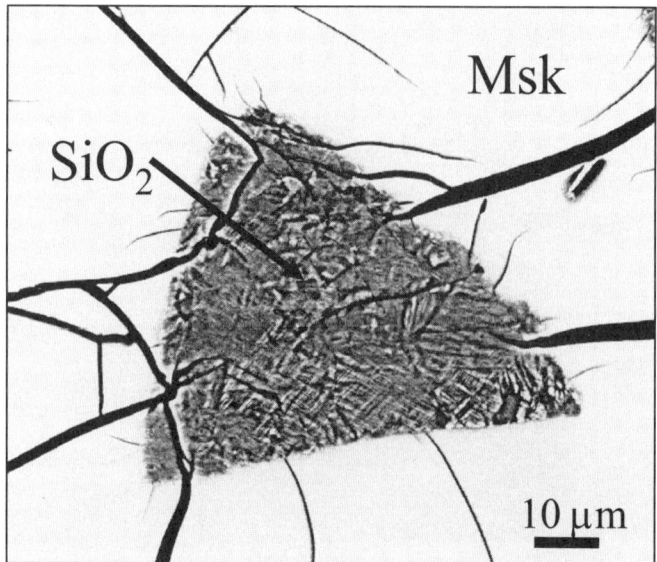

Figure 9. The post-stishovite polymorph seifertite in the Shergotty meteorite (from Sharp et al., 1999). Backscattered electron image of a triangular silica grain embedded in "maskelynite" (Msk). The grain of seifertite polymorph (α-PbO_2) consists of several domains, each depicting two sets of fine-scale orthogonal intergrowth of bright and dark lamellae. The silica grain is surrounded by pervasive radiating fractures that initiate at its surface and penetrate deep in "maskelynite."

ing at their edges, evidencing flow in a dense maskelynite melt. This is ample evidence that the maskelynite liquid was acting as a pressure medium in which former cristobalite or tridymite transformed by a solid-state reaction to the post-stishovite polymorphs. The cristobalite or tridymite precursors are observed in less-shocked meteorites as interstitial phases. Diamond anvil experiments on cristobalite have revealed that this low-pressure silica polymorph transforms directly to the α-PbO_2 polymorph at pressures in excess of 40 GPa (Dubrovinsky et al., 2001a, 2001b). We infer that the natural post-stishovite polymorphs in Shergotty may have formed at similar pressure conditions.

Other pervasive mineral transformations are observed outside the shock melt vein and shocked melt pockets of shocked SNC meteorites. Malavergne et al. (2001) reported majorite crystals of composition $(Mg_{1.60}Fe_{1.00}Ca_{0.88}Al_{0.52})Si_4O_{12}$ embedded in an amorphous phase close to untransformed pyroxenes of similar compositions. In Chassigny, which is a SNC meteorite mainly composed of olivine of composition $(Mg_{0.65}Fe_{0.35})SiO_4$, Malavergne et al. (2001) reported traces of wadsleyite crystals. They had the same composition as the olivines and were localized at grain boundaries between olivine crystals. The wadsleyite crystals resulted from a partial solid-state transformation of olivine crystals. The chemical composition of these crystals was different from what is expected from the phase diagram of $MgSiO_4$-$FeSiO_4$ and might reflect metastable crystallization and the nonequilibrium conditions during the shock event.

Transformation in the Shock Melt Veins

Detailed investigations of the shock melt vein of Zagami, (Langenhorst and Poirier, 2000b) and other shergottites have revealed the presence of hollandite-structured feldspars of plagioclase and K-feldspar compositions, liquidus stishovite, akimotoite, amorphized $(Mg,Fe)SiO_3$-perovskite, and two new high-pressure silicates: a ferromagnesian silicate titanite with the chemical formula $(Mg_{0.4}Fe_{0.4}Ca_{0.2})Si_2O_5$ (Langenhorst and Poirier, 2000b) and a hexaluminosilicate with the formula $(Ca_x,Na_{1-x})Al_{3+x}Si_3O_{11}$ (Beck et al., 2004).

Where the veins cut only pyroxenes, amorphized $(Mg,Fe)SiO_3$-perovskite is observed in association with the new ferromagnesian silicate titanite and akimotoite (Langenhorst and Poirier, 2000b). Perovskite is partly converted at the rims to an assemblage of $(Mg,Fe)O$-magnesiowüstite + SiO_2-stishovite. From simple calculations of shock melt vein cooling, Langenhorst and Poirier (2000b) proposed that perovskite, akimotoite, and the silicate titanite had crystallized from a high-pressure melt of almost pyroxene composition.

Other high-pressure mineral assemblages are observed in shock melt veins that occur at the interface between ferromagnesian silicates and plagioclase. They are dominated by stishovite, Na- and Al-bearing pyroxenes (jadeite and omphacite) and aluminosilicates (hollandites and the new hexaluminosilicate phase). The plagioclases adjacent to the veins or pockets are usually completely converted to maskelynite. In the veins, two chemically different hollandites have been identified. K-rich hollandite is mainly intimately associated with tiny SiO_2-stishovite (Fig. 10). Such an assemblage is in agreement with melting and crystallization relations at pressures higher than 20 GPa in feldspathic or basaltic systems, which show that at these pressures, both K-hollandite and stishovite are liquidus phases (Wang and Takahashi, 1999). The absence of garnet compared to what is observed in high-pressure melting of terrestrial basalts can be accounted for by a significant depletion in Al in Martian basalts.

Other hollandites are rich in Ca and Na and have compositions close to that of the maskelynite (former plagioclase) (Chen and El Goresy, 2000). They represent fragments of plagioclase entrained in the shock melt vein and transformed by a solid-state mechanism into hollandite.

Mineralogy of the Melt Pockets

Shock melt pockets are nearly spherical or amoeboid in shape and have dimensions ranging between 50 and 1000 µm (Fig. 11). Two types of melt pockets can be described: those depicting partial melting, and those with complete melting. The first category shows high abundance of pyrrhotite $(Fe_{1-x}S)$, which occurs in part as spherules in a flow texture, thus indicating partial melting (Fig. 12). They contain large (up to 15 µm in diameter), anhedral fragments of stishovite, with no sign of a tweed pattern characteristic of the post-stishovite polymorph, along with hollandite-structured labradorite composition and the fine-grained assemblage stishovite + $(Ca_x,Na_{1-x})Al_{3+x}Si_3O_{11}$ (Beck et al., 2004). Abundance of silica in these melt pockets is high (>8% by volume) relative to its abundance in the bulk Shergotty (<1% by volume). This strongly suggests that the melt pockets formed preferentially by high-pressure melting of mesostasis, which contained high abundance of pyrrhotite and silica (El Goresy et al., 2004). The second category has maskelynite bulk composition and consists of an intergrowth of idiomorphic $(Ca_x,Na_{1-x})Al_{3+x}Si_3O_{11}$ equant crystals overgrown by acicular stishovite (Beck et al., 2004) that

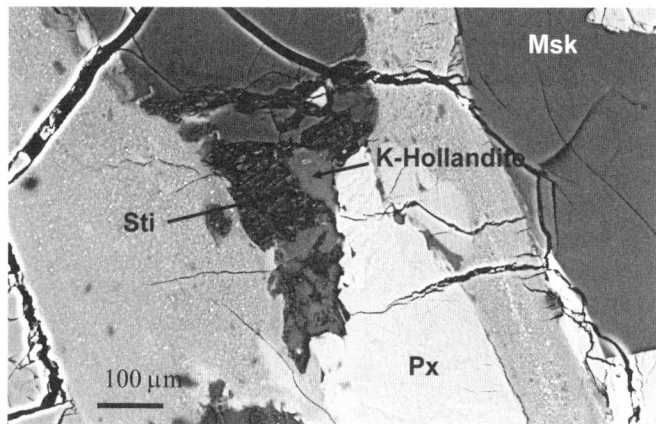

Figure 10. Scanning electron micrographs (SEM) showing the assemblage K-hollandite + stishovite (Sti) occurring between two shock melt veins of the Zagami SNC meteorite. Px—pyroxene; Msk—maskelynite.

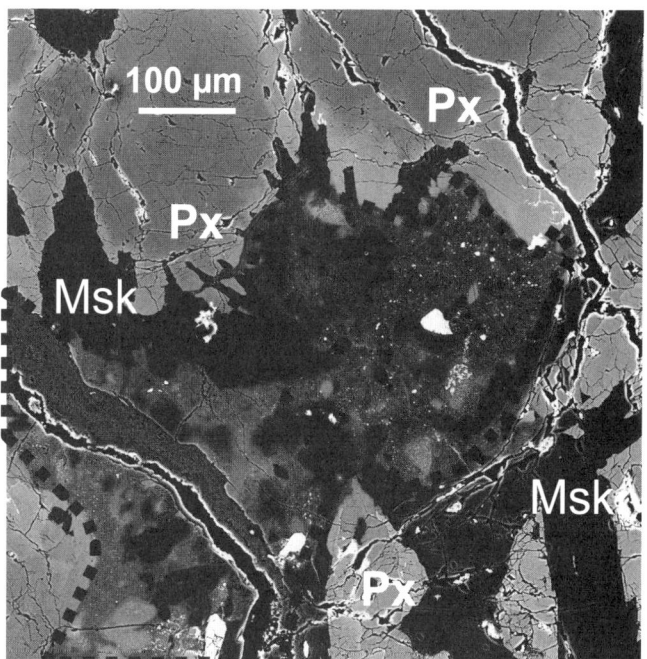

Figure 11. Scanning electron micrograph of a melt pocket (MP) in the NWA 856 SNC meteorite. The dashed line delimits the zone in which partial melting has taken place. Outside of the melt pockets, maskelynites (Msk) and pyroxenes (Px) are observed.

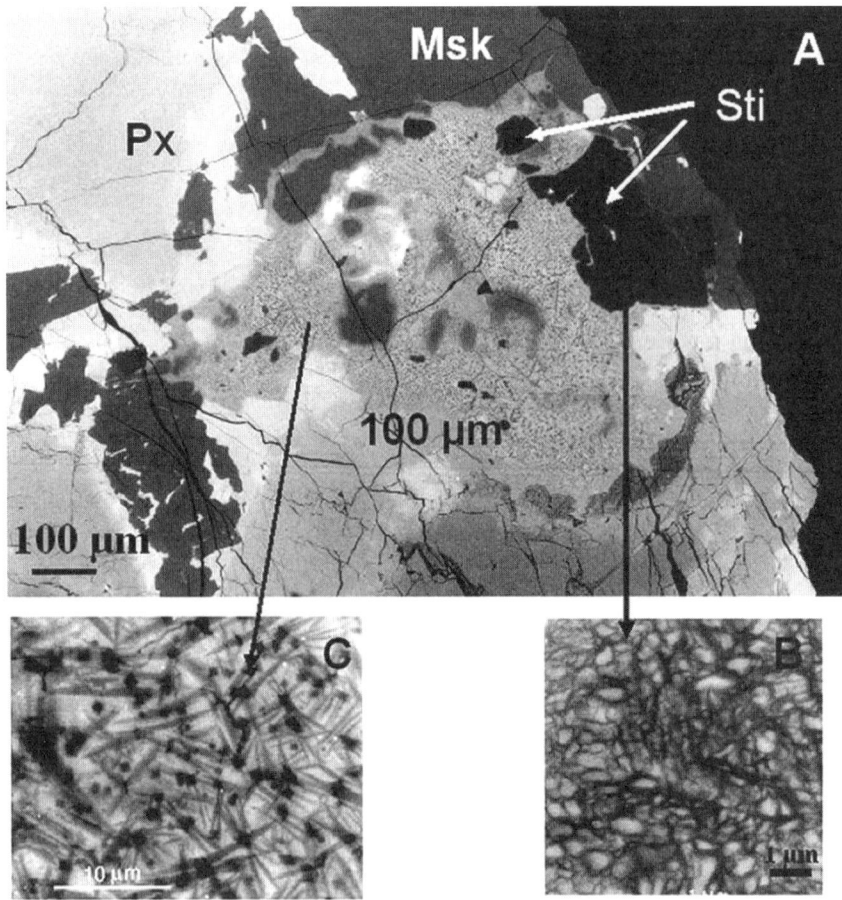

Figure 12. Scanning electron micrographs (SEM) of a melt pocket (MP) in the NWA 480 SNC meteorite. (A) General view of the melt pockets with large blocks of stishovite (Sti). The material surrounding the melt pockets is dominated by pyroxene (Px) and maskelynite (Msk). (B) Detail of the stishovite block showing an aggregate of small-sized crystals. (C) Detail of the groundmass of the melt pockets. Tiny needles of liquidus stishovite are observed embedded in a glass that resulted from melt quenching. The squared crystals are needles sectioned perpendicular to the c-axis of stishovite.

crystallized from the liquid at high pressures (Fig. 13). Table 1 gives a summary of the observed high-pressure minerals in shocked meteorites.

New Shock-Induced Phase Transitions in Terrestrial Rocks

When impacted by large meteorites, target rocks on Earth crust undergo dynamically induced deformations and phase transitions similar to those observed in shocked meteorites. We will only focus on a few recent discoveries of high-pressure minerals in terrestrial rocks.

Since the discovery of the high-pressure polymorphs of silica, coesite, and stishovite in shocked Cocconino sandstones at the Meteor Crater impact site (Chao et al., 1962; Kieffer et al., 1976), only very few new minerals have been identified in shocked terrestrial rocks. Langenhorst and Dressler (2003) have recently described the first terrestrial occurrence of a high-pressure hollandite-structured silicate in a strongly shocked anorthosite from the Manicouagan impact crater in Canada. In fact, this finding is not a surprise since terrestrial rocks suffer similar shock conditions as those revealed in shocked meteorites. Detailed investigations of these rocks are thus needed. In the following, we will describe the occurrences of diamond and diamond-related materials as well as two TiO_2 polymorphs in shocked rocks.

Natural elemental carbon occurs mainly in the form of graphite, diamond, and lonsdaleite. Besides these three major forms, traces of fullerenes have been reported in some carbonaceous chondrites and terrestrial rocks (Becker et al., 1999; Buseck, 2002). In meteorites, diamond has two origins. It can result from the shock-induced transformation of pre-existing graphite like in ureilites (Nakamuta and Aoki, 2000). Interstellar diamond is also present in traces as nanometer-sized grains in some primitive ordinary and carbonaceous chondrites. In that case, the nanodiamonds were formed by synthesis at low pressure by chemical vapor deposition (CVD) around stars (Daulton et al., 1996).

On Earth, most diamonds are formed within the mantle at depths greater than 150 km. Some of them contain mineral inclusions that attest to a formation in the transition zone or in the lower mantle (Gillet et al., 2002; Harte and Harris, 1994). Diamonds are also found in shocked continental rocks and result from the solid-state transformation of pre-existing graphite crystals. Such an origin has been clearly demonstrated by reflected-light microscopy and fine-scale laser micro-Raman spectroscopy of shocked garnet-cordierite-sillimanite gneisses in suevites of the Ries meteorite impact crater, Germany (El Goresy et al, 2001a). Graphite-diamond textural relations permit a clear determination of the solid-state nature of the formation of diamond from graphite, which is estimated to occur at an equilibrium

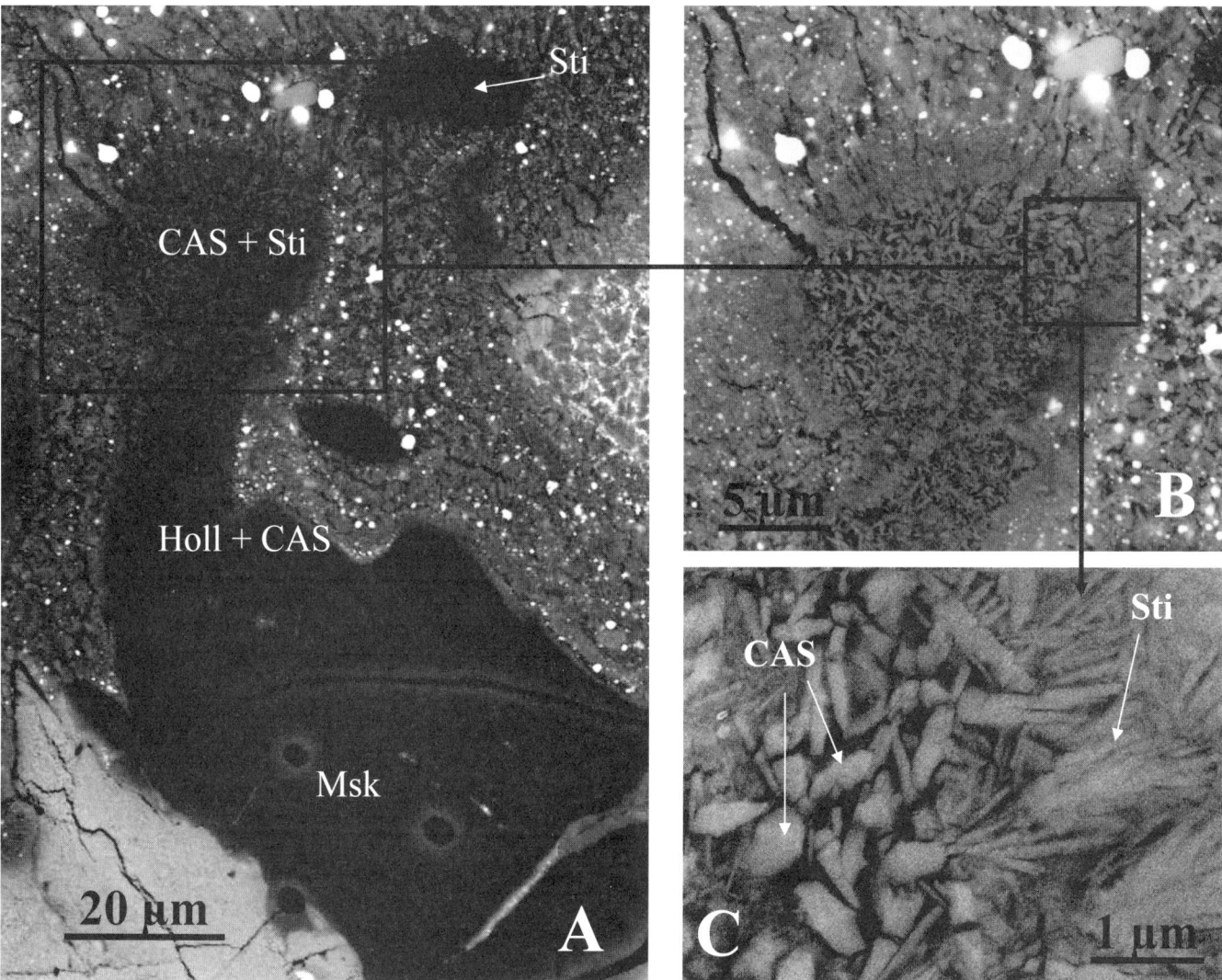

Figure 13. Detail of a melt pocket in SNC meteorite. (A) At the rim of the melt pocket, maskelynite is transformed into hollandite (Holl), stishovite (Sti), and a new hexaluminosilicate (CAS) of formula $(Ca_x,Na_{1-x})Al_{3+x}Si_3O_{11}$ (after Beck et al., 2004). (B) The scalloped surface of former hollandite indicates melting and the formation of a fine-grained polycrystalline assemblage. (C) Detail of the polycrystalline assemblage showing distinct crystals of CAS and stishovite.

shock pressure around 30 GPa. More recently, a new polymorph of carbon, showing an unusually high polishing hardness, has been found in the gneisses of the Popigai impact crater (Siberia) (El Goresy et al., 2003) (Fig. 14). The new species occupies the interior of a multiphase carbon assemblage and is entirely enveloped by lonsdaleite and graphite. Polishing hardness of this new phase is greater than that of lonsdaleite and secondary graphite. Microbeam synchrotron X-ray diffraction, imaging, and fluorescence studies have revealed a pure transparent carbon phase. The diffraction pattern is indexed in terms of a cubic cell (a = 14.697 Å, space group Pm-3m). This species was neither been encountered in static or dynamic high-pressure experiments nor has it been predicted by theoretical calculations. Impact diamonds have also previously been reported in demineralization residues from Popigai, Lapijärvi, and Ries craters, (Koeberl et al., 1997; Langenhorst et al., 1998; Masaitis et al., 1972; Rost et al., 1978).

Rutile (TiO_2) is an important accessory mineral in igneous and metamorphic rocks, and it hosts a number of trace elements. High-pressure experiments have revealed several dense polymorphs, which, until recently, have never been observed in natural rocks. Two of them were encountered in shocked garnet-cordierite-sillimanite gneisses from the Ries crater: one with a α-PbO_2 structure and 2% more dense then rutile (El Goresy et al., 2001b); the second, an ultradense phase with a ZrO_2 structure and 11% more dense than rutile (El Goresy et al., 2001c). Graphite in these gneisses had been partially transformed to diamond in the Ries shock event (El Goresy et al., 2001a). The coexistence of the shock-induced graphite-diamond and the rutile-ZrO_2-struc-

TABLE 1. HIGH-PRESSURE PHASES OBSERVED IN SHOCKED METEORITES

Mineral name	Structure	Formula	Ref	Solid state	From melt	Meteorite	Meteorite type
Wadsleyite	Modified spinel	$(Mg,Fe)_2SiO_4$	A	+			Lunar, Ch, SNC
Ringwoodite	Spinel	$(Mg_{0.75}Fe_{0.25})_2SiO_4$	B	+		Sixiangkou (Ch)	Lunar, Ch
		$(Mg_{0.65}Fe_{0.35})_2(Si_{0.98}Al_{0.02})O_4$	C		+	Tenham (Ch)	
Majorite	Garnet	$(Mg_{0.77}Fe_{0.21}Ca_{0.014}Mn_{0.06})SiO_3$	B	+		Sixiangkou (Ch)	Ch, SNC
		$(Mg_{0.77}Fe_{0.15}Ca_{0.06}Na_{0.03}Al_{0.03})(Al_{0.06}Si_{0.94})O_3$	C		+	Tenham (Sharp)	
		$(Mg_{0.67}Fe_{0.09}Ca_{0.21}Na_{0.03}Al_{0.02})SiO_3$	C	+		Tenham (Sharp)	
Akimotoite	Ilmenite	$(Mg_{0.76}Fe_{0.20}Ca_{0.010}Na_{0.03})SiO_3$	D	+		Tenham (Ch)	Ch, SNC
		$(Mg_{0.89}Fe_{0.12})(Si_{0.95}Al_{0.05})O_3$	E		+	Tenham (Ch)	
	Perovskite	$Mg_{0.76}Fe_{0.21}Ca_{0.005}Na_{0.03}Al_{0.01}Si_{0.99}O_3$	D	+		Tenham (Ch)	Ch
		$Mg_{0.82}Fe_{0.16}Ca_{0.03}Na_{0.08}Al_{0.05}Si_{0.93}O_3$	E		+	Tenham (Ch)	
	Perovskite	$Mg_{0.23}Fe_{0.05}Ca_{0.72}SiO_3$	F	+		Y751000 (Ch)	Ch
	Hollandite	$Na_{0.83}Ca_{0.09}K_{0.06}Al_{1.10}Si_{2.89}O_8$	G	+	+	Sixiangkou (Ch)	Ch, SNC
		$K_{0.74}Ca_{0.06}Na_{0.05}Al_{0.82}Si_{3.08}O_8$	H		+	Zagami (SNC)	
	Hexaluminosilicate	$(Ca_x,Na_{1-x})Al_{3+x}Si_3O_{11}$	I		+	NWA 856 (SNC)	SNC
Magnesiowüstite	NaCl	$Mg_{0.54}Fe_{0.46}O$	B	+		Sixiangkou (Ch)	Ch
		$Mg_{0.68}Fe_{0.32}O$	C	+		Tenham (Ch)	
	Titanite	$Mg_{0.4}Fe_{0.4}Ca_{0.2}Si_2O_5$	J		+	Zagami (SNC)	SNC
Stishovite	Rutile	SiO_2 with traces of Al and Na	I	+	+	NWA 856 (SNC)	SNC
Seifertite	α-PbO_2	SiO_2 with traces of Al and Na	K	+		Shergotty (SNC)	SNC
	Baddeleyite	SiO_2 with traces of Al and Na	L	+		Shergotty (SNC)	SNC
Diamond	Diamond	C	M	+		ALH78019(Ur)	Ureilites
	Si-clathrate?	C	N	+			Ureilites
	$CaTi_2O_4$	$(Fe,Mg,Mn)(Cr,Al,V,Fe)_2O_4$	O	+		Suizhou (Ch)	Ch
	$CaFe_2O_4$	$(Fe,Mg,Mn)(Cr,Al,V,Fe)_2O_4$	O	+		Suizhou (Ch)	Ch
	γ-$Ca_3(PO_4)_2$	$Ca_3(PO_4)_2$	P	+		Sixiangkou (Ch)	Ch

Note: They can be the result either of solid-state transformation or from the crystallization at high pressure from a melt. The chemical formulas are taken from data on specific meteorites. The occurrence in the various types of meteorites is given in the last column. Ch—chondrite. A—Putnis and Price (1979); B—Chen et al. (1996); C—Xie and Sharp (2003); D—Tomioka and Fujino (1997); E—Sharp et al. (1997); F—Tomioka and Kimura (2003); G—Gillet et al. (2000); H—Langenhorst and Poirier (2000a); I—Beck et al. (2004); J—Langenhorst and Poirier (2000b); K—Sharp et al. (1999); L—El Goresy et al. (2004); M—Nakamuta and Aoki (2000); N—El Goresy et al. (personal commun.); O—Chen et al. (2003a, 2003b); P—Xie et al. (2002).

tured polymorph allowed researchers to stringently constrain the equilibrium shock pressure to <28 GPa, a value much lower than deduced from the shock model by (Stöffler, 1988, 1991). These two dense polymorphs have allowed us to constrain the equilibrium shock pressures in the rocks in which they occur (El Goresy et al., 2001b, 2001c).

Pressure-Temperature-Time (*P-T-t*) Characteristics of Shocks in Meteorites

In shock-wave experiments, pressure increases as a wave front passes through the sample as a result of high-velocity impact. Extremely high pressures can be generated (up to 50 TPa), much higher than those reachable by static experiments (up to 400 GPa), but they last only hundreds of nanoseconds to a few microseconds (Ahrens, 1987; Asay and Shahinpoor, 1993). This short duration often inhibits phase transformations, which are observed in static experiments where the pressure and temperature conditions can be maintained for several seconds or minutes. High-pressure and high-temperature conditions in natural shocks can last from a few tens of milliseconds to a few seconds, permitting the incipient to total phase transitions that are also currently observed in static experiments.

Three stages are recognized during a shock: a pressure increase during a few nanoseconds, an initially heterogeneous peak shock pressure lasting 10^{-6} s to tens of seconds, depending on the size and speed of the impactor, and finally pressure heterogeneity, which dampens rapidly to an equilibrium or continuum shock pressure that lasts a few seconds to tens of seconds (Xie et al., 2006). The bulk sample temperature increases up to 1500 K for peak shock pressures of 70–100 GPa due to adiabatic compression, remains constant during the whole continuum pressure, and then decreases prior to decompression, preserving, in most cases, the high-pressure assemblages (Chen et al., 1996; Benzerara et al., 2002; Xie et al., 2006). A progressive shock classification and calibration by Stöffler et al. (1988, 1991) allows a rough estimate of the shock stage (S1 to S6) but does not afford a meaningful constraint of the equilibrium shock pressures. This classification is based on laboratory experiments on rocks and minerals for which the peak shock duration is extremely short (nanoseconds to few microseconds) and which have failed to reproduce the very phase transition–specific features observed in naturally shocked meteorites, the shock melt veins and their related mineralogical high-pressure phase transformations, e.g., olivine-ringwoodite or olivine-wadsleyite, pyroxene-majorite, pyroxene-akimotoite, or pyroxene-perovskite.

It has been extensively demonstrated that in these veins, high-pressure minerals form either by solid-state transformation of former low-pressure polymorphs or by crystallization and quenching assemblages from melts at high pressures (Fig. 6) (Chen et al., 1996; Sharp et al., 1997; Langenhorst and Poirier, 2000a; Gillet et al., 2000; Xie et al., 2006). These minerals are only very occasionally encountered in the bulk of the sample (Sharp et al., 1999; Malavergne et al., 2001). The shock melt veins are formed during the pressure increase by high shear stresses due to pressure heterogeneities related to differences in the shock impedance between minerals or to pores or flaws already present in the rocks before the shock (Baer, 2000). The

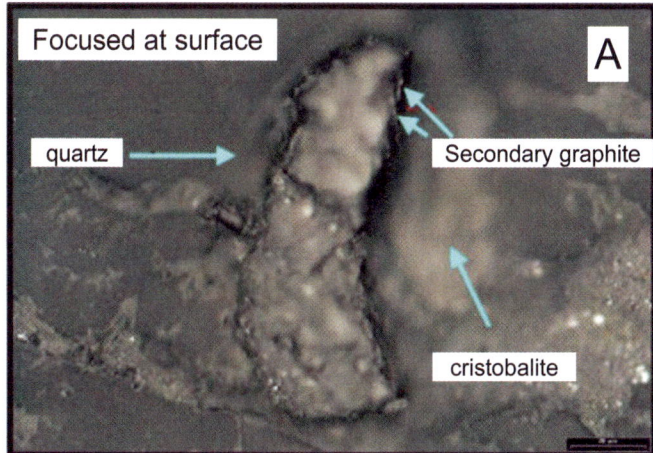

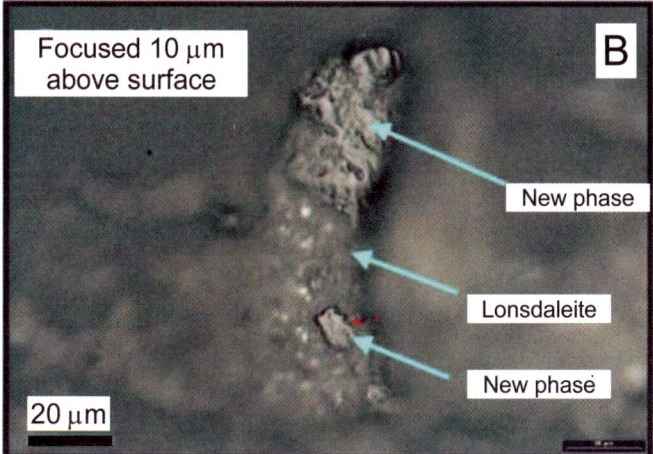

Figure 14. Reflected-light photographs of a carbon composite grain from a polished section of the shocked gneiss in Popigai (after El Goresy et al., 2003). (A) Photograph focused at the surface of the polished thin section showing the hard high-relief part out of focus. (B) Photograph focused 10 µm above the surface, thus depicting the surface features of the new carbon phase with gouges. These features suggest a very high hardness of the new phase with respect to diamond (Lonsdaleite) and graphite polymorphs surrounding the new phase.

unusually high-temperature increase in the shock melt veins (up to 2500 K), which cannot be explained by adiabatic compression unless the peak pressures reach values higher than 150 GPa (Benzerara et al., 2002), results in fact from high-strain rates and frictional heating during the pressure increase (Heider and Kenkmann, 2003). The high-temperature generated within the shock melt veins provides a unique opportunity for precisely bracketing the equilibrium shock pressure. Once formed, the silicate melt acts as a pressure medium. Solid-state phase transformations in solid porphyroclastic mineral inclusions in the melt and the crystallization from the melt of high-pressure liquidus phases take place in the majority of the cases under constant pressure conditions. In fact, once the vein is formed, the pressure inside the vein is similar to that in the host rock due to stress continuity, and it is hydrostatic since the vein is melted. In both the shock melt vein and melt pockets, melting is evidenced by the presence of liquidus phases with interstitial glass and rounded crystal or crystal aggregates (Fig. 5B). The high temperatures generated in the veins permits, for a given equilibrium pressure, the nucleation and growth of high-pressure minerals within the vein and in minerals heated by conduction outside but near the vein (Chen et al., 2004a; Beck et al., 2005; El Goresy et al., 2005).

Comparison of naturally shocked samples with samples processed in static experiments must be done with some caution due to time limitations during dynamic events, which could impede phase transitions. However, in any case, high-pressure dynamic experiments can never reveal a realistic estimate of the pressure-temperature conditions that prevail in shock melt veins since such dynamic experiments have never produced any of the dense phases reported from shock melt veins, e.g., ringwoodite, majorite, akimotoite, or magnesiowüstite. As previously mentioned, the duration of the continuum shock pressure experienced by meteorites can vary over at least three orders of magnitude (10^{-2} s to 10 s). These durations are longer than those obtained in laboratory shock experiments (10^{-6} s). They are, however, in a similar or lower range of duration of static experiments performed in diamond anvil cells or multianvil apparatus (a few seconds to several minutes). It can also be noticed that in static experiments, melts are partially quenched, and the high-pressure liquid usually crystallizes during quenching. In shocked meteorites, glasses are found within melt pockets, testifying to the efficiency of quenching during postshock decompression. The melting textures observed in the shock melt vein of shocked meteorites are also quite similar to those observed in high-pressure assemblages synthesized in high-pressure and high-temperature experiments, in terms of crystal size and shape (Agee et al., 1995; Chen et al., 1996, 2004b).

The mineralogy of the shock melt vein is usually heterogeneous, reflecting the diversity of the minerals it cuts during its propagation in the rock and the temperature gradient in large porphyroclastic mineral fragments floating in the high-pressure silicate melt. The low temperature induced by the adiabatic compression far from the veins inhibits the mineralogical transformations for kinetic reasons. The survival of the high-pressure phases encountered inside and outside the veins gives evidence for the fast temperature drop due to thermal conduction to the cold meteorite matrix before onset of the decompression stage. It is therefore possible to use petrological phase diagrams and the mineral assemblages observed within the shock melt veins to constrain the peak shock pressure undergone by a meteorite (Zhang and Herzberg, 1994; Agee et al., 1995; Chen et al., 1996, 2004a; Xie and Sharp, 2004; Ohtani et al., 2004; El Goresy et al., 2005; Xie et al., 2006).

The shock durations can be inferred from the kinetics of high-pressure mineral transformations, nucleation, and growth rates of high-pressure polymorphs and diffusion of trace elements within the veins (Ohtani et al., 2004; Chen et al., 2004a; Beck et al., 2005; El Goresy et al., 2005).

In the following, the pressure-temperature-time history of shocks events will be illustrated for two different meteorites (Beck et al., 2005).

The P-T-t Shock History of the Tenham Chondrite

The high-pressure mineralogy of shock melt veins of Tenham, which can reach 2 mm in width, is dominated by the $(Mg,Fe)_2SiO_4$ and $(Mg,Fe)SiO_3$ high-pressure polymorphs (Binns, 1970; Putnis and Price, 1979; Tomioka and Fujino, 1997). Plagioclase was also densified during the shock event and transformed to $(CaAl_2Si_2O_8-NaAlSi_3O_8)$ hollandite (lingunite) (Tomioka et al., 2000; Gillet et al., 2000). Most of these high-pressure minerals represent former grains entrained in the shock melt veins that transformed due to the high-pressure and high-temperature conditions that prevailed in the shock melt veins. Beside these minerals, the chondritic melt in the shock melt veins underwent partial crystallization into the assemblage magnesiowüstite + majorite (Chen et al., 1996). The textures, crystal sizes, and mineral compositions in the centers of the shock melt veins are very similar to those produced in static experiments (Zhang and Herzberg, 1994; Agee et al., 1995), strongly suggesting that the phase relationships established in static high-pressure experiments can be used to constrain the pressure and temperature conditions and thus the peak equilibrium pressure undergone by the meteorite.

The phase diagram of melting of chondrites can be used to constrain the P-T conditions of crystallization of the shock melt vein. The Allende phase diagram (Agee et al., 1995; Asahara et al., 2004) is the most comprehensive diagram available at the moment and thus serves as an adequate reference (Fig. 15). It can be used to infer the crystallization conditions of the majorite-magnesiowüstite assemblage observed in the central part of the shock melt vein of the Tenham chondrite at around 24–25 GPa and 2500 K (Chen et al., 1996). These conditions must be lowered by 2 GPa, as shown by Asahara et al. (2004). The assemblage akimotoite + $(Mg,Fe)SiO_3$-perovskite + ringwoodite + majorite observed at the edge of the shock melt vein of the Tenham meteorite leads to the same pressure and temperature conditions if one assumes that akimotoite results from a metastable crystallization during the extremely rapid temperature drop following the shock (Xie and Sharp, 2003; Xie et al., 2006). The fact that the olivine porphyroclasts are only transformed into ringwoodite and not into the assemblage perovskite + magnesiowüstite confirms that the equilibrium shock pressure was not in excess of 25 GPa (Wang et al., 1997). In addition, there is no textural evidence that the ringwoodite in the porphyroclasts formed by recombination of perovskite + magnesiowüstite formed by dissociation of olivine or ringwoodite at $P > 26$ GPa (Wang et al., 1997). It should be noted that this equilibrium shock pressure deduced from the assemblages and textures encountered in shock melt veins is half the value of the low-pressure threshold for S6 shock stage inferred by comparing experimentally induced shock features with deformation effects and transformations observed in the bulk meteorite by Stöffler et al. (1991). This discrepancy casts some doubt on the pressure estimates obtained through this classical shock-pressure scale.

The mechanism of $(Mg,Fe)_2SiO_4$-ringwoodite partially replacing $(Mg,Fe)_2SiO_4$-olivine is one of the most stringent constraints of high-pressure transformation in the shock melt veins of Tenham and other shocked L6 chondrites (Fig. 16). Intracrystalline ringwoodite lamellae have been recently observed along (100) and {101} planes in olivine grains entrained in shock melt veins or adjacent to them (Ohtani et al., 2004b; Chen et al., 2004a; Beck et al., 2005). The kinetics of the intracrystalline transformation along (100) are well known from multianvil experiments on single olivine crystals (Rubie and Ross, 1994; Kerschhofer et al., 1996, 1998, 2000; Mosenfelder et al., 2000; Kubo et al., 2002). The experimentally determined growth of ringwoodite lamellae in olivine has been used to estimate the equilibrium shock pressure duration in the Yamato 791384 and the Sixiangkou chondrites (Ohtani et al., 2004; Chen et al., 2004a). It has been shown that these chondrites experienced pressures of around 20 GPa for at least 3–4 s. Similar calculations for the Tenham meteorite (Beck et al., 2005) show that for similar pressures the equilibrium shock pressure lasted between 0.2 s and 5 s. The latter result is also confirmed by quantitative trace-element mapping and diffusion-time calculations (Fig. 16). In the shock melt veins of Tenham, the Ca contents are identical in both phases and ringwoodite is depleted in Mn with respect to olivine. Knowing the diffusion coefficients of Ca and Mn, it is possible to calculate the diffusion lengths of Ca and Mn (i.e., the expected width of the ringwoodite lamellae) as a function of the cooling rate within the shock melt vein and the shock duration.

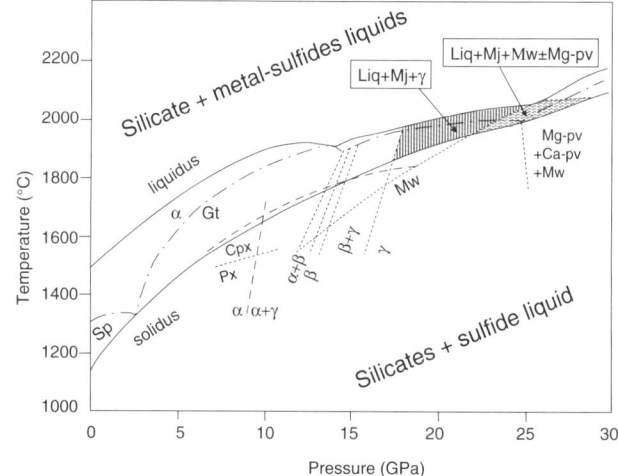

Figure 15. Phase diagram of the Allende meteorite after Agee et al. (1995). The shaded areas encompass the pressure and temperature conditions of the crystallization region of shocked melt veins in heavily shocked meteorites. Liq + Mj + γ—silicate liquid + majorite + γ-$(Mg,Fe)_2SiO_4$; Liq + Mj + Mw ± Mg-pv—silicate liquid + majorite + $(Mg,Fe)O$ ± $(Mg,Fe)SiO_3$-perovskite; Ca-pv—$CaSiO_3$-perovskite; γ—γ-$(Mg,Fe)_2SiO_4$; α—α-$(Mg,Fe)_2SiO_4$; Gt—garnet; Px—orthopyroxenes; Cpx—clinopyroxenes; Sp—spinel. The phase diagram above 20 GPa has been reinvestigated by Asahara et al. (2004), and the pressure estimates of Agee et al. (1995) must be decreased by 2 GPa. β—b-b-$(Mg,Fe)_2SiO_4$.

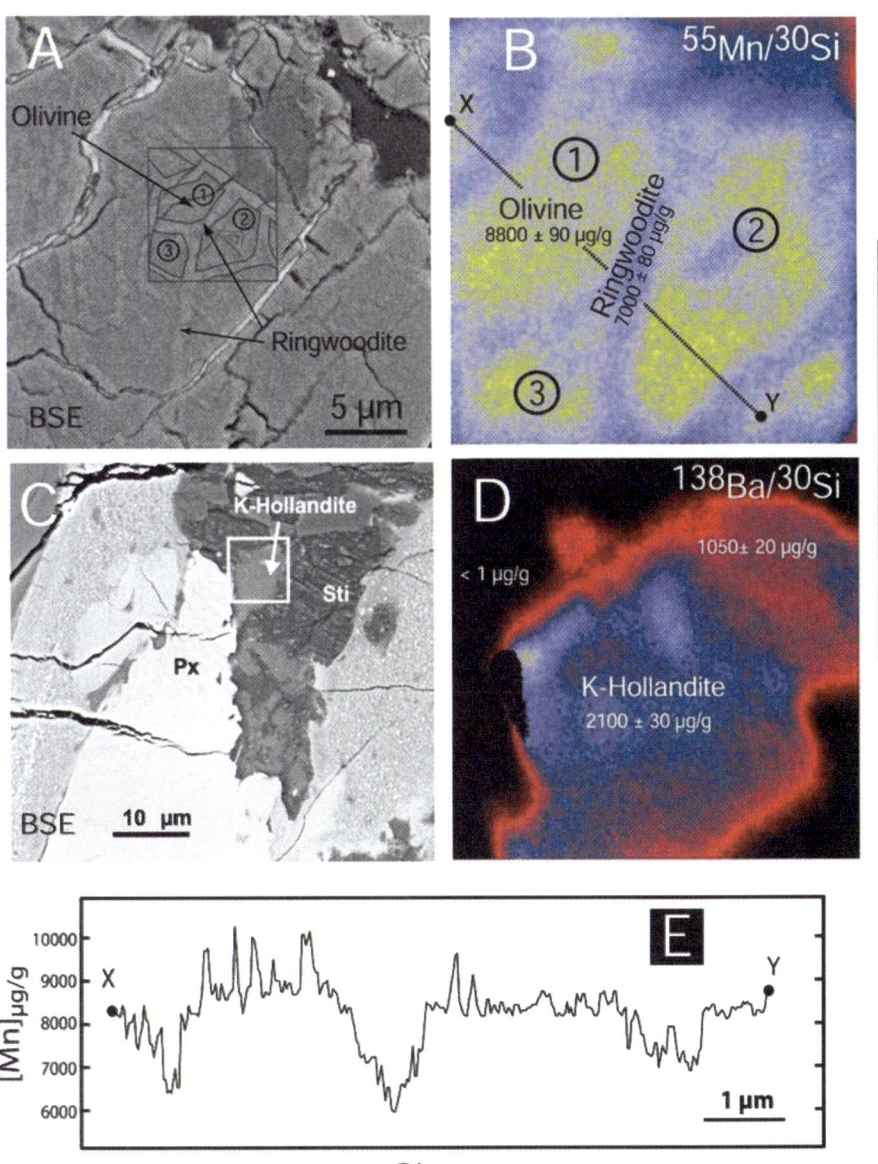

Figure 16. Trace-element maps of high-pressure minerals in Zagami and Tenham meteorites (after Beck et al., 2005). (A) Backscattered electron (BSE) image of an area in a shock melt vein of the Tenham meteorite where intracrystalline ringwoodite (indicated brighter gray) lamellae were formed in an olivine grain (darker gray). The numbers refer to the olivine zones also depicted on the chemical NanoSIMS $^{53}Mn/^{30}Si$ map. The gray square denotes the zone where the NanoSIMS trace-element mapping was conducted. (B) Corresponding $^{55}Mn/^{30}Si$ elemental image. Ringwoodite is clearly depleted in Mn with respect to olivine. The Mn concentration drop between ringwoodite and olivine is not as sharp as the olivine/ringwoodite interface is in E. This observation is indicative of interdiffusion of Mn between olivine and ringwoodite. (C) BSE image showing the K-hollandite aggregate in the Martian meteorite Zagami in which trace-element (Cs, Ba, and Rb) abundances were measured. The K-hollandite aggregate is surrounded by stishovite (Sti). Px—pyroxene. (D) NanoSIMS $^{138}Ba/^{30}Si$ elemental image of the K-hollandite aggregate in Zagami. The Ba concentrations in the grain are high in comparison to the surrounding material, in agreement with a pronounced Ba diffusion during the shock. Similar measurements have been carried out for two other trace elements, Cs and Rb.

The *P-T-t* Shock History of the Martian Meteorite Zagami

In the Zagami shergottite, both shock melt vein and melt pockets are present. Their mineralogy is dominated by K-rich hollandite, (Ca-Na)-hollandite, SiO_2-stishovite, omphacite, silicate-titanite, akimotoite, and amorphous grains of silicate perovskite composition (Langenhorst and Poirier, 2000a). The K-rich hollandite associated with stishovite has been interpreted as a liquidus assemblage (Langenhorst and Poirier, 2000a). The matrix is mostly composed of quenched glassy material associated with the liquidus majorite + magnesiowüstite assemblage, the latter pair observed for the first time in an SNC meteorite and very similar in texture to the one occurring in shock melt veins of chondrites. The liquidus assemblage $(Ca_xNa_{1-x})Al_{3+x}Si_3O_{11}$ + stishovite is also observed (Beck et al., 2004). The presence of this assemblage and hollandite-structured plagioclase tightly constrains the pressure range for crystallization of the melt pockets. The $(Ca_xNa_{1-x})Al_{3+x}Si_3O_{11}$ + stishovite pair presumably formed through dissociation of Ca-rich plagioclase above 16 GPa (Gautron et al., 1996). Above 23 GPa, the hollandite structure dissociates to stishovite + Ca ferrite–structured $NaAlSiO_4$ (Xie et al., 2002; Yagi et al., 1994). Consequently, the crystallization of the hollandite-structured phase sets this pressure as the upper bound for melt pocket solidification. The absence of dense silica glass among these silica grains indicates that stishovite did not form by back transformation of a post-stishovite polymorph. The crystallization temperature of K-hollandite from the high-pressure melt is constrained by the phase diagram of basaltic rocks (Wang and Takahashi, 1999) at $T = 2500$ K and $P = 23$ GPa.

The shock duration in the Zagami shergottite can be obtained from studying trace-element concentrations in coexisting grains in a shock melt pocket. These were determined for 10-μm-sized liquidus aggregates of K-hollandite surrounded by stishovite grains using NanoSIMS (secondary ion mass spectrometry) (Fig. 16) (Beck et al., 2005). The concentrations of Cs, Ba, and Rb in K-hollandite are higher than those for the host rock, or those measured outside the shock melt veins in maskelynite grains (Fig. 16). These observations are consistent with the interpretation that the trace-element partitioning took place between silicate melt and K-hollandite aggregates during the high-pressure shock event. The crystallization of the K-hollandite would have begun as soon as the shock-induced silicate melt was established and the pressure reached 23 GPa. At the end of the peak shock pulse, as pressure dropped below 23 GPa, the vein would remain molten, and lower-pressure mineral phases would be formed (Langenhorst and Poirier, 2000b). The time required for trace elements to diffuse into the K-hollandite thus provides a means for calculating a maximum value for the duration of the equilibrium shock pressure. Using these assumptions, the equilibrium shock pressure duration calculated for all three trace elements is on the order of ~10 ms.

CONCLUSIONS AND PERSPECTIVES

In this review, we have tried to cover the phase transformations, partial melting, and chemical exchanges induced in planetary materials during intense shocks. We have emphasized that most of the minerals synthesized in static experiments are very common in the shear melt veins and melt pockets of heavily shocked meteorites. The short durations of the equilibrium shock pressure and temperature conditions as well as the efficient temperature quenching during natural shocks freeze the initial stages of the changes that minerals and rocks undergo at high pressure and temperature. We emphasize that a comprehensive study of naturally shocked meteorites and terrestrial rocks at a microscopic scale of a few microns down to a few nanometers must be undertaken in order to uncover details of the shock-induced changes. Further studies should be able to reveal high-pressure new minerals and provide further insights into the onset of partial melting and phase transitions. This should lead to a better understanding of the role of high deviatoric stresses in promoting these changes and their time scales.

Since the equilibrium pressure durations are short with respect to the time scales of static experiments, the interpretations and conclusions relevant to Earth's interior must be undertaken with some caution. Nevertheless, the mineralogy of shocked rocks provides valuable insights into the mineral associations that could be present in the deep Earth, including the transition zone and the lower mantle.

The combination of observations in shocked rocks with data from high-pressure static experiments must also be used with the aim of constraining the pressure-temperature-time history of shock events. The combination of exact characterization of the natural assemblages, results of high-pressure static experiments, and trace-element partitioning between high-pressure phases should improve our knowledge of the shock durations. This could be used to scrutinize the time scales of shock processes in the solar system at different stages of its history. This point is briefly discussed below.

The shock pulse durations in Zagami and Tenham are ~0.01 s and ~1 s, respectively, and are correlated with the size of the impactors (Beck et al., 2005). For the Martian meteorite Zagami, the calculated impactor size is 0.1 km, whereas for Tenham, it is ~5 km. Moreover the size of the impactors is related to that of the craters formed on the target planet or asteroid. Using such a relation, which links the shock duration, the impactor velocity, and the crater size, one could deduce that the Martian meteorite Zagami was probably extracted from a crater with a 1500–5000 m diameter on Mars. Moreover, the predicted sizes of the ejected fragments from of such a crater, i.e., the sizes of Martian meteorites arriving on Earth, could range from a few centimeters to a few tens of centimeters, in excellent agreement with the actual size distribution of shergottites. Many shergottites possess similar ejection ages to Zagami, suggesting that they originated from the same shock event on Mars. Given the small expected crater size, it can be suggested that all these rocks of distinct petrology, magmatic ages (180–350 Ma), chemical and isotopic compositions could have coexisted on Mars within a few square kilometers of each other, to a depth of a few hundred meters. Tenham is representative of one of the most heavily shocked chondrites known so far. The age of its shock metamorphism is unknown.

REFERENCES CITED

Agee, C.B., 1998, Phase transformations and seismic structure in the upper mantle and transition zone, in Hemley, R.J., ed., Ultra-High Pressure Mineralogy: Physics and Chemistry of the Earth's Deep Interior: Reviews in Mineralogy, v. 37, p. 165–203.

Agee, C.B., Li, J., Shannon, M.C., and Circone, S., 1995, Pressure-temperature phase diagram for the Allende meteorite: Journal of Geophysical Research, v. 100, p. 17,725–17,740, doi: 10.1029/95JB00049.

Ahrens, T.J., 1987, Shock wave techniques for geophysics and planetary physics, in Sammis, C.G., and Heyney, T.L., eds., Methods of Experimental Physics: San Diego, Academic Press, p. 185–235.

Akaogi, M., Kojitani, H., Matsuzaka, K., Suzuki, T., and Ito, E., 1998, Post-spinel transformations in the system Mg_2SiO_4-Fe_2SiO_4: Element partitioning, calorimetry, and thermodynamic calculations, in Manghnani, M.H., and Yagi, T., eds., Properties of Earth and Planetary Materials at High Pressure and Temperature: American Geophysical Union., Geophysical Monograph, v. 101, p. 373–384.

Akaogi, M., Tanaka, A., and Ito, E., 2002, Garnet-ilmenite-perovskite transitions in the system $Mg_4Si_4O_{12}$-$Mg_3Al_2Si_3O_{12}$ at high pressures and high temperatures: Phase equilibria, calorimetry and implications for mantle structure: Physics of the Earth and Planetary Interiors, v. 132, no. 4, p. 303–324, doi: 10.1016/S0031-9201(02)00075-4.

Akaogi, M., Yano, M., Tejima, Y., Iijima, M., and Kojitani, H., 2004, High-pressure transitions of diopside and wollastonite: Phase equilibria and thermochemistry of $CaMgSi_2O_6$, $CaSiO_3$ and $CaSi_2O_5$-$CaTiSiO_5$ system: Physics of the Earth and Planetary Interiors, v. 143–144, p. 145–156, doi: 10.1016/j.pepi.2003.08.008.

Aoki, I., and Takahashi, E., 2004, Density of MORB eclogite in the upper mantle: Physics of the Earth and Planetary Interiors, v. 143–144, p. 129–143, doi: 10.1016/j.pepi.2003.10.007.

Artemieva, N., and Ivanov, B.A., 2004, Launch of Martian meteorites in oblique impacts: Icarus, v. 171, p. 84–101, doi: 10.1016/j.icarus.2004.05.003.

Asahara, Y., Kubo, T., and Kondo, T., 2004, Phase relations of a carbonaceous chondrite at lower mantle conditions: Physics of the Earth and Planetary Interiors, v. 143–144, p. 421–432, doi: 10.1016/j.pepi.2003.10.011.

Asay, J.R., and Shahinpoor, M., 1993, High-Pressure Shock Compression of Solids: New York, Springer-Verlag, 389 p.

Baer, M.R., 2000, Computational modelling of heterogeneous materials at the mesoscale, in Furnish, M.D., Chabildas, L.C., and Hixon, R.S., eds., CP505: Shock compression of Condensed Matter-1999: American Institute of Physics Proceedings No. 505, p. 27–33.

Bass, J.D., and Anderson, D.L., 1984, Composition of the upper-mantle: Geophysical test of two petrological models: Geophysical Research Letters, v. 11, p. 237–241.

Beck, P., Gillet, P., Gautron, L., Danielle, I., and El Goresy, A., 2004, A new natural high-pressure (Na-Ca)-hexaluminosilicate ([Ca_xNa_{1-x}]$Al_{3+x}Si_{3-x}O_{11}$) in shocked Martian meteorites: Earth and Planetary Science Letters, v. 219, p. 1–12, doi: 10.1016/S0012-821X(03)00695-2.

Beck, P., Gillet, P., El Goresy, A., and Mostefaoui, S., 2005, Timescales of shock processes in the solar system from mineralogical transformations in chondrites and Martian meteorites: Nature, v. 435, p. 1071–1074, doi: 10.1038/nature03616.

Becker, L., Bunch, T.E., and Allamandola, L.J., 1999, Higher fullerenes in the Allende meteorite: Nature, v. 400, p. 227–228, doi: 10.1038/22250.

Belonoshko, A.B., Dubrovinski, L.S., and Dubrovinski, N.A., 1996, A new high-pressure silica phase obtained by molecular dynamics: The American Mineralogist, v. 81, p. 785–788.

Benzerara, K., Barrat, J.A., Guyot, F., Gillet, P., and Lesourd, M., 2002, Cristobalite inclusions in the Tatahouine achondrite: Implications for the shock conditions: The American Mineralogist, v. 87, p. 1250–1265.

Binns, R.A., 1970, $(Mg,Fe)_2SiO_4$ spinel in a meteorite: Earth and Planetary Science Letters, v. 76, p. 109–122.

Bogard, D.D., 1995, Impact ages of meteorites: A synthesis: Meteoritics, v. 30, p. 244–268.

Brearley, A.J., and Jones, R.H., 1998, Chondritic meteorites, in Papike, J.J., ed., Planetary Materials: Reviews in Mineralogy, v. 36, p. 1–398.

Buseck, P.R., 2002, Geological fullerenes: Review and analysis: Earth and Planetary Science Letters, v. 203, p. 781–792, doi: 10.1016/S0012-821X(02)00819-1.

Chambers, J.E., and Wetherill, G.W., 1998, Making the terrestrial planets: N-body integrations of planetary embryos in three dimensions: Icarus, v. 136, p. 304–327, doi: 10.1006/icar.1998.6007.

Chao, E.C.T., Fahey, J.J., Littler, J., and Milton, D.J., 1962, Stishovite, SiO_2, a very high-pressure new mineral from Meteor Crater, Arizona: Journal of Geophysical Research, v. 67, p. 419–421.

Chen, M., and El Goresy, A., 2000, The nature of maskelynite in shocked meteorites: Not diaplectic glass but a glass quenched from shock-induced dense melt at high pressures: Earth and Planetary Science Letters, v. 179, p. 489–502, doi: 10.1016/S0012-821X(00)00130-8.

Chen, M., Wopenka, B., Xie, X., and El Goresy, A., 1995, A new high-pressure polymorph of chlorapatite in the shocked Sixiangkou(L6) chondrite: Lunar and Planetary Sciences, v. 26, p. 237–238.

Chen, M., Sharp, T.G., El Goresy, A., Wopenka, B., and Xie, X.D., 1996, The majorite-pyrope-magnesiowüstite assemblage: Constraints on the history of shock veins in chondrite: Science, v. 271, p. 1570–1573.

Chen, M., Shu, J., Xie, X., and Mao, H.K., 2003a, Natural $CaTi_2O_4$-structured $FeCr_2O_4$ polymorph in the Suizhou meteorite and its significance in mantle mineralogy: Geochimica et Cosmochimica Acta, v. 67, p. 3937–3942, doi: 10.1016/S0016-7037(03)00175-3.

Chen, M., Shu, J., Mao, H.K., Xie, X., and Hemley, R.J., 2003b, Natural occurrence and synthesis of two new post-spinel polymorphs of chromite: Proceedings of the National Academy of Sciences of the United States of America, v. 100, p. 14,651–14,654, doi: 10.1073/pnas.2136599100.

Chen, M., El Goresy, A., and Gillet, P., 2004a, Ringwoodite lamellae in olivine: Clues to olivine-ringwoodite phase transition mechanisms in shocked meteorites and subducting slabs: Proceedings of the National Academy of Sciences of the United States of America, v. 101, p. 15,033–15,037, doi: 10.1073/pnas.0405048101.

Chen, M., El Goresy, A., Frost, D., and Gillet, P., 2004b, Melting experiments of a chondritic meteorite between 16 and 25 GPa: Implication for Na/K fractionation in a primitive chondritic Earth's mantle: European Journal of Mineralogy, v. 16, p. 203–211, doi: 10.1127/0935-1221/2004/0016-0203.

Chudinovskikh, L., and Boehler, R., 2004, $MgSiO_3$ phase boundaries measured in the laser-heated diamond cell: Earth and Planetary Science Letters, v. 219, p. 285–296, doi: 10.1016/S0012-821X(04)00005-6.

Daniel, I., Cardon, H., Fiquet, G., Guyot, F., and Mezouar, M., 2001, Equation of state of Al-bearing perovskite to lower mantle pressure conditions: Geophysical Research Letters, v. 28, p. 3789–3792, doi: 10.1029/2001GL013011.

Daulton, T.L., Eisenhour, D.D., Bernatowicz, T.J., Lewis, R.S., and Buseck, P.R., 1996, Genesis of presolar diamonds: Comparative high-resolution transmission electron microscopy study of meteoritic and terrestrial nanodiamonds: Geochimica et Cosmochimica Acta, v. 60, p. 4853–4872, doi: 10.1016/S0016-7037(96)00223-2.

Dera, P., Prewitt, C.T., Boctor, N.Z., and Hemley, R.J., 2002, Characterization of a high-pressure phase of silica in the Martian meteorite Shergotty: The American Mineralogist, v. 87, p. 1018–1023.

Dewhurst, J.K., and Lowther, J.E., 1996, High-pressure structural phases of titanium oxide: Physical Review B: Condensed Matter and Materials Physics, v. 54, p. 3673–3675.

Dubrovinskaia, N.A., Dubrovinsky, L.S., Saxena, S.K., Tutti, F., Rekhi, S., and Le Bihan, T., 2001, Direct transition from cristobalite to a post-stishovite PbO_2-like silica phase: European Journal of Mineralogy, v. 13, p. 479–483, doi: 10.1127/0935-1221/2001/0013-0479.

Dubrovinsky, L.S., Saxena, S.K., Lazor, P., Ahuja, R., Erikson, O., Wills, J.M., and Johansson, B., 1997, Experimental and theoretical identification of a new high-pressure phase of silica: Nature, v. 388, p. 362–365, doi: 10.1038/41066.

Dubrovinsky, L.S., Dubrovinskaia, N.A., Saxena, S.K., Tutti, F., Rekhi, S., Le Bihan, T., Shen, G., and Hu, J., 2001a, Pressure-induced transformation of cristobalite: Chemical Physics Letters, v. 333, p. 264–270, doi: 10.1016/S0009-2614(00)01147-7.

Dubrovinsky, L.S., Dubrovinskaia, N.A., Swamy, V., Muscat, J., Harrison, M., Ahuja, R., Holm, B., and Johansson, B., 2001b, Cotunnite-structured titanium dioxide: The hardest known oxide: Nature, v. 410, p. 653–654, doi: 10.1038/35070650.

Dubrovinsky, L.S., Dubrovinskaia, N.A., Prakapenka, V., Seifert, F., Langenhorst, F., Dmitriev, V., Weber, H.P., and Le Bihan, T., 2004, A class of new high-pressure silica polymorphs: Physics of the Earth and Planetary Interiors, v. 143–144, p. 231–240, doi: 10.1016/j.pepi.2003.06.006.

El Goresy, A., 1964, Die Erzmineralien in den Ries- und Bosumtwi-Kraterglässern und ihre genetische Deutung: Geochimica et Cosmochimica Acta, v. 28, p. 1881–1891, doi: 10.1016/0016-7037(64)90135-8.

El Goresy, A., 1965, Baddeleyite and its significance in impact glasses: Journal of Geophysical Research, v. 70, p. 3455–3456.

El Goresy, A., Dubrovinsky, L., Sharp, T.G., Saxena, S.K., and Chen, M., 2000, A monoclinic post-stishovite polymorph of silica in the Shergotty meteorite: Science, v. 288, p. 1632–1635, doi: 10.1126/science.288.5471.1632.

El Goresy, A., Gillet, P., Chen, M., Künstler, F., Graup, G., and Stähl, V., 2001a, In situ discovery of shock-induced graphite diamond phase transition in gneisses from the Ries Crater, Germany: The American Mineralogist, v. 86, p. 611–621.

El Goresy, A., Chen, M., Gillet, P., Dubrovinsky, L., Graup, G., and Ahuja, R., 2001b, A natural shock-induced dense polymorph of rutile with α-PbO_2 structure in the suevite from the Ries crater in Germany: Earth and Planetary Science Letters, v. 192, p. 485–495, doi: 10.1016/S0012-821X(01)00480-0.

El Goresy, A., Chen, M., Dubrovinsky, L., Gillet, P., and Graup, G., 2001c, An ultradense polymorph of rutile with seven-coordinated titanium from the Ries crater: Science, v. 293, p. 1467–1470, doi: 10.1126/science.1062342.

El Goresy, A., Dubrovinsky, L., Gillet, P., Mostefaoui, S., Graup, G., Drakopoulos, M., Simionovici, A.S., Swamy, V., and Masaitis, V.L., 2003, A new natural, super-hard, transparent polymorph of carbon from the Popigai impact crater, Russia: Comptes Rendus de l'Académie des Sciences—Geosciences, v. 335, p. 889–898.

El Goresy, A., Dubrovinsky, L., Sharp, T.G., and Chen, M., 2004, Stishovite and post-stishovite polymorphs of silica in the Shergotty meteorite: Their nature, petrographic settings versus theoretical predictions and relevance to Earth's mantle: Physics and Chemistry of Solids, v. 65, p. 1597–1608.

El Goresy, A., Chen, M., Gillet, P., and Dubrovinsky, L., 2005, Two distinct olivine-ringwoodite phase transition mechanisms in shocked L6-chondrites: Genetic implications: Meteoritics & Planetary Science, v. 40, p. 5010.

Fei, Y., and Bertka, C.M., 1999, Phase transitions in the Earth's mantle and mantle mineralogy, in Fei, Y., Bertka, C.M., and Mysen, B.O., eds., Mantle Petrology: Field Observations and High Pressure Experimentation: A Tribute to Francis R. (Joe) Boyd: Geochemical Society Special Publication 6, p. 189–207.

Fei, Y., and Hirose, K., 1999, Majorite-perovskite transformation in the system $MgSiO_3$-$Mg_3Al_2Si_3O_{12}$: Eos (Transactions, American Geophysical Union), v. 78, p. 761.

Ferroir, T., Onozawa, T., Yagi, T., Merkel, S., Miyajima, N., Nishiyama, N., Irifune, T., and Kikegawa, T., 2006, Equation of state and phase transition in $KAlSi_3O_8$ hollandite at high pressure: The American Mineralogist, v. 91, p. 327–332.

Fiquet, G., 2001, Mineral phases of the Earth's mantle: Zeitschrift für Kristallography, v. 216, p. 248–271, doi: 10.1524/zkri.216.5.248.20374.

Funamori, N., Jeanloz, R., Miyajima, N., and Fujino, K., 2000, Mineral assemblages of baslat in the lower mantle: Journal of Geophysical Research, v. 105, p. 26,037–26,043, doi: 10.1029/2000JB900252.

Gasparik, T., 1990, Phase relations in the transition zone: Journal of Geophysical Research, v. 95, p. 15,751–715,769.

Gasparik, T., 1992, Melting experiments on the enstatite-pyrope join at 80–152 kbar: Journal of Geophysical Research, v. 97, p. 15,181–15,188.

Gautron, L., Kesson, S.E., and Hibberson, W.O., 1996, Phase relations for $CaAl_2Si_2O_8$ (anorthite composition) in the system $CaO-Al_2O_3-SiO_2$ at 14 GPa: Physics of the Earth and Planetary Interiors, v. 97, p. 71–81, doi: 10.1016/0031-9201(96)03161-5.

Gautron, L., Fitz Gerald, J.D., Kesson, S.E., Eggleton, R.A., and Irifune, T., 1997, Hexagonal Ba-ferrite: A good model for the crystal structure of a new high-pressure phase $CaAl_4Si_2O_{11}$?: Physics of the Earth and Planetary Interiors, v. 102, p. 223–229, doi: 10.1016/S0031-9201(97)00007-1.

Gautron, L., Angel, R.J., and Miletich, R., 1999, Structural characterization of the high pressure phase $CaAl_4Si_2O_{11}$: Physics and Chemistry of Minerals, v. 27, p. 47–51, doi: 10.1007/s002690050239.

Gillet, P., Chen, C., Dubrovinsky, L., and El Goresi, A., 2000, Natural $NaAlSi_3O_8$-hollandite in the shocked Sixiangkou meteorite: Science, v. 287, p. 1633–1636, doi: 10.1126/science.287.5458.1633.

Gillet, P., Sautter, V., Harris, J.W., Reynard, B., Harte, B., and Kunz, M., 2002, Raman spectroscopy study of garnet inclusions in diamonds from the mantle transition zone: The American Mineralogist, v. 87, p. 312–317.

Gladman, B., 1997, Destination: Earth; Martian meteorite delivery: Icarus, v. 130, p. 228–246, doi: 10.1006/icar.1997.5828.

Glass, B.P., and Liu, S., 2001, Discovery of high-pressure $ZrSiO_4$ polymorph in naturally occurring shock-metamorphosed zircons: Geology, v. 29, p. 371–373, doi: 10.1130/0091-7613(2001)029<0371:DOHPZP>2.0.CO;2.

Grady, D.E., 1980, Shock deformation of brittle solids: Journal of Geophysical Research, v. 85, p. 913–924.

Grady, D.E., Murri, W.J., and De Carli, P.S., 1975, Hugoniot sound velocities and phase transformations in 2 silicates: Journal of Geophysical Research, v. 80, p. 4857–4861.

Grossman, L., 1975, Petrography and mineral chemistry of Ca-rich inclusions in the Allende meteorite: Geochimica et Cosmochimica Acta, v. 39, p. 433–454, doi: 10.1016/0016-7037(75)90099-X.

Grossman, L., Ebel, D.S., and Simon, S.B., 2002, Formation of refractory inclusions by evaporation of condensate precursors: Geochimica et Cosmochimica Acta, v. 66, p. 145–161, doi: 10.1016/S0016-7037(01)00731-1.

Haines, J., and Léger, J.M., 1997, X-ray diffraction of the phase transitions and structural evolution of tin dioxide at high pressure: Relationships between structure types and implications for other rutile-type dioxides: Physical Review B: Condensed Matter and Materials Physics, v. 55, p. 1–11.

Harte, B., and Harris, J.W., 1994, Lower mantle mineral associations preserved in diamonds: Mineralogical Magazine, v. 58A, p. 384–386.

Hartmann, W.K., and Neukum, G., 2001, Cratering chronology and the evolution of Mars: Space Science Reviews, v. 96, p. 165–193, doi: 10.1023/A:1011945222010.

Head, J.W., Melosh, H.J., and Ivanov, B.A., 2002, Martian meteorite launch: High-speed ejecta from small craters: Science, v. 298, p. 1752–1756, doi: 10.1126/science.1077483.

Heider, N., and Kenkmann, T., 2003, Numerical simulation of temperature effects at fissures due to shock loading: Meteoritics & Planetary Science, v. 38, p. 1451–1460.

Hemley, R.J., Prewitt, C.T., and Kingma, K.J., 1994, High-pressure behaviour of silica, in Heaney, P.J., Prewitt, C.T., and Gibbs, G.V., eds., Silica: Physical Behavior, Geochemistry and Material Applications: Mineralogical Society of America Reviews in Mineralogy, v. 29, p. 41–81.

Herzberg, C., and Zhang, J.Z., 1996, Melting experiments on anhydrous peridotite KLB-1: Compositions of magmas in the upper mantle and transition zone: Journal of Geophysical Research, v. 101, p. 8271–8295, doi: 10.1029/96JB00170.

Hirose, K., and Fei, Y., 2002, Subsolidus and melting phase relations of basaltic composition in the uppermost lower mantle: Geochimica et Cosmochimica Acta, v. 66, p. 2099–2108, doi: 10.1016/S0016-7037(02)00847-5.

Hirose, K., Fei, Y., Ma, Y., and Mao, H.K., 1999, The fate of subducted basaltic crust in the Earth's lower mantle: Nature, v. 397, p. 53–56, doi: 10.1038/16225.

Hirose, K., Fei, Y., Ono, S., Yagi, T., and Funakoshi, K.-I., 2001, In situ measurements of the phase transition boundary in $Mg_3Al_2Si_3O_{12}$: Implications for the nature of the seismic discontinuities in the Earth's mantle: Earth and Planetary Science Letters, v. 184, p. 567–573, doi: 10.1016/S0012-821X(00)00354-X.

Inaba, S., Tanaka, H., Nakazawa, K., Wetherill, G.W., and Kokubo, E., 2001, High-accuracy statistical simulation of planetary accretion, II: Comparison with N-body simulation: Icarus, v. 149, p. 235–250, doi: 10.1006/icar.2000.6533.

Irifune, T., and Ringwood, A.E., 1993, Phase transformations in subducted oceanic crust and buoyancy relationships at depths of 600–800 km in the mantle: Earth and Planetary Science Letters, v. 117, p. 101–110, doi: 10.1016/0012-821X(93)90120-X.

Irifune, T., Koizumi, T., and Ando, J.I., 1996, An experimental study of the garnet-perovskite transformation in the system $MgSiO_3-Mg_3Al_2Si_3O_{12}$: Physics of the Earth and Planetary Interiors, v. 96, p. 147–157, doi: 10.1016/0031-9201(96)03147-0.

Jeanloz, R., 1980, Shock effects in olivine and implications for Hugoniot data: Journal of Geophysical Research, v. 85, p. 3163–3176.

Karki, B.B., Warren, S.C., Stixrude, L., Ackland, G.J., and Crain, J., 1997, Ab initio studies of high-pressure structural transformations in silica: Physical Review B: Condensed Matter and Materials Physics, v. 55, p. 3465–3472.

Katsura, T., and Ito, E., 1989, The system $Mg_2SiO_4-Fe_2SiO_4$ at high pressures and temperature: Precise determination of stabilities of olivine, modified spinel and spinel: Journal of Geophysical Research, v. 94, p. 15,663–15,670.

Katsura, T., Yamada, H., Shinmei, T., Kubo, A., Ono, S., Kanzaki, M., Yoneda, A., Walter, M.J., Ito, E., and Urakawa, S., 2003, Post-spinel transition in Mg_2SiO_4 determined by high P-T in situ X-ray diffractometry: Physics of the Earth and Planetary Interiors, v. 136, p. 11–24, doi: 10.1016/S0031-9201(03)00019-0.

Kerschhofer, L., Sharp, T.G., and Rubie, D.C., 1996, Intracrystalline transformation of olivine to wadsleyite and ringwoodite under subduction zone conditions: Science, v. 274, p. 79–81, doi: 10.1126/science.274.5284.79.

Kerschhofer, L., Dupas, C., Liu, M., Sharp, T.G., Durham, W.B., and Rubie, D.C., 1998, Polymorphic transformations between olivine, wadsleyite and ringwoodite: Mechanisms of intracrystalline nucleation and the role of elastic strain: Mineralogical Magazine, v. 62, p. 617–638, doi: 10.1180/002646198548016.

Kerschhofer, L., Rubie, D.C., Sharp, T.G., McConnell, J.D.C., and Dupas-Bruzek, C., 2000, Kinetics of intracrystalline olivine-ringwoodite transformation: Physics of the Earth and Planetary Interiors, v. 121, p. 59–76, doi: 10.1016/S0031-9201(00)00160-6.

Kieffer, S.W., Phakey, P.P., and Christie, J.M., 1976, Shock processes in porous quartzite: Transmission electron microscope observations and theory: Contributions to Mineralogy and Petrology, v. 59, p. 41–93, doi: 10.1007/BF00375110.

Kingma, K.J., Cohen, R.E., Hemley, R.J., and Mao, H.K., 1995, Transformation of stishovite to a denser phase at lower-mantle pressures: Nature, v. 374, p. 243–245, doi: 10.1038/374243a0.

Koeberl, C., Masaitis, V.L., Shafranovsky, G.I., Gilmour, I., Langenhorst, F., and Shrauder, M., 1997, Diamonds from the Popigai impact structure, Russia: Geology, v. 25, p. 967–970, doi: 10.1130/0091-7613(1997)025<0731:MSAALI>2.3.CO;2.

Kubo, A., and Akaogi, M., 2000, Post-garnet transitions in the system $Mg_4Si_4O_{12}-Mg_3Al_2Si_3O_{12}$ up to 28 GPa: Phase relations of garnet, ilmenite and perovskite: Physics of the Earth and Planetary Interiors, v. 121, p. 85–102, doi: 10.1016/S0031-9201(00)00162-X.

Kubo, T., Ohtani, E., Kato, T., Urakawa, S., Suzuki, A., Kanbe, Y., Funakoshi, K., Utsumi, W., Kikegawa, T., and Fujino, K., 2002, Mechanisms and kinetics of the post-spinel transformation in Mg_2SiO_4: Physics of the Earth and Planetary Interiors, v. 129, p. 153–171, doi: 10.1016/S0031-9201(01)00270-9.

Kurashina, T., Hirose, K., Onob, S., Sata, N., and Ohishi, Y., 2004, Phase transition in Al-bearing $CaSiO_3$ perovskite: Implications for seismic discontinuities in the lower mantle: Physics of the Earth and Planetary Interiors, v. 145, p. 67–74, doi: 10.1016/j.pepi.2004.02.005.

Kusaba, K., Syono, M., Kikushi, M., and Fukuoka, K., 1985, Shock behavior of zircon: Phase transition to scheelite structure and decomposition: Earth and Planetary Science Letters, v. 72, p. 433–439, doi: 10.1016/0012-821X(85)90064-0.

Kusaba, K., Kikuchi, M., Fukuoka, K., and Syono, Y., 1988, Anisotropic phase transition of rutile under shock compression: Physics and Chemistry of Minerals, v. 15, p. 238–245, doi: 10.1007/BF00307512.

Langenhorst, F., and Dressler, B., 2003, First observation of silicate hollandite in a terrestrial rock: Nördlingen, in Large Meteorite Impact: Lunar and Planetary Institute contribution No. 1167, Abstract 4046. Langenhorst, F., and Poirier, J.P., 2000a, Anatomy of black veins in Zagami: Clues to the formation of high-pressure phase: Earth and Planetary Science Letters, v. 184, p. 37–55, doi: 10.1016/S0012-821X(00)00317-4.

Langenhorst, F., and Poirier, J.P., 2000b, "Eclogitic" minerals in a shocked basaltic meteorite: Earth and Planetary Science Letters, v. 176, p. 259–265, doi: 10.1016/S0012-821X(00)00028-5.

Langenhorst, F., Shafranovsky, G., and Masaitis, V.L., 1998, A comparative study of impact diamonds from the Popigai, Ries, Sudbury, and Lappajärvi craters: Meteoritics & Planetary Science, v. 33, p. A90–A91.

Leroux, H., 2001, Microstructural shock signatures of major minerals in meteorites: European Journal of Mineralogy, v. 13, p. 253–272, doi: 10.1127/0935-1221/01/0013-0253.

Linde, K., and DeCarli, P.S., 1969, Polymorphic behavior of titania under dynamic loading: The Journal of Chemical Physics, v. 50, p. 319–325, doi: 10.1063/1.1670796.

Liu, L.G., 1978a, High-pressure phase transformation of albite, jadeite and nepheline: Earth and Planetary Science Letter, v. 37, p. 438–444, doi: 10.1016/0012-821X(78)90059-6.

Liu, L.G., 1978b, A fluorite isotype of SnO_2 and a new modification of TiO_2: Implications for the Earth's lower mantle: Science, v. 199, p. 422–425.

Madon, M., and Poirier, J.P., 1983, Transmission electron microscope observation of alpha, beta and gamma $(Mg,Fe)_2SiO_4$ in shocked meteorites—Planar defects and polymorphic transitions: Physics of the Earth and Planetary Interiors, v. 33, p. 31–44, doi: 10.1016/0031-9201(83)90005-5.

Madon, M., Castex, J., and Peyronneau, J., 1989, A new hollandite-type structure as a possible host for calcium and aluminium in the lower mantle: Nature, v. 342, p. 422–425, doi: 10.1038/342422a0.

Malavergne, V., Guyot, F., Benzerara, K., and Martinez, I., 2001, Description of new-shock-induced phases in the SNC meteorites: Shergotty, Nakhla, Chassigny and Zagamy: Meteoritics & Planetary Science, v. 36, p. 1297–1305.

Masaitis, V.L., Futergendler, S.I., and Gnevyshev, M.A., 1972, The diamonds in the Popigai meteoritic crater: Zapiski Vsesouznogo Mineralogicheskogo Obstchestva, v. 101, p. 108–113.

Matas, J., 1999, Modélisation thermochimique des propriétés de solides à hautes températures et pression: Applications géophysiques [Ph.D. dissertation]: Lyon, Ecole Normale Supérieure de Lyon, 175 p.

McCammon, C.A., 1997, Perovskite as a possible sink for ferric iron in the lower mantle: Nature, v. 387, p. 694–696, doi: 10.1038/42685.

McConville, P., Kelley, S., and Turner, G., 1988, Laser probe ^{40}Ar-^{39}Ar studies of the Peace River shocked L6 chondrite: Geochimica et Cosmochimica Acta, v. 52, p. 2487–2499, doi: 10.1016/0016-7037(88)90307-9.

McQueen, R.G., Jamieson, J.C., and Marsh, S.P., 1967, Shock-wave compression and X-ray studies of titanium dioxide: Science, v. 155, p. 1401–1404.

McSween, H.Y.J., 2002, What we learned about Mars from SNC meteorites: Meteoritics & Planetary Science, v. 29, p. 757–779.

Melosh, H.J., 1989, Impact Cratering: A Geologic Process: New York, Oxford University Press, 245 p.

Miura, H., Hamada, Y., Suzuki, T., Akaogi, M., Miyajima, N., and Fujino, K., 2000, Crystal structure of $CaMg_2Al_6O_{12}$, a new Al-rich high-pressure form: The American Mineralogist, v. 85, p. 1799–1803.

Miyajima, N., Yagi, T., Hirose, K., Kondo, T., Fujino, K., and Miura, H., 2001, Potential host phase of aluminium and potassium in the Earth's lower mantle: The American Mineralogist, v. 86, p. 740–746.

Mosenfelder, J.L., Connolly, J.A.D., Rubie, D.C., and Liu, M., 2000, Strength of $(Mg,Fe)_2SiO_4$ wadsleyite determined by relaxation of transformation stress: Physics of the Earth and Planetary Interiors, v. 120, p. 63–78, doi: 10.1016/S0031-9201(00)00142-4.

Murakami, M., Hirose, K., Ono, S., and Ohishi, Y., 2003, Stability of $CaCl_2$-type and α-PbO_2-type SiO_2 at high pressure and temperature determined by in-situ X-ray measurements: Geophysical Research Letters, v. 30, p. 1207–1210, doi: 10.1029/2002GL016722.

Murakami, M., Hirose, K., Kawamura, K., Sata, N., and Ohishi, Y., 2004, Post-perovskite phase transition in $MgSiO_3$: Science, v. 304, p. 855–858, doi: 10.1126/science.1095932.

Murayama, J.K., Nakai, S., Kato, M., and Kumazawa, M., 1986, A dense polymorph of $Ca_3(PO_4)_2$: A high pressure phase of apatite decomposition and its geochemical significance: Physics of the Earth and Planetary Interiors, v. 44, p. 293–303, doi: 10.1016/0031-9201(86)90057-9.

Muscat, J., Swamy, V., and Harrison, N.M., 2002, First principles calculations on the phase stability of TiO_2: Physical Review B: Condensed Matter and Materials Physics, v. 65, p. 224112.

Nakamuta, Y., and Aoki, Y., 2000, Mineralogical evidence for the origin of diamond in ureilites: Meteoritics & Planetary Science, v. 35, p. 487–493.

Nishiyama, N., and Yagi, T., 2003, Phase relation and mineral chemistry in pyrolite to 2200°C under the lower mantle pressures and implications for dynamics of mantle plumes: Journal of Geophysical Research, v. 108, p. 2255–2266, doi: 10.1029/2002JB002216.

Nittler, L.R., 2003, Presolar stardust in meteorites: Recent advances and scientific frontiers: Earth and Planetary Science Letters, v. 209, p. 259–273, doi: 10.1016/S0012-821X(02)01153-6.

Ohtani, E., Kimura, Y., Kimura, M., Takata, T., Kondo, T., and Kubo, T., 2004, Formation of high-pressure minerals in shocked L6 chondrite Yamato 791384: Constraints on shock conditions and parent body size: Earth and Planetary Science Letters, v. 227, p. 505–515, doi: 10.1016/j.epsl.2004.08.018.

Olsen, J.S., Gerward, L., and Jiang, J.Z., 1999, On the rutile/α-PbO_2-type phase boundary of TiO_2: Journal of Physics and Chemistry of Solids, v. 60, p. 229–233, doi: 10.1016/S0022-3697(98)00274-1.

Ono, S., Ito, E., and Katsura, T., 2001, Mineralogy of subducted basaltic crust (MORB) from 25 to 37 GPa, and chemical heterogeneity of the lower mantle: Earth and Planetary Science Letters, v. 190, p. 57–63, doi: 10.1016/S0012-821X(01)00375-2.

Price, G.D., 1983, The nature and significance of stacking faults in wadsleyite, natural beta-$(Mg,Fe)_2SiO_4$ from the Peace River meteorite: Physics of the Earth and Planetary Interiors, v. 33, p. 137–147, doi: 10.1016/0031-9201(83)90146-2.

Putnis, A., and Price, G.D., 1979, High-pressure $(Mg,Fe)_2SiO_4$ phases in the Tenham chondritic meteorite: Nature, v. 280, p. 217–218, doi: 10.1038/280217a0.

Reid, A.F., and Ringwood, A.E., 1969, Newly observed high pressure transformations in Mn_3O_4, $CaAl_2O_4$, and $ZrSiO_4$: Physics of the Earth and Planetary Interiors, v. 6, p. 205–208.

Ringwood, A.E., 1996, Phase transformations in the Earth's mantle: Physics of the Earth and Planetary Interiors, v. 96, p. 79–84, doi: 10.1016/S0031-9201(96)90021-7.

Ringwood, A.E., Reid, A.F., and Wadsley, A.D., 1967, High-pressure $KAlSi_3O_8$, an aluminosilicate with sixfold coordination: Acta Crystallographica, v. 23, p. 1093–1095, doi: 10.1107/S0365110X6700430X.

Rost, R., Dolgov, Y.A., and Vishnevsky, S.A., 1978, Gases in inclusions of impact glass in the Ries Crater, West Germany, and finds of high-pressure carbon polymorphs: Dokladi Akademii Nauk Sssr, v. 241, p. 695–698.

Rubie, D.C., and Ross, C.R., 1994, Kinetics of the olivine-spinel transformation in subducting lithosphere: Experimental constraints and implications for deep slab processes: Physics of the Earth and Planetary Interiors, v. 86, p. 223–241, doi: 10.1016/0031-9201(94)05070-8.

Sharp, T.G., and DeCarli, P.S., 2006, Shock effects in meteorites, in Lauretta, D.S., and McSween, H.Y., eds., Meteorites and the Early Solar System: Tucson, The University of Arizona Press, p. 653–678.

Sharp, T.G., Lingemann, C.M., Dupas, C., and Stöffler, D., 1997, Natural occurrences of $MgSiO_3$-ilmenite and evidence for $MgSiO_3$-perovskite in a shocked L chondrite: Science, v. 277, p. 352–355, doi: 10.1126/science.277.5324.352.

Sharp, T.G., El Goresy, A., Wopenka, B., and Chen, M., 1999, A post-stishovite polymorph in the meteorite Shergotty: Science, v. 284, p. 1511–1513, doi: 10.1126/science.284.5419.1511.

Shim, S.H., Duffy, T.S., and Shen, G., 2001, Stability and structure of $MgSiO_3$ perovskite to 2300-kilometer depth in Earth's mantle: Science, v. 293, p. 2437–2440, doi: 10.1126/science.1061235.

Stöffler, D., Bischoff, A., Buchwald, V., and Rubin, A.E., 1988, Shock effects in meteorites, in Kerridge, J.F., and Matthews, M.S., eds., Meteorites and the Early Solar System: Tucson, University of Arizona Press, p. 165–202.

Stöffler, D., Keil, K., and Scott, E.R.D., 1991, Shock metamorphism of ordinary chondrites: Geochimica et Cosmochimica Acta, v. 55, p. 3845–3867, doi: 10.1016/0016-7037(91)90078-J.

Sueda, Y., Irifune, T., Nishiyama, N., Rapp, R.P., Ferroir, T., Onozawa, T., Yagi, T., Merkel, S., Miyajima, N., and Funakoshi, K., 2004, A new high-pressure form of $KAlSi_3O_8$ under lower mantle conditions: Geophysical Research Letters, v. 31, p. L23612, doi: 10.1029/2004GL021156.

Takahashi, E., and Ito, E., 1987, Mineralogy of mantle peridotite along a model geotherm up to 700 km depth, in Manghnani, M.H., and Syono, Y., eds., High-Pressure Research in Mineral Physics: Tokyo, Terra Scientific Publishing Company (Terrapub), p. 427–437.

Tange, Y., and Takahashi, E., 2004, Stability of the high-pressure polymorph of zircon ($ZrSiO_4$) in the deep mantle: Physics of the Earth and Planetary Interiors, v. 143–144, p. 223–229, doi: 10.1016/j.pepi.2003.10.009.

Teter, D.M., Hemley, R.J., Kresse, G., and Hafner, J., 1998, High pressure polymorphism in silica: Physical Review Letters, v. 80, p. 2145–2148, doi: 10.1103/PhysRevLett.80.2145.

Tomioka, N., and Fujino, K., 1997, Natural $(Mg,Fe)SiO_3$-ilmenite and perovskite in the Tenham meteorite: Science, v. 277, p. 1084–1086, doi: 10.1126/science.277.5329.1084

Tomioka, N., and Fujino, K., 1999, Akimotoite, $(Mg,Fe)SiO_3$, a new silicate mineral of the ilmenite group in the Tenham chondrite: The American Mineralogist, v. 84, p. 267–271.

Tomioka, N., and Kimura, M., 2003, The breakdown of diopside to Ca-rich majorite and glass in a shocked H chondrite: Earth and Planetary Science Letters, v. 208, p. 271–278.

Tomioka, N., Mori, H., and Fujino, K., 2000, Shock-induced transition of $NaAlSi_3O_8$ feldspar into hollandite structure in a L6 chondrite: Geophysical Research Letters, v. 24, p. 3997–4000.

Treiman, A.H., Gleason, J.D., and Bogard, D.D., 2000, The SNC meteorites are from Mars: Planetary Space Sciences, v. 48, p. 1213–1230, doi: 10.1016/S0032-0633(00)00105-7.

Tutti, F., Dubrovinsky, L., and Saxena, S.K., 2000, High pressure phase transformation of jadeite and stability of $NaAlSiO_4$ with calcium-ferrite structure in the lower mantle conditions: Geophysical Research Letters, v. 27, p. 2025–2028, doi: 10.1029/2000GL008496.

Tutti, F., Dubrovinsky, L., and Saxena, S.K., 2001, Stability of $KAlSi_3O_8$ hollandite-type structure in the Earth's lower mantle conditions: Geophysical Research Letters, v. 28, p. 2735, doi: 10.1029/2000GL012786.

Voegelé, V., Cordier, P., Langenhorst, F., and Heinemann, S., 2000, Dislocations in meteoritic and synthetic majorite garnets: European Journal of Mineralogy, v. 12, p. 695–702.

Wang, W., and Takahashi, E., 1999, Subsolidus and melting experiment of a K-rich basaltic composition to 27 GPa: Implication for the behaviour of potassium in the mantle: The American Mineralogist, v. 84, p. 357–361.

Wang, Y., Martinez, I., Guyot, F., and Liebermann, R.C., 1997, The breakdown of olivine to perovskite and magnesiowüstite: Science, v. 275, p. 510–513, doi: 10.1126/science.275.5299.510.

Withers, A.C., Essene, E., and Zhang, Y., 2003, Rutile/TiO_2 II phase equilibria: Contributions to Mineralogy and Petrology, v. 145, p. 199–204.

Xie, X., and Sharp, T.G., 2002, Fayalite-spinel + stishovite in shocked Umbarger L6 chondrite: Lunar and Planetary Science, v. XXXIII, p. 1859.

Xie, X., Minitti, M.E., Chen, M., Mao, H.K., Wang, D., Shu, J., and Fei, Y., 2002, Natural high-pressure polymorph of merrillite in the shock veins of the Suizhou meteorite: Geochimica et Cosmochimica Acta, v. 66, p. 2439–2444, doi: 10.1016/S0016-7037(02)00833-5.

Xie, Z., and Sharp, T.G., 2003, TEM observations of amorphized silicate-perovskite, akimotoite and Ca-rich majorite in a shock-induced melt vein in the Tenham L6 chondrite: Lunar and Planetary Science, v. XXXIV, p. 1469.

Xie, Z., and Sharp, T.G., 2004, High-pressure phases in shock-induced melt veins of the Umbarger L6 chondrite: Constraints of shock pressure: Meteoritics & Planetary Science, v. 39, p. 2043–2054.

Xie, Z., Sharp, T.G., and DeCarli, P.S., 2006, High-pressure phases in a shock-induced melt vein of the Tenham meteorite: Constraints on shock pressure and duration: Geochimica et Cosmochimica Acta, v. 70, p. 504–515.

Yagi, A., Suzuki, T., and Akaogi, M., 1994, High pressure transitions in the system $KAlSi_3O_8$-$NaAlSi_3O_8$: Physics and Chemistry of Minerals, v. 21, p. 12–17, doi: 10.1007/BF00205210.

Yagi, T., Okabea, K., Nishiyama, N., Kubo, A., and Kikegawa, T., 2004, Complicated effects of aluminum on the compressibility of silicate perovskite: Physics of the Earth and Planetary Interiors, v. 143–144, p. 81–91, doi: 10.1016/j.pepi.2003.07.020.

Zanda, B., 2004, Chondrules: Earth and Planetary Science Letters, v. 224, p. 1–17, doi: 10.1016/j.epsl.2004.05.005.

Zhang, J.Z., and Herzberg, C., 1994, Melting experiments on anhydrous peridotite KLB-1 from 5.0 to 22.5 GPa: Journal of Geophysical Research, v. 99, p. 17,729–17,742, doi: 10.1029/94JB01406.

Manuscript Accepted by the Society 14 August 2006

High-pressure mineralogy of diamond genesis

Yuriy A. Litvin[†]

Institute of Experimental Mineralogy, Russian Academy of Sciences, Chernogolovka, Moscow District, 142432, Russia

ABSTRACT

Diamond genesis can be clarified by estimating the common parental media for diamond and its syngenetic inclusions. Formation of diamond and diamondite in carbonatitic melts with garnets, clinopyroxenes, carbonates, iron-chromium alloys, and other minerals was confirmed in experiments using diamond-bearing Kokchetav (Kazakhstan) and Chagatai (Uzbekistan) carbonatitic rocks as the starting materials. Experiments on melting equilibrium of an eclogitic garnet-pyrrhotite join at 7 GPa revealed the existence of a nearly complete silicate-sulfide liquid immiscibility. Very low solubility of silicate components in the sulfide melt implies that the melt is not so efficient for syngenesis of diamonds and silicate inclusions, whereas carbonatitic (carbonate-silicate) parental melts can provide syngenesis of diamond and their primary inclusions more viably. The major components of the parental media for diamond syngenesis are carbonates and silicates, and the minor components are oxides, sulfides, phosphates, haloids, carbon dioxide, water, etc. These media are partially or completely molten during diamond formation, and they have compositionally variable major and minor component contents. It is obvious that the parental media for diamond is closely related to the genesis of carbonatitic magmas in the Earth's mantle.

Keywords: mantle-derived diamonds, parental media, high-pressure experiment, diamond genesis.

INTRODUCTION

Natural mantle-derived diamond is mainly associated with peridotite, pyroxenite, eclogite, lamproite, and kimberlite rocks (Sobolev, 1977; Dawson, 1980; Jaques et al., 1984; Mitchell, 1986; Harris, 1992). However, there is considerable mineralogical and geochemical evidence that these rocks and their melts do not serve as a parental medium for diamond formation. The problems connected with the chemical compositions, phase relations, and origin of the diamond-forming parental media, and the physicochemical mechanisms of nucleation and growth of diamond are poorly understood, in spite of their importance for the genesis of mantle-derived diamond.

The most important data about the chemical properties of parental media for mantle-derived diamond come from syngenetic inclusions of strongly compressed minerals, rocks, melts, and fluids in diamond (see Sobolev, 1977; Meyer, 1987; Navon, 1999; Taylor and Anand, 2004, for review). The materials that form the inclusions are trapped and packed inside diamond single crystals when they grow at mantle conditions. The inclusions retained in diamond are then transported to Earth's crust. The current speculations on the natural growth media for dia-

---

[†]E-mail: litvin@iem.ac.ru.

mond are based on information about these inclusions. Parental media for natural diamond proposed to date include silicate (Mitchell and Crocket, 1971), sulfide (Bulanova et al., 1998), CO_2 and CO_2-H_2O fluid media (Haggerty, 1986; Sobolev and Shatsky, 1990; Schrauder and Navon, 1994). However, these appear to be controversial and do not lead to any solution to the problem.

Experiments at high pressure and temperature provide useful constraints on diamond genesis. Diamond can be synthesized from the carbon-dissolved melts of carbonate, alkaline silicate, carbonate-silicate, and sulfide and CO_2-H_2O fluids (Akaishi et al., 1990a, 1990b, 2000; Arima et al., 1993; Akaishi, 1996; Arima, 1996; Taniguchi et al., 1996; Litvin et al., 1997, 2002; Borzdov et al., 1999; Akaishi and Yamaoka, 2000; Litvin and Zharikov, 2000; Sokol et al., 2001a, 2001b; Pal'yanov et al., 2002a, 2002b). By combining experimental evidence and mineral chemistry data of natural diamond, we can obtain a clue to the physicochemical conditions of diamond formation. The origin of diamond and diamond-bearing rocks is an important part of the chemical evolution of deep-seated materials in the mantle.

This study applied a new approach to the problem of the chemical composition of diamond-forming parental media. This approach focused on the physicochemical potentials for syngenesis of mantle-derived diamond and primary inclusions of minerals, rocks, melts, and fluid. It is apparent that both data on experiments and mineral chemistry should agree with each other.

This paper presents (1) a brief review of the mineral chemistry data available on syngenetic inclusions in natural mantle-derived diamond; (2) a review of the high-pressure experimental data concerning diamond formation; (3) high-pressure experimental data on nucleation and growth of diamond crystals in the multicomponent carbonate-silicate-carbon, carbonatite-carbon, sulfide-carbon, and chloride-carbon systems; (4) high-pressure experimental data on syngenesis of diamond and inclusions in the melts of natural carbonatite rocks (from Kokchetav and Chagatai massifs); (5) high-pressure experimental data on melting equilibrium in the silicate-sulfide system; (6) the physicochemical mechanisms of nucleation and growth of diamond in the parental media with natural chemistry; and (7) realistic models of diamond genesis.

PRELIMINARY STUDIES

Review of Natural Data for Syngenetic Inclusions in Diamonds

Mantle-derived diamond has been grown from the parental media with various compositions. The mineral chemistry data indicate that a wide variety of minerals, melts, and fluid has been captured by natural diamond as primary or syngenetic inclusions. High pressure is retained inside the inclusions after cooling and transportation of the host diamond from the Earth's mantle to the crust (Navon, 1991; Fursenko et al., 2001). Syngenetic inclusions in natural diamond are the primary sources of information on the possible diamond-forming parental media. The media, which were in intimate contact with diamond during its nucleation and growth, have large variations in their phase proportions and chemical compositions. Thus, the compositional variation of primary inclusions and parental media are the characteristic properties of the respective systems, which can be evaluated by both the data from mineral chemistry and high-pressure experiments.

Minerals present in inclusions indicate wide compositional variations with respect to the silicate-oxide constituents. Common mineral assemblages in inclusions (Sobolev, 1977; Meyer, 1987) are peridotitic (olivine, orthopyroxene, clinopyroxene, garnet, spinel, Mg-Fe-ilmenite, phlogopite, magnetite, zircon) and eclogitic (omphacite, garnet, chromite, ilmenite, sanidine, corundum, kyanite, coesite, quartz, rutile, phlogopite, magnetite, zircon, rare glasses). Minerals from the transition zone and the lower mantle (Fe-periclase, Mg-Fe-perovskite, Ca-perovskite, majorite, almandine-pyrope solid solutions) are rarely found as primary inclusions in diamond (Harte and Harris, 1994; Kaminsky et al., 2001, Stachel et al., 2005). Diamond inclusions also contain abundant sulfides, mainly pyrrhotite, pentlandite, chalcopyrite, pyrite, and Cu-Fe-Ni-monosulfide solid solutions (Efimova et al., 1983; Bulanova et al., 1998). Carbonate minerals, calcite, magnesite and dolomite are rare (Wang et al., 1996; Stachel et al., 1998). Metals (Fe and Fe alloys with Cr and Ni) and carbide (SiC) are also rarely included in diamond (Sobolev et al., 1981; Moore et al., 1986; Leung et al., 1990, 1996; Mathez et al., 1995). A large residual compression at room temperature may be retained in some inclusions. For instance, pressure in coesite was 5.5 ± 0.5 GPa at the capture temperature, but it was found to be $\sim 3.62 \pm 0.18$ GPa at room temperature (Fursenko et al., 2001). This demonstrates that the inclusions are captured from their parental media during diamond growth. The residual pressure depends on the thermo-elastic properties of diamond and included phases, and it may be lower than the capture pressure or not retained at all.

The fluid-bearing multiphase primary inclusions with various compositions in fibrous and cloudy diamond (Schrauder and Navon, 1994; Izraeli et al., 1998, 2001; Logvinova et al., 2003; Navon et al., 2003; Klein-BenDavid et al., 2003) comprise the major phases of the diamond-forming parental media. Spectroscopic and microprobe investigations of inclusions have revealed that they are composed of solid phases of oxide, silicate, carbonate, phosphate, sulfide, and chloride minerals, and occasionally carbonate-silicate (carbonatite) quenched melts, together with fluids such as CO_2, CH_4, H_2O, and K-Na-chloride-water brines. The compositions of fluid-bearing Botswanian inclusions are primarily approximated by two end members (Table 1), which are carbonatitic and silicic (Schrauder and Navon, 1994). The average compositions have been calculated for the alkali-chloride brine-bearing Koffifontein inclusions (Izraeli et al., 2001). The fluid-bearing inclusions are

TABLE 1. COMPOSITIONS OF FLUID-BEARING INCLUSIONS IN DIAMOND AND THEIR MODEL COMPOSITIONS USED FOR THE STARTING MATERIALS IN THE EXPERIMENTS

Component	JWN108*	JWN90	JWN91	End members*		Average[†]
				Carbonatitic	Silicic	Alkali-chloride
1. Natural fluid-bearing inclusions in diamond (wt%)						
SiO_2	23.9	25.1	45.1	13.6	58.4	6.1
TiO_2	4.6	4.7	4.9	4.6	4.3	1.4
Al_2O_3	2.5	2.5	5.4	0.8	6.8	0.6
FeO	15.2	16.1	10.7	19.5	7.6	16.6
MgO	10.9	10.1	5.7	13.2	3.3	1.3
CaO	13.5	15.6	5.1	20.5	0.0	6.7
Na_2O	2.2	2.6	1.6	2.2	2.2	3.9
K_2O	22.2	18.3	16.4	20.7	12.3	31.1
P_2O_5	2.2	2.4	1.0	2.4	0.7	1.4
Cl	1.5	1.3	0.8	1.3	0.9	36.9
Total[§]	98.8	98.7	96.7	98.8	96.5	106.0
CO_2/H_2O (m.r.)[#]				9/1	1/9	1/12
2. Starting materials used for the experiments**						
SiO_2	16.7	17.5	36.7	8.8		
TiO_2	3.2	3.3	4.0	3.0		
Al_2O_3	1.7	1.7	4.4	0.5		
FeO[††]	10.6	11.2	8.8	12.7		
MgO	7.6	7.0	4.6	8.6		
CaO	9.4	10.9	4.1	13.3		
Na_2O	1.0	1.0	1.1	0.8		
K_2O	17.7	15.0	14.4	15.6		
P_2O_5[§§]	1.6	1.7	0.8	1.6		
Cl[##]	0.5	0.9	0.6	0.9		
CO_2	30.0	29.8	20.5	34.2		

*Sample numbers and compositions (Schrauder and Navon, 1994).
[†]Composition data from Izraeli et al. (2001).
[§]Without CO_2 and H_2O.
[#]CO_2/H_2O molar ratio.
**CO_2 is added as carbonates such as $FeCO_3$, $MgCO_3$, $CaCO_3$, Na_2CO_3, and K_2CO_3.
[††]Synthetic $FeCO_3$ or natural siderite containing 0.7% $MgCO_3$ and 4.5% $MnCO_3$ by weight were used.
[§§]P_2O_5 was added as $K_4P_2O_7$. K_2O content in carbonate was correspondingly corrected.
[##]Cl was added as NaCl. Na_2O content in carbonate was corrected.

occasionally observed in association with primary inclusions of solid peridotitic and eclogitic minerals (Izraeli et al., 1998; Logvinova et al., 2003). The Fe-Ni-sulfide melts are observed as inclusions in diamond together with quenched carbonatitic melts and peridotitic minerals (Klein-BenDavid et al., 2003). The fluid-bearing inclusions are captured primarily during diamond formation as strongly compressed carbonate-silicate (carbonatite) melts at high pressures and temperatures. The solid and fluid phases or quenched carbonatitic materials may form from the carbonatitic melts during cooling. The capture pressure is ~5–7 GPa at 1000–1400 °C for the carbonatitic melts, and it decreases to 1.5–2.1 GPa at ambient conditions (Navon, 1991).

We emphasize that mineralogists have provided basic information on chemical and phase compositions of natural diamond-forming parental media. Nevertheless, the mineral chemistry data are not sufficient to obtain a direct conclusion on the nature of the parental media for natural mantle-derived diamond. The main reason is that the diamond-forming efficiency of these media cannot be evaluated by mineralogical data alone. An unambiguous estimation of the efficiency of natural media for nucleation and growth of diamond can be made using high-pressure and high-temperature experiments.

Brief Review of Experimental Data for Diamond Formation in Parental Media of Natural Chemical Compositions

The first high-pressure experiments on diamond synthesis (Bundy et al., 1955; Liander, 1955) were based on thermodynamics of carbon-bearing systems and the assumption that metal-carbon systems are preferable for diamond formation (Leipunsky, 1939). The first synthesis was successful by using metallic melts that were effective as solvents for graphite. The solvent model for diamond synthesis was proposed later (Litvin, 1968, 1969). The mechanism of diamond formation in the metal-carbon systems was extended to explain the origin of natural diamond (Wentorf and Bovenkerk, 1961).

Diamond synthesis in the carbonate-carbon systems has been successfully carried out using $CuCO_3 \cdot Cu(OH)_2$ (at 8–9 GPa, 1800–2000 °C), calcite $CaCO_3$ (at 8.5–9.5 GPa, 1900–2100 °C), and carbonates of Tl, In, Ga, Ba, Be, Li, Na, and K (Shul'zhenko and Get'man, 1971, 1972). $CaCO_3$, Na_2CO_3, and K_2CO_3 are the carbonates of interest, since they are observed in nature. Diamond synthesis using $MgCO_3$ and $SrCO_3$ (at 7.7 GPa, 2000–2150 °C) together with carbonates of Li, Na, K, Cs, and Ca has been reported previously (Akaishi et al., 1990a; Akaishi, 1993; Pal'yanov et al., 1999b). The $MgCO_3$-$CaCO_3$ and dolomite $CaMg(CO_3)_2$ systems have also been shown to be efficient diamond-forming media at 7.7 GPa and at 7 GPa and 1700–1750 °C, respectively (Sato et al., 1999; Sokol et al., 2001a). Diamond has also been synthesized effectively at 7.0–11.0 GPa and 1400–1900 °C by double carbonates, such as $K_2Mg(CO_3)_2$ (Taniguchi et al., 1996; Litvin et al., 1997, 1999b), $K_2Ca(CO_3)_2$, $Na_2Mg(CO_3)_2$, $Na_2Ca(CO_3)_2$, $K_2Fe(CO_3)_2$, $Na_2Fe(CO_3)_2$, $K_2Ba(CO_3)_2$, $Na_2Ba(CO_3)_2$, $Li_2Mg(CO_3)_2$, and $Li_2Ca(CO_3)_2$ (Litvin et al., 1998, 1999a, 1999b), using graphite as a carbon source. It has also been shown that carbonates could be effective as a carbon source for diamond formation in strongly reducing conditions (Arima et al., 2002; Yamaoka et al., 2002a; Pal'yanov et al., 2002b, 2005).

Diamond has been successfully synthesized from graphite using fluid components such as hydroxides (Akaishi et al., 1990b) and pure H_2O at 7.7 GPa and 1900 °C (Hong et al., 1999; Sokol et al., 2001b). Diamond nucleation has also been reported in the H_2O-SiO_2 system (Akaishi, 1996; Shaji Kumar et al., 2000, 2001; Akaishi et al., 2001; Yamaoka et al., 2000, 2002b). A few diamond crystals have been nucleated in experiments with carbonates (Na_2CO_3 or K_2CO_3) mixed with oxalic acid dihydrate $(COOH)_2 \cdot 2H_2O$, as a source of the $H_2O + CO_2$ fluid, at 5.7 GPa and 1150–1420 °C (Pal'yanov et al., 1999a, 2002a). Efficient nucleation of diamond has also been demonstrated by the oxalic acid dihydrate at 7.7 GPa and 1400–2000 °C (Akaishi and Yamaoka, 2000).

Diamond formation in the carbonate-silicate systems with graphite was first found for the natural kimberlitic composition at 7.0–7.7 GPa and 1800–2000 °C (Arima et al., 1993). Several carbonate-oxide mixtures of $K_2CO_3 + SiO_2 \pm MgO \pm Al_2O_3$ have also been shown to be effective in runs with long duration for several hours at 7 GPa and 1700–1750 °C, whereas pure K-feldspar $KAlSi_3O_8$ has not resulted in a diamond-forming reaction with graphite (Borzdov et al., 1999). It has been previously reported that no diamond formed in pure SiO_2 with graphite at 7.7 GPa and 1800–2200 °C, but the pure Na_2SiO_3 melt was very efficient for diamond nucleation (Akaishi, 1996). Diamond nucleation was effective at 6.3 GPa and 1650 °C (with a run duration of 40 h) for the carbonate-rich compositions in the carbonate-oxide systems K_2CO_3-Mg_2SiO_4 and K_2CO_3-SiO_2 (Shatcky et al., 2002).

Multicomponent carbonate-carbon and carbonate-silicate-carbon assemblages similar to the primary fluid-bearing inclusions in diamond show high efficiency for diamond nucleation and growth (Litvin and Zharikov, 1999, 2000). The melts of the natural carbonatitic rocks associated with diamonds have also been found to be highly efficient for diamond nucleation (Litvin et al., 2001, 2003). These systems have shown high efficiency characterized by the formation of polycrystalline fine-grained diamond intergrowths (Litvin and Spivak, 2003) similar to those observed in natural diamondite rocks (Kurat and Dobosi, 2000).

The nucleation and growth of diamond have been observed in the sulfide-carbon (Litvin et al., 2002; Litvin and Butvina, 2004), KCl-C (Litvin, 2003), and KCl-K_2CO_3-C (Tomlinson et al., 2004) systems, in addition to all of the above carbonate-carbon, silicate-carbon, carbonate-silicate-carbon, and CO_2-H_2O fluid-carbon systems. This indicates that diamond can be formed efficiently in experiments with in a wide variety of chemically contrasting natural media closely associated with diamond. It means that diamond synthesis experiments can be used for evaluation of diamond-forming efficiency for several parental media. However, such experiments cannot provide an unambiguous conclusion on the problem of the chemical and phase compositions of the parent media for natural diamond. To achieve this goal, one has to reformulate this objective as a problem of syngenesis of diamonds and their primary growth inclusions. The approach as used in this work is to estimate the role of silicates and sulfides in diamond genesis and to substantiate the chemical properties of the natural parent media for diamond genesis.

EXPERIMENTAL METHODS

Starting Materials

Starting materials for the experiments were modeled by two sources: the compositions of fluid-bearing multiphase primary inclusions in diamond and the compositions of diamond-bearing carbonate-silicate rocks.

The averaged compositions of fluid-bearing multiphase inclusions in diamond from Botswana (Schrauder and Navon, 1994) were taken as model compositions for the starting materials for the experiments (Table 1). The model compositions reflected chemical variations of the inclusions, especially SiO_2, FeO, MgO, and CaO contents. Starting materials were prepared as intimate mixtures of chemical reagents: carbonates of K, Na, Ca, and Mg, synthetic Fe-carbonate or natural siderite $FeCO_3$ with 0.7 wt% of $MgCO_3$ and 4.5 wt% of $MnCO_3$, oxides (TiO_2 and Al_2O_3), amorphous SiO_2, chemical reagent $K_4P_2O_7$ and NaCl. The mixtures were stirred with high-purity polycrystalline graphite powder in 1:1 or 3:2 proportions by weight to obtain a large yield of diamond. Homogeneous mixtures of dried powders of spectrographically pure KCl and polycrystalline graphite were used as starting materials in the system KCl-C to investigate the role of alkaline chloride aqueous brines observed as inclusions in diamond formation.

Starting materials for the investigation of the role of sulfides in diamond formation were prepared from powders of natural pyrrhotite $Fe_{1-x}S$ of composition: S = 38.96 wt%; Fe = 60.74 wt%; Co = 0.06 wt%; As = 0.04 wt%; pentlandite $(Fe,Ni)_9S_8$ with ~2 wt% of Co; and chalcopyrite $CuFeS_2$ with negligible Ag and Au impurities. Pyrrhotite and pentlandite were mixed with chalcopyrite in equal proportion. Resultant compositions were as follows (wt%): (1) $Cu_{17}Fe_{35}S_{48}$ and (2) $Cu_{17}Fe_{25}Ni_{10}S_{48}$. Sulfide powders were mixed with 0.5-mm-size polycrystalline graphite grains in a ratio of 3:2 by weight.

Starting materials for the experiments with Kokchetav carbonate-silicate rock were prepared by mixing the powders of a natural rock sample and graphite in a weight ratio of 3:2. The initial rock contained up to 65–75 vol% of dolomite, which formed a matrix of garnet grains (~15–20 vol%), clinopyroxene, and amphibole (both 10–15 vol%). The minor phase (<5 vol%) consisted of mica, K-feldspar, clinozoisite, spinel, corundum, magnetite, rutile, zircon, and calcite. The compositions of the major minerals are given in Table 2. Compositions of Chagatai carbonatite rocks (Djuraev and Divaev, 1999) used as the starting materials are given in Table 3. The carbonatite sample powders were mixed with the polycrystalline graphite grains (0.5 mm in size) in the proportion of 1:1 by weight.

Starting materials used to investigate the sulfide-silicate melting relations were natural pyrrhotite $Fe_{1-x}S$, mentioned already, and Mg-Ca-Fe-garnet of eclogite xenolith from the Udachnaya pipe, Yakutia (the compositions are given in Table 2).

Experimental and Analytical Techniques

Experiments on diamond syntheses and melting-phase equilibrium were conducted under a pressure range of 4.5–8.5 GPa and a temperature range of 1100–2000 °C. The experimental conditions correlate well with pressure and temperature conditions of diamond stability and melting temperatures of materials used for diamond synthesis. We used an "anvil-with-hole" high-pressure apparatus (Litvin, 1991; Eremets, 1996) with a cell made of limestone from Algeti, Georgia. The standard cell assembly with a graphite heater has been described previously (Litvin et al., 1999b). Experiments on diamond synthesis were performed using a graphite tubular furnace that was 7.2 mm long, 5 mm in diameter, and had 0.5 or 1.0 mm wall thickness. The furnace was closed by 1-mm-thick graphite plugs at both ends. In the experiments for melting equilibrium in the sulfide-silicate system, the

TABLE 2. COMPOSITIONS OF MINERALS USED FOR STARTING MATERIALS AND THOSE IN THE RUN PRODUCTS FOR THE EXPERIMENTS IN KOKCHETAV CARBONATITE AND UDACHNAYA PIPE ECLOGITE

Oxides	Dol*	Dol†	Grt*	Grt*	Grt†	Grt†	Cpx*	Cpx*	Cpx†	Cpx†	Grt§
SiO_2	N.D.	N.D.	40.98	40.56	44.64	44.46	54.41	54.55	54.99	53.80	39.16
TiO_2	N.D.	N.D.	0.20	0.18	0.45	N.D.	0.08	0.06	N.D.	N.D.	0.64
Al_2O_3	N.D	N.D.	23.10	23.08	19.97	22.03	1.75	1.89	7.12	6.56	21.01
FeO#	3.23	3.87	5.61	7.67	3.60	2.71	1.74	1.26	1.77	2.33	21.86
MnO	0.72	0.35	0.72	0.67	0.46	N.D.	N.D.	N.D.	N.D.	N.D.	0.44
MgO	20.49	16.42	10.49	10.63	16.69	19.85	17.31	16.66	14.99	15.36	8.55
CaO	28.90	32.24	18.90	17.18	14.20	10.83	23.71	24.19	20.58	21.95	8.50
Na_2O	N.D.	0.68	N.D.	0.03	N.D.	0.13	0.37	0.55	0.22	N.D.	0.09
K_2O	N.D.	N.D.	N.D.	N.D.	N.D.	N.D.	0.55	0.82	0.32	N.D.	N.D.
CO_2**	46.66	46.44	N.D.	N.D.	N.D.	N.D.	N.D.	N.D.	N.D.	N.D.	N.D.
Total	100.0	100.0	100.0	100.0	100.0	100.0	99.92	99.98	99.99	100.0	100.25

Notes: Abbreviations: Dol—dolomite, Gr—garnet, Cpx—clinopyroxene, N.D.—not determined.
*Natural minerals used for the starting materials.
†Minerals syngenetic with diamond observed in the run products.
§Garnet of eclogite paragenesis from Udachnaya pipe, Yakutia, was used as a starting material in the experiments in the garnet-pyrrhotite join.
#Total iron as FeO.
**Calculated value.

TABLE 3. COMPOSITIONS OF THE STARTING MATERIALS OF CHAGATAI AND KOKCHETAV CARBONATITE ROCKS

Oxides	Rock composition				
	Chagatai rocks			Kokchetav rock	
	No. 23	No. 86	No. 88	No. 89	No. 574-1
SiO_2	22.19	12.22	27.86	11.62	25.10
TiO_2	0.60	0.33	0.38	0.60	0.12
Al_2O_3	5.36	2.68	9.64	3.67	3.04
FeO	2.87	3.95	3.52	2.87	2.25
MnO	0.25	0.34	0.28	0.85	0.18
MgO	2.80	0.81	3.58	3.50	14.20
CaO	26.64	36.73	30.14	31.80	28.44
Na_2O	1.62	1.30	0.32	1.15	0.20
K_2O	1.32	0.70	0.90	0.54	0.22
P_2O_5	0.49	4.58	0.15	2.75	0.24
H_2O	0.26	0.20	0.34	0.08	0.25
CO_2	18.70	21.65	10.45	27.92	25.80
Total	98.08	98.52	98.72	98.61	100.04

Notes: Chemical analyses of Chagatai rocks were performed at the Institute of Mineral Resources, Tashkent. Kokchetav rock was analyzed in the Institute of Experimental Mineralogy, Chernogolovka.

homogeneous pyrrhotite and garnet mixtures were pressed into a sample room with 2.0 mm inner diameter and height in the resistance furnace made of a spectroscopic graphite rod of 6.0 mm diameter and 7.5 mm height. Pressure at room temperature was determined using a calibration curve, which was constructed on the basis of the standard reference polymorphic transitions in bismuth: Bi_{I-II} at 2.55, Bi_{II-III} at 2.7, and $Bi_{III-VII}$ at 7.7 GPa (Homan, 1975). Pressure values at high temperatures were corrected based on the diamond-graphite equilibrium curve (Kennedy and Kennedy, 1976). Temperature was measured using a $Pt_{70}Rh_{30}/Pt_{94}Rh_{06}$ thermocouple, 0.3 mm in diameter, and its junction was positioned close to the center of the furnace. The electrode wires and junction of the thermocouple were covered with a MgO ceramic tube and pressed powder layers for protection against contamination from graphite or sulfide. No correction for the effect of pressure on electromotive force of the thermocouple was made. Pressure and temperature in the runs were determined with an accuracy of ±0.1 GPa and ±20 °C, respectively.

Although the graphite heater was in direct contact with the starting material, only a small amount of carbon served as a source of carbon for diamond that crystallized near the wall of the heater. The parental medium efficient for diamond formation was defined by occurrence of spontaneous nucleation as a major criterion. Additionally, in order to check the experimental conditions, a cubic-octahedral metal-synthetic diamond crystal 0.5–0.8 mm in size, which was grown in the Ni-Mn-C melts, was placed in the central part of the packed starting mixture in some runs for diamond synthesis. This was to test for the effects of the growth or partial dissolution and recrystallization of diamond seeds into graphite flakes depending on pressure and temperature conditions. The seeds were useful to elucidate the mechanism of layer growth over different faces of the seeds from carbonate-carbon and other melts. The quenching rate of the experiments was faster than 300 °C/s.

The oxygen fugacity seemed to be buffered by the carbonate-carbon pairs and was close to that of the FeO-Fe buffer (the effects of graphite oxidation or carbonate reduction were not determined by analyzing the experimental products). Solid or molten (quenched) sulfides were also stable in the runs for diamond synthesis in the sulfide-graphite systems.

The experimental run products for diamond synthesis consisted of phases with contrasting mechanical properties, from soft carbonates to super-hard diamonds. Therefore, the scanning electron microscopy and microprobe studies were performed over gold- or carbon-coated surfaces of the broken samples. The silicate-sulfide samples were embedded in epoxy, polished with diamond micropowders, and coated with carbon. The samples were studied by scanning electron microscopy and microprobe analysis (electron microscope CamScan MV2300; VEGA TS 5130MM) with an electron energy-dispersive microprobe Link INCA Energy at the Institute of Experimental Mineralogy, and an electron microscope CamScan with a Link AN 10/85S energy-dispersive system at the Department of Petrology, Moscow State

University, and JEOL Superprobe 733 at University College and Birkbeck College, London.

RESULTS

Carbonatitic Melts Similar to Fluid-Bearing Multiphase Inclusions

The starting compositions, i.e., JWN108, JWN90, and JWN91, and the carbonatitic end member (Table 1), crystallized diamond at 5.5–8.5 GPa and 1200–1600 °C when mixed with graphite (Litvin and Zharikov, 1999, 2000). The graphite content should have been reasonable for producing carbon oversaturation with respect to diamond in the carbonatitic melt and should have maintained the oversaturation during the growth of diamond. The experimental results are summarized in Table 4. Spontaneous nucleation and crystallization of diamond proceeded over a short time interval just after the eutectic melting of a carbonate-silicate-graphite mixture. The present experiments showed that crystallization of diamond is completed within 1–2 min after nucleation. Runs with longer durations were carried out for the seed growth of diamond. The experimental data were applied to determine the pressure and temperature conditions of diamond formation (Fig. 1). The region of diamond crystallization is bounded by the diamond-graphite equilibrium curve (Kennedy and Kennedy, 1976) and the pressure and temperature position of the solidus curve for the carbonate-silicate-graphite system based on our experimental results. The boundary between the fields of spontaneous nucleation and seeded growth of diamond is kinetic and located within the region of diamond formation.

Two major phases of diamond and quenched carbonatite were observed as products of solidification from the melts (Fig. 2A). The white, soft, and loose carbonatite materials were solidified as rounded grains with uniform density. Graphite source was recrystallized into diamond completely in the carbonatitic solutions oversaturated with carbon. No indication of carbonate-silicate liquid immiscibility was observed. Diamond was observed as individual octahedral crystals of 200 μm in size (Fig. 2A), spinel-law (Fig. 2B), and polysynthetic twins (Fig. 2C) similar to the cyclic twin grown in kimberlitic melt (Arima, 1996). Densely intergrown polycrystalline diamond blocks, 1.0–1.5 mm in size, similar to natural diamondite (Kurat and Dobosi, 2000; Litvin and Spivak, 2003) were quickly crystallized within ~10 s (Fig. 2D). It has been estimated from our previous experiments (Spivak and Litvin, 2004) that diamond formation is characterized by dense nucleation, more than $3.0–5.0 \times 10^2$ nuclei/mm^3 for individual crystals and 1.0×10^5 nuclei/mm^3 for diamondite-like materials. Growth rate of spontaneous formation of diamond is strongly variable in the range from several μm/min to mm/min, whereas it was extremely high for diamondite. The diamond

TABLE 4. EXPERIMENTAL CONDITIONS AND RESULTS FOR MODEL CARBONATITE COMPOSITIONS

Starting compositions	Sample No.	Pressure (GPa)	Temperature (°C)	Run duration (min)	Spontaneous crystallization	Seed growth
JWN108	451	7.0	1550	8	+	+
	463	5.8	1420	16	+	+
	463a	5.6	1400	60	–	+
	456	6.0	1330	8	+	+
	475	5.5	1250	70	+	+
	495	5.6	1240	120	+	+
	495a	4.8	1240	80	–	–(dissolved)
	495b	5.5	1160	20	–(not melted)	–
JWN90	449a	7.0	1530	95	+	+
	454	6.0	1450	5	+	no seed
	452	7.0	1440	2	+	no seed
	461	5.7	1380	60	+	+
	461a	5.3	1380	80	–	–dissolved)
	469	5.5	1290	55	–	+
	496	5.2	1220	50	–	+
JWN91	450	7.0	1570	5	+	no seed
	462	5.7	1570	75	–	–(dissolved)
	455	6.0	1340	5	+	no seed
	474	5.5	1320	45	–	+
	492	5.5	1200	90	+	+
Carbonatitic end member	997*	8.0	1990	50	+(diamondite)	no seed
	362	7.5	1700	55	+	+
	363	7.0	1650	60	+	+
	453	6.2	1550	80	+	+
	448	6.5	1460	80	+	+
	460	5.9	1440	65	+	+
	468	5.5	1300	60	–	+
	497	5.8	1220	86	+	+
	1041	5.8	1190	40	–(not melted)	–
	1062	7.0	1300	40	+	+
	1064	7.0	1230	40	+	+
	1065	7.0	1160	40	–(not melted)	–

Note: + or – indicate the presence or absence of spontaneous crystallization or seeded growth.
*Experimental conditions for the formation of diamondite are not shown in Figure 1.

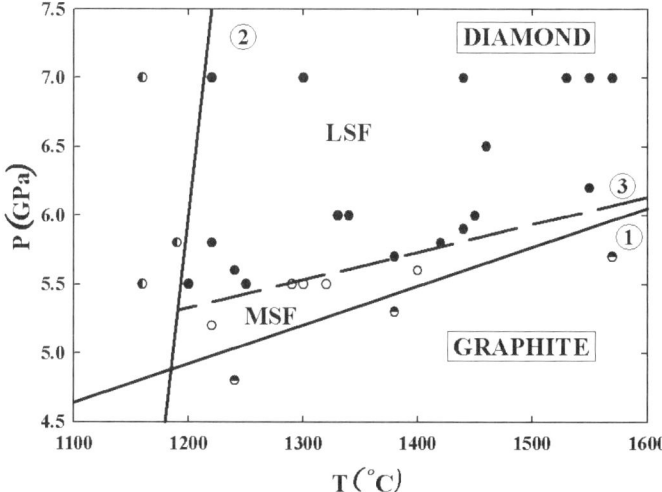

Figure 1. A pressure and temperature diagram of diamond crystallization in carbonate-silicate-carbon melts with compositions similar to fluid-bearing multiphase inclusions in natural diamond: (1) graphite-diamond equilibrium curve (Kennedy and Kennedy, 1976), (2) solidus curve of the carbonate-silicate-graphite starting mixtures, (3) pressure and temperature boundary between the fields of labile (LSF—labile solution field) and metastable (MSF—metastable solution field) solutions, respectively. Run conditions are shown as black circles for spontaneous nucleation within the LSF and as open circles for seeded growth within the MSF. Runs for the solid state starting mixtures are marked by left half–filled circles, and those of dissolution of diamond at conditions of graphite stability are marked by upper half–filled circles.

growth rate is sensitive to variation of pressure and temperature. High nucleation density and large growth rates of spontaneously formed diamond indicate that there is a strong carbon oversaturation in the carbonate-silicate melts. The criterion of the spontaneous nucleation was used as an indicator for the efficiency of parent medium in diamond formation. Diamond crystallization in compositionally simple carbonate-carbon melts composed of K-Mg–, K-Ca–, Na-Mg–, and Na-Ca–double carbonates (Litvin et al., 1997, 1998, 1999a) was found to be relatively less efficient.

The study of diamond growth on the cubic-octahedral seed crystals of the metal-synthetic diamond in the carbonatite-carbon melts revealed that there is a difference in growth of octahedral {111} and cubic {100} faces, similar to that observed in the simpler carbonate-carbon systems (Litvin et al., 1998, 1999b). The growing layers on the {111} faces have octahedral orientations and develop plane faces (Fig. 3A). However, the newly grown layers over cubic {100} faces are formed with octahedral microcrystals and develop rough faces (Fig. 3B). Thus, the newly grown diamond is precipitated on the surfaces parallel to the {111} faces, similar to that observed in natural diamond. Fibrous structures developed in diamond growth on the (100) face (Fig. 3C and 3D). This could result in formation of substantially large cavities, such as those that capture the fluid-bearing Botswanian-type inclusions during natural fibrous diamond growth (Navon et al., 1988; Navon, 1999).

Sulfide Fe-Ni-Cu-S Melts

Our experimental study at high pressures (6.0–8.5 GPa) revealed that diamond is efficiently crystallized from the melts of sulfide-graphite mixtures with different compositions (Litvin et al., 2002). Spontaneous crystallization of diamond from the sulfide melts produced octahedral diamond crystals with flat faces and occurred quickly within 1–2 min after reaching the eutectic melting of a sulfide-graphite mixture (Figs. 4A–4C). Newly grown diamond layers were also observed on the seeds of metal-synthetic diamond (Fig. 4D). Textural patterns and chemical data of the quenched experimental samples showed that diamond was synthesized from the sulfide-carbon melts. When sulfides are molten, the sulfide-carbon solutions that serve as a carbon solvent and a transfer medium for the dissolved carbon during diamond growth are formed in contact with graphite (Litvin et al., 2002). We observed no indications of sulfide breakdown and exsolution of free metal phases, which could form the metal-carbon solutions. This observation indicates that the metal melts are effective solvents in the processes of diamond synthesis (Litvin, 1968), and formation of the melt could cast some doubts on the synthesis of diamonds in the sulfide-carbon solutions. Additionally, only sulfide phases were identified as the quenched melts in intimate contact with newly formed diamond by the microprobe analysis. All of these observations demonstrate high efficiency of sulfide melts for diamond nucleation and growth under the conditions of diamond stability when they are oversaturated with dissolved carbon.

Alkali-Chloride Melts

We also carried out high-pressure experiments on spontaneous nucleation and seeded growth of diamond in the melts of a potassium chloride–graphite system. Diamond was nucleated in the potassium chloride–carbon solutions at 7.0–8.0 GPa and 1500–1700 °C (Litvin, 2003). New diamond layers were grown on single crystal seeds of diamond. The nucleation density of diamond was found to be higher at 8.0 GPa and lower at 7.0 GPa. It should be noted that the rate of spontaneous crystallization of diamond is very high. Single crystals of diamond up to 0.1 mm in size formed within the first 10 s from the onset of crystallization. In most cases, however, experimental conditions were maintained for 40–60 min to provide sufficient time for growth of diamond on the seeds. Examples of diamond crystals spontaneously formed in alkaline chloride (KCl) melts with dissolved carbon are presented in Figure 5A. The KCl melt solidified predominantly in the final cooling stage as fine-grained aggregates. Rhythmic basket-shaped solidified KCl aggregates (Fig. 5D) were also observed around spherules of graphite that formed as a metastable phase at pressure and temperature conditions of diamond stability (Litvin, 2003).

The seeded growth in KCl-C melts oversaturated with dissolved carbon shows dendritic layers both on (111) and (100) faces as shown in Figures 5B and 5C, respectively. The efficiency

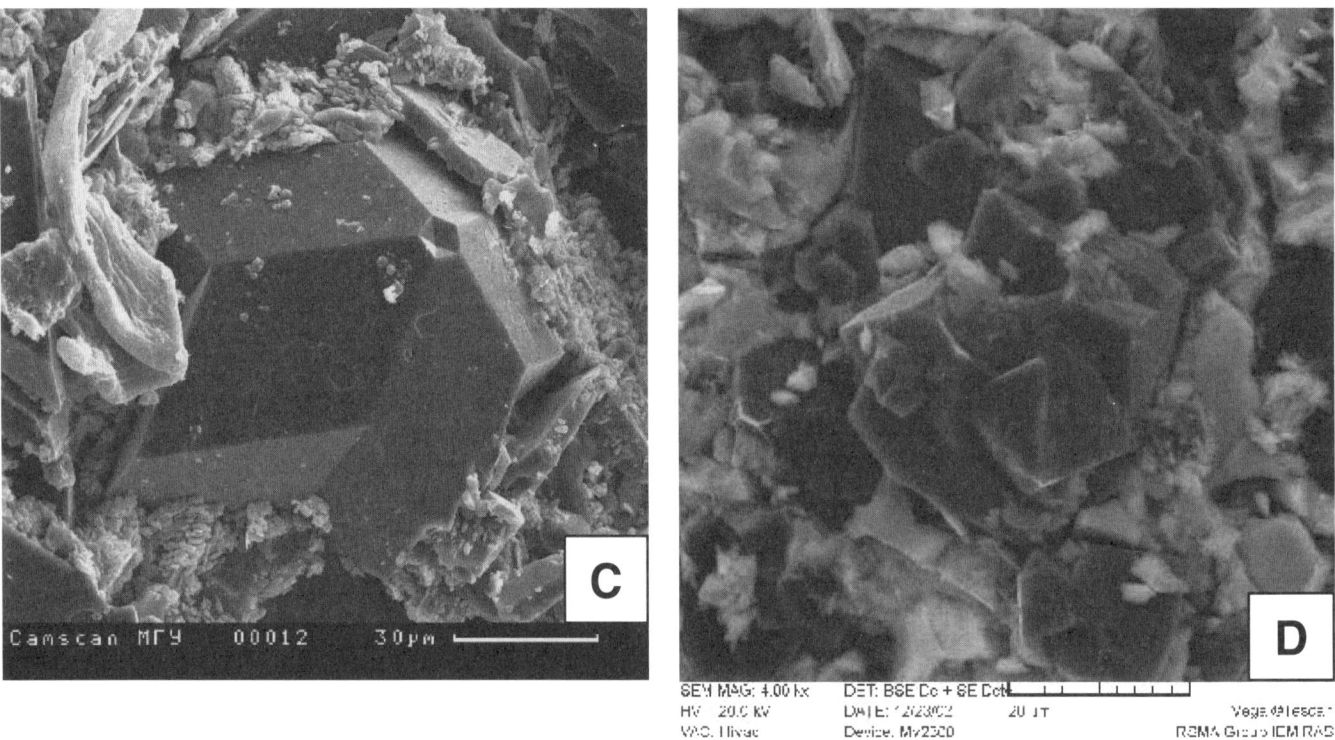

Figure 2. Scanning electron micrograph (SEM) images of diamond and diamondite formed in the multicomponent carbonatite melts: (A) sample 475, individual octahedral crystals; (B) sample 454, spinel-law twins; (C) sample 362, polysynthetic twins; and (D) sample 997, diamondite polycrystals.

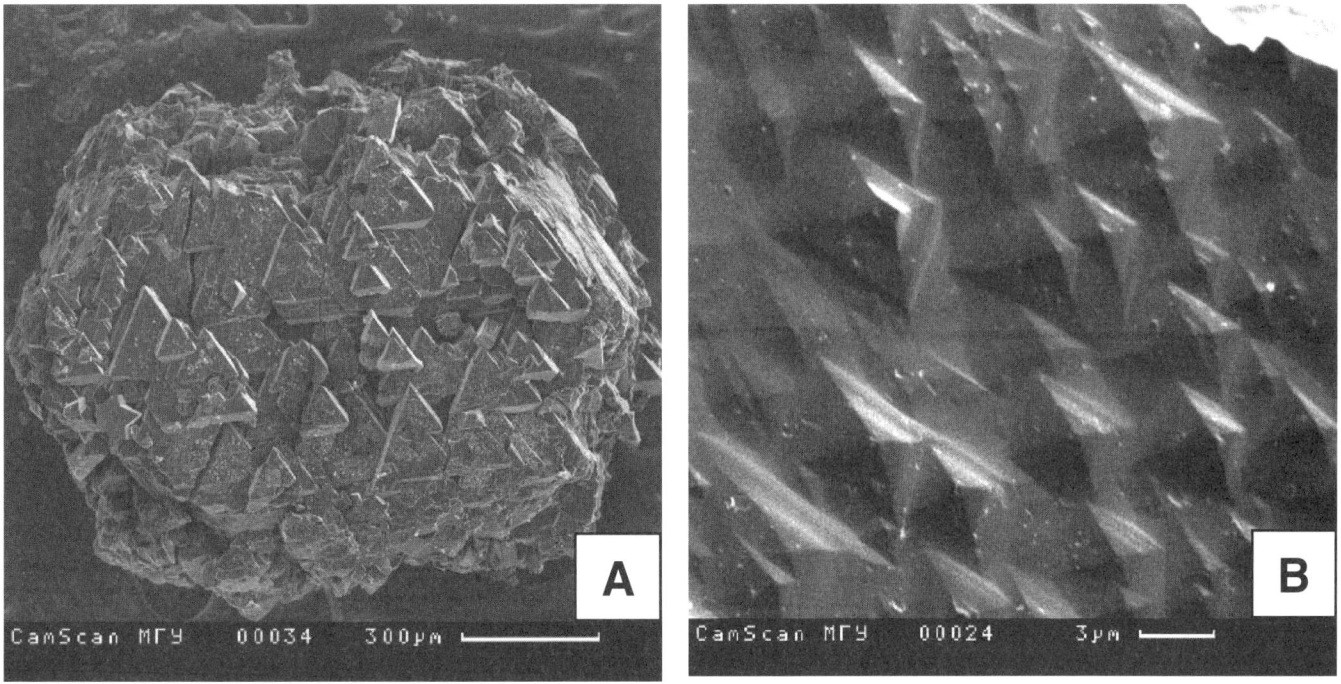
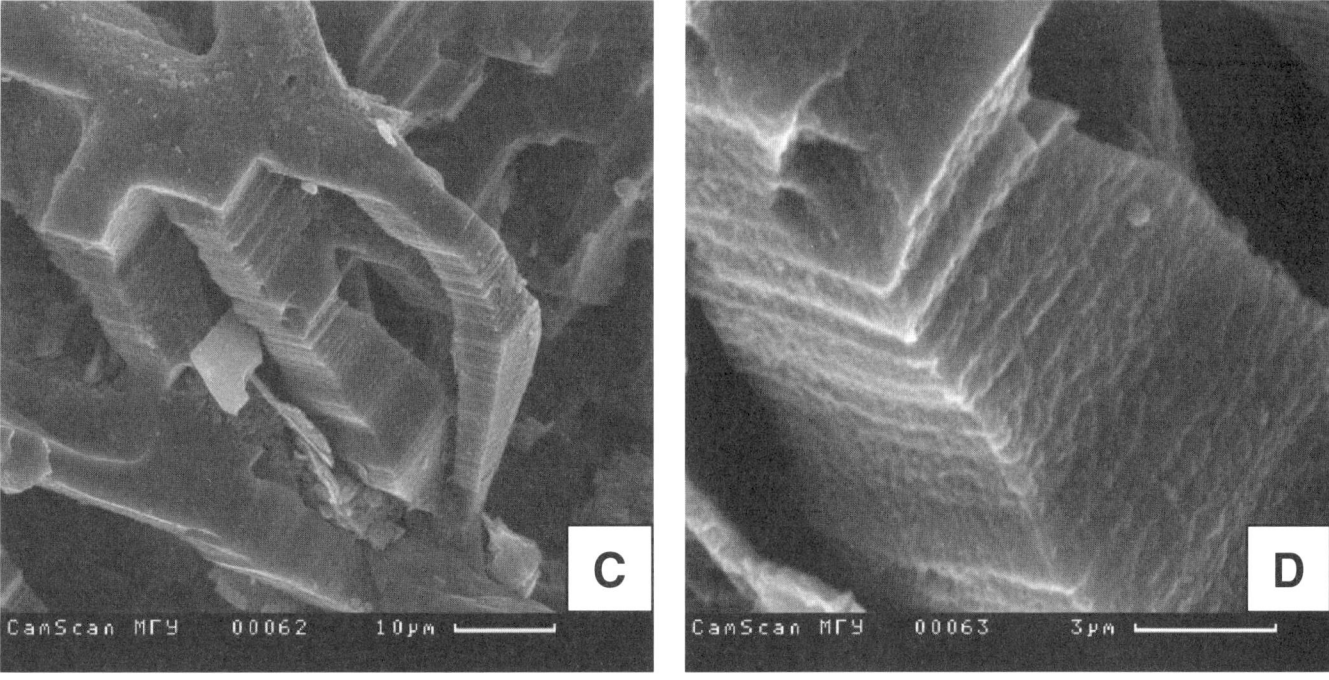

Figure 3. Scanning electron micrograph (SEM) images of diamond overgrowth formed on the seeded crystals: (A) sample 362, flat polycentric growth on octahedral face. The scale bar is 300 μm. (B) Sample 362, micropyramids on cubic face. The scale bar is 3 μm. (C–D) Sample 363, fibrous growth on the cubic face. The scale bars are 10 μm and 3 μm, respectively.

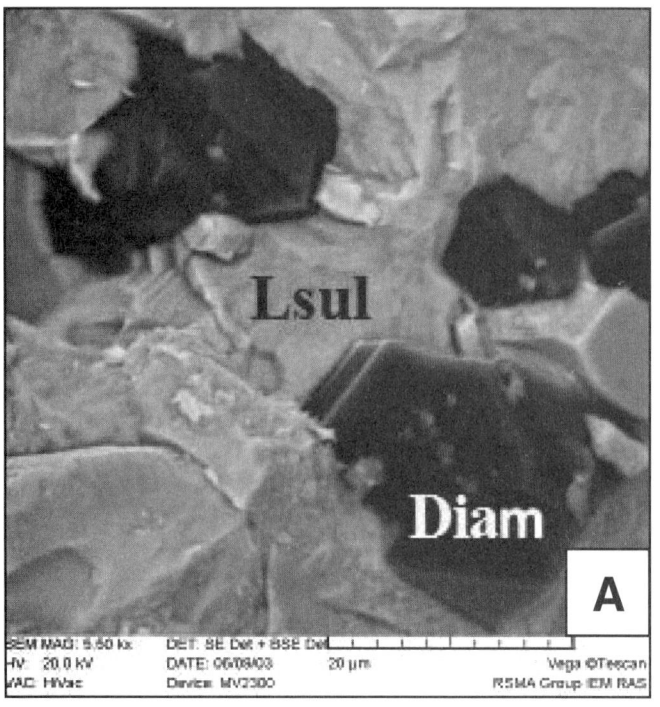

Figure 4. Scanning electron micrograph (SEM) images of diamond crystallized in the sulfide melts: (A) sample 895, octahedral and spinel-law crystals in quenched pyrrhotite melt; (B) sample 667, octahedral crystals in the quenched $Cu_{17}Fe_{35}S_{48}$ melt; (C) sample 697, intergrowth of octahedral and spinel-law crystals in quenched $Cu_{17}Fe_{35}S_{48}$ melt; and (D) sample 658, thin sulfide film covering cubic face of a seed crystal with newly grown micropyramids. Lsul—quenched sulfide melt; Diam—diamond; Sulf—sulfides. The scale bar is 20 μm for (A) and 10 μm for (C–D).

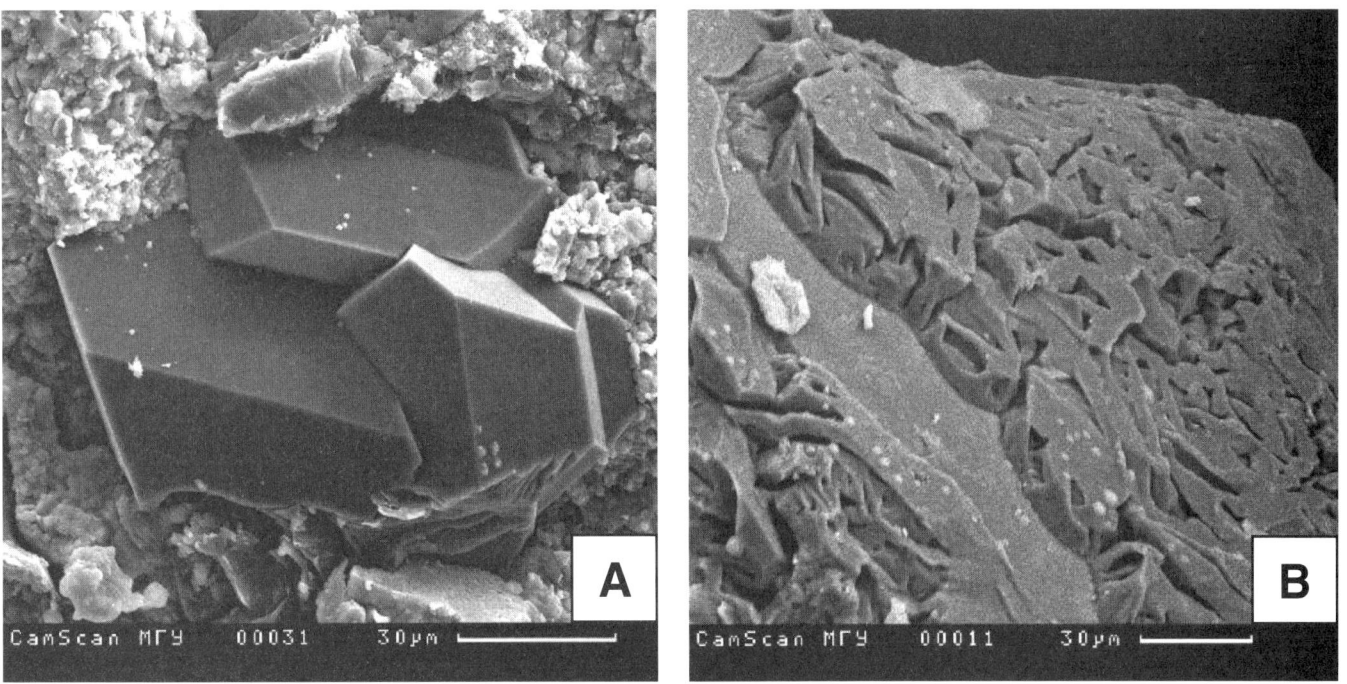
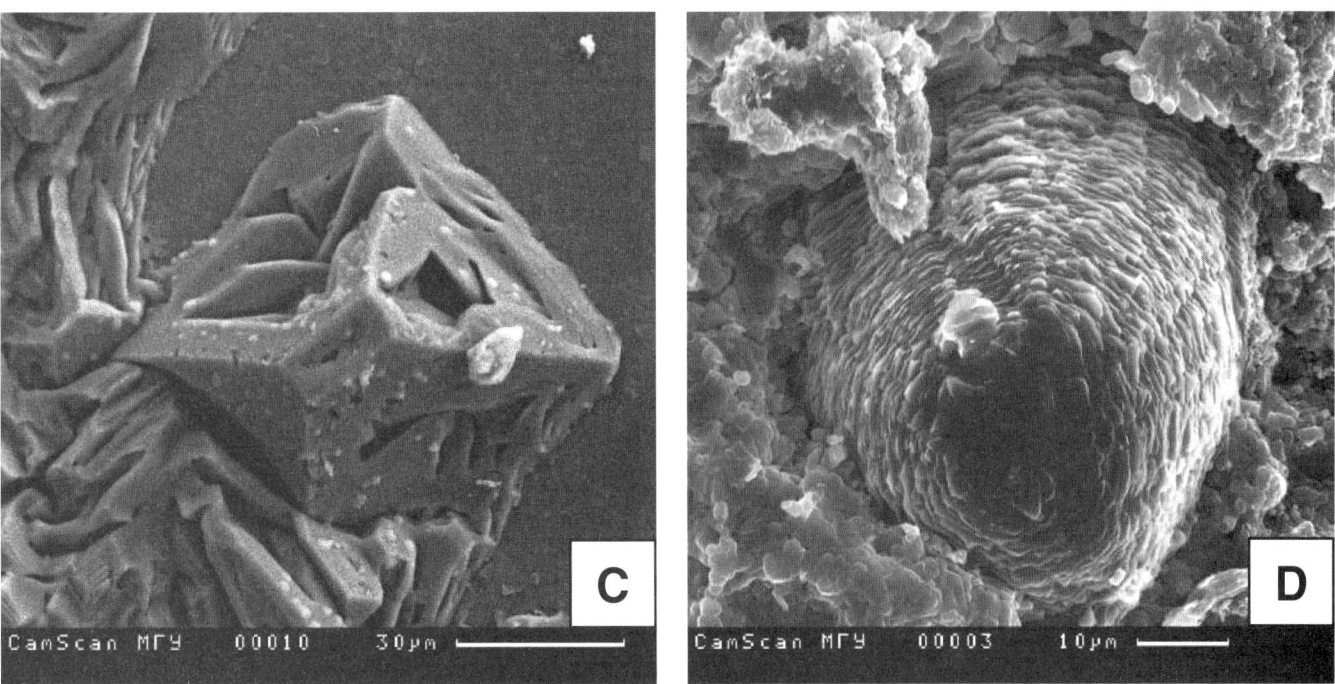

Figure 5. Scanning electron micrograph (SEM) images of diamonds crystallized in KCl melt: (A) sample 876, intergrowth of spinel-law twins in quenched KCl melt; (B–C) sample 877, growth patterns on octahedral and cubic faces of a seeded crystal, respectively; and (D) sample 877, graphite spherule within a basket-shaped shell of the solidified KCl melt.

of the seeded growth of diamond in the alkali chloride–carbon systems has been reported previously (Wang and Kanda, 1998). Microprobe analysis of solidified materials in the contact zone between spontaneously formed diamond single crystals and graphite spherules yields the stoichiometric composition of KCl. All of these observations indicate that both diamond and metastable graphite were crystallized from alkaline chloride KCl melts with dissolved carbon. Since growth of diamond occurs at high pressure in H_2O-C and H_2O-CO_2-C systems (Hong et al., 1999; Akaishi and Yamaoka, 2000; Sokol et al., 2001b), the KCl-H_2O C type melts are undoubtedly efficient in formation of carbon-oversaturated solutions with respect to diamond and spontaneous diamond nucleation. It should be remembered that the KCl-H_2O join is common as the multicomponent brine solutions of fluid-bearing primary inclusions in natural diamond (Izraeli et al., 2001).

Melts of Kokchetav Carbonate-Silicate Rocks

Diamond formation in the melts of the diamond-bearing carbonate-silicate rocks from Kokchetav massif was studied at 5.5–8.5 GPa and 1300–1800 °C with a run duration of 1–40 min (Litvin et al., 2003). Spontaneous nucleation and crystallization of diamond occurred in 1–2 min (Figs. 6A and 6C) after the solidus temperature was attained. Quenching of carbonate-silicate melts at high pressure produces fine-grained carbonate assemblage and fibrous dendritic quench crystals (Fig. 6D). It was found that starting materials of garnet and clinopyroxene were slightly dissolved in a dolomitic melt and involved in the formation of carbonate-silicate melts. Thus, the carbonate (dolomite)-silicate-carbon solutions were formed and served as a parental media not only for diamond, but also for garnet and clinopyroxene. Formation of garnet and clinopyroxene syngenetically with diamond was confirmed in the experiments (Fig. 6B). Garnet crystallized together with diamond from carbonate-silicate-carbon melts has higher pyrope and lower grossular and almandine components, the composition of which is quite different from the starting material. Syngenetic clinopyroxene has Al-rich and alkali-poor compositions (Table 2). These experiments show that the carbonate-silicate-carbon parental melt is capable of providing the syngenesis of diamond, garnet, and clinopyroxene, the latter two of which are abundant primary inclusions in natural diamond.

Melts of Chagatai Carbonatite Rocks

Spontaneous crystallization of diamonds occurred intensively during the first 1–2 min after melting the Chagatai carbonatite-graphite mixture (at 5.5–7.5 GPa and 1250–1550 °C) (Litvin et al., 2001). Diamond was crystallized as octahedral single crystals, spinel-law twins, and their intergrowths (Fig. 7A). The diamond-forming parental carbonatite multicomponent melts were quenched into complex multiphase materials composed of diopside-hedenbergite clinopyroxenes, grossular-almandine garnets, apatite, and calcite. Calcite was formed as the host phase for the other minerals (Fig. 7B). This indicates that these minerals were crystallized syngenetically with diamond from the $CaCO_3$-based carbonate-silicate melts, and calcite was formed in the last stage. Calcite often has an intergrowth with garnet, which seems to be crystallized during rapid solidification from the calcite melt as seen distinctly in Figure 7B. This indicates a high solubility of garnet in the carbonate melt with the calcite composition. It is interesting to note that coesite intergrowths were also observed in the solidified carbonate melt (Fig. 7A), i.e., coesite solubility in the carbonate melt was very high. It was also found that the carbonate-silicate melt of the Chagatai carbonatite is very efficient for the rapid formation of diamondite (Fig. 7C). The diamondite samples contained solidified fragments of carbonatite parental melt, including garnet, clinopyroxene, carbonate, sulfide, and other accessory minerals. Figure 7D shows a pure carbonate-synthetic diamondite. The diverse basic and accessory mineralogy of natural Chagatai carbonatites revealed in high-pressure experiments (Litvin et al., 2001; Bobrov et al., 2004) implies that the rocks are useful in modeling natural parent media for syngenetic formation of diamond and minerals observed as primary inclusions in diamond. Pyrrhotite, wüstite, moissanite, alkaline carbonate $(Na,K)_2Ca(CO_3)_2$, and native Fe-Cr-alloys with variable Cr contents were identified in the experiments as the accessory minerals. It is interesting to note that Fe-based alloys are usually formed in the region of immediate contact with diamond, presumably formed from Fe oxides because of the highly reducing nature of carbon.

Melting Equilibrium between Silicate and Sulfide in the Model Garnet-Pyrrhotite Join

Melting equilibrium of the eclogitic garnet-pyrrhotite mixture was studied at a pressure of 7.0 GPa and a temperature interval of 1400–1900 °C (Litvin et al., 2005), which correspond to the conditions of the diamond stability. A melting phase diagram of the system was constructed, and it is shown in Figure 8. The subsolidus assemblage consists of garnet and pyrrhotite. Pyrrhotite is molten at a solidus temperature of 1530 °C, and a two-phase assemblage of solid garnet and pyrrhotite melt forms. Garnet, as a multicomponent solid solution, is molten partially between 1675 and 1735 °C, and a three-phase assembly consisting of garnet, silicate partial melt, and sulfide melt (Fig. 9A) forms. In this case, the silicate and sulfide melts are immiscible. Garnet has a higher Mg content by removal of iron into the silicate melt. Complete melting of garnet proceeds at temperatures above 1735 °C, and a two-phase field of immiscible silicate and sulfide melts forms (Fig. 9B). Although low solubility, 0.5–0.8 wt%, of the FeS component in the silicate melts was recognized, solubility of the garnet component in the sulfide melt was found to be negligible below the microprobe detection limits for all the "silicate" components. The "silicate" boundary for the two liquid immiscible field is shown by a dashed line in Figure 8. The "sulfide" boundary is nearly coincident with the temperature ordinate. It is important

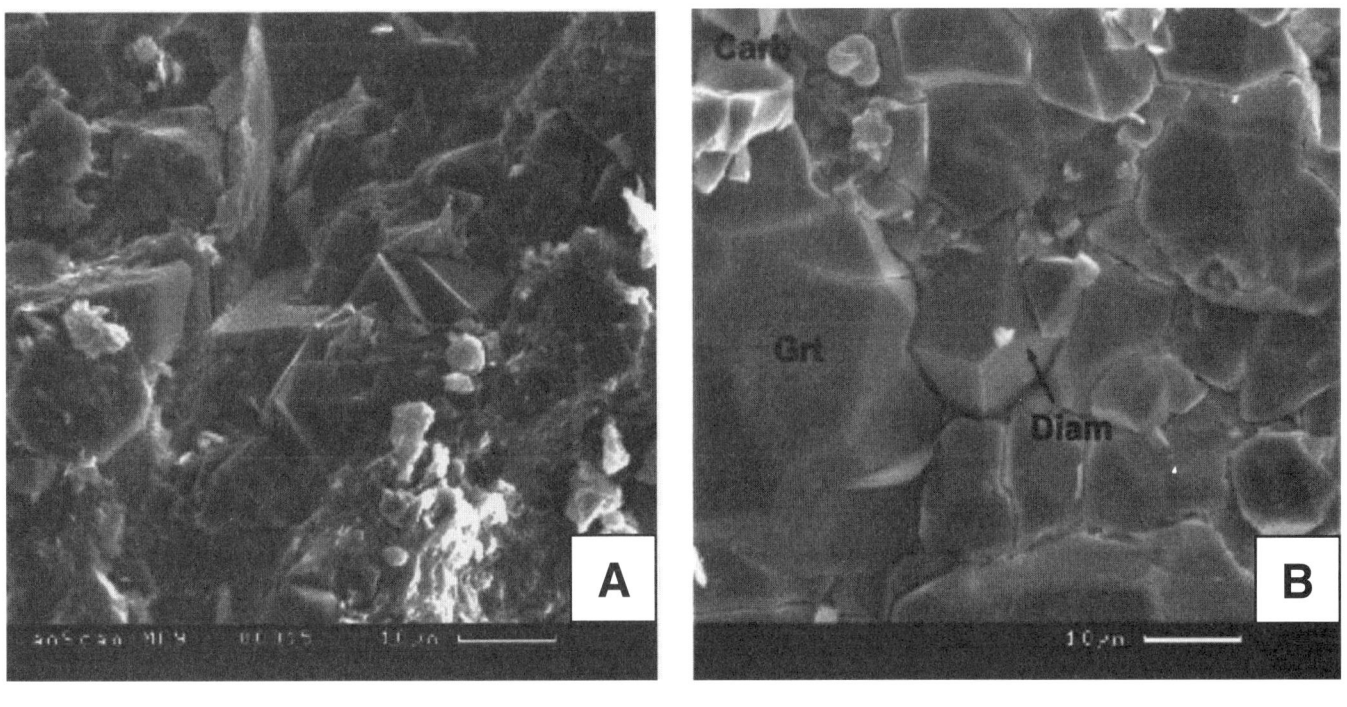

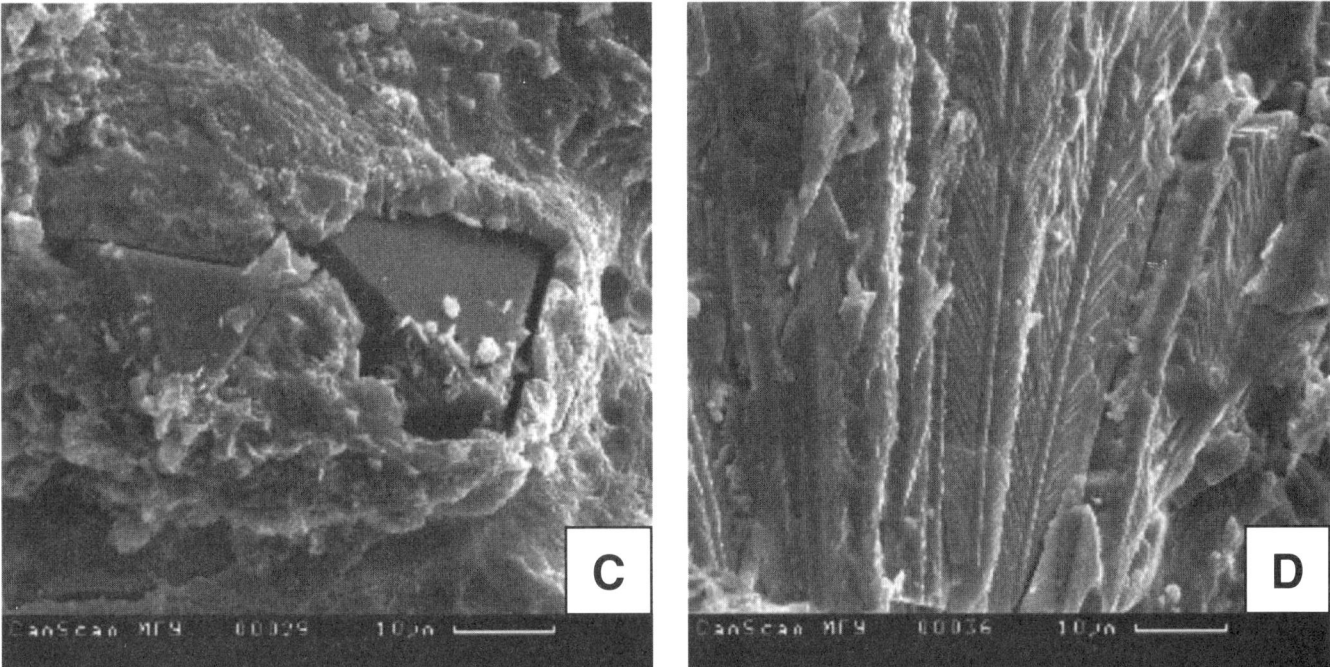

Figure 6. Scanning electron micrograph (SEM) images of diamond and other phases formed in experiments with Kokchetav carbonate-silicate rock: (A) sample 731, extensive diamond crystallization; (B) sample 731, syngenetic diamond and garnet formation from carbonate-silicate melt; Diam—diamond; Carb—carbonate; Grt—garnet; (C) sample 733, octahedral diamond single crystals in carbonate-silicate melt (view after quenching); and (D) sample 735, fibrous dendritic crystallization of parental carbonate-silicate melt during quenching.

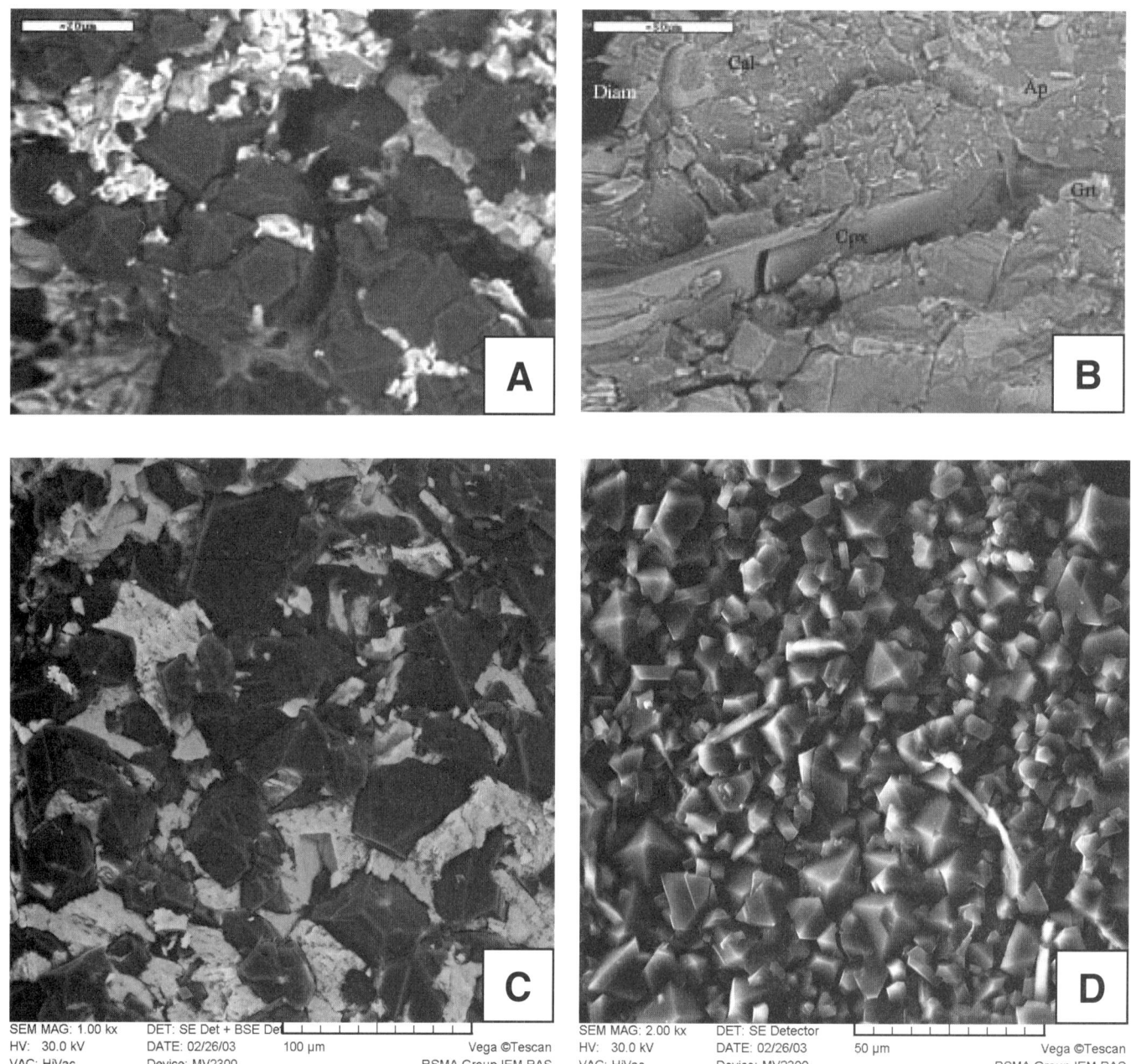

Figure 7. Scanning electron micrograph (SEM) images of diamond, diamondite, and other phases formed in experiments with Chagatai carbonatite rocks: (A) sample 597, diamond (black) and carbonate (gray) with coesite intergrowths (white); (B) sample 599, diamond (Diam) and syngenetic garnet (Grt), clinopyroxene (Cpx), apatite (Ap), and host calcite (Cal) formed during cooling of carbonatitic melt. Calcite (gray) contains garnet intergrowth (white). (C) Sample 997, diamondite formed in carbonatite melt (a view after quenching), and (D) sample 937, view of diamondite, formed in carbonate melt, after chemical treatment.

to note that both solid and molten garnets are not dissolved in the sulfide melt. This implies that the silicate phase is not soluble in the sulfide melt, and, consequently, garnet cannot be crystallized from the melt. Therefore, a syngenetic formation of diamond and garnet inclusions from sulfide melts with dissolved carbon is not possible, although it was shown that the pyrrhotite melt with dissolved carbon is very efficient for diamond nucleation (Fig. 4A). This seems to be applicable also to the other silicate mineral inclusions. The liquid immiscibility of sulfide melt and carbonate or carbonate-silicate melt was also observed in the model silicate-carbonate-sulfide-carbon system (Litvin and Butvina, 2004).

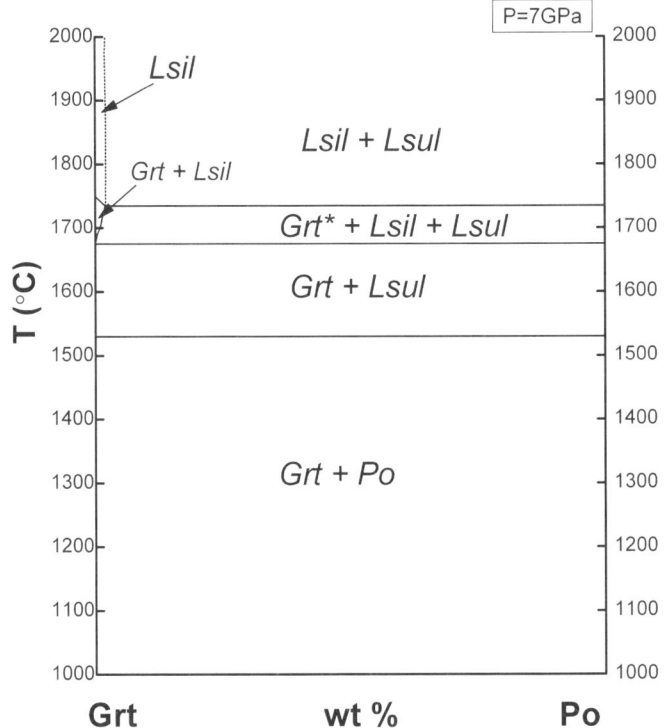
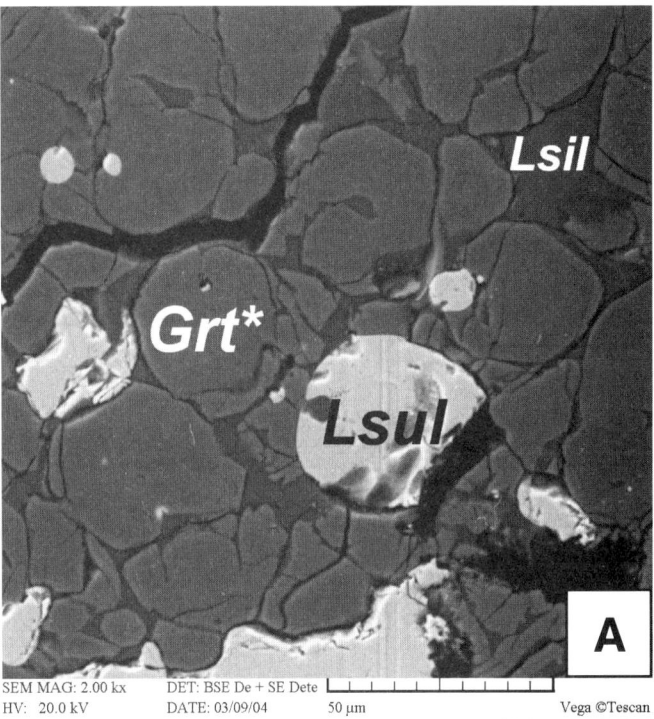
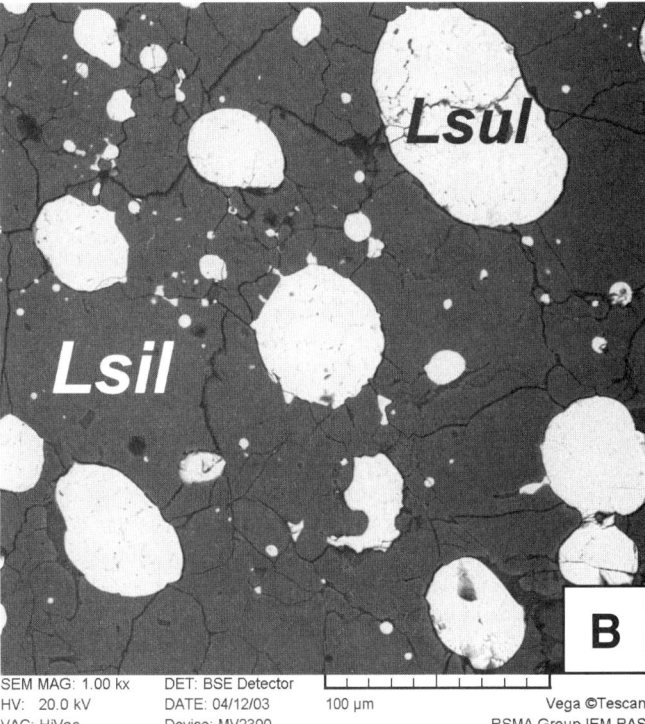

Figure 8. Melting phase diagram of the eclogitic garnet-pyrrhotite (Po) join at a pressure of 7.0 GPa. Lsil—silicate liquid; Lsul—quenched sulfide melt; Grt—garnet.

Figure 9. Scanning electron micrograph (SEM) images of phase relations on the eclogitic garnet-pyrrhotite join: (A) sample 998: 7.0 GPa, 1730 °C, 5 min; three-phase assemblage garnet + silicate melt + sulfide melt is formed by partial melting of garnet. (B) Sample 873: 7.0 GPa, 1760 °C, 5 min; two-phase assemblage of silicate melt + sulfide melt is formed by complete melting of garnet. Lsil—silicate liquid; Lsul—quenched sulfide melt; Grt—garnet.

DISCUSSION

Physicochemical Conditions of Diamond Formation in High-Pressure Experiments

The effect of pressure on the processes of diamond formation in carbonate and carbonate-silicate melts with dissolved carbon is of primary importance for several reasons. First of all, high pressure is required to maintain the pressure and temperature conditions of thermodynamic stability of diamond. It is also important that carbonates melt congruently at high pressure (without breakdown as observed at ambient pressure), and the melts are stable in the course of diamond formation. The high solubility of diamond and graphite in the carbonatite (carbonate or carbonate-silicate) melts is also produced at high pressure. It is important for the physicochemical mechanism of diamond formation in the carbonatite melts with dissolved carbon for graphite to be retained as a thermodynamically unstable phase at pressure and temperature conditions of the diamond stability field. Therefore, an unstable graphite, such as that used as a carbon source in the experiments, has a higher solubility than diamond in the carbonatite melts. As a consequence of the difference in solubility of graphite and diamond, the dissolution of graphite in the carbonate or carbonate-silicate melts leads to the generation of molten carbonatite solutions that are oversaturated with respect to the diamond phase and are responsible for diamond crystallization. Under pressure and temperature conditions of diamond formation, carbonate and silicate melts are completely miscible, which is in contrast to the effect of carbonate-silicate liquid immiscibility reported in the lower-pressure range (Baker and Wyllie, 1990; Kogarko et al., 1995).

These interrelated factors play an important role in the various processes of diamond nucleation and growth in the carbonate-carbon and carbonate-silicate-carbon solutions. The diamond formation can be simulated by the difference in thermodynamic potentials, higher for unstable graphite than diamond, and consequently difference in the solubility of graphite and diamond in the carbonatitic melts. This peculiarity leads to formation of molten carbonate-carbon solutions with carbon concentrations higher than the solubility of diamond. Hence, the carbon solutions automatically become oversaturated with respect to diamond when starting graphite is dissolved in the carbonate melts.

For pressure and temperature conditions of diamond formation, the oversaturated carbon solutions in the molten carbonates may be labile and metastable. Diamonds can spontaneously nucleate in the labile carbon oversaturated solutions, which have higher carbon concentrations relative to the metastable carbon solutions. The carbon concentrations in metastable solutions, which are also oversaturated with respect to diamond, are too low for the spontaneous nucleation but high enough for the seeded growth of diamonds. The critical carbon concentration for spontaneous nucleation of diamond depends on pressure, temperature, and solvent compositions. The pressure and temperature conditions for both the labile and metastable solutions are shown in Figure 1 for the carbonate-silicate solvents labeled as LSF (labile solution field) and MSF (metastable solution field). The position of the LSF/MSF boundary was also estimated and shown in the pressure-temperature diagram together with the diamond-graphite equilibrium boundary (Kennedy and Kennedy, 1976) and the melting curve of the carbonate-silicate-graphite mixtures. It should be remembered that the LSF/MSF boundary is kinetic and corresponds to a lower carbon concentration barrier for spontaneous nucleation of diamond. The positions of both the melting curve of the solvent and the LSF/MSF boundary depend on the chemical composition of the solvent. This is in contrast to the diamond-graphite equilibrium boundary, which has a fixed position. The pressure and temperature conditions for the formation of diamond in carbonatitic end-member composition from fluid-bearing inclusions in the Botswanian diamonds (Schrauder and Navon, 1994) have been reported previously (Spivak and Litvin, 2004). From theoretical and experimental studies, the physico-chemical mechanisms of diamond formation in the sulfide-carbon, chloride-carbon, and metal-carbon solutions were found to be similar to those in carbonate-carbon and carbonate-silicate-carbon solutions.

The diffusion mechanism, together with convection, seems to be responsible for the carbon mass transfer when diamond grows in carbonate-carbon and carbonate-silicate-carbon and other solutions with natural compositions. The rate of carbon mass transfer and diamond growth depends on the chemical composition of a solvent, the carbon solubility therein, pressure, and temperature. The diamond growth rate in carbonate-silicate solutions is strongly variable within the range of several mm/min to several μm/min. In the case of seeded growth, this leads to the formation of both roughly blocked polycentric morphology and smooth layers with nanodimensional steps (Spivak and Litvin, 2004). In the case of spontaneous crystallization, the estimated density of nucleation is no less than $3.0–5.0 \times 10^2$ nuclei/mm^3 for octahedral crystals up to 200 μm in size. In the case of crystallization of the dense polycrystalline diamond aggregates similar to natural diamondite, it reaches a number of 1.0×10^5 nuclei/mm^3.

The spontaneous nucleation and growth of diamond crystals up to 50–100 μm in size in the multicomponent carbonate-carbon and carbonate-silicate-carbon systems proceed without any perceptible induction period over a short time interval of 1–2 min, immediately after melting under pressure and temperature conditions of the labile solution field. The carbon oversaturation value rises generally with increasing pressure and decreasing temperature. The carbon oversaturation in carbonatitic and other melts is the most important factor that influences the density of diamond nucleation and growth rate, which rise with increasing pressure and decreasing temperature. Spontaneous recrystallization of starting graphite into a diamond is rapid when the carbon oversaturation is labile. Seeded growth of diamond is time-dependent and controlled by the carbon concentrations of metastable oversaturation. The boundary conditions and rates of spontaneous and seeded growth of diamond depend on the chemical compositions of molten solvents, which have been shown previously (Litvin et al., 1999b; Spivak and Litvin, 2004). The conditions for the model carbonatitic solvent are limited at pressures above 4.8 GPa and temperatures above 1200 °C (Fig. 1). This is in agreement with natural conditions, e.g., 4.5–7.5 GPa, 950–1350 °C estimated by Navon (1999). The lower temperature of diamond formation may be due to the higher concentrations of C-O-H fluid in the parental carbonatitic melts.

The redox conditions during diamond formation in the high-pressure and high-temperature experiments are controlled by the diamond-forming carbonate-carbon system where carbon is in the form of either graphite and/or diamond. The microprobe analyses data indicate that carbonates are always stable and do not show any indications of reduction by carbon or silicate components. Similarly, solid carbon phases are not oxidized by carbonate components. The most plausible explanation is that the oxygen fugacity of the parent carbonate-carbon and carbonate-silicate-carbon melts is buffered by the carbonate-carbon pairs during diamond formation. The oxygen fugacity seems to be close to that of the FeO-Fe buffer, which is supported by the existence of both Fe oxide and native Fe in the products of diamond formation experiments in the Chagatai carbonatitic melts with dissolved carbon. These results are applicable to characterize the oxidation state of the natural carbonatitic parent melts. Analytical evidence implies that solid or molten sulfides are stable in the runs of diamond synthesis in sulfide-graphite systems, and the oxygen fugacity seems to be regulated by the sulfide-carbon buffer pair during diamond formation. These experimental data indicate that the oxygen fugacity plays an essential role in diamond crystallization (LaTourrette and Holloway, 1994; Chepurov et al., 1999; Arima et al., 2002; Pal'yanov et al, 2005; Stachel et al., 2005).

Implications for Diamond Genesis in Mantle Conditions

The present work can be viewed as an experimental restoration of diamond genesis in mantle carbonate-silicate melts with carbonatitic compositions. The chemical composition of the parental diamond-forming medium is an important factor for providing nucleation and growth of diamond in the Earth's mantle conditions.

The problem of carbon source for mantle-derived diamonds is also important geochemically and was recently reviewed by Navon (1999). The asthenospheric or mantle carbon seems to have a role in the process of diamond formation. Carbonate-silicate-carbon parent melts with labile carbon oversaturation with respect to diamond is required for diamond formation. Carbon seems to be abundant in the mantle as solid phases such as metastable graphite, early formed diamond, and amorphous carbon, which is present in mantle xenoliths (Pineau et al., 1987). It has also been shown in experiments with the $MgCO_3$-SiO_2-H_2 and $MgCO_3$-Al_2O_3-H_2 systems at 6.0–7.0 GPa and 1350–1800 °C that free carbon phases such as diamond and graphite could be formed from carbonate under a strong hydrogen flux (Pal'yanov et al., 2002b). Thus, the formation of carbon from carbonates should occur under strongly reducing conditions.

Speculations on the plausible parental media for diamond have been suggested repeatedly based on mineralogical data (Mitchell and Crocket, 1971; Boyd and Finnerty, 1980; Haggerty, 1986; Harris, 1992; Boyd et al., 1994; Taylor et al., 1996; Bulanova et al., 1998; Navon, 1999; Pal'yanov et al., 2002a; Taylor and Anand, 2004). C and N stable isotopic data and trace-elements studies (see Navon, 1999; Dobosi and Kurat, 2002; Stachel et al., 2004; Cartigny, 2005, for review) can provide indirect estimations of the chemical and phase compositions of the parental media in a mantle-derived diamond as additional sources. It has been proposed that several mantle materials are responsible for diamond formation, such as peridotite, eclogite, sulfide, CO_2-bearing fluid, carbonate, metal, silicate-sulfide, silicate-sulfide-carbonatite-fluid, and some other mantle substances.

The new data on mineral chemistry and high-pressure experiments support the view that carbonatitic (carbonate-silicate) melts are responsible for the formation of most mantle-derived diamond. The diamond-forming parental carbonatitic magmas belong mainly to the multicomponent carbonate-silicate-sulfide-phosphate-chloride-H-O-C system with variable chemistries. Strongly compressed carbonate-silicate melts would efficiently dissolve carbon from the mantle sources and thus could be considered as parental media for diamond formation. These carbonatitic melts can be responsible for the syngenesis of diamonds and their primary inclusions. The carbon concentrations in carbonatitic magmas, when oversaturated with carbon, should be labile for diamond nucleation.

The major and minor components in the composition of the parental carbonatitic melts with dissolved carbon may be involved in diamond formation. The carbonate and silicate components are the major contributors in the parental melts, which are characterized by carbonate-silicate complete miscibility at pressure and temperature conditions of diamond formation.

The minor components are diverse: oxides, phosphates, sulfides, haloids, carbon dioxide, water, and others. On the basis of experiments on synthesis of diamonds, it has been demonstrated indirectly that carbon solubility in the sulfide and potassium chloride melts under high pressure is sufficiently high for spontaneous nucleation of diamonds. Diamond nucleation in the pure water-carbon and water–carbon dioxide–carbon systems generated by the decomposition of oxalic acid dehydrate ($H_2C_2O_4 \cdot 2H_2O$) has been reported by several authors (Hong et al., 1999; Akaishi and Yamaoka, 2000; Sokol et al., 2001b), as has diamond crystallized from Na_2CO_3, K_2CO_3 (Pal'yanov et al., 1999a), and $CaCO_3$ (Yamaoka et al., 2002a) carbonates mixed with the oxalic acid dehydrate as a source for H_2O and CO_2. Thus, the high-pressure and high-temperature experiments on diamond synthesis have revealed a wide variety of natural media (which are closely associated with diamond and diamond-bearing rocks) efficient for diamond nucleation and growth.

Mineral inclusion analyses in diamond indicate that minor components can be concentrated substantially and become the major components locally. For example, locally high concentrations of sulfide inclusions suggest an additional sulfide-related media for diamond formation under mantle conditions (Bulanova et al., 1998). However, high-pressure experiments contradict a role of sulfide melts alone in serving as an additional and accidental media for diamond formation and the formation of silicates, carbonates, oxides, and the other primary inclusions. A similar conclusion may be inferred for the metal-carbon melts based on iron and some other alloys. The appearance of a strongly concentrated alkaline-chloride aqueous solution is also related to local concentration of the minor components in the parental melts. Variation in the chemical compositions of the parental carbonate-silicate melts during the growth of natural diamond can be explained by fractional crystallization of silicate and carbonate minerals and the subsequent accumulation of the components, including K and Na chlorides and water, that are poorly concentrated in the parental melts. A generalized diagram for the composition of primary fluid-bearing multiphase inclusions (Schrauder and Navon, 1994; Izraeli et al., 2001) revealed compositional variation trends for parent melts with the accumulation of alkaline-chloride components and water during fractionation of silicate and carbonate minerals. Fractionation in the carbonate-silicate melts is supported by the high-pressure and high-temperature experiments, as discussed above, in which the simultaneous crystallization of diamond and syngenetic minerals was clearly demonstrated for the melts of Chagatai carbonatite with dissolved carbon.

Preliminary Model for Origin and Evolution of the Carbonatitic Parental Melts

The origin of carbonatite magmas may be related to an interaction of mantle material with chemically active alkaline-fluid agents of mantle plumes, such as those that metasomatize peridotites (Green and Wallace, 1980; Hauri et al., 1993; Rudnick et al., 1993; Wyllie, 1996; Litvin, 1998; Gasparik and Litvin, 2002). Chemically active alkaline CO_2-rich agents may be formed in the process of chemical and thermal evolution of the plume. The plume agents interact with mantle peridotite, when the latter is in solid state or partially molten. The plume-mantle interface reactions lead to carbonatization of olivine, orthopyroxene, and ultrabasic melts, and formation of carbonate minerals and "primary" carbonatitic melts. These melts are efficient solvents for the C-H-O-N fluid components, including CO_2 and H_2O, which are released mainly by plume materials. The newly formed carbonatitic melts dissolve peridotitic minerals, melts, and fluid components due to carbonate-silicate liquid complete miscibility to form silicate-rich carbonatitic (carbonate-silicate) magmas. In this stage, elementary carbon phases associated with peridotite are dissolved into the carbonatitic magmas together with other soluble phases. The sulfide minerals and melts join the carbonatitic magma as separate nonsoluble phases, since the carbonate-silicate melt is completely immiscible with sulfide melt. As a result, the carbonatitic magma would possess the properties of the parental carbonate-silicate media for natural diamond. At the beginning, carbon concentration in the magma is most likely undersaturated in carbon, and because of this, diamond is not nucleated. Diamond may nucleate in the labile carbon solution oversaturated with respect to the diamond phase. For the conditions of diamond stability at depths greater than 150 km, such mantle carbonatite melts become effective parent media for diamond. Concentration of carbon in the carbonatitic magma can grow and reach up to the labile level under the combined factors of decreasing temperature and fractional crystallization of predominant silicate minerals. Carbonates crystallize at practically minimum solidus temperatures. The sulfide melts and minerals, if abundant in the carbonatitic magma, are not involved into the process of silicate-carbonate fractional crystallization. However, sulfide melts are capable of dissolving carbon and nucleating diamond when they reach labile carbon oversaturation, and they are able to grow diamond without silicate, carbonate, and other primary inclusions. The carbonatitic (carbonate-silicate) melts, when oversaturated with carbon, are solely responsible for the syngenesis of diamond and primary inclusions, including the common sulfides.

CONCLUSIONS

Formation of mantle-derived diamonds in melts with chemical compositions similar to fluid-bearing multiphase and sulfide primary inclusions in natural diamonds was studied experimentally at high pressures and temperatures. Diamonds and diamondites were synthesized at 5.5–8.5 GPa in multicomponent carbonate-silicate-carbon melts with a model carbonatitic chemistry similar to the compositions of fluid-bearing multiphase inclusions in diamonds from Botswana. On the basis of the present experimental results, it is argued that sulfide-graphite and potassium chloride–graphite mixtures are also efficient for diamond formation when these melts are formed under pressure-temperature conditions of the diamond stability field. Diamonds and diamondites were formed with garnets, clinopyroxenes, carbonates, phosphates, sulfides, iron-chromium alloys, and some other minerals, which have also been observed as primary inclusions in natural diamonds, in the experiments using carbonate-silicate melts with diamond-bearing Kokchetav (Kazakhstan) and Chagatai (Uzbekistan) carbonatitic rock compositions. Experimental study of the model garnet-pyrrhotite system at pressures up to 7 GPa revealed a complete silicate-sulfide liquid immiscibility and also showed that sulfide melts are not effective for the syngenesis of diamonds and their primary silicate inclusions. Thus, the carbonatitic (carbonate-silicate) nature of mantle melts responsible for the formation of natural diamonds is supported by new experimental data on the formation of diamonds and syngenetic phases. The physicochemical mechanism for the formation of diamond was revealed in the experiments using the starting materials with natural chemistry, and it may be applied to the genesis of diamond under conditions prevalent in the mantle. This mechanism includes formation of oversaturated carbon solutions with respect to diamonds in the natural carbonate-silicate and some other melts. The degree of oversaturation of carbon in the melt solutions may be as high as labile for spontaneous nucleation of diamond and as low as metastable for the seeded growth of diamond. It is appropriate to recognize the major components of carbonatitic parental media for the natural diamond as carbonate and silicate and the minor components as oxides, sulfides, phosphates, haloids, carbon dioxide, water, etc. The compositions of the parent media, which are partially or completely molten during the time of diamond formation, are variable with respect to the contents of the major and minor components. Compositional variation in natural carbonate-silicate parental melts is caused by fractional crystallization of silicate, carbonate, and other minerals during diamond formation. Minor components including sulfides, potassium chloride, and water may increase in concentration and become comparable to major components. The origin of the parental media for mantle-derived diamond can be explained by a more general phenomenon, i.e., formation and evolution of carbonatitic magma in the Earth's mantle. The origin of carbonatitic magma may be related to carbonatization of mantle peridotite by chemically active alkali-fluid agents associated with plumes.

Results of high-pressure experiments to determine the chemical and mineral compositions of carbonatitic parental melts of mantle-derived diamonds were also discussed briefly in this review. Additional essential data would be obtained by experimental studies on multicomponent parental carbonate-silicate melts with C-O-H fluid and other minor components. These experiments would be helpful to solve the problem of syngenesis of mantle-derived diamonds and their primary inclusions.

ACKNOWLEDGMENTS

This work was supported by the Russian Foundation for Basic Research grants 05-05-64101 and 04-05-97220 (jointly with the Moscow District Ministry of Science and Technology) and the Russian Academy of Sciences Project P9 for Material Study at Extreme Conditions.

REFERENCES CITED

Akaishi, M., 1993, Non-metallic catalysts for synthesis of high pressure, high temperature diamond: Diamond and Related Materials, v. 2, p. 183–189, doi: 10.1016/0925-9635(93)90050-C.

Akaishi, M., 1996, Effect of Na_2O and H_2O addition to SiO_2 on the synthesis of diamond from graphite: Tsukuba, Ibaraki, Japan, Proceedings of the 3rd NIRIM (National Institute for Research in Inorganic Materials) International Symposium on Advanced Materials (ISAM'96), p. 75–80.

Akaishi, M., and Yamaoka, S., 2000, Crystallization of diamond from C-O-H fluids under high-pressure and high-temperature conditions: Journal of Crystal Growth, v. 209, p. 999–1003, doi: 10.1016/S0022-0248(99)00756-3.

Akaishi, M., Kanda, H., and Yamaoka, S., 1990a, Synthesis of diamond from graphite-carbonate systems under very high temperature and pressure: Journal of Crystal Growth, v. 104, p. 578–581, doi: 10.1016/0022-0248(90)90159-I.

Akaishi, M., Kanda, H., and Yamaoka, S., 1990b, High pressure synthesis of diamond in the systems of graphite-sulfate and graphite-hydroxide: Japanese Journal of Applied Physics, v. 29, p. L1172–L1174, doi: 10.1143/JJAP.29.L1172.

Akaishi, M., Shaji Kumar, M.D., Kanda, H., and Yamaoka, S., 2000, Formation process of diamond from supercritical H_2O-CO_2 fluid under high pressure and high temperature conditions: Diamond and Related Materials, v. 9, p. 1945–1950, doi: 10.1016/S0925-9635(00)00366-6.

Akaishi, M., Shaji Kumar, M.D., Kanda, H., and Yamaoka, S., 2001, Reactions between carbon and a reduced C-O-H fluid under diamond-stable HP-HT conditions: Diamond and Related Materials, v. 10, p. 2125–2130, doi: 10.1016/S0925-9635(01)00490-3.

Arima, M., 1996, Experimental study of growth and resorbtion of diamond in kimberlitic melts at high pressures and temperatures: Tsukuba, Ibaraki, Japan, Proceedings of the 3rd NIRIM (National Institute for Research in Inorganic Materials) International Symposium on Advanced Materials (ISAM'96), p. 223–228.

Arima, M., Nakayama, K., Akaishi, M., Yamaoka, S., and Kanda, H., 1993, Crystallization of diamond from silicate melt of kimberlite composition in high-pressure high-temperature experiments: Geology, v. 21, p. 968–970, doi: 10.1130/0091-7613(1993)021<0968:CODFAS>2.3.CO;2.

Arima, M., Kozai, Y., and Akaishi, M., 2002, Diamond nucleation and growth by reduction carbonate melts under high-pressure and high-temperature conditions: Geology, v. 30, p. 691–694, doi: 10.1130/0091-7613(2002)030<0691:DNAGBR>2.0.CO;2.

Baker, M.B., and Wyllie, P.J., 1990, Liquid immiscibility in a nephelinite-carbonate system at 25 kbar and implications for carbonatite origin: Nature, v. 346, p. 168–177, doi: 10.1038/346168a0.

Bobrov, A.V., Litvin, Yu.A., and Divaev, F.K., 2004, Phase relations and diamond synthesis in the carbonate-silicate rocks of the Chagatai complex, western Uzbekistan: Results of experiments at P = 4–7 GPa and T = 1200–1700°C: Geochemistry International, v. 42, p. 39–48.

Borzdov, Yu.M., Sokol, A.G., Pal'yanov, Yu.N., Kalinin, A.A., and Sobolev, N.V., 1999, Study of diamond crystallization in alkaline silicate, carbonate and carbonate-silicate melts: Doklady Akademii Nauk, v. 366, p. 530–533.

Boyd, F.R., and Finnerty, A.A., 1980, Conditions of origin of natural diamonds of peridotitic affinity: Journal of Geophysical Research, v. 85, p. 6911–6918.

Boyd, S.R., Pineau, F., and Javoy, M., 1994, Modeling the growth of natural diamonds: Chemical Geology, v. 116, p. 29–42, doi: 10.1016/0009-2541(94)90156-2.

Bulanova, G.P., Griffin, W.L., and Ryan, C.G., 1998, Nucleation environment of diamonds from Yakutian kimberlites: Mineralogical Magazine, v. 62, p. 409–419, doi: 10.1180/002646198547675.

Bundy, F.P., Hall, H.T., Strong, H.M., and Wentorf, R.H., 1955, Man-made diamond: Nature, v. 176, p. 51–54, doi: 10.1038/176051a0.

Cartigny, P., 2005, Stable isotopes and the origin of diamond: Elements, v. 1, p. 79–84.

Chepurov, A.I., Fedorov, I.I., Sonin, V.M., Bagryantsev, D.G., and Osorgin, N.Yu., 1999, Diamond formation during reduction of oxide-and silicate-carbon systems at high P-T conditions: European Journal of Mineralogy, v. 11, p. 355–362.

Dawson, J.B., 1980, Kimberlites and Their Xenoliths: Berlin, Springer-Verlag, 252 p.

Djuraev, A.D., and Divaev, F.K., 1999, Melanocratic carbonatites—New type of diamond-bearing rocks, Uzbekistan, in Stanley, S.N., ed., Mineral Deposits: Processes to Processing: Rotterdam, Balkema, p. 639–642.

Dobosi, G., and Kurat, G., 2002, Trace element abundances in garnets and clinopyroxenes from diamondites—A signature of carbonatitic fluids: Mineralogy and Petrology, v. 76, p. 21–38, doi: 10.1007/s007100200030.

Efimova, E.S., Sobolev, N.V., and Pospelova, L.N., 1983, Inclusions of sulphides in diamonds and peculiarities of their paragenesis: Zapiski Vsesoyuznogo Mineralogicheskogo Obshchestva, v. 92, p. 300–309.

Eremets, M., 1996, High Pressure Experimental Methods: New York, Tokyo, Oxford University Press, 390 p.

Fursenko, B.A., Goryainov, S.B., and Sobolev, N.V., 2001, High pressures in coesite inclusions in diamond: Combination scattering spectroscopy: Doklady Akademii Nauk, v. 379, p. 812–815.

Gasparik, T., and Litvin, Yu.A., 2002, Experimental investigation of the effect of metasomatism by carbonatitic melt on the composition and structure of the deep mantle: Lithos, v. 60, p. 129–143, doi: 10.1016/S0024-4937(01)00078-0.

Green, D.H., and Wallace, M.E., 1980, Mantle metasomatism by ephemeral carbonatite melts: Nature, v. 336, p. 459–462.

Haggerty, S.E., 1986, Diamond genesis in a multiply constrained model: Nature, v. 320, p. 34–38, doi: 10.1038/320034a0.

Harris, J.W., 1992, Diamond geology, in Field, J.E., ed., The Properties of Natural and Synthetic Diamond: London, Academic Press, p. 345–393.

Harte, B., and Harris, J.W., 1994, Lower mantle mineral associations preserved in diamonds: Mineralogical Magazine, v. 58A, p. 384–385.

Hauri, F.N., Shimizu, N., Dieu, J.J., and Hart, S.R., 1993, Evidence for hotspot-related carbonatite metasomatism in the oceanic upper mantle: Nature, v. 365, p. 221–227, doi: 10.1038/365221a0.

Homan, C.G., 1975, Phase diagram of Bi up to 140 kbars: Journal of Physics and Chemistry of Solids, v. 36, p. 1249–1254, doi: 10.1016/0022-3697(75)90199-7.

Hong, S.M., Akaishi, M., and Yamaoka, S., 1999, Nucleation of diamond in the system of carbon and water under very high pressure and temperature: Journal of Crystal Growth, v. 200, p. 326–328, doi: 10.1016/S0022-0248(98)01288-3.

Izraeli, E.S., Schrauder, M., and Navon, O., 1998, On the connection between fluid and mineral inclusions in diamonds: Cape Town, VII International Kimberlite Conference Extended Abstracts, p. 352–354.

Izraeli, E.S., Harris, J.H., and Navon, O., 2001, Brine inclusions in diamonds: A new upper mantle fluid: Earth and Planetary Sciences Letters, v. 187, p. 323–332, doi: 10.1016/S0012-821X(01)00291-6.

Kaminsky, F.V., Zakharchenko, O.D., Davies, R., Griffin, W.L., Khachatryan-Blinova, G.K., and Shiryaev, A.A., 2001, Superdeep diamonds from the Juina area, Mato Grosso State, Brazil: Contributions to Mineralogy and Petrology, v. 140, p. 734–753.

Kennedy, C.J., and Kennedy, G.C., 1976, The equilibrium boundary between graphite and diamond: Journal of Geophysical Research, v. 81, p. 2467–2470.

Klein-BenDavid, O., Logvinova, A.M., Izraeli, E.S., Sobolev, N.V., and Navon, O., 2003, Sulphide melt inclusions in Yubileinaya (Yakutia) diamonds: Victoria, Canada, VIII International Kimberlite Conference Long Abstract: Mineralogical Society of America, CD-ROM.

Kogarko, L.N., Henderson, C.M.B., and Pacheco, H., 1995, Primary Ca-rich carbonatite magma and carbonate-silicate-sulphide liquid immiscibility in the upper mantle: Contributions to Mineralogy and Petrology, v. 121, p. 267–274.

Kurat, G., and Dobosi, G., 2000, Garnet and diopside-bearing diamondites (framesites): Mineralogy and Petrology, v. 69, p. 143–159, doi: 10.1007/s007100070018.

LaTourrette, T., and Holloway, J.R., 1994, Oxygen fugacity of the diamond + C-O fluid assemblage and CO_2 fugacity at 8 GPa: Earth and Planetary Science Letters, v. 128, p. 439–451, doi: 10.1016/0012-821X(94)90161-9.

Leipunsky, O.I., 1939, On the artificial diamonds: Uspekhi Khimii, v. 8, p. 1520–1534.

Leung, I.S., Guo, W., Friedman, I., and Gleason, J., 1990, Natural occurrence of silicon carbide in diamondiferous kimberlite from Fuxian: Nature, v. 346, p. 352–354, doi: 10.1038/346352a0.

Leung, I.S., Taylor, L.A., Tsao, C.S., and Han, Z., 1996, SiC in diamond and kimberlites: Implications for nucleation and growth of diamond: International Geological Reviews, v. 37, p. 483–496.

Liander, H., 1955, Diamond synthesis: Allmana Svenska Elektriska Aktiebolaget Journal, v. 28, p. 97–98.

Litvin, Yu.A., 1968, On the mechanism of diamond formation in metal-carbon systems: Izvestiya Akademii Nauk SSSR, Neorganicheskie Materialy, v. 4, p. 175–181.

Litvin, Yu.A., 1969, On the problem of diamond origin: Zapiski Vsesoyuznogo Mineralogicheskogo Obshchestva, v. 98, p. 116–123.

Litvin, Yu.A., 1991, Physico-Chemical Study of Melting Relations of the Deep-Seated Earth's Substance: Moscow, Publishing House "Nauka," 312 p. (in Russian).

Litvin, Yu.A., 1998, Hot spots of the mantle and experiment to 10 GPa: Alkaline reactions, lithosphere carbonatization, and new diamond-generating systems: Russian Geology and Geophysics, v. 39, p. 1760–1768.

Litvin, Yu.A., 2003, Alkaline-chloride components in processes of diamond growth in the mantle and high-pressure experimental conditions: Doklady Earth Sciences, v. 389A, p. 388–391.

Litvin, Yu.A., and Butvina, V.G., 2004, Diamond-forming media in the system eclogite-carbonatite-sulphide-carbon: Experiments at 6.0–8.5 GPa: Petrology, v. 12, p. 377–387.

Litvin, Yu.A., and Spivak, A.V., 2003, Rapid growth of diamondite at the contact between graphite and carbonate melt: Experiments at 7.5–8.5 GPa: Doklady Earth Sciences, v. 391A, p. 888–891.

Litvin, Yu.A., and Zharikov, V.A., 1999, Primary fluid-carbonatite inclusions in diamond: Experimental modeling in the system K_2O-Na_2O-CaO-MgO-FeO-CO_2 as a diamond formation medium at 7–9 GPa: Doklady Earth Sciences, v. 367A, p. 801–805.

Litvin, Yu.A., and Zharikov, V.A., 2000, Experimental modeling of diamond genesis: Diamond crystallization in multicomponent carbonate-silicate melts at 5–7 GPa and 1200–1570°C: Doklady Earth Sciences, v. 373, p. 867–870.

Litvin, Yu.A., Chudinovskikh, L.T., and Zharikov, V.A., 1997, Experimental crystallization of diamond and graphite from alkali-carbonate melts at 7–11 GPa: Doklady Earth Sciences, v. 355, p. 669–672.

Litvin, Yu.A., Chudinovskikh, L.T., and Zharikov, V.A., 1998, Crystallization of diamond in the $Na_2Mg(CO_3)_2$-$K_2Mg(CO_3)_2$-C system at 8–10 GPa: Doklady Earth Sciences, v. 359A, p. 464–466.

Litvin, Yu.A., Aldushin, K.A., and Zharikov, V.A., 1999a, Diamond synthesis at 8.5–9.5 GPa in the $K_2Ca(CO_3)_2$–$Na_2Ca(CO_3)_2$ system modeling compositions of fluid-carbonatite inclusions in kimberlitic diamonds: Doklady Earth Sciences, v. 367, p. 529–532.

Litvin, Yu.A., Chudinovskikh, L.T., Saparin, G.V., Obyden, S.K., Chukichev, M.V., and Vavilov, V.S., 1999b, Diamonds of new alkaline carbonate-graphite HP syntheses: SEM morphology, CCL-SEM and CL spectroscopy studies: Diamond and Related Materials, v. 8, p. 267–272, doi: 10.1016/S0925-9635(98)00318-5.

Litvin, Yu.A., Jones, A.P., Beard, A.D., Divaev, F.K., and Zharikov, V.A., 2001, Crystallization of diamond and syngenetic minerals in melts of diamondiferous carbonatites of the Chagatai Massif, Uzbekistan: Experiment at 7.0 GPa: Doklady Earth Science, v. 381A, p. 1066–1069.

Litvin, Yu.A., Butvina, V.G., Bobrov, A.V., and Zharikov, V.A., 2002, The first syntheses of diamond in sulphide-carbon system: The role of sulphides in diamond genesis: Doklady Earth Sciences, v. 382, p. 40–43.

Litvin, Yu.A., Spivak, A.V., and Matveev, Yu.A., 2003, Crystallization of diamond in the molten carbonate-silicate rocks of the Kokchetav metamorphic complex at 5.5–7.5 GPa: Geochemistry International, v. 11, p. 1090–1098.

Litvin, Yu.A., Shushkanova, A.V., and Zharikov, V.A., 2005, Immiscibility of sulfide-silicate melts in the mantle: Role in the syngenesis of diamond and inclusions (based on experiments at 7.0 GPa): Doklady Earth Sciences, v. 403, p. 719–722.

Logvinova, A.M., Klein-BenDavid, O., Izraeli, E.S., Navon, O., and Sobolev, N.V., 2003, Microinclusions in fibrous diamonds from Yubileinaya kimberlite pipe (Yakutia): Victoria, Canada, VIII International Kimberlite Conference Long Abstract: Mineralogical Society of America, CD-ROM.

Mathez, E.A., Fogel, R.A., Hutcheon, I.D., and Marshintsev, V.K., 1995, Carbon isotopic composition and origin of SiC from kimberlites of Yakutia, Russia: Geochimica et Cosmochimica Acta, v. 59, p. 781–791, doi: 10.1016/0016-7037(95)00002-H.

Meyer, H.O.A., 1987, Inclusions in diamond, in Nixon, H.P., ed., Mantle Xenoliths: New York, Wiley, p. 501–523.

Mitchell, R.H., 1986, Kimberlites: Their Mineralogy, Geochemistry and Petrology: New York, Wiley, 436 p.

Mitchell, R.H., and Crocket, J.H., 1971, Diamond genesis—A synthesis of opposing views: Mineralium Deposita (Berlin), v. 6, p. 392–403.

Moore, R., Otter, M.L., Richardson, R.S., Harris, J.W., and Gurney, J.J., 1986, The occurrence of moissanite and ferro-periclase as inclusions in diamond: Geological Society of Australia Abstract, v. 16, p. 409–411.

Navon, O., 1991, High internal pressures in diamond fluid inclusions determined by infrared-absorption: Nature, v. 353, p. 746–748, doi: 10.1038/353746a0.

Navon, O., 1999, Diamond formation in the Earth's mantle, in Gurney, J.J., Gurney, J.L., Pascoe, M.D., and Richardson, S.H., eds., Proceedings of the VII International Kimberlite Conference: Cape Town, Red Roof Design, v. 2, p. 584–604.

Navon, O., Hutcheon, I.D., Rossman, G.R., and Wasserburg, G.L., 1988, Mantle-derived fluids in diamond micro-inclusions: Nature, v. 335, p. 784–789, doi: 10.1038/335784a0.

Navon, O., Izraeli, E.S., and KleinBen-David, O., 2003, Fluid inclusions in diamonds—The carbonatitic connection: Victoria, Canada, VIII International Kimberlite Conference Long Abstract: Mineralogical Society of America, CD-ROM.

Pal'yanov, Yu.N., Sokol, A.G., Borzdov, Yu.M., Khokhryakov, A.F., and Sobolev, N.V., 1999a, Diamond formation from mantle carbonatite fluid: Nature, v. 400, p. 417–418, doi: 10.1038/22678.

Pal'yanov, Yu.N., Sokol, A.G., Borzdov, Yu.M., Khokhryakov, A.F., and Sobolev, N.V., 1999b, The diamond growth from Li_2CO_3, Na_2CO_3, K_2CO_3, and Cs_2CO_3 solvent-catalysts at $P = 7$ GPa and $T = 1700$–$1750°C$: Diamond and Related Materials, v. 8, p. 1118–1124, doi: 10.1016/S0925-9635(99)00098-9.

Pal'yanov, Yu.N., Sokol, A.G., Borzdov, Yu.M., and Khokhryakov, A.F., 2002a, Fluid-bearing alkaline carbonate melts as the medium for the formation of diamonds in the Earth's mantle: An experimental study: Lithos, v. 60, p. 145–159, doi: 10.1016/S0024-4937(01)00079-2.

Pal'yanov, Yu.N., Sokol, A.G., Borzdov, Yu.M., Khokhryakov, A.F., and Sobolev, N.V., 2002b, Diamond formation through carbonate-silicate interaction: The American Mineralogist, v. 87, p. 1009–1013.

Pal'yanov, Yu.N., Sokol, A.G., Tomilenko, N.V., and Sobolev, N.V., 2005, Conditions of diamond formation through carbonate-silicate interaction: European Journal of Mineralogy, v. 17, p. 207–214.

Pineau, F., Javoy, M., and Kornprobst, J., 1987, Primary igneous graphite in ultramafic xenoliths: II. Isotopic composition of the carbonaceous phases present in xenoliths and host lava at Tissemt (Eggere, Algerian Sahara): Journal of Petrology, v. 28, p. 313–332.

Rudnick, R.L., McDonough, W.F., and Chappel, B.W., 1993, Carbonatite metasomatism in the northern Tanzanian mantle: Petrographic and geochemical characteristics: Earth and Planetary Sciences Letters, v. 114, p. 463–475, doi: 10.1016/0012-821X(93)90076-L.

Sato, H., Akaishi, M., and Yamaoka, S., 1999, Spontaneous nucleation of diamond in the system $MgCO_3$-$CaCO_3$-C at 7.7 GPa: Diamond and Related Materials, v. 8, p. 1900–1905, doi: 10.1016/S0925-9635(99)00157-0.

Schrauder, M., and Navon, O., 1994, Hydrous and carbonatitic mantle fluids in fibrous diamonds from Jwaneng, Botswana: Geochimica et Cosmochimica Acta, v. 58, p. 761–771, doi: 10.1016/0016-7037(94)90504-5.

Shaji Kumar, M.D., Akaishi, M., and Yamaoka, S., 2000, Formation of diamond from supercritical H_2O-CO_2 fluid at high pressure and high temperature: Journal of Crystal Growth, v. 213, p. 203–206, doi: 10.1016/S0022-0248(00)00352-3.

Shaji Kumar, M.D., Akaishi, M., and Yamaoka, S., 2001, Effect of fluid concentration on the formation of diamond in the CO_2–H_2O-graphite system under HP-HT conditions: Journal of Crystal Growth, v. 222, p. 9–13, doi: 10.1016/S0022-0248(00)00921-0.

Shatcky, A.F., Borzdov, Yu.M., Sokol, A.G., and Pal'yanov, Yu.N., 2002, Peculiarity of phase formation and diamond crystallization in ultra potassium carbonate-silicate systems with carbon: Geologia i Geofizika, v. 43, p. 936–946 (in Russian).

Shul'zhenko, A.A., and Get'man, A.F., 1971, Diamond synthesis: German Patent 2032083, 04 November 1971.

Shul'zhenko, A.A., and Get'man, A.F., 1972, Diamond synthesis: German Patent 2124145, 16 March 1972.

Sobolev, N.V., 1977, The Deep-Seated Inclusions in Kimberlites and the Problem of the Composition of the Upper Mantle: Washington, D.C., American Geophysical Union, 304 p.

Sobolev, N.V., and Shatsky, V.S., 1990, Diamond inclusions in garnets from metamorphic rocks: Nature, v. 343, p. 742–746, doi: 10.1038/343742a0.

Sobolev, N.V., Efimova, E.S., and Pospelova, L.N., 1981, Native iron in diamond from Yakutia and its paragenesis: Geologia i Geofizika, v. 22, p. 25–29 (in Russian).

Sokol, A.G., Borzdov, Yu.M., Pal'yanov, Yu.N., Khokhryakov, A.F., and Sobolev, N.V., 2001a, An experimental demonstration of diamond formation in the dolomite-carbon and dolomite-fluid-carbon systems: European Journal of Mineralogy, v. 13, p. 893–900, doi: 10.1127/0935-1221/2001/0013/0893.

Sokol, A.G., Pal'yanov, Yu.N., Pal'yanova, G.A., Khokhryakov, A.F., and Borzdov, Yu.M., 2001b, Diamond and graphite crystallization from C-O-H fluids under high pressure and high temperature conditions: Diamond and Related Materials, v. 10, p. 2131–2136, doi: 10.1016/S0925-9635(01)00491-5.

Spivak, A.V., and Litvin, Yu.A., 2004, Diamond syntheses in multi-component carbonate-carbon melts of natural chemistry: Elementary processes and properties: Diamond and Related Materials, v. 13, p. 482–487, doi: 10.1016/j.diamond.2003.11.104.

Stachel, T., Harris, J.W., and Brey, G.P., 1998, Rare and unusual mineral inclusions in diamonds from Mwadui, Tanzania: Contributions to Mineralogy and Petrology, v. 132, p. 34–47, doi: 10.1007/s004100050403.

Stachel, T., Aulbach, S., Brey, G.P., Harris, J.W., Loest, I., Tappert, R., and Viljoen, K.S., 2004, The trace element composition of silicate inclusions in diamonds: A review: Lithos, v. 77, p. 1–19, doi: 10.1016/j.lithos.2004.03.027.

Stachel, T., Brey, G.P., and Harris, J.W., 2005, Inclusions in sublithospheric diamonds: Glimpses of deep mantle: Elements, v. 1, p. 73–78.

Taniguchi, T., Dobson, D., Jones, A.P., Rabe, R., and Milledge, H.J., 1996, Synthesis of cubic diamond in the graphite-magnesium carbonate and graphite-$K_2Mg(CO_3)_2$ systems at high pressure of 9–10 GPa region: Journal of Materials Research, v. 11, p. 2622–2632.

Taylor, L.A., and Anand, M., 2004, Diamonds: Time capsules from the Siberian mantle: Chemie der Erde Geochemistry, v. 64, p. 1–74, doi: 10.1016/j.chemer.2003.11.006.

Taylor, L.A., Snyder, G.A., Crozaz, G., Sobolev, V.N., and Sobolev, N.V., 1996, Eclogitic inclusions in diamonds: Evidence of complex mantle processes over time: Earth and Planetary Sciences Letters, v. 142, p. 535–551, doi: 10.1016/0012-821X(96)00106-9.

Tomlinson, E., Jones, A.P., and Milledge, J.H., 2004, High-pressure experimental growth of diamond using $C-K_2CO_3-KCl$ as an analogue of Cl-bearing carbonate fluid: Lithos, v. 77, p. 287–294, doi: 10.1016/j.lithos.2004.04.029.

Wang, Y., and Kanda, H., 1998, Growth of HTHP diamonds in alkali haloids: Possible effect of oxygen contamination: Diamond and Related Materials, v. 7, p. 57–63, doi: 10.1016/S0925-9635(97)00183-0.

Wang, A., Pasteris, J.D., Meyer, H.O.A., and Dele-Duboi, M.L., 1996, Magnesite-bearing inclusion assemblage in natural diamond: Earth and Planetary Sciences Letters, v. 141, p. 293–306, doi: 10.1016/0012-821X(96)00053-2.

Wentorf, R.H., and Bovenkerk, H.P., 1961, On the origin of natural diamonds: Journal of Astrophysics, v. 134, p. 995–1005, doi: 10.1086/147227.

Wyllie, P.J., 1996, Carbonate and carbonate-rich liquids in the Earth's interior, and critical fluids in diamond inclusions: Tsukuba, Ibaraki, Japan, Proceedings of the 3rd NIRIM (National Institute for Research in Inorganic Materials) International Symposium on Advanced Materials, p. 69–74.

Yamaoka, S., Shaji Kumar, M.D., Akaishi, M., and Kanda, H., 2000, Reactions between carbon and water under diamond-stable high pressure and high temperature conditions: Diamond and Related Materials, v. 9, p. 1480–1486, doi: 10.1016/S0925-9635(00)00274-0.

Yamaoka, S., Shaji Kumar, M.D., Kanda, H., and Akaishi, M., 2002a, Formation of diamond from $CaCO_3$ in a reduced C-O-H fluid at HP-HT: Diamond and Related Materials, v. 11, p. 1496–1504, doi: 10.1016/S0925-9635(02)00053-5.

Yamaoka, S., Shaji Kumar, M.D., Kanda, H., and Akaishi, M., 2002b, Crystallization of diamond from CO_2 fluid at high pressure and high temperature: Journal of Crystal Growth, v. 234, p. 5–8, doi: 10.1016/S0022-0248(01)01678-5.

MANUSCRIPT ACCEPTED BY THE SOCIETY 14 AUGUST 2006

Melting of ice VII and new high-pressure, high-temperature amorphous ice

Leonid Dubrovinsky
Bayerisches Geoinstitut, Universität Bayreuth, D-95440 Bayreuth, Germany

Natalia Dubrovinskaia
Mineralphysik und Strukturforschung, Mineralogisches Institut, Universität Heidelberg, 69120 Heidelberg, Germany
and *Lehrstuhl für Kristallographie, Physikalisches Institut, Universität Bayreuth, D-95440 Bayreuth, Germany,*
and *Bayerisches Geoinstitut, Universität Bayreuth, D-95440 Bayreuth, Germany*

ABSTRACT

Properties of H_2O at elevated pressure and temperature are of fundamental importance in both condensed matter physics and planetary sciences. We studied behavior of H_2O in externally heated diamond anvil cells (DACs) at pressures up to 50 GPa and temperatures to 1150 K by combining visual observations, Raman spectroscopy, and X-ray powder diffraction. The melting curve of H_2O was found to be well described by the Simon equation $P(\text{GPa}) = 2.2 + 1.31\{[T(\text{K})/364]^{3.3} - 1\}$. Above 30 GPa and 950 K, using visual observations and Raman spectroscopy, we found an X-ray amorphous phase clearly distinct from liquid H_2O. The new material reversibly transforms to ice VII and can be obtained on cooling or compression of liquid H_2O, which suggests that the high-pressure, high-temperature amorphous phase may be thermodynamically stable.

Keywords: diamond anvil cell, amorphous ice, melting, ice VII.

INTRODUCTION

Properties and phase relations of ice and water (H_2O) at high pressures and temperatures are of fundamental interest and importance for a number of physical, chemical, geophysical, and technological problems (Mishima and Stanley, 1998; Mishima and Suzuki, 2002; Hemley et al., 1987; Cavazzoni et al., 1999; Bina and Navrotsky, 2000; Evans, 2004; Klug, 2002; Umemoto et al., 2004; Besson et al., 1997). Information on fluid H_2O at very high pressure and temperature has been obtained largely from shock-wave experiments (Mitchell and Nellis, 1982; Lyzenga et al., 1982). The pressure-temperature phase diagram of H_2O ices is extremely complex (Petrenko and Whitworth, 2002), and various solid forms of H_2O have been and still are the subject of intense experimental and theoretical investigations. Ice Ih was the first system to be observed to amorphize under pressure (Mishima et al., 1984). Several glassy and amorphous ices (including low-density amorphous [LDA] and high-density amorphous [HAD] forms) have been observed to exist metastably at pressures below ~1.5 GPa and low temperatures (Mishima and Stanley, 1998; Mishima and Suzuki, 2002; Petrenko and Whitworth, 2002).

Ice VII is a stable H_2O phase at room temperature above 2.3 GPa (Petrenko and Whitworth, 2002). Melting of ice VII has been studied over the past 40 yr by different techniques (for recent reviews, see, for example, Datchi et al., 2000; Petrenko and Whitworth, 2002; Frank et al., 2004). Fei et al. (1993) studied melting of ice VII in externally electrically heated diamond anvil cell (DAC) to ~16 GPa by monitoring the disappearance of the (110) peak in the X-ray diffraction collected using an energy-dispersive detector. A similar technique was used by Frank et al. (2004). Datchi et al. (2000) followed melting of H_2O at pressure up to 13 GPa (temperatures ~750 K) by visual observations in externally heated DACs. Dubrovinskaia and Dubrovinsky (2003) extended the studied pressure range up to 37 GPa by employing the whole-cell heating assemblage for pressure and temperature generation and angle-dispersive synchrotron-based X-ray diffraction for detection of the ice VII melting event. Lin et al. (2004) applied Raman spectroscopy and visual observations in an externally heated DAC to detect melting of H_2O and reached 22 GPa and 900 K. Keeping in mind different techniques used for recognition of melting and different standards for pressure characterization, Au equation of state (Fei et al., 1993; Dubrovinskaia and Dubrovinsky, 2003) or fluorescence markers (Datchi et al., 2000; Lin et al., 2004, 2005), there is fair agreement between the data (Fei et al., 1993; Dubrovinskaia and Dubrovinsky, 2003; Frank et al., 2004; Datchi et al., 2000; Lin et al., 2004, 2005) obtained in electrically heated DACs (Fig. 1) (at 650 K, difference in pressure is in the order of 2 GPa). In contrast, the melting curve of H_2O measured in a laser-heated diamond anvil cell with visual observation of the laser-speckle pattern (Schwager et al., 2004) is at odds with all other data to date (at 15 GPa, difference is at least 250 K and increases with pressure). Goncharov et al. (2005) reported a melting curve of H_2O measured in laser-heated DAC by means of Raman spectroscopy and found reasonable agreement with results of the experiments in electrically heated DACs (note, however, that Goncharov et al. [2005] did not measure pressure at high temperature, and this fact significantly limited reliability of the data). Resolving such a controversy requires further experimental studies with accurate in situ measurements of pressure and temperature in diamond anvil cells.

EXPERIMENTAL DETAILS

In order to study the behavior of pure doubly distilled deionized H_2O at elevated pressures and temperatures, we employed external electrical heating diamond anvil cell (DAC) technique (Dubrovinskaia and Dubrovinsky, 2003). The high-pressure, high-temperature system was composed of three subunits, namely, the anvil assembly, mechanical loading mechanism, and heaters (Fig. 2). As a prototype for the anvil assembly, we used a so-called Tel Aviv University (TAU) type cell, which, itself is based on the Merrill-Bassett (Bassett, 2003) design. Different alloys were tested as construction materials—Inconel 718, Udimet 700, Republica, and Nimonic. Cells (including backing plates) made of Inconel 718 and Udimet 700 could be safely operated in air at temperature up to 973 K, and in an inert atmosphere (Ar + 2% H_2) up to 1173 K. Cells made of Republica alloy did not oxidize in air up to 1023 K, but the best performance was demonstrated

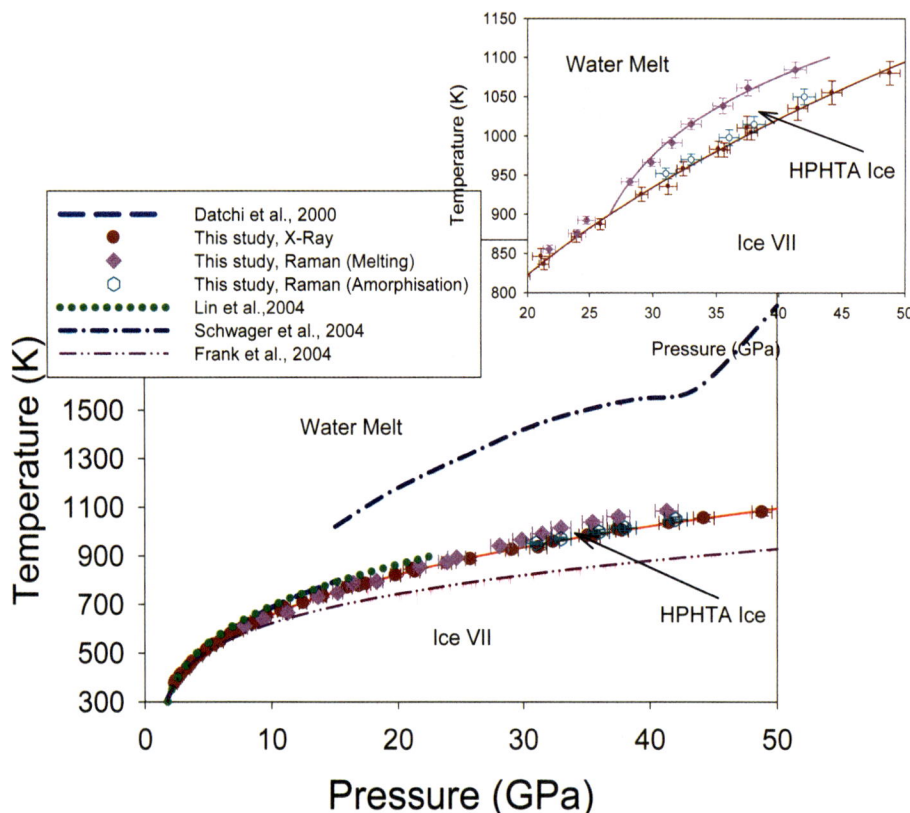

Figure 1. Phase relations among ice VII, H_2O melt, and high-pressure, high-temperature amorphous (HPHTA) ice. Dark-red circles—X-ray data (melting or transformation in noncrystalline phase); pink diamonds—melting according to Raman and visual observations; open hexagons—transformation of ice VII to HPHTA ice by Raman and visual data. Red solid line—curve describing transformation of ice VII to noncrystalline phase (Equation 1, melting curve of ice VII to ~30 GPa). Inset shows enlarged high-pressure part of the phase diagram. Solid pink line on inset describes transformation curve between HPHTA ice and melted H_2O according to equation $P = 28.2 + 4.51[(T/941)^{9.65} - 1]$.

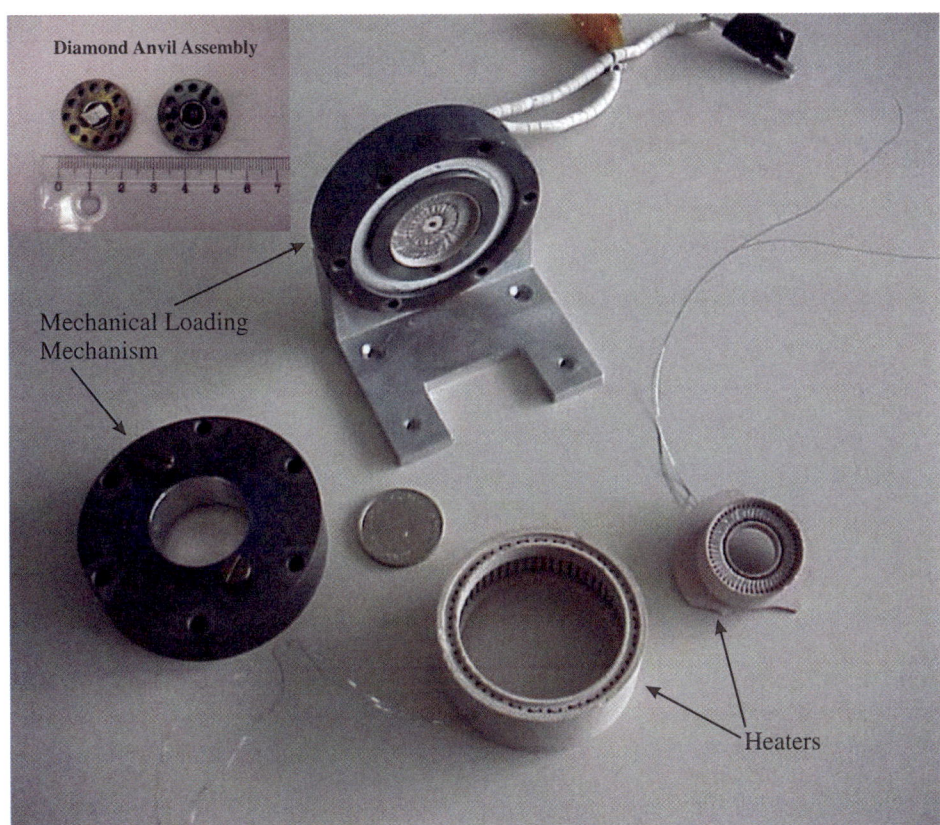

Figure 2. The diamond anvil assembly, mechanical loading mechanism, and heaters.

by cells made of Nimonic—up to 1123 K in air, and up to 1373 K in an inert atmosphere. All the experimental results described below were obtained from cells made of Nimonic.

Two massive cylindrical plates were used to apply mechanical loading to the anvil assembly (Fig. 2). One of the plates had six holes, while the other had six holes with M5 threads for tension screws. Screws made with hexagonal heads were easily accessible and could be adjusted even during heating. Both plates were made of Nimonic. The plates also have cylindrical holes and a trench for heaters.

The system included three heaters—two mounted directly on the outside of the front and back plates, and one placed around the cell (Fig. 2). We have tested heaters made of pyrophyllite and zirconia ceramics and found both of them to work satisfactory to 1473 K. As heating elements, we used either a nichrome wire (working range to 1073 K in air), or a molybdenum wire (tested to 1573 K in an inert atmosphere), or a platinum wire (works equally well in air and in an Ar atmosphere to 1373 K reached in high-pressure experiments).

A thermocouple (S type) was placed near the center of the high-pressure cell, next to the gasket-diamond interface. We found that on heating to 1273 K, a temperature gradient in the center of the high-pressure assemblage was practically absent. In the area of ~5 mm diameter around the diamonds, the temperature varied no more than 2–3 K.

Pressure was determined from the equation of state of gold and/or fluorescence sensors (Dubrovinskaia and Dubrovinsky, 2003) (Sm:YAG and ruby; Sm:YAG was used only at temperatures below 650 K, because it reacts with H_2O at high temperatures; Datchi et al., 2000). Natural low-fluorescence type IIa and high-purity synthetic (for Raman spectroscopy, see following) diamonds with 250–350-µm-diameter culets were used. Re gaskets were used at temperatures to 900 K. At higher temperatures during long-duration heating (dozens of minutes), Re starts to react with H_2O—shape of the hole in the gasket changes, "black clouds" start to grow at gasket-water interface, scanning electron micrograph (SEM) and microprobe analyses reveal contamination by Re (most probably in the form of ReO_3 oxide; exact chemical analysis is difficult because of the low amount and rough surface of the material) of the whole working surface of the diamonds after experiments. So, in experiments aimed to study temperatures approaching or higher than 900 K, we used Ir gaskets, which did not show any sign of reaction with H_2O. Changes in the state of H_2O were monitored by visual observations, Raman spectroscopy, and X-ray powder diffraction.

Raman spectra were recorded using Dilor XY (514 nm Ar laser, 1800 g/mm and 1200 g/mm grating, double-stage spectrometer, 1 cm^{-1} resolution) and LabRam (632 nm He-Ne laser, 1200 g/mm grating, ~2 cm^{-1} resolution) systems. High-pressure Raman study of H_2O ices is a challenging task due to a weak signal from the sample, merging of Raman-active stretching mode of ice VII into the second-order Raman signal from diamond anvils above 25 GPa, as well as the stresses and pressure gradients that develop in the ice. We overcame the latter problems by melting the samples and allowing them to subsequently cool slowly (~10 K/min) to the desired temperature: even at ~50 GPa (highest pressure in this study), the pressure variation in the sample cooled to room temperature was less than 1.5 GPa across the 130-μm-diameter pressure chamber. In order to improve a signal-to-noise ratio and to collect Raman spectra of ice VII and liquid H_2O that are especially weak at high temperature, we followed the methodology proposed by Goncharov et al. (1999): we used confocal optical configuration and spatial filtering with depth of focus down to 2 μm, as well as high-purity thin (1.2–1.5 mm thickness) synthetic diamonds. The background spectra collected from the part of the pressure chamber free of H_2O (it was blocked by a particle of Au or Ir on the surface of the anvil) were subtracted.

X-ray powder diffraction experiments were conducted at the European Synchrotron Radiation Facility (ESRF) on beam line ID30 and at Bayerisches Geoinstitut (BGI). At ESRF, the data were collected using the MAR345 or Bruker CCD detectors and an X-ray beam of 0.3738 Å wavelength with a size of 10 × 15 μm. The detector-to-sample distance varied in different experiments from 170 to 350 mm. At BGI, we used FR-D high-brilliance Rigaku X-ray rotating anode generator (MoKα radiation, operating at 55 kV and 60 mA) with Osmic MaxFlux focusing optics that allowed us to produce a beam of ~40 μm in diameter. Data were collected using a Bruker APEX CCD area detector. The opening angle of 36° in the backing plates allowed us to collect diffraction data down to a d-spacing of 1.1 Å. The collected images were integrated using the Fit2D program in order to obtain a conventional diffraction spectrum.

RESULTS AND DISCUSSION

Melting of Ice VII

Visual observations have been proven to be a reliable method in detecting the melting of ice VII at pressures of ~20 GPa (Datchi et al., 2000; Lin et al., 2004). At higher pressures, however, samples become thinner, boundaries between grains are difficult to distinguish, optical access suffers at higher temperatures (which are necessary in order to melt H_2O at higher pressure), and visual identification of melting becomes problematic (Fig. 3). Angle-dispersive X-ray diffraction with an area detector is a powerful method for studying melting (Dubrovinskaia and Dubrovinsky, 2003; Shen et al., 2004). We monitored melting by the disappearance on heating or appearance on cooling, of reflections from ice VII (Dubrovinskaia and Dubrovinsky, 2003). As the major criterion of loss of crystalline structure by ice VII, we used the appearance of a diffusion halo. As shown in Figure 1, our X-ray data points, extended in this study to 48 GPa, fall along a line that can be describe by Simon equation (Dubrovinskaia and Dubrovinsky, 2003):

$$P = 2.2 + 1.31[(T/364)^{3.3} - 1].$$

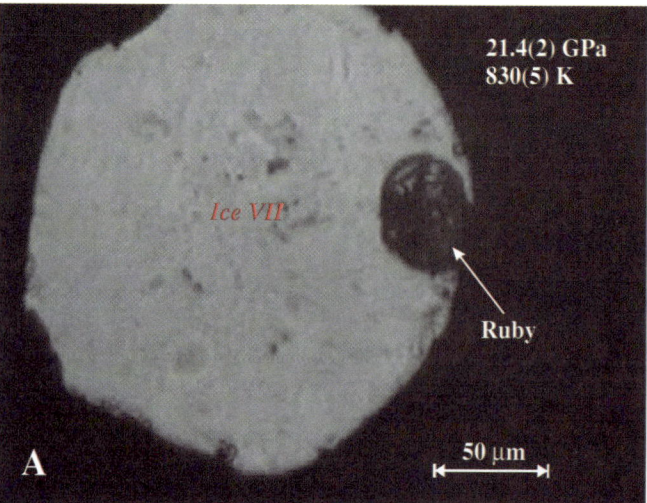
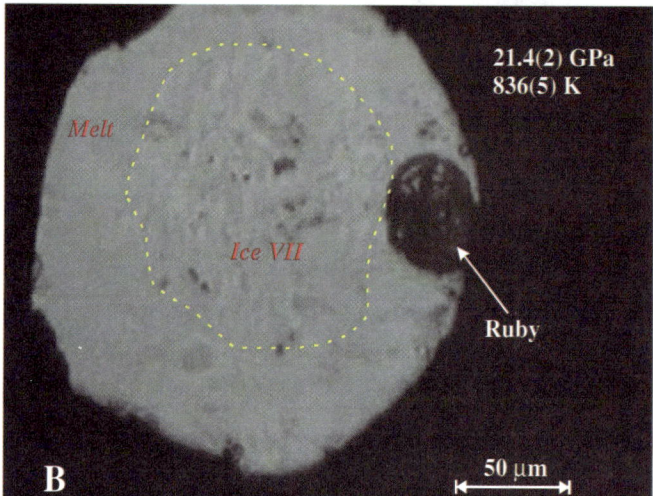
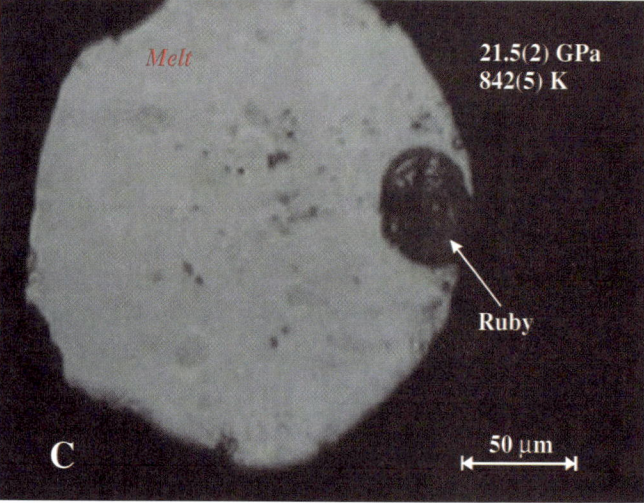

Figure 3. Photographs taken in course of melting of ice VII (upon heating at 21.4[0.2] GPa). Dotted line in B shows an area of ice VII surrounded by H_2O melt. Even at relatively low pressure (~20 GPa), it is difficult to distinguish weak grain boundaries in ice VII (A).

Melting manifested itself in Raman spectra by disappearance of active translational modes, changes in shape, and reduction of intensity of OH-stretching modes (Petrenko and Whitworth, 2002; Lin et al., 2004) (Fig. 4). The results of observations of melting of ice VII using Raman spectroscopy are summarized in Figure 1. Spectroscopic and X-ray observations are in good agreement at pressure up to 30 GPa, although Raman data are more scattered.

Relatively small difference in the common pressure range (e.g., to 15 GPa in Fei et al. [1993], Datchi et al. [2000], and 22 GPa in Lin at al. [2004]) between our curves and previously reported melting curves of ice VII measured in externally heated diamond anvil cells can be attributed to the difference in methods of registration of the melting event (e.g., X-ray diffraction with energy-dispersive detector used by Fei et al. [1993] and Frank et al. [2004] versus angle-dispersive in the present study), to the difference in pressure measurements (Au and ruby in direct contact with H_2O sample in our work versus Sm:YAG isolated from H_2O in Datchi et al. [2000]), or to the presence of additional substances (particularly MgO in experiments of Frank et al. [2004]) in a pressure chamber, which could dissolve in H_2O and affect melting temperatures. Significant discrepancies with data obtained in laser-heated DAC (Schwager et al., 2004) remain unresolved and can be related, as proposed by Lin et al. (2004), to methodology of temperature measurements in the sample while a metallic hot-plate is used as an absorber of laser radiation.

Goncharov et al. (2005) found an abrupt increase in the melting temperature of H_2O at pressures above 47 GPa by measurements of Raman spectra in laser-heated DACs, while Lin et al. (2005) reported significant increase in the temperature stability of ice VII at pressures above 35 GPa (in fact, Lin et al. [2005] were not able to melt H_2O above 35 GPa). Moreover, Lin et al. (2005) reported partial dissociation of H_2O and formation of pure oxygen detected by Raman spectroscopy. We measured the melting curve of ice VII to ~40 GPa by means of Raman spectroscopy and by X-ray powder diffraction to 49 GPa and did not confirm sudden changes in the melting temperature of ice in the studied pressure range or H_2O dissociation. We observed, however, significant increase in reactivity of the H_2O at temperatures above 1000 K (see experimental section) and suggest that chemical contamination could be responsible for inconsistence in the all previous studies of the melting curve of ice at pressures exceeding 35 GPa.

Phase Transition of Ice VII to High-Pressure, High-Temperature Amorphous Ice

At pressures above 30 GPa and heating at temperatures approaching those of the expected melting, we observed the formation of a distinct pattern that apparently looked like grain boundaries (we called it "cell pattern") (Fig. 5). Note, that we did not observe recrystallization of ice VII at lower pressure before melting. Raman spectra corresponding to the cell pattern are clearly different from those of ice VII and from the spectra of "usual" H_2O melted at high pressure and temperature (Fig. 6). For example, upon heating to 980 K at 33 GPa, the cell pattern forms, while in Raman spectra, translational modes disappear and OH-stretching modes broaden and shift to higher frequencies (Fig. 6). On further heating to 1015 K, the "cells" melt, the intensity of Raman OH-stretching modes rapidly decreases, and the spectra look like those typical for H_2O melt at corresponding pressures and temperatures. The diffusion halo in X-ray diffraction images collected at 33 GPa and temperatures above 980 K unambiguously reveals a noncrystalline state of the material at corresponding conditions (Fig. 7). Our observations imply that the cell pattern forms when ice VII transforms to an amorphous phase (which we term high-pressure high-temperature amorphous [HPHTA] ice). We think that the cells are observed due to formation of cracks as a result of stresses generated by the difference in densities of ice VII and the amorphous phase (similar to the system of cracks that forms at the phase transition between low- and high-density amorphous ices at low pressure; see Mishima and Suzuki, 2002).

The formation of the cell pattern appears visually in the course of transformation from ice VII to HPHTA ice (e.g., on heating or decompression of ice VII at appropriate pressures and temperatures; Fig. 1). On reverse paths (by compression or cooling melted H_2O toward ice VII), we did not observe clear cell patterns, although perturbation of the images and occasional cracks were observed. At the same time, Raman spectra (Fig. 6) provide clear evidence that ice VII and liquid H_2O reversibly transform into HPHTA ice. The reversibility of the transformations suggests that HPHTA ice may be a thermodynamically stable phase, especially taking into account that the appearance of the amorphous phase does not depend on

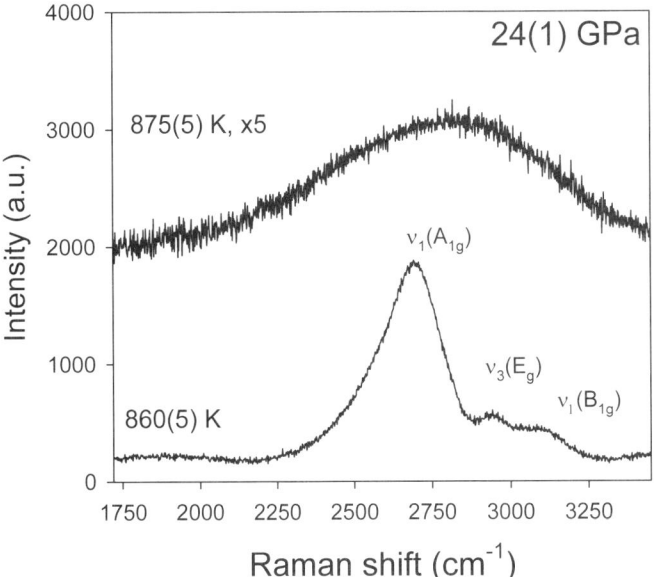

Figure 4. Raman spectra of ice VII at 24($\pm$1) GPa and 860($\pm$5) K and melted H_2O at the same pressure and 875($\pm$5) K. a.u.—arbitrary units.

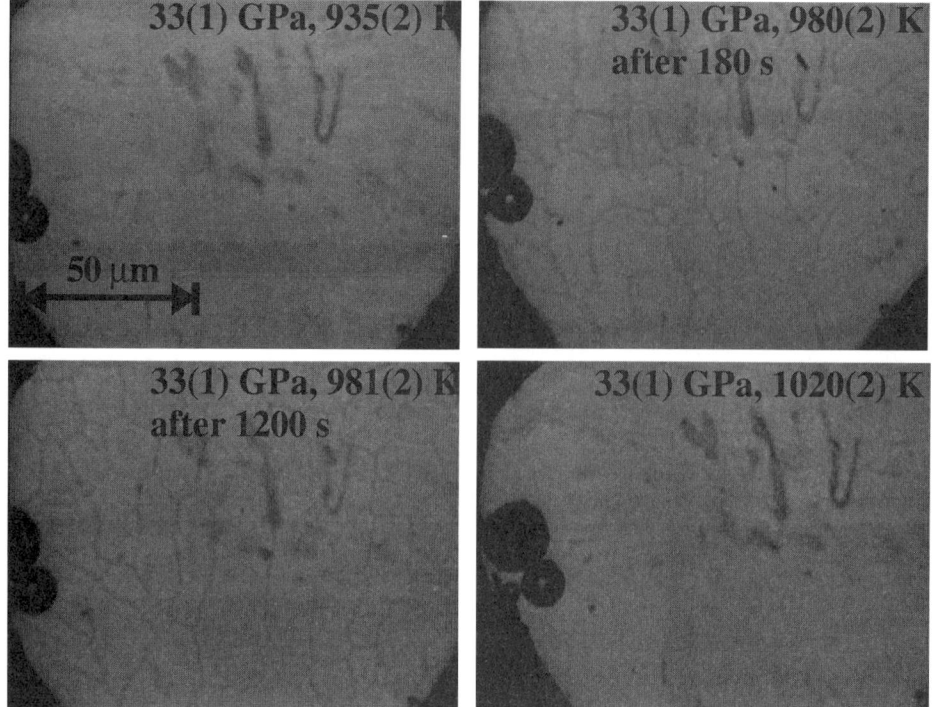

Figure 5. Photographs taken upon heating of ice VII at 33(±1) GPa. At 935(±2) K, solid phase is ice VII (upper left). After increasing temperature to ~980 K with a speed of 2 K/min, we observed formation of the "cell pattern" (upper right and bottom left; time was counted since temperature reached 980 K). Cell pattern is preserved for dozens of minutes if temperature and pressure remain constant, although the shape of individual cells changes. Upon further heating above 1015 K, cells disappear (bottom right). Small spheres in the left part of photographs are ruby crystals.

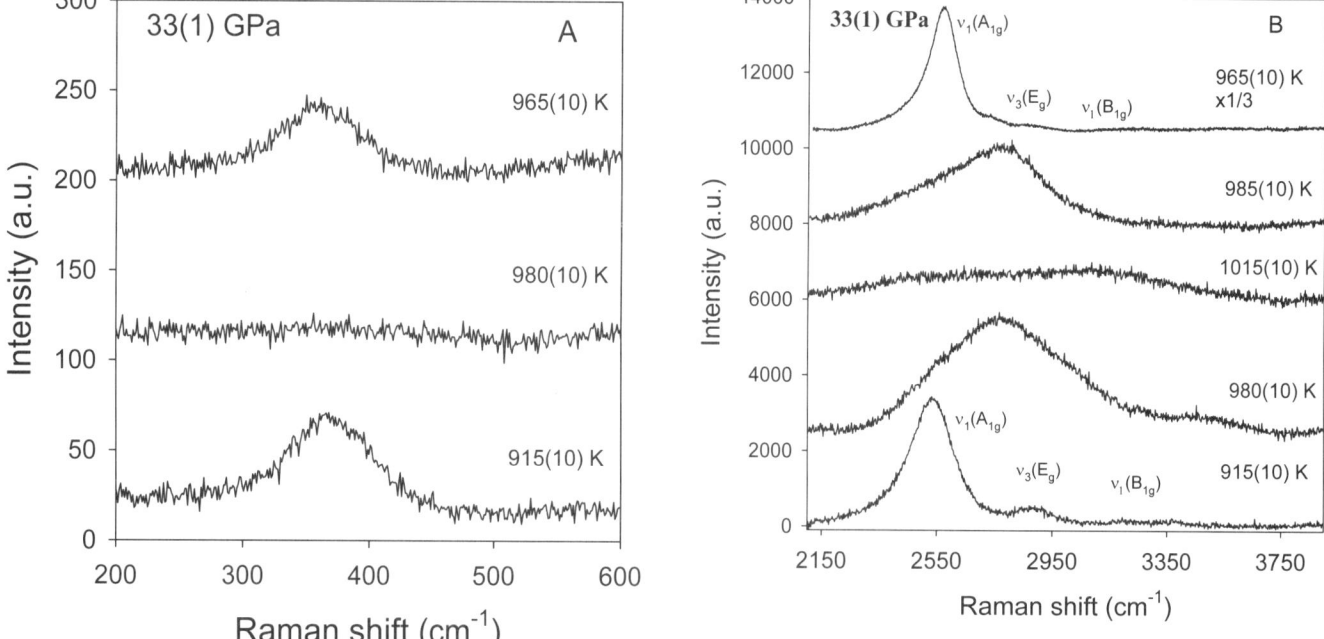

Figure 6. Raman spectra of H_2O collected at 33(±1) GPa and different temperatures in (A) low- and (B) high-frequency regions. Spectra at 915(±10) K on heating and 965(10) K on cooling correspond to ice VII. At 980(±10) K on heating and 985(±10) K on cooling, we observed spectra of the new high-pressure, high-temperature amorphous ice. At 1015(±10) K, the spectrum is from melted H_2O. For all spectra, collection time was 600 s (except upper one in B, which was collected for 1800 s). a.u.—arbitrary units.

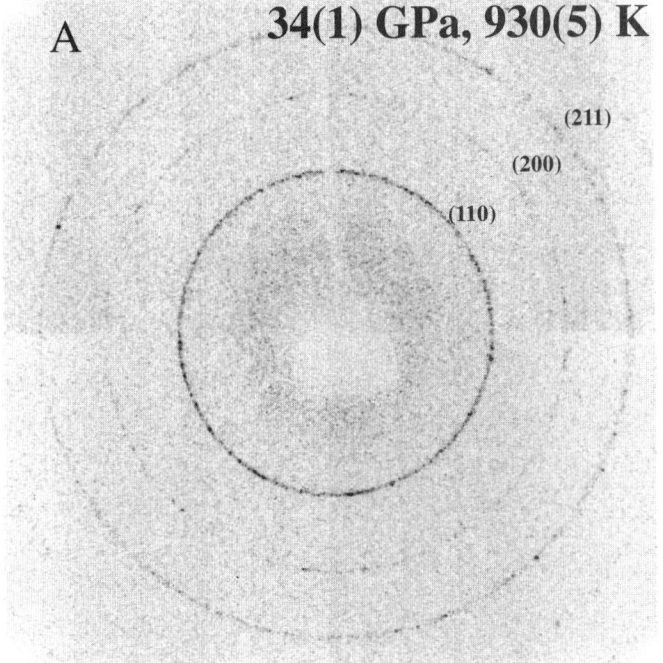
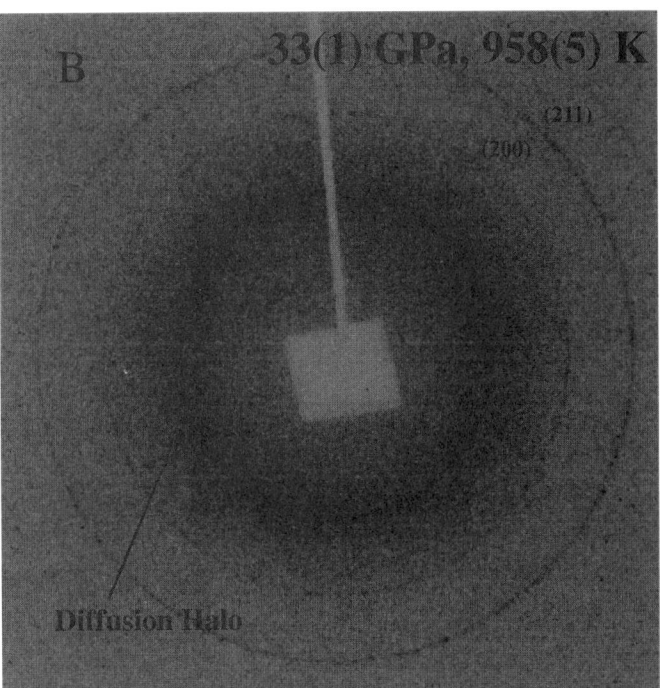
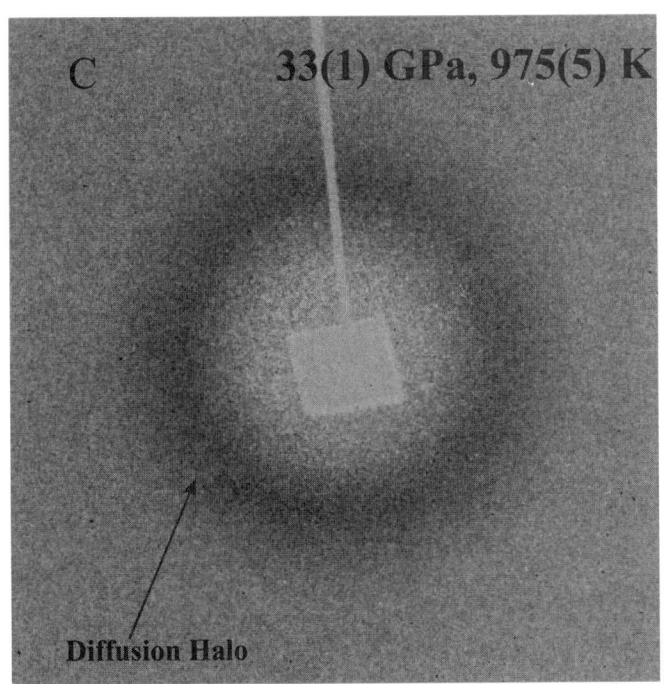
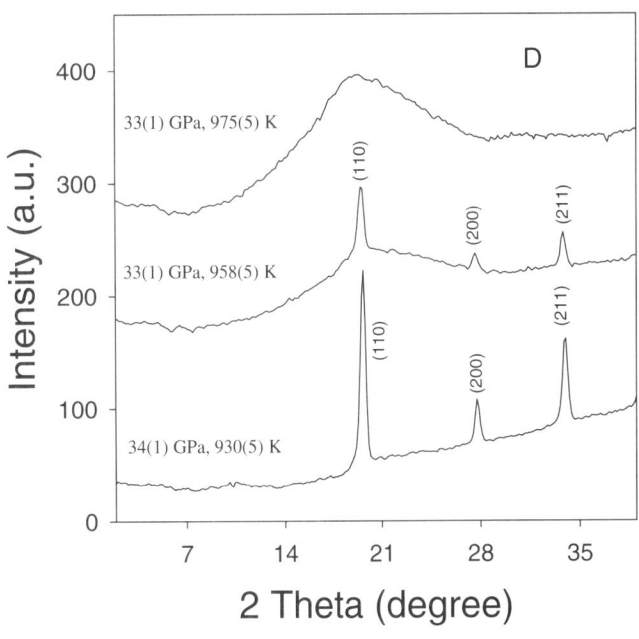

Figure 7. Example of two-dimensional diffraction images collected on heating of ice at high pressure and temperature. At 34(±1) GPa and 930(±5) K, (A) the pattern corresponds to the cubic ice VII phase with lattice parameter 2.938(±3) Å. At 33(±1) GPa and 958(±5) K, (B) the crystalline phase coexists with the X-ray amorphous phase. At 975(±5) K and 33(±1) GPa, (C) only the diffusion halo from noncrystalline phase is observed. (D) Integrated diffraction patterns of the images shown in A–C. a.u.—arbitrary units.

the rate of cooling or heating (ranging in our study from 1 °/h to 10 °/min), and it can be preserved for dozens of minutes if temperature and pressure remain constant.

Apart from very rich polymorphism of crystalline phases, H_2O is known to have several X-ray amorphous phases at low pressures and temperatures, including low- and high-density amorphous ices (Mishima and Stanley, 1998; Mishima and Suzuki, 2002; Petrenko and Whitworth, 2002). Even liquid H_2O is thought to exist in certain thermodynamic conditions in two distinct (low- and high-density) forms (Mishima and Stanley, 1998; Mishima and Suzuki, 2002). New HPHTA ice is clearly distinct from all known H_2O phases as shown by its Raman signature (Fig. 6), as well as by the pressure and temperature conditions at which it was observed. Moreover, all amorphous ices known to date are metastable and cannot reversibly transform to phases stable at corresponding conditions.

Pressure-induced amorphization of solids is a complex phenomenon not fully understood yet. Experimental and theoretical studies (Umemoto et al., 2004; Umemoto et al., 2004; Besson et al., 1997) relate phase transformations and amorphization of ices VIII and XI to the nonlinear behavior and softening of some stretching and transitional modes, which, in turn, lead to collapse of the hydrogen-bonded networks and mechanical instabilities. Ab initio molecular dynamic simulations (Cavazzoni et al., 1999) suggest that at pressures above 30 GPa and temperatures above 1000 K, H_2O becomes superionic with the oxygen Body centred cubic (bcc) sublattice preserved, and hydrogen atoms jump among equivalent sites along the O-O separations. Apparent coincidence of the thermodynamic conditions at which theory (Cavazzoni et al., 1999) predicts formation of superionic H_2O and at which we found HPHTA ice indicates that these two phenomena may be related. However, further detailed combined experimental and theoretical studies are necessary in order to the uncover atomistic nature of the new high-pressure, high-temperature amorphous ice.

CONCLUSIONS

We followed the melting curve of H_2O to ~50 GPa by combining visual observations, Raman spectroscopy, and angle-dispersive X-ray powder diffraction techniques. At 49($\pm$1) GPa, melting occurred at 1080($\pm$10) K. Our data support the proposal of Bina and Navrotsky (2000) that high-pressure ice could be present in cold subducting slabs. Above 30 GPa and 950 K, using visual observations and Raman spectroscopy, we found an X-ray amorphous phase clearly distinct from liquid H_2O. The new material reversibly transforms to ice VII and can be obtained on cooling or compression of liquid H_2O, suggesting that the high-pressure, high-temperature amorphous phase may be thermodynamically stable. From a methodological point of view, our observations imply that the X-ray diffraction technique alone may not be sufficient to study melting curves of solids due to their possible amorphization prior to melting.

REFERENCES CITED

Bassett, W.A., 2003, High pressure-temperature aqueous systems in the hydrothermal diamond anvil cell (HDAC): European Journal of Mineralogy, v. 15, p. 773–780, doi: 10.1127/0935-1221/2003/0015-0773.

Besson, J.M., Klotz, S., Hamel, G., Marshall, W.G., Nelmes, R.J., and Loveday, J.S., 1997, Structural instability in ice VIII under pressure: Physical Review Letters, v. 78, p. 3141–3144, doi: 10.1103/PhysRevLett.78.3141.

Bina, C.R., and Navrotsky, A., 2000, Possible presence of high-pressure ice in cold subducting slabs: Nature, v. 408, p. 844–847, doi: 10.1038/35048555.

Cavazzoni, C., Chiarotti, G.L., Scandolo, S., Tosatti, E., Bernasconi, M., and Parrinello, M., 1999, Superionic and metallic states of water and ammonia at giant planet conditions: Science, v. 283, p. 44–46, doi: 10.1126/science.283.5398.44.

Datchi, F., Loubeyre, P., and LeToullec, R., 2000, Extended and accurate determination of the melting curves of argon, helium, ice (H_2O), and hydrogen (H_2): Physical Review B: Condensed Matter and Materials Physics, v. 61, p. 6535–6546.

Dubrovinskaia, N., and Dubrovinsky, L., 2003, Whole-cell heater for the diamond anvil cell: The Review of Scientific Instruments, v. 74, p. 3433–3477, doi: 10.1063/1.1578151.

Evans, R., 2004, Just add more water: Nature, v. 429, p. 356–357, doi: 10.1038/429356a.

Fei, Y., Mao, H.-K., and Hemley, R.J., 1993, Thermal expansivity, bulk modulus, and melting curve of H_2O–ice VII to 20 GPa: The Journal of Chemical Physics, v. 99, p. 5369–5373, doi: 10.1063/1.465980.

Frank, M.R., Fei, Y.W., and Hu, J.Z., 2004, Constraining the equation of state of fluid H_2O to 80 GPa using the melting curve, bulk modulus, and thermal expansivity of ice VII: Geochimica et Cosmochimica Acta, v. 68, p. 2781–2790, doi: 10.1016/j.gca.2003.12.007.

Goncharov, A., Struzhkin, V.V., Mao, H., and Hemley, R.J., 1999, Raman spectroscopy of dense H_2O and the transition to symmetric hydrogen bonds: Physical Review Letters, v. 83, p. 1998–2001, doi: 10.1103/PhysRevLett.83.1998.

Goncharov, A.F., Goldman, N., Fried, L.E., Crowhurst, J.C., Kuo, I.-F.W., Mundy, C.J., and Zaug, J.M., 2005, Dynamic ionization of water under extreme conditions: Physical Review Letters, v. 94, p. 125508 (1–4).

Hemley, R.J., Jephcoat, A.P., Mao, H.K., Zha, C.S., Finger, L.W., and Cox, D.E., 1987, Static compression of H_2O-ice to 128 GPa: Nature, v. 330, p. 737–740, doi: 10.1038/330737a0.

Klug, D.D., 2002, Dense ice in detail: Nature, v. 420, p. 749–750, doi: 10.1038/420749a.

Lin, J.-F., Militzer, B., Struzhkin, V.V., Gregoryanz, E., Hemley, R.J., and Mao, H., 2004, High pressure-temperature Raman measurements of H_2O melting to 22 GPa and 900 K: The Journal of Chemical Physics, v. 121, p. 8423–8424, doi: 10.1063/1.1784438.

Lin, J.-F., Gregoryanz, E., Struzhkin, V.V., Somayazulu, M., Mao, H.K., and Hemley, R.J., 2005, Melting behaviour of H_2O at high pressures and temperatures: Geophysical Research Letters, v. 32, p. L11306, doi: 10.1029/2005GL022499.

Lyzenga, G.A., Ahrens, T.J., Nellis, W.J., and Mitchell, A.C., 1982, The temperature of shock-compressed water: The Journal of Chemical Physics, v. 76, p. 6282–6286, doi: 10.1063/1.443031.

Mishima, O., and Stanley, H.E., 1998, The relationship between liquid, supercooled and glassy water: Nature, v. 396, p. 329–335, doi: 10.1038/24540.

Mishima, O., and Suzuki, Y., 2002, Propagation of the polyamorphic transition of ice and the liquid-liquid critical point: Nature, v. 419, p. 599–603, doi: 10.1038/nature01106.

Mishima, O., Calvert, L.D., and Whaley, E., 1984, 'Melting' ice I at 77 K and 10 kbar: A new method of making amorphous solids: Nature, v. 310, p. 393–395, doi: 10.1038/310393a0.

Mitchell, A.C., and Nellis, W.J., 1982, Equation of state and electrical conductivity of water and ammonia shocked to the 100 GPa (1 Mbar) pressure range: The Journal of Chemical Physics, v. 76, p. 6273–6281, doi: 10.1063/1.443030.

Petrenko, V.F., and Whitworth, R.W., 2002, Physics of Ice: New York, Oxford University Press, 373 p.

Schwager, B., Chudinovskich, L., Gavriliuk, A., and Boehler, R., 2004, Melting curve of H_2O to 90 GPa measured in a laser-heated diamond cell: Journal of Physics Condensed Matter, v. 16, p. S1177–S1179, doi: 10.1088/0953-8984/16/14/028.

Shen, G.Y., Prakapenka, V.B., Rivers, M.L., and Sutton, S.R., 2004, Structure of liquid iron at pressures up to 58 GPa: Physical Review Letters, v. 92, p. 5701–5704.

Umemoto, K., Wentzcovitch, R.M., Baroni, S., and de Gironcoli, S., 2004, Anomalous pressure-induced transition(s) in ice XI: Physical Review Letters, v. 92, p. 105502 (1–4).

MANUSCRIPT ACCEPTED BY THE SOCIETY 14 AUGUST 2006

Effect of water on the phase relations in Earth's mantle and deep water cycle

Konstantin D. Litasov[†]
Eiji Ohtani

Institute of Mineralogy, Petrology and Economic Geology, Faculty of Sciences, Tohoku University, Sendai 980-8578, Japan

ABSTRACT

Water is transported by subducting slabs into the transition zone and lower mantle. Important water carriers to the deep mantle may be serpentine, chlorite, phase A, and superhydrous phase B in peridotite; zoisite, lawsonite, and phengite in basalt; and topaz-OH and phase Egg in sediments. Phase D, stable in peridotite, and the δ-AlOOH phase in the sedimentary component may transport water to at least 1200–1500 km depth. Phase relations in hydrous peridotite show that the phase boundaries of olivine to wadsleyite and ringwoodite to Mg-perovskite + ferropericlase phase transitions shift to lower and higher pressures, respectively. Thus, elevation of the 410 km discontinuity and depression of the 650 km discontinuity in subduction zones may be partially affected by water. Water may also control the separation of the basaltic layer of the slab near the 650 km discontinuity. The density crossover that occurs under dry conditions between peridotite and basalt components of slab near 650 km disappears under hydrous conditions due to a significant shift to lower pressures of post-garnet transformation in basalt. Due to high water solubility in wadsleyite and ringwoodite, the transition zone may be a water reservoir in Earth's interior. Recent data show that the transition zone at least locally may contain 0.2–1.5 wt% H_2O. In addition, the transition zone may serve as a water absorber for the water circulation system of the mantle. The upper mantle appears to be largely degassed through the action of mid-ocean-ridge and hotspot volcanism. The water storage capacity of the lower mantle as well as hydrogen storage potential of the core are still uncertain. There are several potential dehydration sites in the mantle that may control water circulation through plate tectonics: the mantle wedge above subduction slabs, a region above the 410 km discontinuity, the top of the lower mantle, and the deep lower mantle.

Keywords: mantle, water, phase transition, high pressure.

[†]E-mail: klitasov@ganko.tohoku.ac.jp.

Litasov, K.D., and Ohtani, E., 2007, Effect of water on the phase relations in Earth's mantle and deep water cycle, *in* Ohtani, E., ed., Advances in High-Pressure Mineralogy: Geological Society of America Special Paper 421, p. 115–156, doi: 10.1130/2007.2421(08). For permission to copy, contact editing@geosociety.org. ©2007 Geological Society of America. All rights reserved.

INTRODUCTION

Although many comprehensive reviews have been published on water circulation in the Earth (e.g., Thompson, 1992; Gasparik, 1993; Mysen et al., 1998; Frost, 1999; Prewitt and Parise, 2000; Williams and Hemley, 2001), over the last couple of years, a significant amount of new information has emerged to indicate that the role of water in Earth's interior might be underestimated.

Hydrogen is the most abundant element in our solar system (90%), while oxygen ranks third (0.08%). In Earth, instead, oxygen is the second most abundant element (after Fe), composing 30 wt% of bulk earth, whereas hydrogen abundance is uncertain. Estimates of the hydrogen abundance in Earth have spanned a range from approximately one ocean/hydrosphere mass to on the order of several hydrospheres if hydrogen is a dominant light element in Earth's core.

Understanding of the global history of Earth and its evolution strongly depends on the fundamental question: how efficiently is Earth's interior degassed and regassed? From the "hydrous" point of view, this problem is related to the following processes: (1) the initial amount of primordial water in Earth's parental body; (2) the water dynamics during early Earth's evolution, the influence of possible impact inputs, the differentiation of a magma ocean and segregation of the core; and (3) the efficiency of water exchange between the interior and the surface during plate tectonic history. With the initiation of plate tectonics, subduction of oceanic and some continental material and volcanism at island arcs, mid-ocean ridges and hotspots work to transport volatiles including water or hydrogen in and out Earth's interior.

Water has played many important roles in Earth's evolution because of its strong influence on the chemical and physical properties of minerals, melts, and fluids. It produces major distinguishing topographic features of the surface of our planet and plays an active role in the generation of continents (Campbell and Taylor, 1983). Water has a remarkable effect on weakening of rocks and minerals and displacement of phase transition boundaries. Water-induced reduction of the viscosity and strength of mantle materials may be crucial for understanding the driving forces of plate tectonics (e.g., Stevenson, 2004). It also dramatically depresses the melting temperature of silicate minerals.

The effect of water on phase transformations in the mantle transition zone and the core-mantle boundary may lead to a significant revision of temperature distribution inside the Earth and its spatial and temporal energy balance. The uncertainty in the abundance of hydrogen in core material gives rise to variations by thousand of degrees in the inferred temperature profile of Earth's core (e.g., Williams and Hemley, 2001).

Several studies have indicated that water took part in the generation of voluminous mantle melting and formation of continental flood basalts (Gallagher and Hawkesworth, 1992; Turner and Hawkesworth, 1995) and Archean komatiites (Stone et al., 1997; Parman et al., 1997; Shimizu et al., 2001; Wilson et al., 2003). It is most likely that water played an important role during solidification of a magma ocean and core segregation during early evolution of Earth (e.g., Abe et al., 2000).

Recent analytical and experimental studies have demonstrated that many hydrous phases may be stable at mantle conditions. In particular, several dense hydrous magnesium silicates like phase A, D (G), and superhydrous B, are stable at upper- and lower-mantle conditions (e.g., Frost, 1999; Ohtani et al., 2001b, 2004). Moreover, significant amounts of water, in a volume of at least several current hydrospheres, could be stored in nominally anhydrous minerals, such as olivine, garnet, and stishovite (e.g., Bell and Rossman, 1992a; Pawley et al., 1993; Ingrin and Skogby, 2000; Bolfan-Casanova et al., 2000).

Many recent geophysical studies have provided information about the possible existence of water in Earth's mantle, and especially in its transition zone, where wadsleyite and ringwoodite can incorporate up to 3 wt% of H_2O in their structures (e.g., Inoue et al., 1995; Kohlstedt et al., 1996). Low-velocity zones detected on top of the 410 km discontinuity may indicate the existence of trapped high-density melt (e.g., Revenaugh and Sipkin, 1994; Song et al., 2004), which is most likely to be hydrous according to differences in water solubility between olivine and wadsleyite and melting temperature of anhydrous peridotite mantle. Electrical conductivity anomalies in the upper mantle and transition zone (e.g., Fukao et al., 2004) can also be interpreted as water-bearing zones rather than high-temperature anomalies.

This paper reviews the aforementioned interrelated problems, and special attention is paid to the influence of water on phase relations in the mantle and the potential of different mantle phases to store water. We also discuss the origin of water and water cycling in Earth's mantle since the time of formation, which is schematically illustrated in Figure 1.

HYDROUS PHASES IN THE MANTLE

Mantle Composition and Subduction Zones

Ringwood (e.g., Ringwood, 1975) proposed the pyrolite model for averaging Earth's mantle composition, based on observation of peridotite samples worldwide, that combines primitive basaltic composition with refractory dunite in a 1:3 proportion (Table 1). This model was modified with time, when several other pyrolite-like primitive-mantle compositions were proposed (Jagoutz et al., 1979; McDonough and Sun, 1995; McDonough and Rudnick, 1998). Anhydrous pyrolite contains 50%–60% olivine by volume and minor pyroxenes and garnet in the uppermost mantle. The phase relations of olivine (α-$[Mg,Fe]_2SiO_4$) imply sharp progressive high-pressure phase transitions from olivine to wadsleyite (modified spinel or β-$[Mg,Fe]_2SiO_4$), then to ringwoodite (silicate spinel or γ-$[Mg,Fe]_2SiO_4$) and to a mixture of silicate perovskite and ferropericlase (magnesiowüstite). The remaining nonolivine components, such as orthopyroxene, clinopyroxene, and garnet, undergo gradual high-pressure transitions of dissolution of pyroxenes into garnet, resulting in the pyrope-majorite solid solution, and they eventually transform into a

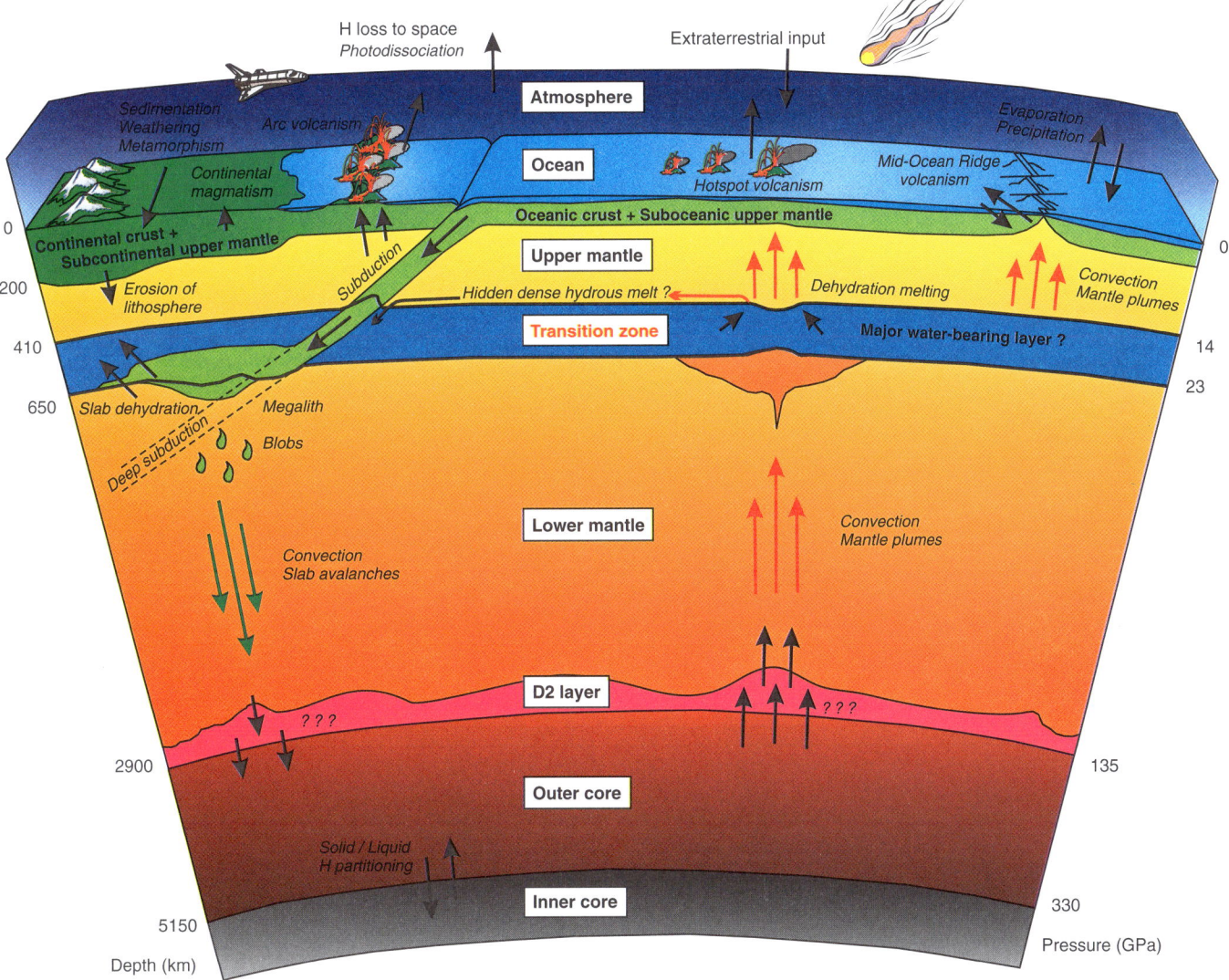

Figure 1. Schematic model for global water circulation and exchange between the mantle and other spheres using data by Williams and Hemley (2001), Ohtani et al. (2001b, 2004, 2005), Poli and Schmidt (2002), Ohtani (2002, 2005), and Bercovici and Karato (2003). Arrows indicate directions of water or hydrogen movement. See text for more explanations.

perovskite phase at higher pressures (Figs. 2 and 3). Akimotoite ($MgSiO_3$-ilmenite phase) and CaMg-perovskite (e.g., Asahara et al., 2005) may exist at depths close to the 650 km discontinuity under lower-temperature conditions. These two phases, as well as some hydrous minerals (see next section), may introduce complexities into seismic velocity structures near the 650 km depth, as reported by many authors (e.g., Niu and Kawakatsu, 1996; Simmons and Gurrola, 2000; Ai et al., 2003). The major part of the lower mantle consists of 78–80 wt% Mg-perovskite, 15–16 wt% ferropericlase, and 5–6 wt% Ca-perovskite (e.g., Wood, 2000). The lowermost part of the mantle, D″ layer, may contain a "post-perovskite phase" (e.g., Murakami et al., 2004), post-stishovite oxide components, such as $CaCl_2$ or α-PbO_2–type SiO_2 and $(Mg,Fe)O$, or some Fe-alloys (e.g., Knittle and Jeanloz, 1991; Garnero and Helmberger, 1995; Luo et al., 2002).

Subduction of the lithosphere to the deep mantle may be the most important source for mantle refertilization and, therefore, supply of water into deep reservoirs. Oceanic slabs may penetrate as deep as the 650 km seismic discontinuity. Some authors have argued that slabs may sink to the core-mantle boundary (e.g., Grand et al., 1997; Van der Hilst et al., 1998; Fukao et al., 2001). Magmatic activities in mid-ocean ridges, island arcs, and hotspots coupled with whole-mantle or layered convection act in the opposite direction to stir heterogeneities produced by subduction and to deliver water back to the surface.

A simplified lithologic section of the oceanic lithosphere, i.e., the subducting slab, includes a thin upper layer of sediments (0–1 km), a layer of mafic rocks (7–10 km), pillow basalts, gabbro, troctolites, which have mid-ocean-ridge basalt (MORB) bulk composition on average, and a lower layer of

TABLE 1. MODEL COMPOSITION (WT%) OF MAJOR MATERIALS RELEVANT TO EARTH'S MANTLE AND ITS SUBDUCTION ZONES

Sample	Pyrolite (1)	Pyrolite (2)	Piclogite (3)	C1 (4)	Lherzolite (5)	Harzburgite (5)	Dunite (5)
SiO_2	45.1	45.0	47.0	22.01	39.53	40.39	38.37
TiO_2	0.2	0.20	0.4	0.07	0.32		0.11
Al_2O_3	4.3	4.45	8.6	1.55	2.62	0.85	1.06
Cr_2O_3	0.4	0.38	0.2	0.37	0.20	0.40	0.24
FeO	8.0	8.05	10.8	22.84	11.15	9.65	11.54
MnO	0.1	0.135	0.1	0.24	0.20	0.14	0.13
MgO	38.1	37.80	24.0	15.34	32.95	36.71	36.28
NiO	0.2	0.25	0.1	1.32	0.25	0.25	0.25
CaO	3.1	3.55	8.0	1.26	1.95	0.93	0.60
Na_2O	0.4	0.36	1.0	0.65	0.30	0.04	0.05
K_2O	0.03	0.029	0.1	0.06	0.03	0.01	0.01
P_2O_5	0.02	0.021		0.20	0.02		0.03
H_2O		0.090		17.51†	9.49	9.76	10.33
CO_2		0.044		11.36			
Total	99.55	100.37	100.3	100.0	99.01	99.13	99.00

Sample	MORB (6)	Altered MORB (7)	Fe-gabbro (8)	Gabbro (5)	Troctolite (5)	Volcanoclastic (9)	Red clay (9)
SiO_2	50.45	45.80	46.78	44.28	41.09	55.97	52.60
TiO_2	1.62	1.18	4.47	0.28	0.08	1.03	0.67
Al_2O_3	15.26	15.53	11.30	16.00	25.89	12.51	14.01
FeO	10.43	9.01	17.02	4.05	4.73	7.36	7.81
MnO	0.17	0.17	0.26	0.10	0.11	0.26	1.68
MgO	7.58	6.66	6.72	10.05	10.14	5.39	3.05
CaO	11.30	12.88	9.18	18.28	9.83	5.25	3.14
Na_2O	2.68	2.07	3.85	0.60	1.62	3.21	2.79
K_2O	0.11	0.56	0.10	0.05	0.06	1.67	2.84
P_2O_5	0.15	0.11	0.32	0.05	0.05	0.24	0.66
H_2O	0.25	2.68		5.71	5.86	6.25	10.04
CO_2		2.95				0.73	0.39
Total	100.0	99.6	100.0	99.45	99.46	99.87	99.68

Sample	Silicic (9)	Terrigenous (9)	Shale (9)	Carbonate (9)	Ophicarbonate (10)	GLOSS (9)	Upper cont. crust (11)	Cont. crust (12)
SiO_2	71.56	59.26	62.0	32.79	28.51	58.57	66.0	59.1
TiO_2	0.41	0.74	1.0	0.18	0.02	0.62	0.5	0.7
Al_2O_3	8.07	15.53	15.9	3.99	0.43	11.91	15.2	15.8
FeO	3.99	5.88	6.5	2.40	1.27	5.21	4.5	6.6
MnO	0.38	0.17	0.1	0.25	0.04	0.32		0.11
MgO	1.69	2.46	2.2	1.43	25.61	2.48	2.2	4.4
CaO	1.08	2.78	1.3	30.41	13.41	5.95	4.2	6.4
Na_2O	2.39	2.78	1.2	1.58	0.33	2.43	3.9	3.2
K_2O	1.52	2.11	3.7	0.61	0.28	2.04	3.4	1.9
P_2O_5	0.22	0.19	0.1	0.17	0.04	0.19		0.2
H_2O	8.49	8.22	6.0	2.82	18.27	7.29		
CO_2	0.08			23.31	11.41	3.01		
Total	99.88	100.12	100.0	99.94	99.62	100.0		98.39

Note: Numbers in parentheses are references: 1—Ringwood (1975); 2—McDonough and Sun (1995); 3—Anderson and Bass (1986); 4—average C1 carbonaceous chondrite after Palme and Jones (2003); 5—from Garret transform (Constantin, 1999); 6—average mid-ocean-ridge basalt (MORB) after Hofmann (1988); 7—Staudigel et al. (1996); 8—Mottana et al. (1990); 9—average sediments from data in Plank and Langmuir (1998); 10—Sciurto and Ottonello (1995); GLOSS—average global subductiong sediment (Plank and Langmuir, 1998; Poli and Schmidt, 2002); 11—upper continental crust (anhydrous basis, Taylor and McLennan, 1985); 12—average continental crust after Rudnick and Fountain (1995).
†Presented not only in the form of H_2O or OH in minerals (H_2O = ~10 wt%).

depleted and fertile peridotites (30–100 km). However, cross sections of modern oceanic floors and ophiolites reveal large compositional and structural variations in the form of layers and pods. Some complexities appear during subduction of seamounts and oceanic plateaus, such as hydrothermally altered rocks, ophicarbonatites, phosphorites, Fe-Mn nodules, and metalliferous sediments (e.g., Plank and Langmuir, 1998; Li and Schoonmaker, 2003). In addition, significant contribution of sediments from continental materials may also be expected. Table 1 summarizes the compositions of the most important materials entering Earth's mantle in subduction zones. Experimental studies on phase transformations in the major components of subducting slabs are important for understanding the density relationship between slab components and the surrounding mantle and rheological behavior controlling the deformation and fragmentation of the slab in the deep mantle. Ringwood and co-workers (e.g., Irifune and Ringwood, 1993; Ringwood, 1994; Irifune et al., 1994) determined the anhydrous modal mineralogy of sediment, basalt, and peridotite layers of the model slab (Fig. 3) and the density changes with depth. The hydrous phase relations in the major slab components are discussed in detail next.

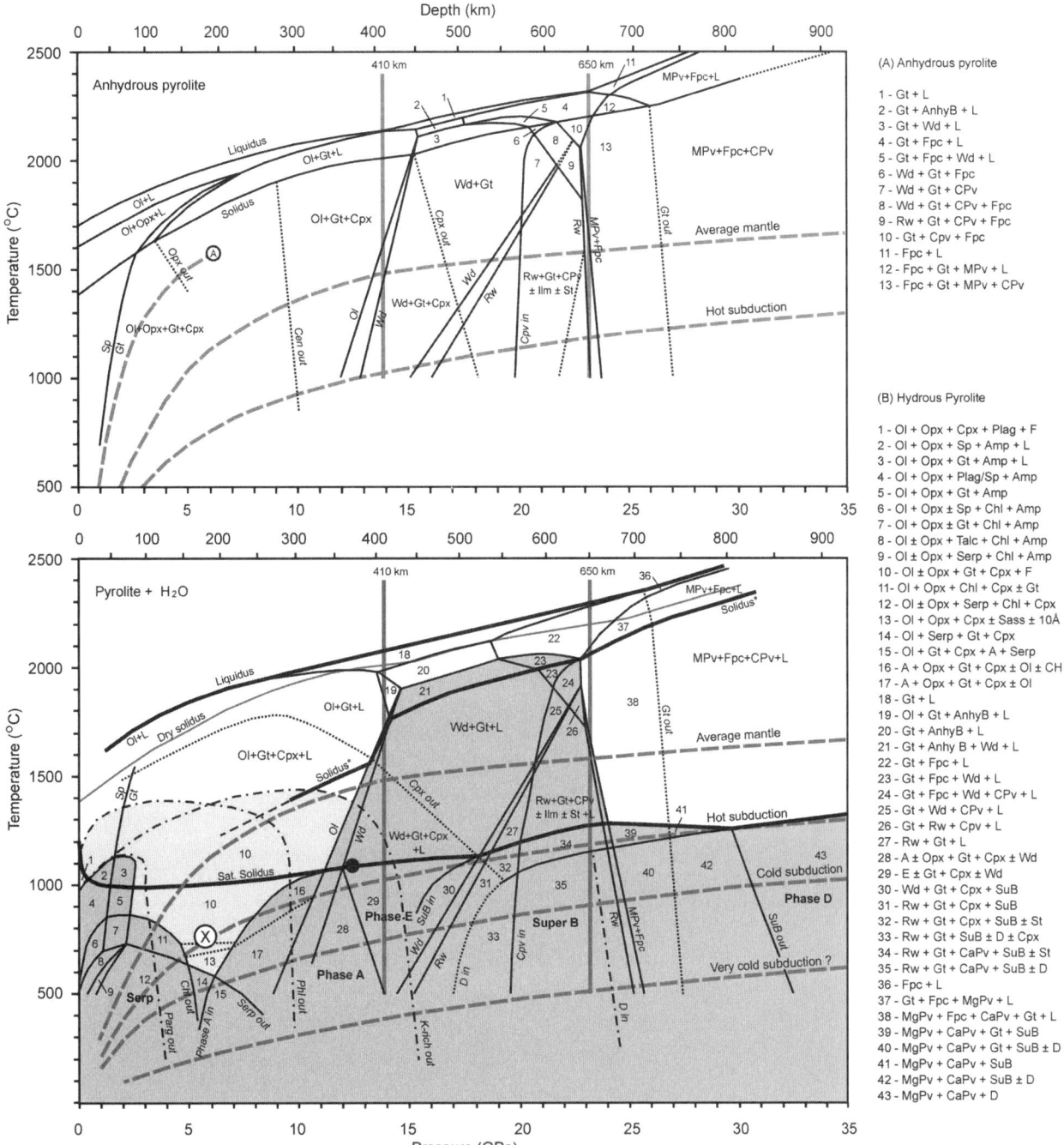

Figure 2. Phase diagram of the anhydrous and H$_2$O-saturated pyrolite (average mantle peridotite) deduced from experimental results (Takahashi, 1986; Inoue, 1994; Zhang and Herzberg, 1994; Kawamoto et al., 1995, 1996; Herzberg and Zhang, 1996; Burnley and Navrotsky, 1996; Schmidt and Poli, 1998; Wunder, 1998; Irifune et al., 1998; Ulmer and Trommsdorff, 1999; Artioli et al., 1999; Ohtani et al., 2000, 2001b, 2003; Asahara and Ohtani, 2001; Litasov et al., 2001; Litasov and Ohtani, 2002, 2003a; Hirose, 2002; Fumagalli et al., 2001; Bromiley and Pawley, 2002; Poli and Fumagalli, 2003; Kawamoto, 2004; Fumagalli and Poli, 2005). Solidus* indicates apparent solidus for pyrolite + 2 wt% H$_2$O (Litasov and Ohtani, 2002). Black circle indicates location of second critical end point for Mg$_2$SiO$_4$-MgSiO$_3$-H$_2$O system after Stalder et al. (2001). "X" indicates choke point of dehydration in hot subduction zone (Tatsumi, 1989; Bose and Ganguly, 1995; Kawamoto et al., 1995). Dotted lines show results for nonpyrolite systems. Mantle geotherm is after Akaogi et al. (1989). Subduction slab geotherms are after Kirby et al. (1996). "A" indicates oceanic geotherm. Chained lines show stability of K-bearing phases: Parg—pargasite stability field after Niida and Green (1999); Phl—phlogopite after Konzett and Ulmer (1999) and Trønnes (2002); K-rich—K-richterite after Inoue et al. (1998a) and Konzett and Ulmer (1999). Abbreviations: Ol—olivine; Opx—orthopyroxene; Cpx—clinopyroxene; Plag—plagioclase, Sp—spinel; Amp—amphibole; Chl—chlorite; Serp—serpentine; Gt—garnet; Sur—Mg-sursassite; 10Å—10 Å phase; A—phase A; E—phase E; AnhyB—anhydrous phase B; SuB—superhydrous phase B; Wd—wadsleyite; Rw—ringwoodite; MPv—Mg-perovskite; CPv—Ca-perovskite; Fpc—ferropericlase; Ak—akimotoite; St—stishovite; D—phase D; Cen—clinoenstatite; Phen—phengite; F—fluid; L—liquid.

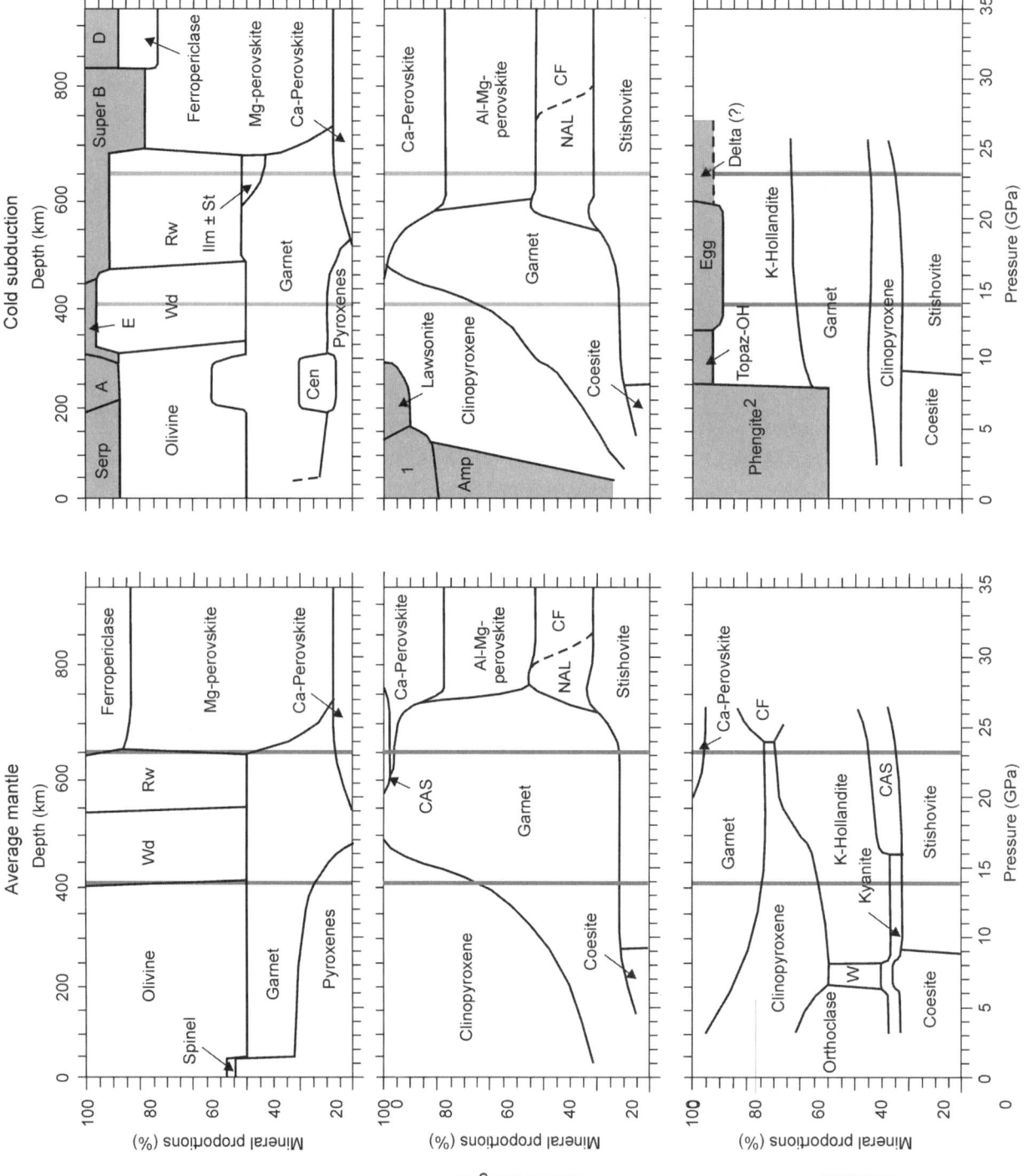

Figure 3. Mineral proportions in dry and hydrous pyrolitic mantle, mid-ocean-ridge basalt (MORB), and sediments estimated based on phase relations given by Irifune and Ringwood (1993); Irifune et al. (1994); Yasuda et al. (1994); Kawamoto et al. (1995); Schmidt and Poli (1998); Ono (1998); Litasov and Ohtani (2002, 2003a); Hirose and Fei (2002); Poli and Schmidt (2002); Ohtani et al. (2004); Kawamoto (2004); and Litasov and Ohtani (2005). Water content is calculated along cold mantle geotherm (see Fig. 2). 1—multiphase field including chlorite, epidote, staurolite, chloritoid, zoisite, and some other phases; 2—field including biotite, lawsonite, zoisite, chloritoid, talc, and amphibole. CAS—Ca-Al phase; NAL—Na-Al phase; Amp—amphibole; W—wadeite; Egg—phase Egg; Delta—δ-AlOOH; CF—Al-rich CF phase; see Figure 2 for other abbreviations.

Phase Relations in Hydrous Peridotite

Phase Relations to 6 GPa

Although numerous experimental works have been conducted on the melting of the pyrolite/peridotite composition, including anhydrous, hydrous, and carbonate-bearing systems, the subsolidus phase relations in undersaturated hydrous pyrolite are still poorly constrained at pressures above 2–3 GPa. Most mineralogical models of hydrated subducting slabs are based on the systematic studies of simplified systems, such as $MgO-SiO_2-H_2O$ (MSH) or $MgO-Al_2O_3-SiO_2-H_2O$ (MASH) (e.g., Ulmer and Trommsdorff, 1999; Ohtani et al., 2000, 2001b; Stalder and Ulmer, 2001), CMASH (e.g., Asahara and Ohtani, 2001; Litasov et al., 2001; Litasov and Ohtani, 2002, 2003a), or H_2O-saturated peridotite compositions (e.g., Kawamoto et al., 1996; Kawamoto and Holloway, 1997; Poli and Fumagalli, 2003; Kawamoto, 2004). The phase relations and stability of hydrous phases in pyrolite with ~2 wt% H_2O are shown in Figure 2, and the density and composition of major hydrous phases relevant to the mantle are summarized in Table 2 and Figure 4.

TABLE 2. MAJOR HYDROUS MINERALS IN THE PERIDOTITE MANTLE AND MAJOR PARTS OF A SUBDUCTING SLAB

Mineral	Formula (Mg end member)	H_2O* (wt%)	Density[†] (g/cm^3)	Depth[§] (km)	Temp.[§] (°C)	Paragenesis	References[§]
Pargasite	$NaCa_2Mg_4Al_3Si_6O_{22}(OH)_2$	2.2	3.02	0–120	1150	P B S	Mengel and Green (1989), Niida and Green (1999)
K-richterite	$K_2CaMg_5Si_8O_{22}(OH)_2$	2.1	3.01	0–450	1450	P B S	Inoue et al. (1998a), Konzett and Ulmer (1997, 1999)
Phlogopite[#]	$KMg_3AlSi_3O_{10}(OH)_2$	4.0	2.7–3.3	0–300	1400	P B S	Sudo and Tatsumi (1990), Liu (1993), Luth (1997), Trønnes (2002)
Talc	$Mg_3Si_4O_{10}(OH)_2$	4.8	2.6–2.8	0–150	700	P B S	Yamamoto and Akimoto (1977), Bose and Ganguly (1995), Chinnery et al. (1999)
Chlorite	$Mg_5Al_2Si_3O_{10}(OH)_8$	13.0	2.6–3.3	0–150	850	P B S	Deer et al. (1992)
Serpentine	$Mg_3Si_2O_5(OH)_4$	13.0	2.6	0–200	700	P	Ulmer and Trommsdorff (1995), Wunder and Schreyer (1997), Irifune et al. (1998)
10 Å phase	$Mg_3Si_4O_{14}H_6$	13.0	2.7	100–250	700	P ?	Sclar et al. (1965), Yamamoto and Akimoto (1977), Chinnery et al. (1999), Fumagalli et al. (2001)
Mg-sursassite**	$Mg_5Al_5Si_6O_{21}(OH)_7$	7.0	3.30	0–200	750	P ? S	Fockenberg (1998), Artioli et al. (1999), Bromiley and Pawley (2002)
Humite	$Mg_7Si_3O_{12}(OH)_2$	3.7	3.20	0–100	900	P ?	Liu (1993), Wunder et al. (1995)
Clinohumite[††]	$Mg_9Si_4O_{16}(OH)_2$	3.0	3.3–3.3	0–300	1050	P ?	Yamamoto and Akimoto (1977), Kanzaki (1991), Wunder (1998)
Chondrodite	$Mg_5Si_2O_8(OH)_2$	5.3	3.16–3.26	0–300	1100	P ?	Akimoto and Akaogi (1980), Wunder (1998)
Norbergite	$Mg_3SiO_4(OH)_2$	9.8	3.18	0–200	1100	P ??	Yamamoto and Akimoto (1977)
Phase A	$Mg_7Si_2O_8(OH)_6$	11.8	2.96	200–400	1050	P	Ringwood and Major (1967), Horiuchi et al. (1979), Wunder (1998)
Phase E[§§]	$Mg_{2.27}Si_{1.26}H_{2.4}O_6$	11.4	2.78–2.92	350–500	1100	P	Kanzaki (1991), Kudoh et al. (1993), Inoue (1994), Frost (1999)
Superhydrous phase B	$Mg_{10}Si_3O_{14}(OH)_4$	5.8	3.21–3.33	450–800	1300	P	Gasparik (1990, 1993), Pacalo and Parise (1992)
Phase B	$Mg_{24}Si_8O_{38}(OH)_4$	2.4	3.32–3.38	350–650	1000	P ??	Ringwood and Major (1967), Finger et al. (1989, 1991)
Phase D (F, G)	$Mg_{1.14}Si_{1.73}H_{2.81}O_6$	10–14	3.50	550–1200	2400	P	Liu (1987), Ohtani et al. (1995, 1997, 2001b), Yang et al. (1997), Irifune et al. (1998), Frost and Fei (1998), Shieh et al. (1998, 2000)
Brucite	$Mg(OH)_2$	31	2.37	>1800[##]	1300	P ?	Duffy et al. (1991, 1995), Fei and Mao (1993), Johnson and Walker (1993)
Hydrous wadsleyite I	$Mg_7Si_4O_{14}(OH)_2$	3.3	3.47	350–600	2000	P	Smyth et al. (1997)
Hydrous wadsleyite II	$Mg_7Si_4O_{14}(OH)_2$	3.3	3.51	350–600	2000	P	Smyth and Kawamoto (1997)
Hydrous ringwoodite	$Mg_7Si_4O_{14}(OH)_2$	2.6	3.47–3.65	500–700	1900	P	Kohlstedt et al. (1996)
Zoisite, clinozoisite***	$Ca_2Al_3Si_3O_{12}(OH)$	2.0	3.15–3.37	0–100	800	B S	Poli and Schmidt (1998)
Chloritoid	$Mg_2Al_4Si_2O_{10}(OH)_4$	8.2	3.46–3.80			B S	Schmidt and Poli (1998)
Staurolite	$Mg_2Al_9Si_4O_{22}(O,OH)_2$	2.2	3.74–3.83			B ? S	Deer et al. (1992)
Lawsonite	$CaAl_2Si_2O_7(OH)_2·H_2O$	11.0	3.09	0–300	850	B S	Pawley (1994), Schmidt (1995), Okamoto and Maruyama (1999)
Phengite	$K_2Al_2Mg_2Si_8O_{20}(OH)_4$	4.0	2.9	0–300	1050	B S	Schmidt (1996), Domanik and Holloway (1996, 2000)
Phase X	$K_4Mg_8Si_8O_{25}(OH)_2$	1.8	3.01	500–600	1300	B ? S ?	Luth (1997), Inoue et al. (1998a), Yang et al. (2001), Konzett and Fei (2000)
Muscovite	$K_2Al_6Si_6O_{20}(OH)_4$	4.5	2.77–2.88	0–60	800	S	Domanik and Holloway (1996, 2000)
Phase Pi	$Al_3Si_2O_7(OH)_3$	9.0	3.23	200	700	S	Wunder et al. (1993a, 1993b)
Topaz-OH	$Al_2SiO_4(OH)_2$	10.7	3.37	250–350	1400	S	Wunder et al. (1993a, 1993b, 1999)
Phase Egg	$AlSiO_3(OH)$	7.5	3.84	350–600	1700	S	Eggleton et al. (1978), Schmidt et al. (1998), Ono (1998), Sano et al. (2004)
Diaspore	$AlOOH$	15.0	2.38	0–450	900	S ?	Wunder et al. (1993a, 1993b)
δ-AlOOH phase	$AlOOH$	15.0	3.53	>450	1100	S ?	Suzuki et al. (2000a), Ohtani et al. (2001a)

Note: P—peridotite, B—basalt, S—sediment, ?—occasionally, ??—unlikely to be stable.
*Maximum H_2O content based on stoichiometry.
[†]Includes range reported in cited references, see also Deer et al. (1992) and Smyth and McCormick (1995).
[§]Approximate range of depth stability and maximum temperature at which the mineral is stable. The references include citations of the initial characterization of phase and major studies of stability range.
[#]Fe end member biotite is stable in sediments at the depth 0–60 km.
**MgMgAl-pumpellyite (Schiffman and Liou, 1980; Schreyer et al., 1986).
[††]Including Ti-clinohumite, $Mg_{9-x}Ti_xSi_4O_{2x}(OH)_{2-2x}$.
[§§]Related Fe-bearing phase, called preliminary phase H, has been characterized (Williams and Hemley, 2001).
[##]Including high-pressure polymorphs (Duffy et al., 1995).
***Including epidote.

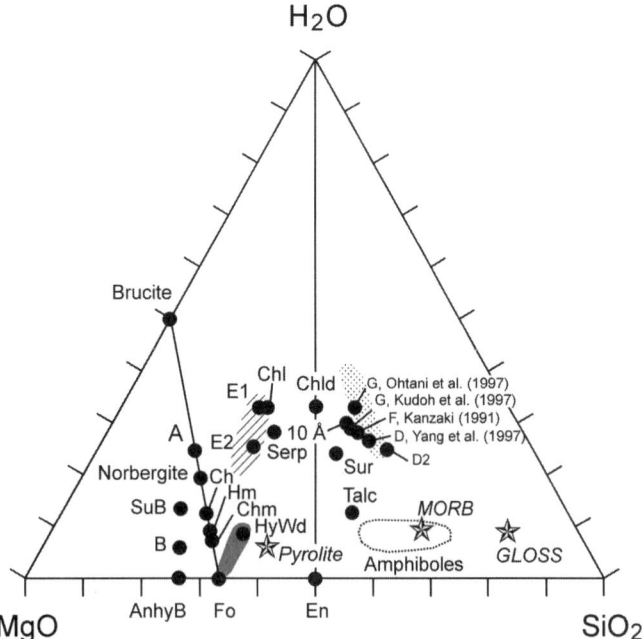

Figure 4. Ternary diagram illustrating the compositions of dense hydrous magnesium silicates and other phases related to mantle assemblages. Striped field—range of compositions of phase E; Dotted field—composition of phase D/G/F. Stars—pyrolite; mid-ocean-ridge basalt (MORB) and GLOSS—global average sediment composition (Table 1). Ch—chondrodite; Hm—humite; Chm—clinohumite; B—hydrous phase B; Cld—chloritoid; Fo—forsterite; En—enstatite; see Figure 2 for other abbreviations.

Serpentine is the major hydrous phase in the lower-pressure assemblages, corresponding to depths of 0–200 km along the temperature gradients (Wunder, 1998; Ulmer and Trommsdorff, 1999). In the alumina excess systems, chlorite is expected to occur throughout greenschist, amphibolite, and eclogite facies of metamorphism. Ca-amphibole (tremolite at low temperature and pargasite at high temperature) might be present in pyrolite composition to 3 GPa and 1150 °C, depending on the proportions of CaO and Na_2O in the system (Niida and Green, 1999).

Other hydrous phases, which might be stable in the pyrolite composition, at pressures to 5 GPa, are talc and K-bearing phases, like phlogopite. Talc is restricted to lower pressures (2 GPa); however, Tschermak-substituted talc might occur at somewhat higher pressures (Poli and Schmidt, 2002). Phlogopite and K-richterite might exist as an accessory phase in a wide pressure and temperature range to ~1400 °C and 10 and 15 GPa, respectively (Sudo and Tatsumi, 1990; Luth, 1997; Konzett et al., 1997; Inoue et al., 1998a; Konzett and Ulmer, 1999; Trønnes, 2002). The potassium content of the average mantle peridotite is very low; however, phlogopite and amphibole are the most abundant hydrous phases in mantle xenoliths worldwide (e.g., Carswell, 1975; Ionov et al., 1997; Pearson et al., 2003), suggesting significant local enrichment of potassium in the upper mantle.

Experiments in the MASH system and peridotite reveal that chlorite breaks down at ~5 GPa to a mineral assemblage that is critical to transfer H_2O further to extreme depths. At these conditions (5–7 GPa, 600–650 °C), the existence of a so-called choke point has been proposed, a minimum in hydrous phase stability between the stability fields of antigorite and phase A, where peridotite loses most of its water in hot subduction zones (Tatsumi, 1989; Bose and Ganguly, 1995; Kawamoto et al., 1995; field X in Fig. 2). However, some recent studies indicate that a 10 Å phase may form at temperatures up to 700 °C at this pressure (Fumagalli et al., 2001; Poli and Fumagalli, 2003) and Mg-sursassite may occur up to 750 °C (Fockenberg, 1998; Artioli et al., 1999; Bromiley and Pawley, 2002). Thus, subducting slabs with geotherms passing above the choke point may transport bonded water further into the deep mantle (Fig. 2). It should be noted that a 10 Å phase was recently observed as inclusions in mantle olivines (Khisina and Wirth, 2004).

Whatever reaction or mechanism transfers volatiles from serpentine/chlorite-bearing assemblages to pressures above 6 GPa, once phase A appears together with enstatite, there is no more hindrance for water transport to the deep subduction environment (Poli and Schmidt, 2002).

Alphabet Phases

At pressures above 6 GPa, major hydrous phases in peridotite compositions are the so-called alphabet phases: A, B, C, D, E, F, G, and superhydrous phase B or dense hydrous magnesium silicates (DHMS) (Table 2). Reviews on the stability field and the terminology of different DHMS have appeared in several papers (e.g., Gasparik, 1993; Frost, 1999; Ohtani et al., 2001b; Williams and Hemley, 2001; Komabayashi et al., 2004). Superhydrous phase B is the same as phase C of Ringwood and Major (1967), and the phases named as phase D, F, and G are identical (Liu, 1987; Kanzaki, 1991; Ohtani et al., 1995, 1997; Kudoh et al., 1997; Yang et al., 1997; Shieh et al., 1998; Frost and Fei, 1998; Irifune et al., 1998). All these hydrous phases are stable at low temperatures in the peridotite composition and can be stable only within subduction-zone environments.

Phase A is stable in different systems from 3 to 17 GPa, and its stability expands to a temperature of 1100 °C near 11 GPa (Yamamoto and Akimoto, 1977; Burnley and Navrotsky, 1996; Pawley and Wood, 1996; Frost, 1999). It forms by the reaction of forsterite + H_2O = A + enstatite at 8 GPa and 800 °C (Luth, 1995) or by the breakdown of serpentine at 6 GPa and 600 °C (Ulmer and Trommsdorff, 1995). The composition of phase A falls on a brucite-forsterite line, along with humite- and chondrodite-series minerals (Fig. 4). Phase A was detected at 10 GPa and below 900 °C together with phase E, clinohumite, and enstatite by Kawamoto et al. (1995). The data on the stability of phase A, chondrodite, and clinohumite (Yamamoto and Akimoto, 1977; Akimoto and Akaogi, 1980; Luth, 1995; Wunder, 1998; Stalder and Ulmer, 2001) show that the assemblage of phase A + olivine (+ garnet + clinopyroxene + clinoenstatite) could be stable at 7–10 GPa and below 800 °C. In peridotite systems (CMAS + 2 wt% H_2O, Litasov and Ohtani, 2003a; and H_2O-saturated KLB-1 peridotite, Kawamoto et al., 1995;

Kawamoto, 2004), phase A is stable at 6–13 GPa and temperatures up to 1050 °C (Fig. 2). In systems with low-H_2O contents, phase A plays a limited role in the upper-mantle assemblages, since a significant amount of water may be stored in olivine and wadsleyite.

Occasionally, chondrodite and clinohumite group phases are observed in natural environments. Ti-clinohumite was found in ultramafic massifs (e.g., Evans and Trommsdorff, 1983; Dymek et al., 1988; Zhang et al., 1995) and peridotite nodules in kimberlites (e.g., McGetchin et al., 1970; Aoki et al., 1976; Smith, 1979). Ti-chondrodite is less common than Ti-clinohumite, but it appears in the same environment (e.g., Aoki et al., 1976; Smith, 1995; Scambelluri and Rampone, 1999). Stalder and Ulmer (2001) showed that, in the MSH system, even minor amounts of fluorine enhances the low-pressure and low-temperature stability of clinohumite for a wide range of Mg/Si bulk ratios, and Kawamoto et al. (1995) found Ti-rich clinohumite and Ti-chondrodite at 7–9 GPa and 600–900 °C in the water-saturated peridotite system with or without phase A. Therefore, phase relations at pressures above the serpentine stability field (6–10 GPa) are still ambiguous and strongly affected by presence of minor components such as Ti, F, and Cl.

Phase E has been observed in the system Mg_2SiO_4 + 20 wt% H_2O at 13–17 GPa and 800–1000 °C (Kanzaki, 1991). It has relatively broad compositional variations, and the compositions fall approximately along a forsterite-water line (Fig. 4). The stability field of phase E partially overlaps with wadsleyite. At low temperatures, wadsleyite can accommodate up to 3 wt% of H_2O (Inoue et al., 1995). Phase E may not be stable when wadsleyite is undersaturated with H_2O (Frost, 1999). Litasov and Ohtani (2003a) observed phase E coexisting with H_2O-saturated wadsleyite at 12 GPa and 1050 °C in the CMASH pyrolite system containing 2 wt% H_2O.

Superhydrous phase B coexisting with stishovite at pressures of 16–24 GPa and temperatures between 800 and 1400 °C was recognized by Gasparik (1990, 1993) using a starting composition Mg/Si = 1.5 and 3.6 wt% of H_2O. In the systems containing more H_2O, superhydrous phase B has been observed coexisting with phase D/G at pressures of 16–24 GPa and temperatures of 900–1500 °C (Ohtani et al., 1995; Frost and Fei, 1998). A limited H_2O solubility in ringwoodite may stabilize superhydrous phase B in pyrolite + 2 wt% H_2O at 18–24 GPa. Superhydrous phase B has been observed in the CMASH system and water-saturated peridotite at temperatures below 1100 °C at 18.5 GPa and below 1300 °C at 25 GPa (Litasov and Ohtani, 2003a; Kawamoto, 2004). Appearance of superhydrous phase B in the transition zone may depend on the water content. When the bulk water content is <1 wt%, hydrous ringwoodite may contain no more than 2 wt% H_2O, and superhydrous phase B may appear only as a decomposition product of hydrous ringwoodite at pressures close to the 650 km discontinuity. Superhydrous phase B decomposes into Mg-perovskite + periclase + phase D in the uppermost lower mantle at pressures ~30–32 GPa and temperatures below 1250 °C (Ohtani et al., 2003).

The stability field of DHMS phases may expand to higher temperatures with halogen content, as is observed in amphiboles or micas. The substitution of F for the OH group appears to enhance stability of superhydrous phase B and phase E to 1500 °C at 15 GPa (Gasparik, 1993). This observation may be important because micas and amphiboles reported in the upper mantle contain up to 2 wt% fluorine (e.g., Smith and Dawson, 1981; Ionov et al., 1997; Litasov and Litasov, 1999).

Phase D is potentially the most important hydrous phase among DHMS, because it has the highest pressure stability and can represent a major water container in the lower mantle. It is the only DHMS with Si entirely in octahedral coordination (e.g., Kudoh et al., 1997; Yang et al., 1997). In the MSH system, phase D has been observed at pressures above 15 GPa and temperatures up to 1100 °C (Liu, 1987; Ohtani et al., 1995, 1997, 2001b; Frost and Fei, 1998; Irifune et al., 1998). In the model pyrolite in the CMASH system, phase D was not observed up to 25 GPa pressure (Litasov and Ohtani, 2002); however, in the hydrous multicomponent peridotite system containing FeO, it was observed coexisting with superhydrous phase B at 20–22 GPa (Kawamoto, 2004). The composition of phase D varies widely (Fig. 4), and there is a correlation between Mg/Si ratio and H_2O content. According to Shieh et al. (1998), phase D can be synthesized at elevated pressures of 30–45 GPa from a bulk serpentine composition.

Williams and Hemley (2001) discussed a possible metastability of DHMS in the mantle and noted that existence of hydrous phases in water-bearing systems need not be coincident with their thermodynamic stability fields: nucleation of metastable phases is a well-known phenomenon in experiments (e.g., Yoder, 1952). For instance, in cold subducting slabs, metastable hydrous phases might be carried to depths well beyond their stability fields (e.g., Scott and Williams, 1999; Daniel et al., 2000).

On the contrary, phases A, B, E, and D have been observed to decompose or amorphize rapidly at ambient pressure and temperatures of 440, 300–400, 200, and 100 °C, respectively (Liu et al., 1997a, 1997b, 1998a, 1998b). Thus, we may not observe these phases as inclusions in diamonds and other mantle minerals; and the possible existence of the deepest DHMS remains enigmatic.

Water in Olivine Polymorphs and Seismic Discontinuities

Seismic studies have clarified variations in depth and sharpness of the 410, 520, and 650 km seismic discontinuities, which have usually been attributed to the phase changes in $(Mg,Fe)_2SiO_4$ from olivine to wadsleyite, wadsleyite to ringwoodite, and ringwoodite to Mg-perovskite + ferropericlase (postspinel transformation), respectively (Fig. 2, e.g., Ringwood, 1975; Agee, 1998). According to the seismological observations, the 410 and 650 km discontinuities are sharp, and the changes in physical properties associated with them occur over small depth intervals. For the 650 km discontinuity, a width of 5 km is consistent with seismic data, and a width of <10 km has been suggested for the 410 km discontinuity (e.g., Benz and Vidale, 1993; Helffrich and

Wood, 1996). Studies of topography of the discontinuities indicate elevation of the 410 km discontinuity and depression of the 650 km discontinuity in subduction zones and an opposite effect in hotspots and mantle plumes (e.g., Flanagan and Shearer, 1998; Helffrich, 2000; Collier et al., 2001; Gu and Dziewonski, 2002). Until recently, these variations were considered to be consistent with the experimentally observed Clapeyron slopes of the olivine to wadsleyite and postspinel transformations (e.g., Ito and Takahashi, 1989; Katsura and Ito, 1989; Akaogi et al., 1989; Bina and Helffrich, 1994).

Newest data for Mg_2SiO_4 and average peridotite (pyrolite) systems obtained by in situ X-ray diffraction studies (Katsura et al., 2003; Fei et al., 2004; Litasov et al., 2005a) indicate that the postspinel transformation boundary has a gentle negative Clapeyron slope, which varies between −0.4 and −1.3 MPa/K (Figs. 2 and 5). These studies create significant complexities in the effort to explain the topography of the 650 km discontinuity, and they indicate that the postspinel phase transition may account for less than half of the variations in the depth of the 650 km discontinuity. Higo et al. (2001) reported that water shifts the postspinel phase boundary insignificantly (+0.2 GPa). However, very recently, Litasov et al. (2005b) observed a shift of the postspinel transformation boundary in hydrous pyrolite for 0.6 GPa at 1200 °C (Fig. 5), which may account for the missing half of 30–40 km depressions of the 650 km discontinuity in subduction zones (e.g., Flanagan and Shearer, 1998; Collier et al., 2001; Gu and Dziewonski, 2002). It is possible that the transformation boundary is not linear; therefore, this shift should be stronger at lower temperatures (cold subduction), even if the water content in the system is significantly less than the 2 wt% used in the experiments by Litasov et al. (2005b). It should be noted that Katsura et al. (2003) also suggested a 0.6 GPa shift of the postspinel transition boundary to the higher pressure in hydrous Mg_2SiO_4 based on their preliminary investigations at 1400 °C.

On the other hand, recent quench multianvil experiments (Smyth and Frost, 2002; Chen et al., 2002; Litasov and Ohtani, 2003a) have shown a shift of the olivine-wadsleyite boundary to lower pressures by 1–2 GPa with added water (Fig. 5). In addition, in Fe-bearing systems, the two-phase interval appears to broaden from ~0.4 GPa (12 km) in the anhydrous system to as much as 1.3 GPa (40 km) in the hydrous system (Smyth and Frost, 2002; Frost, 2003). This is consistent with calculations by Wood (1995), who estimated that the presence of 500 ppm H_2O would necessarily expand the olivine-wadsleyite loop interval from a minimum of ~7 km for a dry system to at least 22 km. He argued that the observed sharpness of the 410 km discontinuity would be inconsistent with H_2O contents greater than ~200 ppm. On the other hand, Smyth and Frost (2002) suggested that hydrogen diffusion and gravitational stratification narrow the phase transition interval between olivine and wadsleyite.

The observed shifts of the olivine-wadsleyite and postspinel transition boundaries are consistent with different water solubility in wadsleyite and olivine and ringwoodite and Mg-perovskite, respectively. Wadsleyite may accommodate 5–40 times more

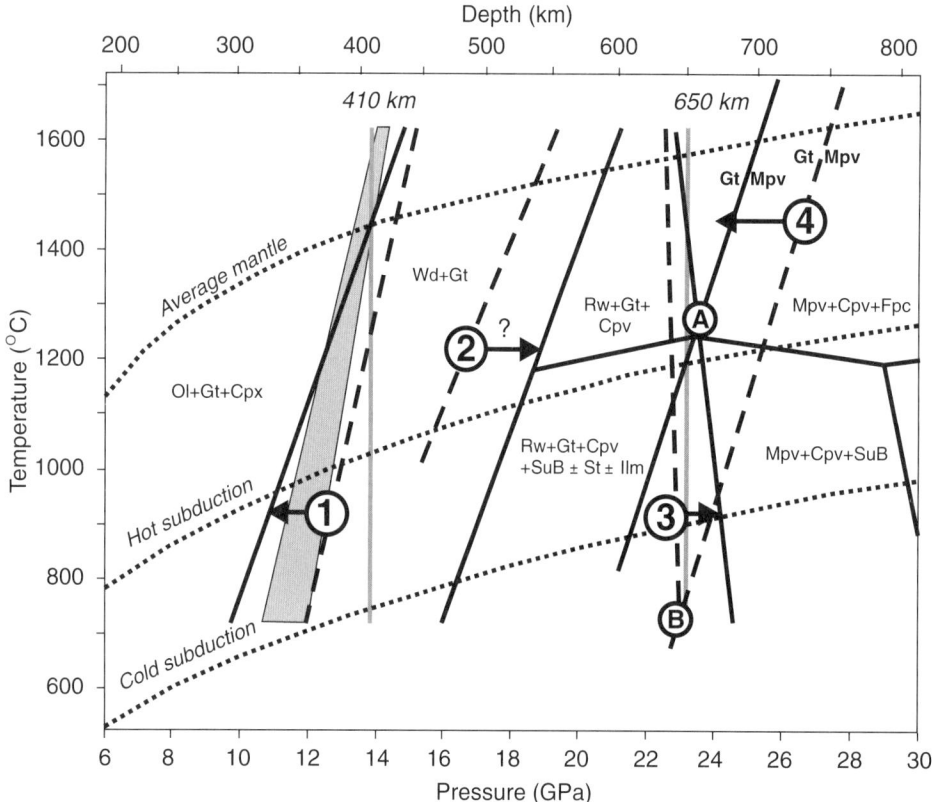

Figure 5. Shift of phase transition boundaries by adding water and comparison with mantle geotherms (modified after Litasov et al., 2006b). Solid lines indicate phase transitions under hydrous conditions (2 wt% H_2O); dashed lines indicate anhydrous transitions. 1—Olivine-wadsleyite transition after Morishima et al. (1994) and Litasov and Ohtani (2003a); gray field shows olivine-wadsleyite transition in $(Mg,Fe)_2SiO_4$ system after Katsura and Ito (1989). 2—Wadsleyite-ringwoodite transition after Suzuki et al. (2000b), Litasov and Ohtani (2003a), and Kawamoto (2004). 3—Postspinel transition in pyrolite after Litasov et al. (2005a, 2005b). 4—Post-garnet transition in mid-ocean-ridge basalt (MORB) (Gt/Mpv) after Litasov et al. (2004) and Sano et al. (2006). A and B indicate the points of equal density of basalt and pyrolite under hydrous and dry conditions, respectively. See Figure 2 for other abbreviations. Stability of superhydrous phase B (SuB) is after Litasov and Ohtani (2003a) and Ohtani et al. (2003).

water relative to olivine at 1100 °C; and ringwoodite contains much more water (up to 2.6 wt%) than Mg-perovskite and ferropericlase (<100 ppm water; see Chapter 3, this volume). In addition, Mg/Fe partitioning between the phases (e.g., Irifune and Isshiki, 1998; Chen et al., 2002) and composition of the fluid phase (e.g., Kawamoto et al., 2004) may affect the phase transition boundaries, but these effects have not yet been systematically studied.

Litasov and Ohtani (2003a) and Kawamoto (2004) reported a significant shift of the wadsleyite to ringwoodite phase transition to higher pressures caused by addition of water to the peridotite system (Fig. 5), although the reasons for significant displacement of the wadsleyite-ringwoodite phase boundary are not clear. These observations are consistent with depth variations of the discontinuity at 520 km recently observed by Deuss and Woodhouse (2001).

Melting Relations of Hydrous Peridotite

Melting relations of hydrous peridotite at low pressures (0–3 GPa) were reviewed by Green and Falloon (1998). Melting of amphibole and phlogopite controls the composition of low-degree partial melts in experiments on pyrolite and Tinaquillo peridotite compositions (Mengel and Green, 1989; Wallace and Green, 1991; Niida and Green, 1999). They determined the solidus temperature of water-saturated pyrolite as 1000–1030 °C at 3 GPa. The experiments also showed the expansion of the liquidus field of olivine at the expense of pyroxene under hydrous conditions. Hirose and Kawamoto (1995) determined the composition of the low-degree partial melt in hydrous KLB-1 peridotite at 1 GPa and found that it is enriched in SiO_2 (quartz-normative) and depleted in MgO relative to the anhydrous melt (Hirose and Kushiro, 1993).

Melting relations of hydrous CMASH peridotite containing 0.5–5.0 wt% H_2O at higher pressures (4–25 GPa) were studied by Asahara and Ohtani (2001), Litasov et al. (2001), and Litasov and Ohtani (2002, 2003a, 2003b). Bureau and Keppler (1999) and Stalder et al. (2001) reported recently that the disappearance of the immiscibility between aqueous fluid and melt along the solidus occurs at 12–13 GPa. Therefore, it is difficult to recognize the solidus temperature at higher pressures. However, in the study by Litasov and Ohtani (2002), we defined "apparent solidus" and distinguished between the existence of fluid at low temperatures and melt at high temperatures based on the texture of the quench run products (see also Mantle Fluids section). The apparent solidus is at ~1600 °C at 13.5 GPa in the upper mantle, whereas we observed a drastic change of the apparent solidus to 1850 °C at 15.5 GPa at the top of the transition zone. The apparent solidus is 1900 °C at 20 GPa and 2150 °C at 25 GPa, i.e., ~180 °C lower at 20 GPa and 120 °C lower at 25 GPa than the dry solidus temperature, respectively. In Figure 2, we assumed that the decrease of the solidus temperature in hydrous Fe-bearing peridotite relative to the CMAS system is the same as that under dry conditions (e.g., Herzberg and Zhang, 1996; Herzberg et al., 2000; Litasov and Ohtani, 2002). The drastic change of the apparent solidus temperature may be caused by restricted solubility of water in olivine compared to that of wadsleyite, resulting in a difference in water activity in the silicate melt, which may have important implications for melting processes in ascending hot mantle plumes.

Litasov and Ohtani (2002) observed expansion of liquidus majorite in hydrous pyrolite relative to dry pyrolite toward both low and high pressures and expansion of the stability field of anhydrous phase B ($Mg_{14}Si_5O_{24}$). Majorite was a liquidus phase from 10 to 25 GPa (Fig. 2). Asahara and Ohtani (2001) observed expansion of the enstatite stability field under hydrous conditions, and based on both studies, we argued that Al-depleted komatiite (deepest among komatiite series rocks) may be formed by partial melting of hydrous garnet peridotite at pressures up to 10 GPa. At higher pressures, compositions of the melt become more magnesian and contain less Al_2O_3 compared to komatiite melts. Studies of melting relations in CMASH pyrolite supported a model of komatiite genesis by dehydration melting of rising wet plumes at the base of the upper mantle.

Phase Relations in Hydrous Basalt

Subsolidus Phase Relations

Intensive experimental studies on metamorphic mafic rocks in basaltic systems have demonstrated that the phase relations involve continuous reactions of different hydrous minerals (e.g., Thompson et al., 1982). Phase relations in basaltic systems with application to water transport in subduction zones have been studied by many authors (e.g., Pawley and Holloway, 1993; Poli and Schmidt, 1995, 2002; Schmidt and Poli, 1998; Liu et al., 1996; Vielzeuf and Schmidt, 2001). At lower pressures (<2.5 GPa), phase relations in hydrous basalt are quite complicated (Fig. 6), and they have been considered in detail by Schmidt and Poli (1998, 2003) and Poli and Schmidt (2002). In the MORB system, sodic amphibole, chlorite, and clinopyroxene/plagioclase coexist with epidote, zoisite, or lawsonite along pressure-temperature (*P-T*) paths of subduction zones. In hot subduction zones, Na-Ca-amphibole is stable. Experimental studies show that amphibole is stable in the basaltic composition to ~2.5 GPa (Fig. 6), and its stability field expands slightly to higher pressures with increasing Mg number. Breakdown of amphibole leads to formation of Mg-chloritoid: Amphibole + Zoisite = Chloritoid + Clinopyroxene + Quartz + H_2O (Poli, 1993). Talc, paragonite, and Mg-staurolite are also strongly controlled by the rock composition. Phengite is the ubiquitous K-phase present at any conditions up to the pressures of 10 GPa. Lawsonite and phengite may be competitive phases in basaltic composition at pressures up to 10 GPa (Poli and Schmidt, 1995; Schmidt and Poli, 1998; Okamoto and Maruyama, 1999), whereas K-rich phlogopite and K-richterite may be stable as accessory phases to 10 and 15 GPa, respectively, similar to the peridotite system (Figs. 2 and 6). Unlike in peridotite, there are no stable, highly hydrous phases in basaltic composition above 10–15 GPa, even at lower temperatures corresponding

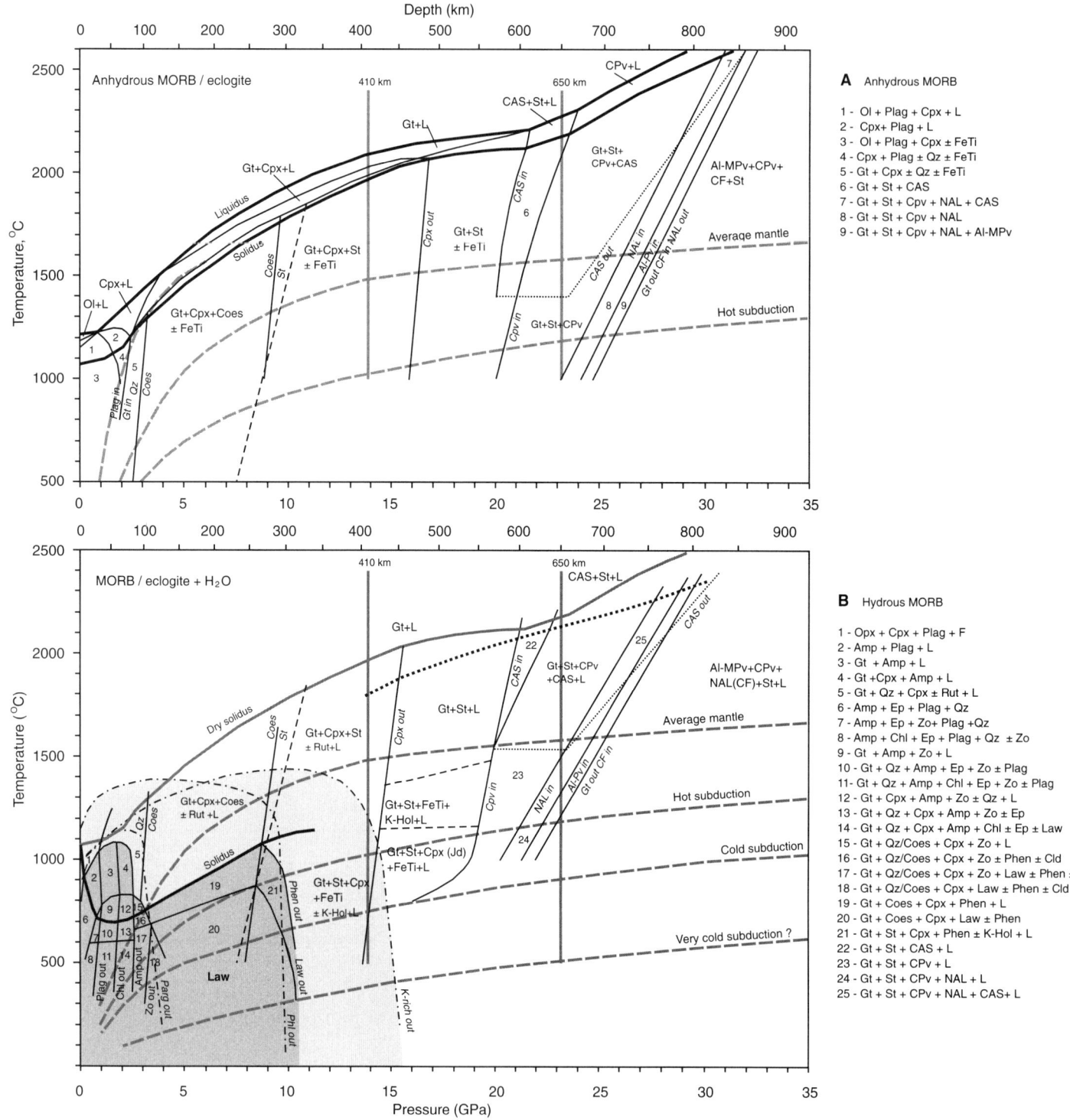

Figure 6. Phase diagram of the anhydrous and H$_2$O-saturated mid-ocean-ridge basalt (MORB) deduced from experimental results (Yasuda et al., 1994; Poli and Schmidt, 1995, 2002; Schmidt and Poli, 1998; Hirose and Fei, 2002; Litasov et al., 2004; Litasov and Ohtani, 2005; Sano et al., 2006). Bold dotted line indicates apparent solidus for MORB + 2 wt% H$_2$O (Litasov and Ohtani, 2005). Chained lines show stability of K-bearing phases (see Fig. 2). Quartz/coesite and coesite/stishovite transition lines are after Bohlen and Boettcher (1982) and Zhang et al. (1996), respectively. Qz—quartz; Coes—coesite; FeTi—Fe-Ti oxides; Rut—rutile; Ep—epidote; Zo—zoisite; Cld—chloritoid; Law—lawsonite; K-Hol—K-hollandite; Jd—jadeite; Al-Pv—Al-rich Mg-perovskite. See Figures 2 and 3 for other symbols.

to subducting slabs (Okamoto and Maruyama, 2004; Litasov and Ohtani, 2005). Inoue et al. (1998a) observed K-bearing hydrous phase X as a decomposition product of K-amphibole, which was stable at pressures above 15 GPa and 1000–1300 °C. Phase X was stable up to 20 GPa in the experiments by Konzett and Fei (2000). Yang et al. (2001) described the structure of phase X (Table 2) and its anhydrous equivalent as trigonal and suggested that it is isostructural with the $Na_2Mg_2Si_2O_7$ phase synthesized by Gasparik and Litvin (1997).

High-pressure mineralogy of the hydrous basalt in the transition zone and lower mantle is virtually the same as that of anhydrous basalt: eclogite changes to garnetite and then to a Mg-perovskite–bearing lithology with pressure (e.g., Yasuda et al., 1994; Hirose et al., 1999; Hirose and Fei, 2002). K-hollandite and Fe-Ti-oxide phases are stable at low temperatures (Fig. 6). Mineral assemblages of MORB above 28 GPa consist of Al- and Fe-rich Mg-perovskite, Ca-perovskite, stishovite, and Al-rich phases NAL (Na-Al bearing phase) or CF (Ca-ferrite structured Al-phase) (Hirose and Fei, 2002; Litasov and Ohtani, 2005). However, water may have effects on post-garnet (garnet- to perovskite-bearing lithology) transformation.

Litasov and Ohtani (2005) suggested a shift of the post-garnet phase transition boundary in hydrous MORB (2 wt% H_2O) to lower pressures, by 1.0–1.5 GPa, based on quench multianvil experiments at pressures up to 28 GPa and observed clear differences between hydrous and anhydrous charges in the simultaneous double-capsule experiments. Recently, we performed in situ X-ray diffraction experiments on the post-garnet transformation in anhydrous and hydrous MORB systems (Litasov et al., 2004; Sano et al., 2006) and confirmed the results obtained by Litasov and Ohtani (2005). Sano et al. (2006) observed more than a 2 GPa shift of the post-garnet transition in MORB containing 5–10 wt% H_2O. In addition, Litasov et al. (2004) found that the Clapeyron slope of the post-garnet phase transformation obtained by in situ experiments is much steeper (+4.1 MPa/K) than that observed in quench experiments (+0.8 MPa/K) (Figs. 5 and 6). Litasov et al. (2004) and Sano et al. (2006) concluded that there is no density crossover between peridotite and basaltic parts of the slabs descending to the lower mantle along the cold subduction geotherm (Fig. 5), and therefore accumulation of basaltic crust near the 650 km discontinuity (e.g., Gasparik, 1997) is unlikely under the hydrous conditions of subduction.

Differences in water solubility in garnet and post-garnet phases may shift the post-garnet phase boundary to the higher pressures (garnet may accommodate up to 1250 ppm, and Mg-perovskites and NAL phase contain less than 50 ppm H_2O; see Chapter 3, this volume). Therefore, there must be another explanation for the phase boundary shift to lower pressures. Litasov and Ohtani (2005) suggested that this shift could be explained by the oxidation of a garnet-bearing assemblage by hydrous fluid and formation of Fe^{3+}-bearing aluminous perovskite (McCammon, 1997; McCammon et al., 2004a) at lower pressures relative to the anhydrous system.

Melting Phase Relations of Hydrous Basalt

In the basaltic composition, amphibole and epidote/zoisite are the most important hydrous minerals at the solidus between 1 and 3 GPa. Zoisite is stable to temperatures of ~750 °C, and amphibole decomposes at 850–1000 °C (Vielzeuf and Schmidt, 2001; Fig. 6). At pressures above 3 GPa, phengite is only possible remaining hydrous phase at the water-saturated solidus of basalt. There are no data on melting-phase relations of hydrous basalt in the pressure interval from 3–4 to 20 GPa. Litasov and Ohtani (2005) reported that liquidus phase relations are different from those observed in anhydrous basalt at 20–28 GPa (Fig. 6). Hirose and Fei (2002) reported that Ca-perovskite is a liquidus phase of anhydrous MORB at pressures above 22 GPa; however, Litasov and Ohtani (2005) showed that the Ca-Al-CAS phase is stable at the liquidus together with stishovite at 22 GPa, and stishovite is a liquidus phase at pressures of 23–28 GPa in hydrous basalt. The partial melt of the hydrous basalt at 20–28 GPa is Fe-rich and similar in composition to that of dry basalt (Hirose and Fei, 2002; Litasov and Ohtani, 2005).

Phase Relations in Metasediments

The majority of experimental work on sediment composition related to metamorphic facies has been performed to 3 GPa. The large number of possible mineral assemblages in metapelites and metagraywackes is commonly described in the model system KFMASH (+CaO, Na_2O). Poli and Schmidt (2002) reviewed phase relations to 3 GPa based on their experimental work and found that they were consistent with the topological analyses made by Kepezhinskas and Khlestov (1977). At pressures to 1.5 GPa, this topology shows the following prograde metamorphic sequence with increasing temperature: (1) garnet + chlorite; (2) chloritoid + biotite; and (3) staurolite + biotite-bearing assemblages. At pressures above the staurolite stability field, the characteristic garnet + talc + chloritoid and garnet + biotite assemblies are observed in most sediment compositions (Fig. 7). Quartz, phengite, amphibole, zoisite, and lawsonite are common minerals of the sediment assemblages. Talc may be transformed to the 10 Å phase assemblage at 6 GPa and 500 °C (Chinnery et al., 1999). However, the stability of the 10 Å phase in sediment composition remains untested.

Hermann (2002) determined phase relations in hydrous continental crust at 2.0–4.5 GPa and found that talc (<700 °C, 4.5+ GPa), zoisite and amphibole (<800 °C and <3 GPa), biotite (<900 °C, <3.5 GPa), and phengite (<1000 °C, 2.5–4.5 GPa) are the primary hydrous phases in the assemblage, which are generally consistent with the other sediment systems. However, Mg-rich biotite may be stable up to 5 GPa, as demonstrated by biotite-diamond intergrowths in gneisses from the Kokchetav ultrahigh-pressure rocks, Kazakhstan (e.g., Sobolev and Shatsky, 1990).

In the Al_2O_3-SiO_2-H_2O system, which may be relevant to sediments, topaz-OH phase appears at pressures above 5 GPa, and the phase Pi has been observed at 5 GPa and low temperatures

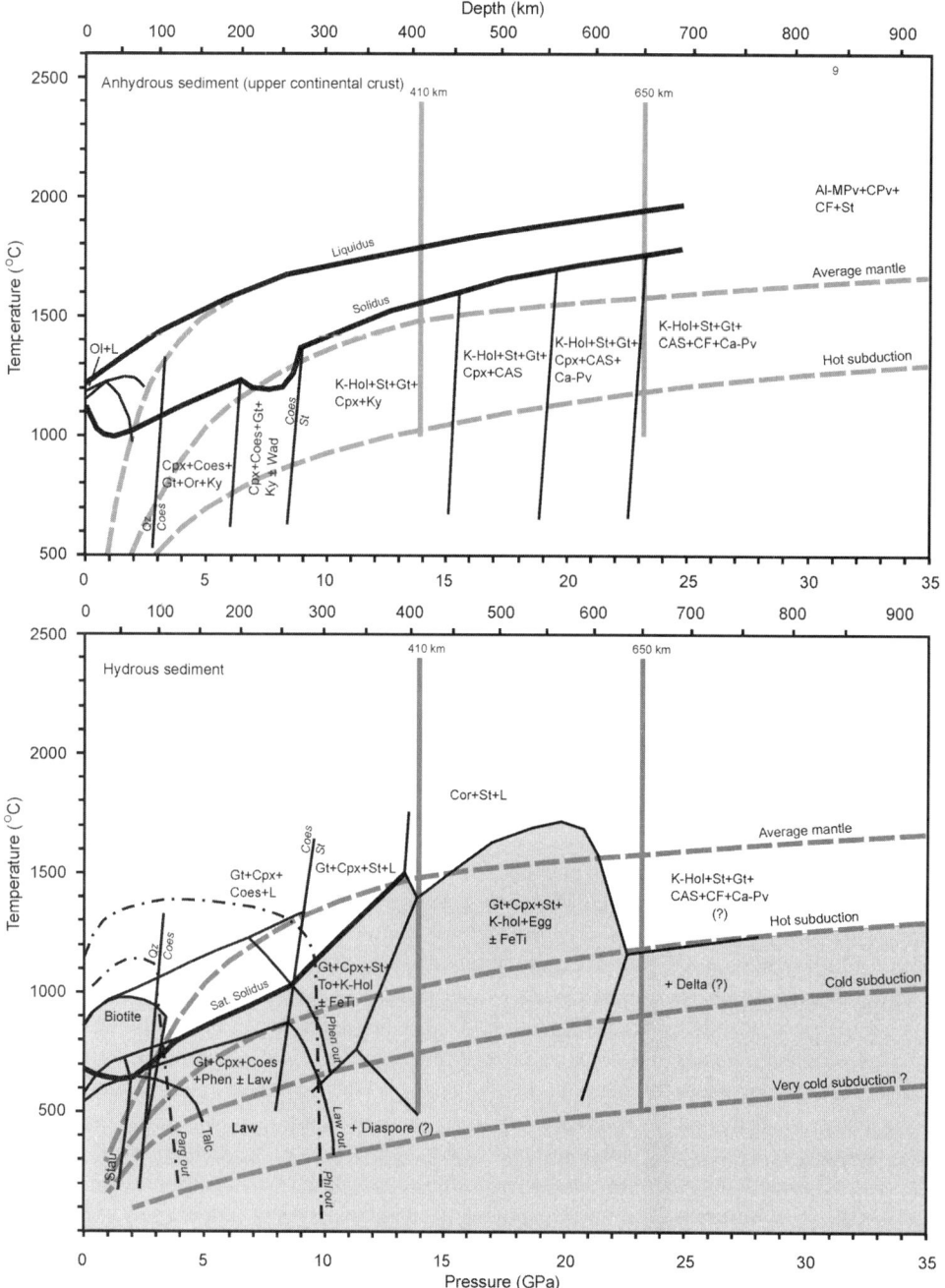

Figure 7. Simplified phase diagram of pelagic sediment (metapelite–upper continental crust) under anhydrous and H$_2$O-saturated conditions based on experimental results (Irifune et al., 1994; Ono, 1998; Poli and Schmidt, 2002; Sano et al., 2004). Low-pressure assemblages to 3 GPa are similar to typical assemblages of metamorphic facies (e.g., Maruyama et al., 1996; Liou et al., 1998). Wad—wadeite-structured K$_2$Si$_4$O$_9$; Ky—kyanite; Stau—staurolite; Egg—phase Egg; Cor—corundum; Delta—δ-AlOOH. See Figures 2 and 6 for other symbols.

(Table 2; Wunder et al., 1993b). Phase Egg has been observed at pressures above 10 GPa (Table 2; Eggleton et al., 1978; Ono, 1999). Domanik and Holloway (1996) determined the stability field of phengitic muscovite and observed phengite, lawsonite, topaz-OH, and Mg-sursassite at 6–8 GPa and 700–900 °C. Lawsonite and sursassite dehydrate at 8–9 GPa, while topaz-OH and phengite remain stable. However, detailed subsolidus phase relations above 3–4 GPa and the transition from low-pressure garnet-chloritoid-talc or garnet-biotite assemblages to high-pressure assemblages containing phengite and topaz-OH have not yet been constrained in complex sediment systems. Ono (1998) determined phase relations in hydrous metapelite and showed that phengite is stable at pressures below 8 GPa, whereas topaz-OH phase is stable at 8–12 GPa and temperatures up to 1200 °C, and phase Egg is stable at higher pressures and temperatures up to 1300 °C (Fig. 7). Recently, Sano et al. (2004) studied stability of phase Egg by in situ X-ray diffraction and found that it is stable to 23 GPa and 1700 °C (Fig. 7). Besides at low temperature, AlOOH phases may be important in sediment compositions. Diaspore is stable to 17 GPa and temperatures below 1000 °C, whereas above this pressure, it transforms to the δ-AlOOH (delta) phase (Table 2; Suzuki et al., 2000a; Ohtani et al., 2001a). Although the stability of the delta phase has not yet been confirmed in multicomponent basaltic or sediment systems, it may play an important role as a water carrier to the deep lower mantle (e.g., Ohtani et al., 2004; Ohtani, 2005). AlOOH phases may also be important if specific types of rocks, such as diasporites (e.g., Feenstra and Wunder, 2002), are subducted.

Irifune et al. (1994) determined subsolidus and liquidus phase relations in the anhydrous upper continental crust composition, which is close to the composition of pelagic sediments (Fig. 7). They found that the solidus temperature decreases at pressures of 6–9 GPa (assumed stability field of wadeite-structured $K_2Si_4O_9$; Yagi et al., 1994). Under subsolidus conditions, the assemblage of garnet–clinopyroxene–K-hollandite–stishovite is stable to 15 GPa. CAS and Ca-perovskite are added to this assemblage at 15 and 20 GPa, respectively, and clinopyroxene is replaced by CF phase at ~24 GPa (Fig. 3). The liquidus phases were coesite at 6 GPa, kyanite at 10 GPa, clinopyroxene at 16 GPa, and stishovite at pressures above 21 GPa.

Melting relations of hydrous sediments have been studied by many authors (e.g., Johnson and Plank, 1999; Vielzeuf and Schmidt, 2001; Poli and Schmidt, 2002; Schmidt and Poli, 2003). At 2 GPa, the solidus assemblage is mostly phengite + biotite + plagioclase + garnet + quartz ± kyanite. At higher pressures, the solidus assemblage, phengite + clinopyroxene + garnet + coesite ± kyanite, is notably identical to mafic basaltic systems (Vielzeuf and Schmidt, 2001; Poli and Schmidt, 2002, see previous). Liquidus phase relations of the hydrous sediments at pressures above 4 GPa are poorly constrained (e.g., Ono, 1998).

Mantle Fluids

Fluid inclusions in mantle minerals in various lithologies observed in mantle xenoliths, peridotite massifs, and ophiolites indicate the existence of hydrous fluid in the mantle. However, typically, inclusions contain more than one volatile component, and CO_2 is one of the most common volatiles in addition to H_2O. Moreover, in most mantle xenoliths, CO_2 is dominant or the only fluid component in fluid inclusions (e.g., Andersen et al., 1984; Schiano and Clocchiatti, 1994; Frezzotti, 2001). In contrast to the fluid inclusions in shallow-mantle minerals, many diamonds contain inclusions of H_2O-bearing fluids (e.g., Chrenko et al., 1967; Navon et al., 1988; Akagi and Matsuda, 1988; Turner et al., 1990; Schrauder and Navon, 1994). Thompson (1992) argued that hydrous silicate melts are a major water reservoir within the mantle, simply as a consequence of the temperature distribution of the upper mantle and transition zone, which lies above the temperature of the hydrous solidus.

Fluid Compositions

Hydrous fluid can dissolve large amounts of silicate components at high pressures (e.g., Kennedy et al., 1962; Manning, 1994; Closmann and Williams, 1995). A pure H_2O fluid is unlikely to be stable in the mantle; Bina and Navrotsky (2000) have suggested that ice VII might be stable in the coolest region of subducted slab. In the MSH system, a maximum 12 wt% of silicates may be dissolved in aqueous fluid at 1–6 GPa and solidus temperature (1150 °C) (e.g., Schneider and Eggler, 1986). Addition of CO_2 to the fluid strongly depresses solubility of silicates by approximately one order of magnitude (Poli and Schmidt, 2002), while addition of chlorine may increase the amount of solutes in the fluid to 50 wt% (e.g., Brenan et al., 1995; Keppler, 1996; Shmulovich et al., 2001). It should be noted that chlorine may be an abundant component of the both subduction-related fluids (e.g., Scambelluri et al., 1997; Scambelluri and Philippot, 2001) and fluids in diamonds (e.g., Navon et al., 1988, 2003). Experimental studies of hydrous fluid compositions in the system $MgO-SiO_2-H_2O$ at higher pressures have indicated that the solubility of silicate components in aqueous fluid increases from 30 wt% at 3 GPa to 70 wt% at 8 GPa at temperatures above the solidus. The Mg/Si ratio of this fluid is 0.2 at 1.5 GPa and almost constant (1.2) at 3–8 GPa (Mibe et al., 2002; Kawamoto et al., 2004). Even higher Mg/Si ratios (up to 3) of the melt/fluid have been observed at higher pressures (e.g., Inoue, 1994; Kawamoto, 2004), while ratios decrease with temperature. Kawamoto et al. (2004) argued that, at pressures of ~3 GPa, Mg/Si ratios change rapidly from SiO_2-rich (reported previously by Ryabchikov et al., 1982) to MgO-rich. Mibe et al. (2002) suggested that aqueous fluid migrating upward through the mantle can modify the Mg/Si of residual rocks and create Mg/Si stratification in the Earth. Stalder (2004) doubted the results of Mibe et al. (2002, 2004a) and suggested that silicate solubility in hydrous fluid is less than 30 wt% at 9 GPa. A recent study by Kessel et al. (2005) showed that the solubility of silicates in hydrous fluid is ~20% at 4 GPa in K-free basaltic composition, and the composition of fluid is essentially Si- and Na-rich.

Water is the most abundant volatile component in subduction-zone fluid. There are numerous works on the composition of fluid phases and dehydration in subduction zones (e.g., Scambelluri and Philippot, 2001; Kerrick and Connolly, 2001a, 2001b; Stern, 2002; Schmidt and Poli, 2003). In his review of the chemistry of subduction-zone fluids, Manning (2004) emphasized that aqueous fluid may be substantially dilute, containing only 2–3 times of the total dissolved solids of seawater (6–10 wt%) at 1–2 GPa. Silica and alkalis are dominant solutes, with significant Al and Ca and low Mg and Fe. In addition, this aqueous fluid contains a significant amount of incompatible elements and chlorine. Understanding the evolution of subduction-zone fluid is very complicated because it requires quantitative reaction-flow modeling of porous and channeled flows.

Fluids preserved in inclusions in diamonds formed at 5–6 GPa may be the deepest existing fluid observed to date. Fibrous diamonds contain significant amounts of H_2O as numerous micrometer-scale micro-inclusions in fibers (e.g., Schrauder and Navon, 1994; Kagi et al., 2000). Schrauder and Navon (1994) and Navon et al. (2003) concluded that the major-element composition of the fluids in diamonds varies widely, but they were able to distinguish two end members, a carbonate-rich one enriched in CaO, FeO, MgO, and P_2O_5; and a hydrous fluid enriched in SiO_2 and Al_2O_3. Most of the fluids are enriched in K and halogens (Johnson et al., 2000). A third type of micro-inclusion represents brine enriched in water, chlorine, carbonates, and alkalis. Izraeli et al. (2001) studied brine inclusions in diamonds from both peridotite and eclogite suites and proposed a brine composition of $(K,Na)_8(Ca,Fe,Mg)_4SiO(CO_3)_4(H_2O)_{28-44}$, from which diamonds were precipitated. Navon et al. (2003) suggested that various fluids trapped in diamonds may have evolved from a parental carbonatitic melt, which may have been derived from a MORB-like source based on the isotopic and trace-element characteristics of diamonds and their inclusions (Schrauder et al., 1996; Cartigny et al., 1998; Johnson et al., 2000; Burgess et al., 2002). Pal'yanov et al. (1999, 2002), Akaishi et al. (2000), and Sokol et al. (2001) demonstrated that, in experiments on various compositions at 5–8 GPa, diamond is grown from the carbonate-water system, which is similar in composition to fluids observed as inclusions in natural diamonds.

Mobility of Hydrous Fluid in the Mantle

Measurements of dihedral angle (θ) between coexisting fluid and mantle minerals allow us to estimate the efficiency of fluid transport. A dihedral angle of 60° marks the transition between an interconnected fluid phase ($\theta < 60°$) and a fluid phase isolated at grain-boundary junctions ($\theta > 60°$). Watson et al. (1990) documented a negative pressure dependence of θ in the olivine-H_2O system at pressures between 0.5 and 2 GPa. In their study, θ became <60° at pressures above 0.7 GPa at 1200 °C. Mibe et al. (1998, 1999) extended these experiments to higher pressures and confirmed the negative pressure and temperature dependence of θ. They proposed that the transition from isolated to interconnected fluid networks in the subducting slab may control the location of the volcanic front in island arcs.

Watson and Lupulescu (1993) reported $\theta > 60°$ in the clinopyroxene-H_2O system at 1.5 GPa and 900–950 °C. Ono et al. (2002) studied pyrope-H_2O at 4–13 GPa and 900–1200 °C and found an increase of θ with pressure. At pressures above 9 GPa, θ was greater than 60°. They suggested that in the basaltic portion of the slab, aqueous fluid released from dehydration of lawsonite and phengite can be trapped in the garnet-rich matrix and transported further to the deep mantle. Mibe et al. (2003) extended these experiments on eclogite composition and found θ of 62–68° for garnet-clinopyroxene-fluid aggregates at 3–5 GPa and 700–800 °C, thus confirming the conclusions obtained by Ono et al. (2002), and they suggested that 1–2 vol% of aqueous fluid may be trapped in eclogite and transported to the deeper mantle.

Fluid/Melt Miscibility in the Deep Mantle

Fluid phases at high pressure can trigger melting. Water-saturated silicate melts become increasingly hydrous with increasing pressure. These trends result in a total convergence of these two phases at a distinct pressure and temperature (e.g., Stalder et al., 2001). This singular point in the given P-T-X-space, which terminates the solidus curve, is called the "second critical end point." At supercritical conditions, we could consider the system as supersolidus at any temperature as long as the fluid phase is present. Therefore, fluid may exist below mantle adiabat at any pressures above the second critical end point. An open question at present concerns the nature and wetting properties of the fluid/melt phase at pressures above the second critical end point. Complete miscibility between the melt and fluid was observed in the albite-H_2O system by using a diamond anvil cell (Shen and Keppler, 1997). Bureau and Keppler (1999) studied the solvus line at 1.5 GPa in different system and found miscibility of nepheline-, jadeite-, and granite-H_2O at 550, 800, and 900 °C, respectively. However, the maximum of the solvus line is not the same as the second critical end point, which is a pressure-defined single point, and at this particular pressure, it coincides with the maximum of the solvus. Stalder et al. (2000) reported the second critical end point in albite-H_2O system above 1.5 GPa using the diamond-trap technique, whereas in silicate (forsterite + enstatite)-H_2O, total convergence between fluid and melt along the solidus probably occurs at 12–13 GPa (Stalder et al., 2001). Wyllie and Ryabchikov (2000) argued that the compositions of fluid inclusions in fibrous diamonds support the occurrence of a second critical end point in lherzolite-CO_2-H_2O system (placed at ~8 GPa). Recently, Kessel et al. (2005) reported the second critical end point between 5 and 6 GPa for K-free eclogite using the diamond-trap technique. Mibe et al. (2004b, 2005) and Kawamoto et al. (2005) developed a new method of direct observations of immiscible fluids using synchrotron X-ray radiography. Using this approach, Kawamoto et al. (2005) and Mibe et al. (2005) reported the second critical end point in calc-alkaline andesite, andesite, and normal MORB at 1, 2.5, and 3 GPa, respectively. Their results for basalt are different from those obtained by Kessel et al. (2005) and need to be clarified.

At pressures above the second critical end point, the solubility of silicate components dissolved in the aqueous fluid/melt increases drastically. There is no H_2O-saturated solidus, and melting occurs somewhere at "practical" or "apparent" solidus at these pressures (e.g., Iwamori, 1998; Litasov and Ohtani, 2002; Kawamoto et al., 2004). At pressures of 10–25 GPa, it is possible to distinguish the existence of "fluid" at low temperatures and "melt" at high temperatures based on the texture and morphology of the quench run products (e.g., Inoue, 1994; Irifune et al., 1998; Litasov and Ohtani, 2002). Although typical bubbles of fluid phase were absent, "fluid" formed at low temperatures was quenched as round-shaped patches with euhedral crystals, whereas the melt formed at high temperatures was observed as irregular-shaped or interconnected interstitial quench crystals. Therefore, it is possible to define the temperature where the melt fraction exceeds a certain threshold and intensive melting occurs. Appearance of quench crystals was defined as an "apparent solidus" (Fig. 2; Litasov and Ohtani, 2002). The textural differences (and differences in dihedral angles?) between liquid phases at high and low temperatures and pressures above the second critical end point need to be clarified in further works.

WATER SOLUBILITY IN NOMINALLY ANHYDROUS MINERALS

Hydrogen Incorporation to High-Pressure Minerals

In near-surface environments, various hydrogen-bearing minerals have been identified, and they contain OH^-, H_2O, and "free" protons bound typically as H_3O^+, $H_5O_2^+$, and NH_4^+ (Hawthorne, 1992, 1994; Prewitt and Parise, 2000). There are also some unusual mechanisms, such as the so-called hydrogarnet substitution, the exchange of Si^{4+} by $4H^+$, or Al^{3+} by $3H^+$ (Lager et al., 1989; Wright et al., 1994; Lager and VonDreele, 1996; see Prewitt and Parise, 2000, for detailed discussion). Molecular water as well as proton bonds cannot be incorporated into the structure of high-pressure minerals. Therefore, in high-pressure phases, we should not use the term "water" but rather hydrohyl group (OH^-) or hydride (H^-). However, for simplicity and following the majority of previous studies, we hereafter express OH^- contents in minerals as H_2O. Incorporation of water as hydrohyl can form relatively dense frameworks that are directly related to anhydrous equivalents by simple substitution. For example, superhydrous phase B is related to humite and ultimately to olivine by simple crystallographic shear (Pacalo and Parise, 1992; Prewitt and Parise, 2000).

The most common method of detecting OH in minerals is Fourier transform infrared spectroscopy (FTIR), which is able to distinguish structurally incorporated OH from molecular water in fluid inclusions and grain boundaries. Determination of the amount of water (OH) in the minerals requires calibration of the intensity of the absorption with the amount of water present. Therefore, an independent technique to determine the amount of water in standard material is required. There are generalized calibrations based on glasses and nominally anhydrous phases (Paterson, 1982; Libowitzky and Rossman, 1997). These relationships are, however, poorly applicable to all nominally anhydrous minerals, and it is most likely that each mineral needs to be calibrated independently (e.g., Bell et al., 1995, 2003).

There are some other techniques with the potential to estimate hydrogen abundance in minerals. Hydrogen manometry is an absolute technique, where hydrogen is extracted from the sample and its volume is precisely measured. It requires, however, large samples (e.g., 4 g of garnet for pyrope calibration; Bell et al., 1995).

Secondary ion mass spectrometry (SIMS) is also an important method (Kurosawa et al., 1997; Hauri, 2002; Koga et al., 2003), although it depends significantly on the matrix correction, and structurally incorporated hydrogen and very small (nanometer-scale) fluid inclusions may not be distinguished by this method.

Another bulk technique is ^1H nuclear magnetic resonance spectroscopy using spinning at the magic angle (^1H MAS NMR) (e.g., Yesinowski et al., 1988; Kohn, 1996; Kohn et al., 2002), which has some advantages, such as the area of the spectrum is directly proportional to the amount of hydrogen in the sample independent of the matrix.

Another technique is nuclear reaction analysis (Lanford et al., 1976; Rossman, 1988; Skogby et al., 1990; Maldener et al., 2001; Bell et al., 2003), which has the advantage of yielding absolute hydrogen concentrations; however, it requires a large sample area (at least 1 mm in diameter). Bell et al. (2003) applied this technique to hydrogen in olivine and found that previous values of OH content determined by using integration of FTIR spectra had been underestimated.

Upper-Mantle Minerals

The presence of hydrohyl point defects in nominally anhydrous phases, such as olivine, pyroxene, and garnet, has long been considered (Sclar, 1970; Fyfe, 1970; Martin and Donnay, 1972; Wilkins and Sabine, 1973). Systematic studies of water content in nominally anhydrous minerals from various geological environments, such as samples from mantle xenoliths, indicate that typical mantle minerals, olivine, pyroxene, and garnet contain variable amounts of structurally bound hydrogen as defects (Table 3; Fig. 8). The amount of water in nominally anhydrous minerals may be as much as thousands of ppm. This leads to an intriguing possibility that the major hosts of water in Earth's interior are nominally anhydrous phases, rather than hydrous minerals stable only at low temperatures (e.g., Bell and Rossman, 1992a).

Olivine

Water solubility in olivine has been studied intensively because it is the most abundant mineral in the upper mantle. Wilkins and Sabine (1973), Beran and Putnis (1983), Kitamura et al. (1987), Miller et al. (1987), Bell and Rossman (1992a), Libowitzky and Beran (1995), Kohn (1996), Kurosawa et al. (1997), Jamtveit et al. (2001), Matsyuk and Langer (2004), Bell et al. (2004a), and Matveev et al. (2005) have all reported the water concentration in natural olivines, whereas Mackwell and Kohlstedt (1990), Bai and Kohlstedt (1992, 1993), Kohlstedt et al. (1996), Matveev et al. (2001), Chen et al. (2002), Smyth et al. (2004), and Litasov et al. (2006a) have measured the water solubility in synthetic olivine samples (Table 3).

In the systematic study by Bell and Rossman (1992a), the maximum water concentration in natural olivine (140 ppm) was observed in megacrysts from Monastery kimberlites. However, recently, Bell et al. (2003) applied nuclear reaction analysis for calibration of hydrogen in olivine and found that some previous estimates of hydrogen concentrations from FTIR spectra need to be revised upward by a factor of 3–3.5. Recently, Matsyuk and Langer (2004) reported maximum water concentrations of 419 ppm in olivine megacrysts from Udachnaya kimberlites in Siberia. The typical range of water content in natural olivines lies between 0 and 60 ppm, including island arc–related environments (e.g., Miller et al., 1987; Kurosawa et al., 1997).

Experimental data on water solubility in olivine indicate that natural samples do not reflect saturation with water and clearly

TABLE 3. WATER SOLUBILITY IN NOMINALLY ANHYDROUS MANTLE MINERALS

Mineral	Mg #	Al$_2$O$_3$ (wt%)	P (GPa)	T (°C)	H$_2$O (ppm)	Method* (cm^{-1})	Reference
Olivine[†]	88–92		1–6	1000–1100	0–140	3430–3640	Bell and Rossman (1992a)
	90		2–7	1100	135–496	3567, 3579, 3598, 3613	Kohlstedt et al. (1996)
	90		10–12	1100	1070–1510	Same	Kohlstedt et al. (1996)
	90		2–4	1300	23–57[§]	3285–3380	Matveev et al. (2001)
	90		2–4	1300	282–738[#]	3430–3640	Matveev et al. (2001)
	87–93		~4–6	~1100	0–419	FTIR (up to 70 bands)	Matsyuk and Langer (2004)
Olivine	96–97		13–14	1200	3700–**7100**	SIMS	Chen et al. (2002)
	100		0.3	1000–1300	7–18	3332–3384, 3568, 3579, 3612	Zhao et al. (2004)
	83		0.3	1000–1300	40–101	3329–3386, 3487–3573	Zhao et al. (2004)
	90		6	1100	1730	3567, 3579, 3598, 3613	Mosenfelder et al. (2006)
	90		12	1100	6400	Same	Mosenfelder et al. (2006)
	95		12	1250	5000[#]	FTIR	Smyth et al. (2004)
	90		9.5	1200	1200	3477, 3533–3579, 3598, 3613	Litasov et al. (2006a)
	90		13.5	1400	5200	Same	Litasov et al. (2006a)
Orthopyroxene[†]	77–90	0.3–5.2	0–6	?	26–350	3410, 3510, 3560	Skogby et al. (1990)
	88–92	1–5	1–6	900–1100	50–650	3410, 3560	Bell and Rossman (1992a)
	88–92	3–6	1.0–2.2	900–1100	39–265	3060, 3210, 3300, 3570	Peslier et al. (2002)
Orthopyroxene	100	0	0.2–10	1100	55–867	3064, 3363, (3605, 3676, 3690)[**]	Rauch and Keppler (2002)
	100	1	1.5	1100	**1100**	3363, 3474, 3520, 3549	Rauch and Keppler (2002)
Clinoenstatite	100		15	1300–1500	580–590	3380, 3460, 3675	Bolfan-Casanova et al. (2000)
Diopside[†]	70–100	0–7	~0–2	?	20–1000	3355, 3460, 3535	Skogby et al. (1990)
	88–92	3–6	1.0–2.2	900–1100	140–528	3460, 3540	Peslier et al. (2002)
Omphacite[†]	88	11	~4	?	1200	3355, 3460, 3535	Peslier et al. (2002)
	75–90	10–19	~2–6	?	1000–1840	3470, 3620	Smyth et al. (1991)
Omphacite[†]	67–78	8–9	2.0–2.7	700	1000–1100	3440, 3500, 3600	Katayama and Nakashima (2003)
	74	9–10	6–8	1000	3020	Same	Katayama and Nakashima (2003)
Diopside	70–90	0–0.2	0	700–900	0–1100	3355, 3450, 3535, 3620–3640	Skogby (1994)
	97	0	0.5–4.0	1000–1100	95–446	3434, 3525, 3558, 3646	Bromiley et al. (2004)
Omphacite	70–90	3–18	1.3–5.5	900–1300	61–886	3445–3465, 3500–3540, 3600–3624	Koch-Müller et al. (2004)
Jadeite	100	24–25	2–10	1000	100–450	3373, 3613	Bromiley and Keppler (2004)
Pyrope[†]	92–98	25	3	900	20–30	3600–3660	Rossman et al. (1989)
	46–87	21–23	~2–6	900–1300	0–135	3512, 3570, 3650	Bell and Rossman (1992b)
Pyrope	100	25	2–5	800–1000	200–700	3604, 3618, 3641, 3651	Geiger et al. (1991)
	92–98	25	1.5–10	800–1000	73–199	3602, 3640–3660	Lu and Keppler (1997)
Pyr.-almandine[†]	25–70	20–23	~3	600–700	92–**1735**	3400–3420, 3560–3580	Xia et al. (2005)
Pyrope-grossular	100	22–25	2–6	1000	140–1010	3630	Withers et al. (1998)
	100	22–25	7–13	1000	0		Withers et al. (1998)
Majorite	100	0	17.5	1500	677	3550	Bolfan-Casanova et al. (2000)
	66–85	22	20	1400–1500	1130–**1250**	3450, 3580	Katayama et al. (2003)
Coesite		0	5–10	1200	43–**212**	3299, 3459, 3523, 3573, 3606	Mosenfelder (2000)
		~0	3.1–7.5	700–1100	4–200	3459, 3516, 3575, SIMS	Koch-Müller et al. (2001)
Stishovite		1.5	10	1200	82	3111, 3240, 3311	Pawley et al. (1993)
		0	15–21	1300–1500	15–72	3111, 3240, 3314	Bolfan-Casanova et al. (2000)
		0.6–0.8	10	1200	210–377	3111	Chung and Kagi (2002)
		1.3	15	1400	570–844	3111	Chung and Kagi (2002)
		?	30–60	2000–4000	80–480	3111	Panero et al. (2003)
		4.6	20	1400	**2900**	2660, 3126, 3320	Litasov et al. (2006c)
Wadsleyite	92		14	1300–1400	600–4000	3329, 3580, 3615	Young et al. (1993)
	100		15.5	1200–1300	1.1–3.1[††]	SIMS	Inoue et al. (1995)
	90		14–15	1100	2.1–2.4[††]	3270, 3305, 3360, 3590, 3605	Kohlstedt et al. (1996)
	87–93		13–15	1200	1.9–**3.4**[††]	SIMS	Chen et al. (2002)
Ringwoodite	90		19.5	1100	0.19–**2.6**[††]	3120, 3220, 3345, 3645	Kohlstedt et al. (1996)
	100		19	1300	2.2[††]	SIMS	Inoue et al. (1998b)
	100		20	1300	2.0[††]	SIMS	Kudoh et al. (2000)
	89–100		18–22	1400–1500	0.2–1.1[††]	3120	Smyth et al. (2004)
Akimotoite	100	0	19–24	1300–1600	352–443	3260, 3300, 3320, 3390	Bolfan-Casanova et al. (2002a)
Mg-perovskite	100	0	27	1500	**60**	3423, 3483	Meade et al. (1994)
	100	0	24	1200	<1	FTIR	Bolfan-Casanova et al. (2000)
	88–89	3.4–5.8	25.5	1600	1900–2400	SIMS	Murakami et al. (2002)
	100	0	25	1300	12	3423, 3448, 3482	Litasov et al. (2003)
	100	2.0	25	1400	<35	3397, 3448	Litasov et al. (2003)
	100	4.4	25	1200	1300[#]	SIMS	Litasov et al. (2003)
	100	7.2	26	1200	1400[#]	SIMS	Litasov et al. (2003)
	91	5.8	25	1600	<20	3397, 3448, 3482	Litasov et al. (2003)
	61	16.8	25	1300	<10	3397, 3423, 3448	Litasov et al. (2003)
	93–98	0–4.9	24	1400	<2	3388	Bolfan-Casanova et al. (2003)
	100	1.4	24	1800	1300	SIMS	Krawczynski et al. (2004)
Ca-perovskite		0.3	25.5	1600	3700	SIMS	Murakami et al. (2002)
		2.0	25	1900	5100	3343, 3607	Litasov and Ohtani (2003c)
Ferropericlase	93		5	1200	2	3320	Bolfan-Casanova et al. (2002b)
	93		25	1200	20	3320, 3480	Bolfan-Casanova et al. (2002b)
	92		25.5	1600	1900	SIMS	Murakami et al. (2002)
	88	2.0	25	1800	**60**	3299, 3308, 3474	Litasov and Ohtani (2003c)
CF phase	100		20	1000	3700–4300	SIMS	Litasov and Ohtani (2004)
NAL phase	71–77	54–57	20–25	1200	0–50	SIMS	Litasov and Ohtani (2004)

*Frequencies of OH absorption bands (in cm^{-1}) are shown for Fourier transform infrared (FTIR) spectra. SIMS—secondary ion mass spectrometry.
[†]Natural samples (experimental samples are not marked).
[§]Silica excess in the sample.
[#]Magnesia excess in the sample.
[**]Minor peaks visible in few samples in parentheses.
[††]Wt% H$_2$O. Well-documented maximum H$_2$O solubilities for particular mineral are in bold.

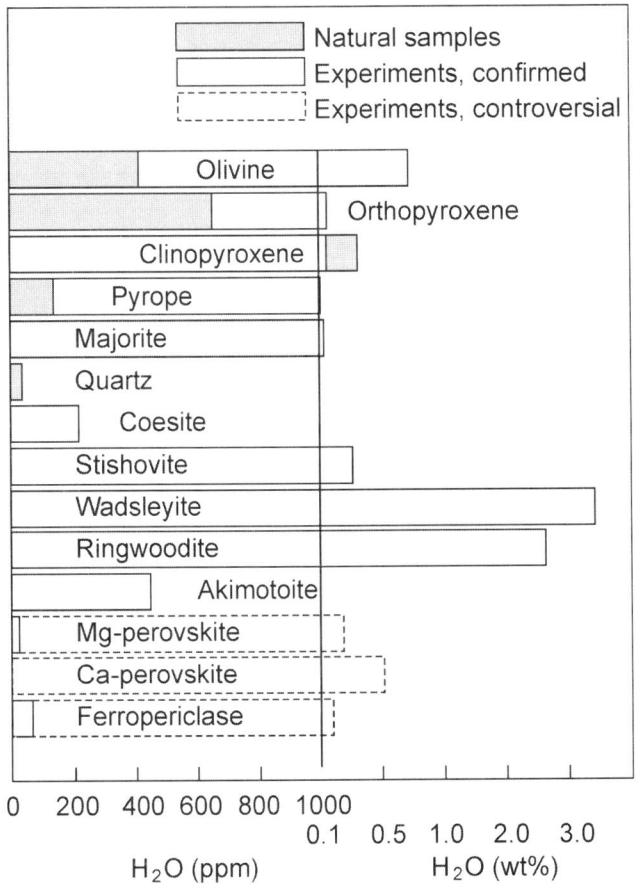

Figure 8. Range of water concentrations measured in natural and synthetic samples. For clinopyroxene (omphacite), concentrations in natural samples exceed those obtained in experiments.

0.37–0.71 wt% H_2O in olivine synthesized at 13–13.5 GPa and 1200 °C, measured by SIMS. Very recently, Smyth et al. (2006), Mosenfelder et al. (2006), and Litasov et al. (2006a) found that 5000–8900 ppm H_2O can be incorporated to olivine (Fo_{90-95}) at 12–14 GPa and 1200–1400 °C (Table 3). It was shown that the H_2O concentration in olivine decreases with increasing temperature at 12–14 GPa from ~8900 ppm at 1250 °C to 4500 ppm at 1400 °C and 1000 ppm at 1600 °C (Smyth et al., 2006).

Most models of hydrogen incorporation to olivine have involved vacancy substitution on the Si and Mg sites (e.g., Mackwell and Kohlstedt, 1990) and some oxygen interstitials. Using molecular dynamic simulation, Wright and Catlow (1994) suggested that incorporation of OH in olivine via a reduction of Fe^{3+} is energetically the most favorable mechanism. Complex infrared (IR) spectra of olivines, however, indicate that all possible substitutions may be realized (Bai and Kohlstedt, 1992, 1993; Kohlstedt et al., 1996). Matsyuk and Langer (2004) emphasized up to 70 absorption bands in natural olivines from kimberlites, including those related to nonintrinsic separate inclusions.

Experimental olivines and major parts of natural olivines show complex IR absorption bands near 3600 cm^{-1}, whereas some other mantle olivines show IR absorption around 3340 cm^{-1}, which may indicate a different mechanism of hydrohyl incorporation (Miller et al., 1987; Kohlstedt et al., 1996). Beran and Putnis (1983) and Libowitzky and Beran (1995) suggested that high-frequency OH bands near 3600 cm^{-1} may be due to hydrohyl-balancing Si vacancies, whereas Kohlstedt et al. (1996) and Kohlstedt and Mackwell (1998) attributed all absorption bands near 3600 cm^{-1} and those at 3170–3445 cm^{-1} indiscriminately to hydroxyl balancing M2 vacancies. Matveev et al. (2001) synthesized olivine with both types of IR spectra and showed that low-frequency bands at 3285–3380 cm^{-1} are assigned to OH substitution in the M2 site, and are indicative of silica-excess olivines, whereas high-frequency bands at 3430–3640 cm^{-1} and a single peak at 3295 cm^{-1} are attributed to OH substitution in the Si site and are indicative of magnesia-excess olivines (Fig. 9).

Khisina et al. (2001) and Khisina and Wirth (2002) observed nanometer-scale OH-bearing inclusions in peridotitic olivines and identified them as "hydrous olivine" phase with the general formula $(Mg_{1-y}Fe^{2+}_y)_{2-x}V_xSiO_4H_{2x}$, where V is a vacancy in the metal sublattice. Kudoh (2002) proposed a possible structure of this hydrous olivine with $x = 0.2$ in this above formula, which can be deduced from the forsterite structure by a simple crystallographic shear. Natural olivines contain many other micro-inclusions of hydrous phases, which may precipitate from hydrated host olivine during decompression and ascend to the surface. In a detailed study by Matsyuk and Langer (2004), nonintrinsic inclusions of serpentine, talc, Mg-edenite, pargasite, and probably wadsleyite were found in olivines from xenoliths and xenocrysts in Yakutia kimberlites. Khisina and Wirth (2002, 2004) also noted lamellar inclusions of the 10Å phase in olivines from Udachnaya kimberlite xenoliths (Yakutia).

show that the water solubility increases with pressure, temperature, and FeO-content, and decreases with silica activity in the system (e.g., Bai and Kohlstedt, 1993; Young et al., 1993; Kohlstedt et al., 1996; Matveev et al., 2001; Zhao et al., 2004; Smyth et al., 2004; Litasov et al., 2006a). Kohlstedt et al. (1996) synthesized Fo_{90}-olivines, the water contents of which increased with pressure from 135 ppm water at 2.5 GPa to 1510 ppm at 12 GPa and 1100 °C. These values were calculated using the calibration of Paterson (1982). It should be noted that the calibration by Bell et al. (2003) allows recalculation of the maximum solubility of water in Kohlstedt's olivines to ~5300 ppm. Zhao et al. (2004) showed that water solubility in olivine at 300 MPa increases in temperature range (1000–1300 °C) from 7 to 18 ppm in pure forsterite and from 40 to 101 ppm in Fo_{83}. Accordingly, water solubility in olivine also increases with FeO content.

Matveev et al. (2001) studied water solubility in olivines at 2–4 GPa and 1300 °C and found strong dependence with silica activity in the system. They reported that olivine coexisting with orthopyroxene (high silica activity) contains 20–50 ppm water, whereas that coexisting with ferropericlase (low silica activity) contains 280–740 ppm H_2O (Fig. 9). Chen et al. (2002) reported

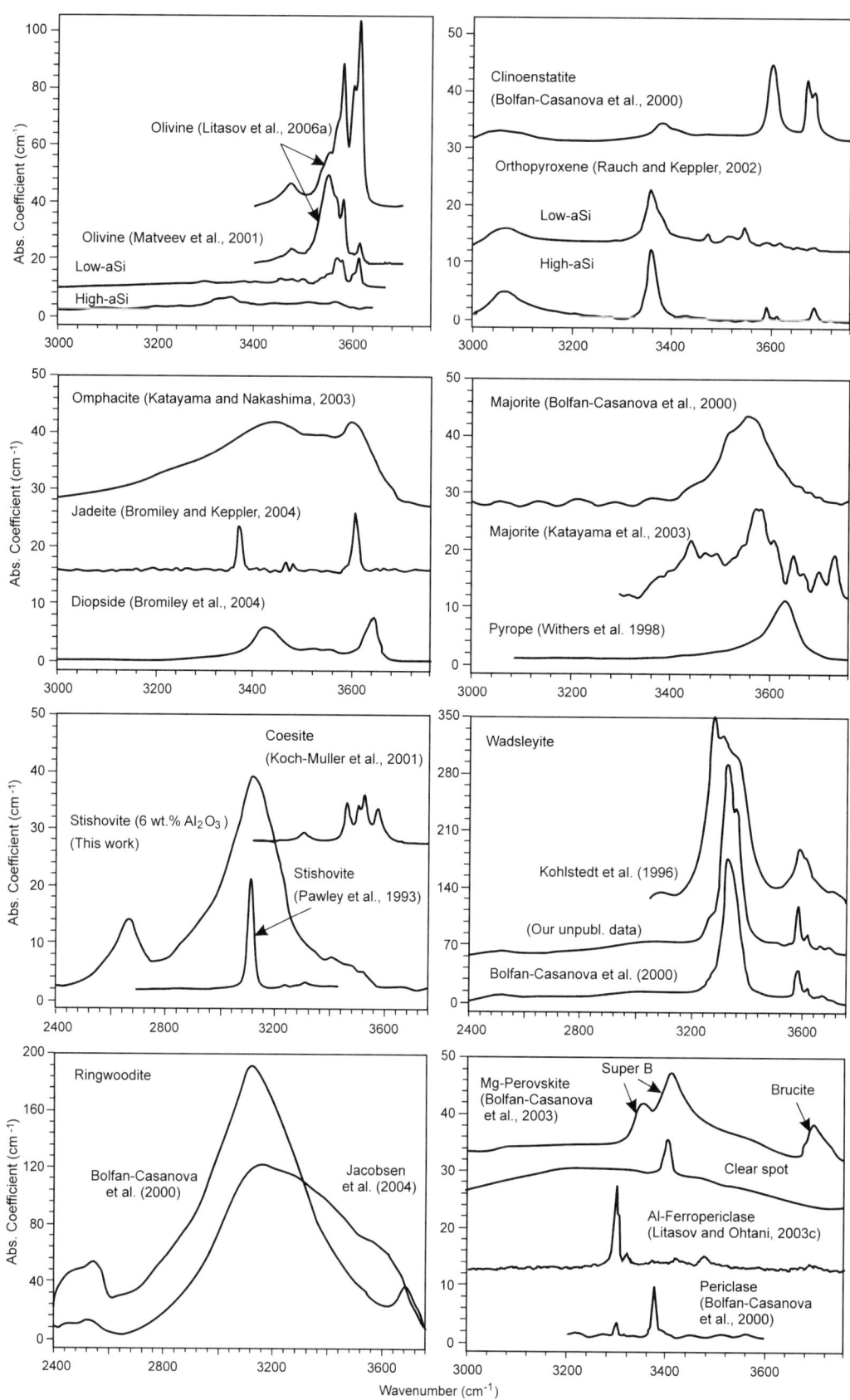

Figure 9. Examples of OH bands present in infrared (IR) spectra of mantle minerals. Note some spectra have not been normalized to common thickness. Low aSi and high aSi are low and high silica activity, respectively.

Orthopyroxene

The water solubility in pure $MgSiO_3$ enstatite is roughly comparable to that in olivine. Bell and Rossman (1992a) and Rossman (1996) found that most mantle orthopyroxenes contain water in the range of 50–650 ppm (Table 3). Highest concentrations were observed in garnet peridotite xenoliths from the Premier kimberlite, South Africa. Peslier et al. (2002) found water contents of 39–265 ppm H_2O in orthopyroxene from spinel peridotite xenoliths in the subarc mantle of North America. They also noted a correlation of major-element composition and oxidation state with water contents.

Experimental data show that water solubility in orthopyroxene increases with pressure, temperature, and Al_2O_3 content. Rauch and Keppler (2002) found that water content in pure enstatite increases from 55 ppm at 0.2 GPa to 867 ppm at 7.5 GPa and 1100 °C. They also reported that addition of 1 wt% Al_2O_3 increases water solubility from 199 ppm to 1100 ppm H_2O at 1.5 GPa and 1100 °C. Linear extrapolation to higher Al_2O_3 contents, typical for mantle orthopyroxenes (2–5 wt%), indicates that the water solubility may be as high as 5000 ppm H_2O. Consequently, Rauch and Keppler (2002) concluded that Al-orthopyroxene may be the major host of water in the uppermost mantle.

Bolfan-Casanova et al. (2000) measured water in clinoenstatite at 15 GPa and 1300–1500 °C and found 580–590 ppm H_2O in clinoenstatite coexisting with wadsleyite or stishovite, i.e., in the samples under Mg- and Si-excess conditions, respectively. SIMS measurements of H_2O in clinoenstatite suggested 0.3–0.5 wt% from experiments at 13–15 GPa and 1300 °C (Yamada et al., 2004).

Most orthopyroxenes show a strong narrow band at 3363 cm^{-1} in the IR spectra. Also, a weak broad band at 3064 cm^{-1} has been observed. Rauch and Keppler (2002) reported weak bands at 3605, 3676, and 3690 cm^{-1} at 10 GPa (Fig. 9). These bands become dominant at 15 GPa in clinoenstatite (Bolfan-Casanova et al., 2000). Beran and Zemann (1986) suggested replacement of Mg^{2+} by two protons; this substitution was confirmed by IR and electron microprobe study by Rauch and Keppler (2002). However, additional peaks in high clinoenstatite may represent different environments of protons in the structure. Several additional bands at 3474, 3520, and 3549 cm^{-1} are indicative of Al_2O_3-bearing orthopyroxene and may be consistent with a coupled substitution of Si^{4+} by $Al^{3+} + H^+$ (Rauch and Keppler, 2002). Another substitution may involve reduction of Fe^{3+} (Skogby and Rossman, 1989).

Clinopyroxene

Water solubility in natural clinopyroxenes (Table 3) has been studied by many authors (Wilkins and Sabine, 1973; Ingrin, 1989; Skogby and Rossman, 1989; Skogby et al., 1990; Smyth et al., 1991; Bell and Rossman, 1992a; Beran et al., 1993; Bell et al., 1995; Peslier et al., 2002; Katayama and Nakashima, 2003; Koch-Müller et al., 2004). Experiments on solubility and diffusivity of hydrogen in clinopyroxene have been conducted by Skogby (1994), Ingrin et al. (1995), Bromiley and Keppler (2004), and Bromiley et al. (2004). Clinopyroxene contains the highest concentration of water among uppermost mantle minerals in natural samples. The water solubility ranges from 100 to 600 ppm in augite and diopside and up to 1200 ppm in omphacite from mantle eclogites (Smyth et al., 1991; Bell and Rossman, 1992a; Ingrin and Skogby, 2000; Koch-Müller et al., 2004). However, maximum concentrations may exceed 3000 ppm as reported by Katayama and Nakashima (2003) for omphacite from ultrahigh-pressure eclogites of the Kokchetav massif, Kazakhstan.

Experimental works show that water solubility in clinopyroxene depends significantly on its composition. In diopside and augite, water content increases with FeO and Al_2O_3 and under oxidized conditions (Skogby et al., 1990; Skogby, 1994; Bromiley et al., 2004). Solid solution of jadeite and diopside and, in particular, a Ca-Escola component leads to a drastic increase of water solubility in clinopyroxene (Katayama and Nakashima, 2003; Bromiley and Keppler, 2004). Skogby (1994) reported up to 1100 ppm water in Fe-bearing diopside synthesized at 1 atm and 700–1000 °C. High solubility of 1160 ppm water in synthetic Al-rich diopside was reported by Kohn (1996) using ^1H MAS NMR spectroscopy. Bromiley et al. (2004) found up to 446 ppm water in Cr-diopside synthesized at 1.5 GPa under the redox state close to the IW (Iron-Wüstite) buffer, whereas water solubility was low (130–242 ppm at 0.5–4.0 GPa) in all Cr-diopsides synthesized in the redox conditions close to the NiNiO buffer. Water solubility in jadeite decreased with pressure from 470 ppm at 2 GPa to 90–100 ppm at 8–10 GPa (Bromiley and Keppler, 2004).

Three groups of OH-stretching bands have been distinguished in clinopyroxenes: (1) 3440–3470 cm^{-1}, (2) 3500–3560 cm^{-1}, and (3) 3600–3640 cm^{-1} (e.g., Ingrin and Skogby, 2000). Bromiley and Keppler (2004) noted also a sharp peak at 3373 cm^{-1} in jadeite. Major OH bands of clinopyroxenes are probably due to hydrogen incorporation into the O2 site in the diopside structure with vibration of the OH dipole in the different directions. There are several possible mechanisms of hydrogen incorporation in clinopyroxenes, including formation of vacancies in both M1 and M2 sites by simple charge-balancing substitution or solid solution with Ca-Escola component and by reduction of Fe^{3+} in the M2 site (e.g., Smyth, 1989; Smyth et al., 1991; Katayama and Nakashima, 2003; Bromiley et al., 2004).

Garnet

There have been many studies on the water contents in natural garnets (e.g., Wilkins and Sabine, 1973; Aines and Rossman, 1984a, 1984b; Rossman, 1988, 1989; Lager et al., 1989; Rossman and Aines, 1991; Bell and Rossman, 1992a, 1992b; Beran et al., 1993; Cho and Rossman, 1993; Langer et al., 1993; Bell et al., 1995; Amthauer and Rossman, 1998; Matsyuk et al., 1998; Bell et al., 2004a) and synthetic garnets (Ackermann et al., 1983; Geiger et al., 1991; Khomenko et al., 1994; Wang et al., 1996; Lu and Keppler, 1997; Withers et al., 1998; Bolfan-Casanova et al., 2000; Katayama et al., 2003). Water concentrations in natural pyrope garnet range from 0 to 200 ppm; however, most samples contain less than 60 ppm (e.g., Bell and Rossman, 1992a, 1992b;

Ingrin and Skogby, 2000). Among xenolith samples, the highest concentrations have been reported in Cr-poor garnet megacrysts in kimberlites. Bell and Rossman (1992b) noted that hydroxyl concentrations correlate positively with FeO content in garnet from kimberlite xenoliths. Xia et al. (2005) reported high concentrations of H_2O in pyrope-almandine garnet from ultrahigh-pressure eclogites of Dabieshan (China). Concentrations ranged from 92 to 1735 ppm and were heterogeneously distributed at 1 cm scale.

Ackermann et al. (1983) and Geiger et al. (1991) reported 200–700 ppm H_2O in pyrope synthesized at 2–5 GPa and 800–1200 °C and suggested that water dissolves by the hydrogarnet substitution. In experiments at 1000 °C, Lu and Keppler (1997) found that water solubility in pyrope increases with pressure from 73 ppm at 1.5 GPa to 199 ppm at 10 GPa. However, Withers et al. (1998) found that water solubility in pyrope reaches 1010 ppm at 4–6 GPa, then the concentration decreases with pressure, and no water was detected at 7–13 GPa. Withers et al. (1998) explained this discrepancy by pressure limitation for hydrogarnet substitution caused by growth of the SiO_4 tetrahedron size (Lager et al., 1989). Lu and Keppler (1997) suggested a charge-balance substitution for oxygen provided by boron and lithium (replacing silicon) in their natural pyrope samples. The data on water solubility in majoritic garnet are limited. Bolfan-Casanova et al. (2000) measured 677 ppm H_2O in $MgSiO_3$ tetragonal garnet, and Katayama et al. (2003) reported 1130–1250 ppm H_2O in majorite related to the basaltic systems. Bolfan-Casanova et al. (2000) noted that hydrogen substitutes for a cation in either the octahedral or the dodecahedral sites of the majorite structure.

Natural kimberlitic pyropes show OH absorption bands at 3570 cm^{-1}, with minor vibrations at 3512 and 3650 cm^{-1}. Pyropes from the Dora-Maira Massif show weak bands at 3602 and 3660 cm^{-1} (Rossman et al., 1989), and the same bands were reproduced in similar samples by Lu and Keppler (1997). Xia et al. (2005) observed two groups of OH absorption bands in garnets from Dabieshan eclogites at 3400–3420 and 3560–3580 cm^{-1}. Synthetic pyropes have an asymmetric band near 3630 cm^{-1}, which is close to the band of grossular at 3622 cm^{-1} attributed to hydrogarnet (4OH → SiO_4) substitution (Withers et al., 1998). Majorite garnets show sharp or broad asymmetric bands at 3450 and 3550–3580 cm^{-1} (Fig. 9; Bolfan-Casanova et al., 2000; Katayama et al., 2003).

SiO_2 Polymorphs

Quartz is rare in mantle rocks; however, it is transported to depth by sediments and eclogites in subduction zones. Water solubility in quartz has long been studied by many authors (e.g., Kats, 1962; Paterson, 1982, 1986; Aines et al., 1984; Mackwell and Paterson, 1985; Kronenberg et al., 1986; Rovetta et al., 1986, 1989; Yurimoto et al., 1989; Gerretsen et al., 1989; Kronenberg, 1994). Paterson (1986) reviewed the experimental data and concluded that hydroxyl solubility in both natural and synthetic quartz is very low (<20 ppm H_2O). Incorporation of trivalent cations may not increase water solubility significantly (e.g., Kats, 1962; Rovetta et al., 1989; Yurimoto et al., 1989). Yurimoto et al. (1989) reported up to 30 ppm water in natural quartz containing Al^{3+} and Li^+ impurities. They observed major absorption bands at 3380 and 3483 cm^{-1}, similar to those observed by Kats (1962).

Coesite can incorporate more hydrogen relative to quartz (Li et al., 1997; Mosenfelder, 2000; Koch-Müller et al., 2001, 2003). Experimental studies have shown that the water solubility in coesite is in the range of 0–212 ppm (Table 3). Mosenfelder (2000) showed that hydroxyl solubility in pure coesite increases with pressure from 43 ppm at 5 GPa to 212 ppm at 10 GPa in experiments at 1200 °C. Similar results were obtained by Koch-Müller et al. (2001). They observed OH absorption bands at 3459, 3520, and 3575 cm^{-1} and argued that OH defects are introduced into the structure by a hydrogarnet-type substitution (4OH → SiO_4).

Solubility of water in stishovite was studied experimentally by Pawley et al. (1993), Bolfan-Casanova et al. (2000), Chung and Kagi (2002), and Panero et al. (2003) and theoretically by Panero and Stixrude (2004). Pawley et al. (1993) observed 82 ppm water in Al-bearing stishovite synthesized at 10 GPa and 1200 °C. Pure stishovite contained almost the same amount of water (up to 72 ppm) in experiments at 15–24 GPa (Bolfan-Casanova et al., 2000). However, Chung and Kagi (2002) recently reported that Al-bearing stishovite synthesized in a MORB system at 10–15 GPa may contain up to 844 ppm water. Panero et al. (2003) studied water solubility in Al-bearing stishovite to 60 GPa using a diamond anvil cell and observed up to 480 ppm water at 2600–3800 °C. Al_2O_3 and water solubility in stishovite notably increases with temperature. Water content of Al-bearing (4.6 wt% Al_2O_3) stishovite synthesized at 20 GPa and 1400 °C may exceed 3000 (±200) ppm, whereas at 20 GPa and 1800 °C, stishovite contains 6.1 wt% Al_2O_3 and 2500 (±200) ppm H_2O (Litasov et al., 2006c). A similar temperature effect was observed by Chung and Kagi (2002) in stishovite synthesized at 10–15 GPa and in stishovite synthesized at 30–60 GPa by Panero et al. (2003). Since the total Al_2O_3 + Fe_2O_3 content of stishovite stable in a MORB system at pressures above 15 GPa may exceed 4 wt% (e.g., Litasov and Ohtani, 2005), stishovite may be a dominant water-bearing phase in the lower mantle and even in the transition zone, where water solubility in wadsleyite and ringwoodite notably decreases with temperature.

Stishovite possesses a major OH absorption band at 3111 cm^{-1}. Pawley et al. (1993) and Bolfan-Casanova et al. (2000) observed also some minor peaks at 3240 and 3311 cm^{-1}. In our Al-rich stishovite, the peak positions were shifted to higher frequencies (3126 and 3320 cm^{-1}), and the major band at 3126 cm^{-1} was very broad. One additional band at 2660 cm^{-1} was also observed. Broadness of the band at 3126 cm^{-1} (Fig. 9) may be explained by deformation of stishovite structure due to incorporation of 4–6 wt% Al_2O_3. Due to a clear correlation between alumina and water contents in stishovite, coupled Al^{3+} + H^+ substitution is the most plausible mechanism for water incorporation in stishovite (Pawley et al., 1993; Smyth et al., 1995; Chung and Kagi, 2002; Panero and Stixrude, 2004).

Other Minerals

Rutile, kyanite, and zircon are the typical accessory minerals relevant to Earth's upper mantle. They are observed as xenoliths and xenocrysts in many kimberlites (e.g., Smyth and Hatton, 1977; Mitchell, 1986; Griffin et al., 2000), and they can be important water containers. Water solubility in rutile containing significant amount of trivalent cations may exceed 1 wt% (Rossman and Smyth, 1990; Hammer and Beran, 1991; Vlassopoulos et al., 1993; Swope et al., 1995). Natural kyanite may contain up to 1800 ppm (0.18 wt%) H_2O (Beran and Götzinger, 1987; Rossman and Smyth, 1990; Beran et al., 1993). However, recent calibration by Bell et al. (2004b) revealed no more than 230 ppm H_2O in various natural kyanites. Bell et al. (2004a) reported 28–34 ppm H_2O in zircon megacrysts from Monastery kimberlites, South Africa, although Woodhead et al. (1991) noted that water solubility in zircons from kimberlites may be 100 ppm.

Transition-Zone Minerals

Water solubility in minerals in the transition zone and lower mantle are studied mainly experimentally because these phases are rare in natural environments. The transition zone may be an important water reservoir in Earth's interior, which is supported by the high water solubility of wadsleyite (β-$[Mg,Fe]_2SiO_4$) and ringwoodite (γ-$[Mg,Fe]_2SiO_4$). Some other phases relevant to the transition zone, such as majorite garnet, clinopyroxene, and stishovite were considered in previous sections.

Wadsleyite

High water solubility in wadsleyite was predicted by Smyth (1987, 1994). He suggested the existence of Si_2O_7 units in the wadsleyite structure, which are absent in the olivine and ringwoodite structures. Accordingly, 0.5 oxygen atoms per formula unit (p.f.u.) are bound to the magnesium ions and are very compliant to protonation. There is a stoichiometric hydrogen-bearing phase $Mg_{1.75}SiH_{0.5}O_4$ (or $Mg_7Si_4O_{12}[OH]_2$ containing 3.3 wt% H_2O), which has been observed to exist in the wadsleyite structure (Inoue et al., 1995). According to experimental data, wadsleyite can accommodate up to 3.4 wt% of water (McMillan et al., 1991; Young et al., 1993; Gasparik, 1993; Cynn and Hofmeister, 1994; Inoue, 1994; Inoue et al., 1995; Kohlstedt et al., 1996; Chen et al., 2002). However, water solubility in wadsleyite is temperature dependent (e.g., Litasov and Ohtani, 2003a; Demouchy et al., 2005). At the temperature of the average mantle, it may contain only 0.1–0.5 wt% H_2O.

Hirschmann et al. (2005) reviewed all published data for hydrous wadsleyite and concluded that estimates of water content in wadsleyite do not require revision by the factor of 3–3.5 that must be applied to FTIR analyses of olivine made with Paterson's (1982) method. Intercalibration with NMR suggests that FTIR gives accurate results for wadsleyite (Kohn et al., 2002). Similarly, comparison of SIMS and FTIR for wadsleyite II yields comparable H_2O concentrations (Smyth et al., 1997, 2005).

Kohlstedt et al. (1996) observed two groups of absorption bands in IR spectra of wadsleyite, i.e., large peaks located near 3300 cm^{-1} (3270 cm^{-1}, 3305 cm^{-1}, and 3360 cm^{-1}) and smaller peaks near 3600 cm^{-1} (3590 cm^{-1} and 3605 cm^{-1}; Fig. 9). Smyth (1987, 1994) suggested that the O1 site (nonsilicate oxygen) is the most likely oxygen position for protonation, whereas Downs (1989) favored O2 as well as the O1 site. Cynn and Hofmeister (1994) argued that the IR absorption bands near 3300 cm^{-1} are associated with protonation of the O2 site and the bands near 3600 cm^{-1} are due to protonation of the O1 site. Kohlstedt et al. (1996) reported that ~2/3 of the hydrogen of their samples was bonded to the O2 site.

Single-crystal X-ray diffraction studies have indicated the existence of two different structures of hydrous β-Mg_2SiO_4 (wadsleyite I and II) (Smyth, 1994; Smyth and Kawamoto, 1997; Smyth et al., 1997). Wadsleyite II is expected to be stable in the peridotite system. The deviation of these two phases from the orthorhombic symmetry of anhydrous wadsleyite results from a combination of ordering of H, Mg/Fe, and Si. The mode of arrangement of the Mg-vacant structural modules defines deviation of symmetry from anhydrous wadsleyite and hydrogen content of hydrous wadsleyite (Kudoh and Inoue, 1999).

Ringwoodite

Ringwoodite can incorporate up to 2.6 wt% H_2O at 1100 °C and 20 GPa (Kohlstedt et al., 1996; Inoue et al., 1998b). Smyth et al. (2003) suggested restricted water solubility (0.2–1.1 wt%) in ringwoodite synthesized at 18–22 GPa and 1400–1500 °C. The IR spectra of ringwoodite display broad absorption bands at 3120, 3345, and 3645 cm^{-1} (Kohlstedt et al., 1996). The maximum major absorption band may be shifted to 3150 cm^{-1} in Fe-bearing (Fo_{89}) ringwoodite (Smyth et al., 2003). Bolfan-Casanova et el. (2000) observed relatively sharp bands at 2540, 3120, and 3700 cm^{-1}.

Kudoh et al. (2000) determined the crystal structure of hydrous ringwoodite $Mg_{1.89}Si_{0.98}H_{0.30}O_4$ containing ~2.0 wt% H_2O and showed the existence of vacancies in both tetrahedral ($V = 0.02$ p.f.u.) and octahedral ($V = 0.11$ p.f.u.) sites, as well as a small amount of cation disorder. However, they did not report the positions of protons. Kleppe et al. (2002) and Smyth et al. (2003) suggested that the principal hydration mechanism involves octahedral cation vacancies. Ohtani et al. (2000) and Smyth et al. (2003) showed that the water content in ringwoodite decreases significantly with temperature, and it can accommodate less water than wadsleyite along the typical mantle geotherm.

Akimotoite

Akimotoite could exist in the harzburgite layer of subducting slabs. Water solubility in the $MgSiO_3$-ilmenite phase (akimotoite) was studied by Bolfan-Casanova et al., 2000, 2002a). They observed sharp absorption bands at 3260, 3300, 3320, and 3390 cm^{-1} (which is similar to those in corundum spectra by Beran, 1991) and estimated water contents of 352–443 ppm in the samples synthesized at 19–24 GPa and 1300–1600 °C.

Lower-Mantle Minerals

Major constituents of the lower mantle are Mg-perovskite, Ca-perovskite, and ferropericlase. Local heterogeneities caused by subducted oceanic crust (MORB) may stabilize stishovite and Al-rich phases NAL and CF.

Mg-Perovskite

Al^{3+} substitution of silicon may create oxygen vacancies in the Si sites within the perovskite structure, which can be replaced by hydrogen (e.g., Navrotsky, 1999). Some ceramic perovskites can absorb water (e.g., Norby, 1990; Baikov and Shalkova, 1992; Kreuer, 1997; 1999). However, water solubility in $MgSiO_3$-perovskite is very controversial and has been studied in experiments by Meade et al. (1994), Bolfan-Casanova et al. (2000, 2003), Murakami et al. (2002), and Litasov et al. (2003).

Studies of Mg-perovskite show that water solubility may increase with increasing Al_2O_3 content, and formation of oxygen vacancies in the Si sites by Al^{3+} incorporation may depend on the SiO_2 and MgO activity (Mg/Si ratio) in the system (Litasov et al., 2003). Meade et al. (1994) reported results of FTIR measurements of water solubility in $MgSiO_3$-perovskite synthesized at 27 GPa and documented two pleochroic hydroxyl absorbance peaks at 3423 cm^{-1} and 3483 cm^{-1}. Calculation of the water content yielded 60–70 ppm H_2O. Similar results were obtained by Lu et al. (1994) for $MgSiO_3$-perovskite synthesized at 26 GPa and 1600 °C. However, Bolfan-Casanova et al. (2000, 2003) showed an absence of water (<1 ppm H_2O) in $MgSiO_3$-perovskite and trace amounts (<2 ppm) of H_2O in Mg-perovskite with 1–4 wt% FeO and 2–6 wt% Al_2O_3 obtained in experiments at 24 GPa. Recent data on water solubility in Mg-perovskites with similar compositions measured by SIMS have revealed up to 4000 ppm H_2O (Murakami et al., 2002; Sanehira et al., 2002; Litasov et al., 2003; Krawczynski et al., 2004). Litasov et al. (2003) also studied Al- and Fe-rich perovskites stable in the MORB system (13–17 wt% Al_2O_3 and 23–27 wt% FeO_{total}) and found water solubility less than 100 ppm.

Bolfan-Casanova et al. (2003) explained minor bands in Mg-perovskites at 3350 and 3690 cm^{-1} and a broad band near 3400 cm^{-1} as inclusions of superhydrous phase B and brucite (Fig. 9). They also showed that clear spots in perovskite crystals reveal no OH absorption bands in IR spectra or minor band at 3388 cm^{-1}. Although, Murakami et al. (2002) and Litasov et al. (2003) reported no visible inclusions in their perovskites, which was also supported by SIMS measurements, the IR spectra of Mg-perovskites show a very broad band near 3400 cm^{-1} similar to those observed by Bolfan-Casanova et al. (2003). IR spectra of Mg-perovskite from Murakami et al. (2002) also contain a clear band of brucite near 3690 cm^{-1}. The IR spectra reported by Litasov et al. (2003) showed that most water may be attributed to the broad background from possible micro-inclusions, and the water content calculated from minor bands at 3397, 3423, 3448, and 3482 cm^{-1} did not exceed 35 ppm (Table 3). However, discrepancy between the FTIR and SIMS results and dependence of water solubility in Mg-perovskite on silica activity (which affects the amount of oxygen vacancies) should be clarified more carefully in future studies.

Ca-Perovskite

Beran et al. (1996) observed up to 70 ppm water in $CaTiO_3$ perovskite from the San Benito mine (California), which showed two OH absorption bands at 3326 and 3394 cm^{-1}. It is difficult to account for water solubility in high-pressure $CaSiO_3$-perovskite due to its amorphization during quenching. Murakami et al. (2002) measured 0.37–0.38 wt% H_2O using SIMS in Ca-perovskite from peridotite samples synthesized at 25.5 GPa and 1600–1650 °C. Litasov and Ohtani (2003c) measured the IR spectra of amorphous Al-bearing (2 wt% Al_2O_3) $CaSiO_3$-perovskites with two strong OH absorption bands at 3343 and 3607 cm^{-1} and determined a water content of 5100 ppm (Table 3). This is consistent with preferential incorporation of Al^{3+} into Si sites in the Ca-perovskite structure. However, IR measurements of quenchable Al-bearing (3 wt% Al_2O_3) $CaGeO_3$ perovskites revealed no water in the samples synthesized at 15 GPa and 1300–1900 °C (our unpubl. data). Based on first principle calculations, Panero and Akber-Knutson (2004) estimated that Ca-perovskite may contain 100 ppm H_2O and 1.7 wt% Al_2O_3 at 1750 °C and 25 GPa.

Ferropericlase

Water solubility in ferropericlase has been studied by Bolfan-Casanova et al. (2000, 2002b), Murakami et al. (2002), and Litasov and Ohtani (2003c). Murakami et al. (2002) observed significant water solubility (up to 0.2 wt%, 2000 ppm) in ferropericlase measured by SIMS. However, IR spectra of these samples showed a broad band at 3400 cm^{-1} similar to those in Mg-perovskite (see previous). Bolfan-Casanova et al. (2002b, 2003) observed clear spots of ferropericlase crystals (Mg# = 91) synthesized at pressures to 25 GPa and concluded that they may contain up to 60 ppm H_2O. Similar data were obtained by Litasov and Ohtani (2003c) for Al-bearing ferropericlase and periclase when the water content was calculated by the calibration of Libowitzky and Rossman (1997). Bolfan-Casanova et al. (2002b) observed an increase of water solubility in ferropericlase with pressure (Table 3), while Litasov and Ohtani (2003c) reported notable increases of water content in Al-ferropericlase with temperature.

Periclase has a strong absorption band at 3372 cm^{-1} and a band at 3295 cm^{-1}, whereas Al-rich periclase and ferropericlase show major bands at 3299 or 3320 cm^{-1} with weak bands at 3308 and 3480 cm^{-1} (Fig. 9; Bolfan-Casanova et al., 2000, 2002b; Litasov and Ohtani, 2003c). The preferential mechanism of hydrogen incorporation in ferropericlase may be a reduction of ferric iron, which creates oxygen vacancies (Bolfan-Casanova et al., 2002b). However, complexities in the IR spectra of Al-bearing ferropericlase indicate that other mechanisms, such as the substitution of two Mg atoms by Al^{3+} and H^+, may be plausible.

Al-Rich Phases CF and NAL

The structure of the Al-rich NAL phase has a high potential to store hydrogen by replacing vacancies related to monovalent cation sites (e.g., Gasparik et al., 2000). However, preliminary measurements of water in NAL phase synthesized at 20–25 GPa and 1200 °C by SIMS indicate that concentrations are less than 50 ppm H_2O (Litasov and Ohtani, 2004). In addition, IR spectra of NAL phase show no bands in the OH stretching interval. SIMS data for CF phase ($NaAlSiO_4$) synthesized at 20 GPa indicate that it may contain up to 0.43 wt% (4300 ppm) of water (Litasov and Ohtani, 2004).

Water Partitioning between Mantle Phases

Water partitioning between mantle phases may be derived roughly from their solubility and mineral/melt partitioning values; however, direct information on coexisting phases is more reliable, and this information is particularly important for understanding the distribution of water within Earth's mantle (Bolfan-Casanova et al., 2000, 2003; Koga et al., 2003; Aubaud et al., 2004; Bell et al., 2004a).

Bell et al. (2004a) calculated solid/solid and solid/melt partition coefficients of water for megacrysts in kimberlites: $D_{Cpx/Ol}$ = 3.0 ± 0.3; $D_{Cpx/Opx}$ = 2.0 ± 0.3; $D_{Opx/Ol}$ = 1.6 ± 0.2; $D_{Ol/Zrn}$ = 3.2 ± 0.9; $D_{Ol/L}$ = 0.0046–0.0053; $D_{Opx/L}$ = 0.0059–0.0093; $D_{Cpx/L}$ = 0.013–0.016; $D_{Gt/L}$ = 0.0003–0.0014 (where Cpx—clinopyroxene, Opx—orthopyroxene, Ol—olivine, Zrn—zircon, Gt—garnet, and L—liquid); and the bulk D of water between garnet lherzolite and melt is 0.0051–0.0063. Aubaud et al. (2004) measured hydrogen partition coefficients in the phases synthesized at 1–2 GPa and 1230–1380 °C using SIMS and found a strong partitioning of water to pyroxenes relative to olivine: $D_{Cpx/Ol}$ = 12.5 ± 1.0; $D_{Cpx/Opx}$ = 1.4 ± 0.3; $D_{Opx/Ol}$ = 9.1 ± 0.5; $D_{Ol/L}$ = 0.0017 ± 0.0005; $D_{Opx/L}$ = 0.019 ± 0.004; $D_{Cpx/L}$ = 0.023 ± 0.005; and the bulk D for peridotite/melt is 0.009. These bulk partition coefficients are consistent with a D value for water between peridotite and melt estimated from the measurements of water in basaltic glasses and correlation with water and trace-element compositions of basalts (e.g., Michael, 1988; Dixon et al., 1988; Sobolev and Chaussidon, 1996; Danyushevsky et al., 2000). Sweeney (1997) reported that $D_{Ol/L}$ increases with pressure from 0.04 at 1.5 GPa to 0.12–0.13 at 6–10 GPa in the experiments coexisting with olivine (Fo_{90}) and potassic silicate melt.

Bolfan-Casanova et al. (2000, 2003) reported partition coefficients of water (Ds) between minerals in the transition zone and lower mantle: $D_{Wd/Cen}$ = 3.8; $D_{Cen/St}$ = 8.2; $D_{Gt/St}$ = 270; $D_{Rw/St}$ = 521; $D_{Rw/Ak}$ = 18 $D_{Ak/St}$ = 18; $D_{Rw/Mpv}$ = 1000–1400; $D_{Fp/Mpv}$ = 60–75 (where Wd—wadsleyite, Cen—clinoenstatite, St—stishovite, Gt—majorite garnet, Rw—ringwoodite, Ak—akimotoite, Mpv—Mg-perovskite, and Fp—ferropericlase).

Partitioning of water between olivine and wadsleyite is particularly important because it may have an effect on the depth of the 410 km discontinuity and change the transport properties at this boundary. However, estimated values of $D_{Wd/Ol}$ vary between 1.5 and 40 in different studies. In particular, uncertainty in calibration methods of FTIR spectra may be the cause. The first measurement of water partitioning between wadsleyite and olivine was estimated to be $D_{Wd/Ol}$ = 20–40 at 1100–1300 °C (Young et al., 1993; Kohlstedt et al., 1996), and the boundary of this transformation was considered to be an important zone of dehydration in the mantle. Recently, Chen et al. (2002) reported results of SIMS measurements of $D_{Wd/Ol}$ and suggested it is 4.6–5.1 at 1200 °C. Our estimations of $D_{Wd/Ol}$ from FTIR measurements (Litasov et al., 2006a) suggest $D_{Wd/Ol}$ = 3.9 at 1200 °C, 2.4 at 1400 °C, and 1.5 at 1500 °C if the calibration by Bell et al. (2003) is used for olivine spectra and the calibration by Paterson (1982) is used for wadsleyite spectra.

WATER TRANSPORT AND STORAGE IN THE MANTLE

Subduction-Zone Water Flux

Major exchange of material between the present mantle and crust occurs in oceanic spreading centers and subduction zones. It is believed that a long-term average production of oceanic crust is nearly balanced by subduction (e.g., Jarrard, 2003). However, the same may not be true for individual elements. Subduction recycling of H_2O as well as some other important volatiles (CO_2, chlorine) is a key issue for understanding global geochemical cycles (Peacock, 1990). Volatile outgassing from MORB and arc magmas was probably the main source of water for early growth of the ocean (Rubey, 1951), but these fluxes are currently balanced or exceeded by a loss due to subduction.

Estimation of the water flux in arc magmatic systems is very complicated, if not impossible. The uncertainties result from: (1) variable amounts of sediments subducted beyond accretionary prisms; (2) the different degree and depth of hydration and hydrothermal alteration of subducting oceanic crust and hydrated portions of the mantle; and (3) an unknown amount of water that is emplaced within the mantle wedge or returned to the surface due to décollement between two plates (e.g., Moore and Vrolijk, 1992; Williams and Hemley, 2001). Therefore, trying to balance an uncertain number with another uncertain number may lead to confusing conclusions. Many authors have tried to estimate water fluxes, and we hold to a moderate view on these estimations.

Peacock (1990) estimated that the amount of water transported by subducting slabs containing pelagic sediments is ~0.7 × 10^{11} kg/yr, whereas basaltic oceanic crust transports 8.0 × 10^{11} kg/yr of water. The total amount of water that is returned to Earth's mantle is estimated to be 8.7 × 10^{11} kg/yr. This quantity is broadly consistent with other calculated values but probably is a lower limit because the amount of trapped water along the grain boundary of minerals in oceanic crust is not included in these estimations (Table 4). In a recent review, Jarrard (2003) estimated nearly twice Peacock's value (1.83 × 10^{12} kg/yr) for the subduction flux and summarized the contribution of different sources for subducting water as follows: 42% for pore water in

TABLE 4. COMPARISON OF VOLUMES FOR MODERN SUBDUCTION-ZONE WATER FLUXES AND THE DEGASSING RATIOS

Reference	F_{sub}	F_{rec}	F_{deg}
Ito et al. (1983)	8.8	0.9–2.0	
Peacock (1990)	8.7	1.4	
Von Huene and Scholl (1991)	10		
Moore and Vrolijk (1992)	18		
Rea and Ruff (1996)	9.0		
Bebout (1996)	9.0–18.0	8.5–9.5	
Maruyama (1999)	11.2		2.3
Bounama et al. (2001)	10.3	5.2	2.1
Jarrard (2003)	18.3		

Note: F_{sub}—subduction flux to the mantle; F_{rec}—recycling flux to the mantle wedge; F_{deg}—degassing flux by volcanic activity; Flux values are in 10^{11} kg/yr.

subducting sediments; 7% for structural water in sediments; 33% for structural water in igneous crust; and 18% for pore water in igneous crust. The greatest source of water entering the subduction zone is seawater. The contribution of primary water from the crust (primary water of unaltered MORB) is 0.9% of total subducted water.

Much less is known about the contribution of water from the lithospheric mantle. Normal faults occurring during plate bending between the outer rise and the trench axis provide a possible mechanism for seawater penetration and reaction with lithospheric mantle. High-resolution seismic-reflection profiles show that normal faults can cut across the crust far enough (>15 km) to reach the mantle (Ranero et al., 2003; Rüpke et al., 2004). In such a scenario, the lithospheric mantle would be highly serpentinized around these faults, while the unfaulted part may remain dry (Christensen, 2004). Thus, the global average of the degree of serpentinization is not at the moment resolvable. Besides, minor hydration of the uppermost mantle may be related to the reaction of MORB melts and fluids with peridotites to form amphibole- and phlogopite-bearing lithologies outside of the ridge axis. Perhaps we can ignore this water in calculation of subduction-zone fluxes, but the water stored in hydrous minerals (~0.1–0.2 wt% of the rock total) in the coldest part of the slab may be almost totally transported to the deep mantle even in relatively hot subduction zones.

The degree of dehydration of subducted slabs in island arcs is also poorly constrained. Seismic tomography and geochemical studies provide some insights into the amount of slab-derived water in the mantle wedge (e.g., Helffrich, 1996; Wiens and Smith, 2003), but so far, thermal modeling of subduction zones based on experimental phase diagrams seems to be the most powerful tool for the study of water release in the mantle wedge. Consequently, numerous studies have been devoted to modeling of slab dehydration (e.g., Peacock, 1990, 1996, 2003; Davies and Stevenson, 1992; Kincaid and Sacks, 1997; Iwamori, 1998, 2004; Peacock and Wang, 1999; Kerrick and Connolly, 2001a, 2001b; Poli and Schmidt, 2002; Funiciello et al., 2003; Hacker et al., 2003a, 2003b; Schmidt and Poli, 2003; Rüpke et al., 2004).

Water release to the mantle wedge largely depends on the temperature structure of the slab, which in turn is controlled by plate speed and age. Accordingly, mature, thick, and cold slabs may transport more water to depth, whereas younger, thin, and hot slabs may be totally dehydrated at the mantle wedge. In the sediment and basalt layers, dehydration is controlled by the stability of lawsonite. Phengite is important, but a less abundant phase. In peridotite, dehydration is controlled by the stability of serpentine and the reaction of serpentine to phase A. If the temperature of the slab is lower than the choke point (Fig. 2), it can transport a significant amount of water to the transition zone. It should be noted that water can be transported by fluid phase (see section Mantle Fluids), but this water is difficult to account for in calculations.

Rüpke et al. (2004) calculated that a young slab of 40 Ma age may retain less than 7% of water, stored mainly in basaltic crust to 8 GPa, while a 120 Ma mature slab may retain 30%–35% of water stored both in basalts and peridotites (water transport by sediments was negligible in their calculation). Bebout (1995, 1996) and many others have estimated that only 5%–15% of subducting water returns to the surface by arc volcanism. This indicates that a huge amount of water resides in the mantle wedge and is likely to be involved in continent growth. Yet, if the estimations of Rüpke et al. (2004) are correct and roughly 10%–20% of water in subducting slabs reaches the stability field of phase A, only ~0.1–0.2 times the ocean's mass of water may be transported to the transition zone in 1 billion years. It is most likely that this value is underestimated, because it is nearly balanced by the average amount of water degassed through volcanism (Table 4, see following). At the same time, it is broadly consistent with boron isotope and B/H_2O data, which suggest that 20% of the water is recycled to the deep mantle (e.g., Chaussidon and Jambon, 1994). For comparison, if all subducting water (1.0–1.8 × 10^{12} kg/yr) were transported to the transition zone, then 0.7–1.3× present ocean mass would be returned to the mantle in 1 billion years.

Knowledge of the present-day water flux in subduction zones is not enough to estimate the flux over plate tectonic history, because it is most likely that earlier Earth was much hotter. This is consistent with decreasing sea level, at least through Phanerozoic time. For the last 600 m.y., Hallam (1992) has estimated a sea-level drop of 500 m, which may correspond to a loss of 11% of the present ocean mass. There are not enough geological data to constrain sea level before this, while rough estimations suggest a gradual increase of sea level since a major ocean-forming event in the early Archean. It should be noted that sea-level drop was not gradual but instead was greatly stepped. For example, Maruyama et al. (2001) described an unusual sea-level drop at 800–700 Ma contemporaneous with the breakup of the Rodinia supercontinent and a change of thermal regime from "older and warmer" to "younger and colder" Earth. After this time, they proposed that sea level decreased due to a change from a degassing-dominated to a regassing-dominated regime. Further, in the Phanerozoic, sea-level change is well-correlated in time with the production rate of oceanic crust (volcanic activity in MORB) (e.g., Haq et al., 1987; Larson, 1991; Kerr, 1998), while also dependent on some other factors, such as catastrophic events and glaciation cycling (e.g., Condie, 2004). Figure 10 shows a model of subduction-zone water recycling through geological time proposed by Rüpke et al. (2004) that shows only 5% sub-Moho (~10 km of oceanic mantle) serpentinization after formation of major continental mass (ca. 2.0 Ga) and increasingly more water recycled to the mantle than returned to the surface by degassing.

Degassing from the Mantle and Water Content in the Basalt Source Regions

Basaltic volcanism in mid-ocean ridges and hotspots is a primary mechanism for extraction of water from the mantle. How much of this water is retained in the oceanic crust or how much enters into the hydrosphere remains unclear. Rough estimation indicates that ~2.1–2.3 × 10^{11} kg/yr of water is returning to the hydrosphere via volcanism in mid-ocean ridges and hotspots (Table 4; Maruyama, 1999; Bounama et al., 2001).

There are two sources of water in the mantle: (1) primordial water in the undegassed mantle and (2) subducted water. In the latter case, water entrapped by basaltic magmas in the ridges or hotspots represents the final stage of circulation of subducted water. Primary MORBs contain 700–6000 ppm H_2O (Delaney et al., 1978; Pineau and Javoy, 1994; Jambon, 1994). Typical amounts of water in island-arc basalts are between 1.0 and 2.9 wt% (Sobolev and Chaussidon, 1996; Walker et al., 2003); this demonstrates the importance of water in generation of subduction-related basalts. The amount of water in MORB correlates with enrichment in incompatible elements, such as Ba, K, Nb, and light rare earth elements (LREEs) (Delaney et al., 1978; Michael, 1988; Dixon et al., 1988; Sobolev and Chaussidon, 1996; Danyushevsky et al., 2000). Water has a bulk solid/liquid partition coefficient around 0.01, thus it is more compatible than K in MORB (Michael, 1995; Danyushevsky et al., 2000). Accordingly, one can estimate the water content in the source region of the basalts. For normal (N-) MORB, water content of the source mantle is on the order of 80–180 ppm, whereas the water abundances estimated in the source region of enriched (E-) MORB are between 200 and 450 ppm (Michael, 1988; Sobolev and Chaussidon, 1996). Sources associated with mantle upwelling (FOZO, Focal Zone, common mantle plume component, e.g. Stracke et al., 2005) have been associated with regions of higher water content (~750 ppm) than the normal MORB source region (e.g., Simons et al., 2002; Dixon et al., 2002). The MORB source region appears to be grossly homogeneous in global scale with respect to its average water content (Michael, 1995). However, it should be noted that some MORB source regions may contain significantly more water (to 1000 ppm at South Atlantic "cold" spot; Ligi et al., 2005), whereas some hotspot sources may contain less water (400 ppm at Hawaii; Dixon and Clague, 2001). These values are generally in agreement with water solubility in the nominally anhydrous minerals under shallow-mantle conditions and indicate that more than 0.1 ocean masses of water can be stored only in nominally anhydrous phases in the upper mantle (to 410 km depth).

The subcontinental lithosphere may also be significantly hydrated. Abundant evidence from mantle xenoliths, kimberlites, and flood basalts indicates that at least some regions of the continental mantle have water contents higher (~0.2 wt%) than the

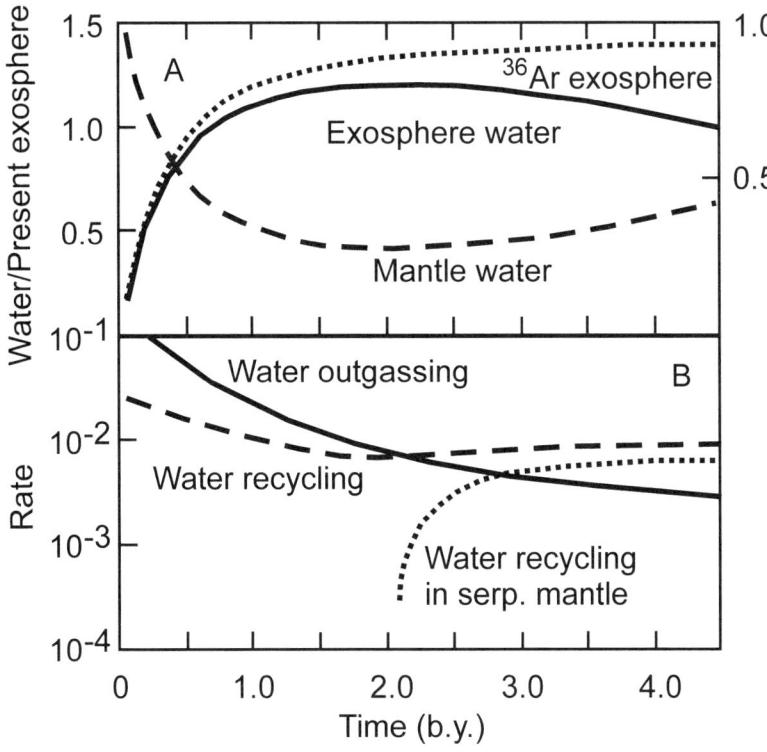

Figure 10. A model for the geological water cycle (after Rüpke et al., 2004). (A) Water contents of Earth's mantle and exosphere are plotted along with ^{36}Ar in exosphere calculated assuming 5% of sub-Moho serpentinization. Present-day water in the exosphere was estimated to be 2.2 × 10^{21} kg (36% in continental crust + 64% in present ocean, e.g., Henderson, 1982). (B) Water recycling rate at subduction zone and water outgassing rate at ridges and hotspots over time. ^{36}Ar fraction of exosphere is plotted as a reference curve for mantle degassing.

source regions of oceanic basalts. Such water may be stored in hydrous phases, most likely amphibole and/or phlogopite and as fluid inclusions, as observed in diamond. However, the source of water degassed by continental volcanism may be related to the subducted water, which was added to the mantle wedge by arc volcanism.

Water Storage Capacity of the Mantle and Dehydration Sites

In the previous sections, we argued that significant amounts of water may be transported by different parts of the slab to the transition zone. Thermal modeling of subduction zones are generally restricted to the shallow mantle up to 8 GPa (e.g., Peacock, 1990; Kerrick and Connolly, 2001a, 2001b; Hacker et al., 2003a, 2003b; Rüpke et al., 2004). Water storage capacity of the deep mantle has been discussed by Iwamori (2004), Komabayashi et al. (2004), and Ohtani et al. (2004) for model peridotite and by Okamoto et al. (2002) for the basaltic component of the slab. Figure 3 summarized the modal proportions of minerals in hydrous and anhydrous peridotite, basalt, and sediment at pressures to 35 GPa along cold subduction and average mantle geotherms given in Figure 2, and potential concentrations of water in the major mantle assemblages are presented in Figure 11 (sediments are ignored in this figure, because their phase relations are poorly known and their contribution for water transport to the transition zone may be negligible relative to basalt and peridotite parts of the slab).

The mineral assemblage of serpentinized peridotite may contain more than 4.5 wt% water in cold slabs, and most of this water can be transported to the deep mantle. However, in the hot subduction environment, very limited water may be transported to the transition zone due to dehydration to the mantle wedge. Basalt can transport a large amount of water to the transition zone but only in the form of fluid (section Mantle Fluids). We would like to emphasize that the noticeable water storage capacity of basalt in the transition zone and lower mantle results from the high water solubility in stishovite. Stishovite may contain more than 4 wt% Al_2O_3, and its water solubility is expected to be 0.2–0.3 wt%. Thus, lower-mantle assemblages with 20% stishovite may contain up to 0.06 wt% H_2O.

We have suggested several dehydration sites related to subducting slabs: (1) uppermost mantle (mantle wedge), (2) top of the lower mantle, and (3) deep lower mantle (Fig. 11) (Ohtani et al., 2004). Dehydration at the top of the lower mantle due to the decomposition reactions of superhydrous phase B and ringwoodite to the lower-mantle assemblage might be very important for water dynamics. Fluids generated at the top of the lower mantle can move upward to saturate overlying transition-zone regions adjacent to subducting slabs or may react with minor metallic Fe. Since all lower-mantle Mg-perovskites, synthesized under the different oxidation states, and all Mg-perovskite inclusions in diamonds are Fe^{3+}-rich ($Fe^{3+}/Fe_{total} > 0.6$; e.g., McCammon et al., 2004a, 2004b), Frost et al. (2004) argued that the lower mantle is either Fe^{3+}-rich or contains ~1 wt% of a metallic Fe-rich alloy, which can be formed due to disproportionation of Fe^{2+} into Fe^{3+} and iron metal. Reaction of this iron metal with water from subduction slabs produces FeO plus FeH (Ohtani et al., 2005). Although, influence of oxidation state on this reaction has not yet been studied, newly formed FeO may react with lower-mantle minerals, whereas FeH can migrate toward the core. If a cold slab is able to penetrate deep into the lower mantle without thermal re-equilibration with the mantle, phase D can transport water to 1200 km depth, where it can dehydrate again and produce hydrous fluid or FeH.

The most important region of dehydration melting in a rising hot plume may also exist at the base of the upper mantle, as shown in Figure 11, due to preferential partitioning of water to wadsleyite relative to olivine. Recent seismic observations suggest the existence of a low-velocity anomaly at depths corresponding to the dehydration sites. Revenaugh and Sipkin (1994) suggested a possible existence of a melt at the base of the upper mantle or top of the 410 km discontinuity beneath eastern China. Recent data on P-wave tomography, including the data beneath Japan (Zhao, 2001, 2004; Fukao et al., 2001), suggest existence of low-velocity anomalies at the base of the upper mantle and the top of the lower mantle. Zhao et al. (1997) reported a low-P-velocity anomaly in the transition zone and the base of the upper mantle beneath the Tonga Trench, which is the coldest among subducted slabs. This low-velocity anomaly cannot be assigned to high temperature, but it may be caused by existence of fluid/melt. Using the estimation of density of the hydrous melt and seismic observations of low-velocity zones above 410 km, Ohtani (2002) and Ohtani et al. (2004) suggested the existence of a density crossover between a hydrous melt and crystals and existence of a gravitationally stable hydrous melt at the base of the upper mantle. Bercovici and Karato (2003) proposed a transition-zone "water filter" hypothesis, suggesting existence of a "hidden" dense hydrous melt (generated from melts arising with the mantle plumes), which may capture water and incompatible trace elements and recycle them back into the transition zone (Fig. 1). Very recently, measurements of density of hydrous basaltic and peridotitic melts have confirmed this possibility. It has been shown that MORB and peridotite melt with H_2O up to 6 wt% (Matsukage et al., 2005; Sakamaki et al., 2006) is denser than the surrounding mantle near 410 km depths.

It is obvious that the transition zone can be considered as a major water reservoir in the mantle. However, this suggestion may be too simple, because of the high mobility of hydrogen in olivine crystals (Mackwell and Kohlstedt, 1990). There are possibilities of heterogeneity in water content in the transition zone even if it has a high water storage capacity, i.e., the transition zone adjacent to the subducting slabs may be enriched in water relative to the normal transition zone. The stratification of the mantle in terms of the water content, i.e., relatively lower water content in the upper and lower mantles and higher water content in the transition zone, may be altered by various processes such as a convective circulation, interaction of the ascending plume

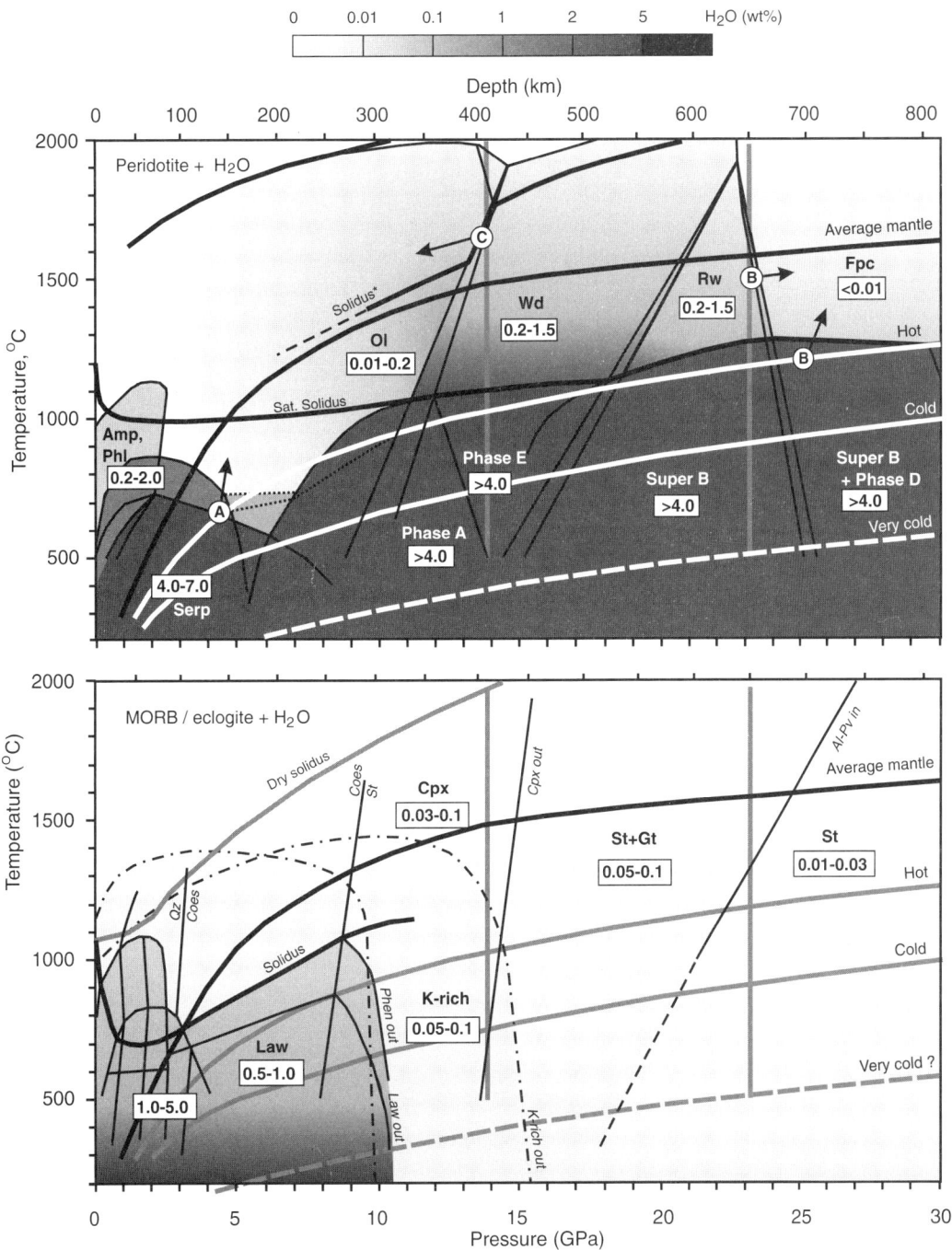

Figure 11. Water storage capacity of the mantle assemblages (based on Figs. 2 and 6). Water contents are calculated based on phase proportions (Fig. 3). Major water-bearing phases and water content of the assemblage are shown in bold. In peridotite composition, sequence serpentine–phase A–phase E–superhydrous phase B–phase D has high water storage capacity over the limit of possible water concentration in the subduction slab. In basaltic composition, lawsonite and phengite stability restrict significant solubility of water in the assemblages. Water contents above the solidus line are based on solubilities in nominally anhydrous minerals. Peridotite: (1) upper mantle (to 10 GPa): Cpx > Opx > Ol > Gt; (2) upper mantle (10–14 GPa): Ol > Cpx > Opx > Gt; (3) transition zone: Wd/Rw >> Cpx > Gt > CaPv (?); (4) lower mantle: Fpc > MgPv, CaPv (?). Mid-ocean-ridge basalt (MORB)/eclogite: (1) upper mantle: K-rich > Cpx > St > Gt > Coes; (2) transition zone: St = Gt; (3) lower mantle: St >> Al-MgPv = CaPv (?) = NAL (?). A–C—potential dehydration sites (A—mantle wedge; B—uppermost lower mantle; C—lowermost upper mantle; see text for explanation). See Figures 2 and 6 for symbols and abbreviations.

with the transition zone, mobility of hydrogen, and the permeable flow of fluids in the transition zone. The lack of basic data, such as the dihedral angle and permeability of water in wadsleyite and ringwoodite matrices, makes it difficult to assess the heterogeneity of water content in the transition zone. As an illustration, we should note that Richard et al. (2002), using numerical simulation, proposed a model of the homogeneous distribution of water throughout the mantle, if advection and diffusion are the only processes involved in water transport.

Summarizing this section, it is possible to estimate reasonable water storage capacity of the different layers of the mantle based on the approved water solubilities in mantle minerals (Table 3). The upper mantle contains ~200 ppm water in MORB source regions, and may be essentially depleted in water to less than 100 ppm, but water content may increase with depth due to increasing water solubility in olivine and pyroxenes and may reach 1000–2000 ppm near the 410 km discontinuity. Thus, along the average mantle geotherm, one can estimate 0.04–0.5 ocean masses of water in the upper mantle. If water content of wadsleyite and ringwoodite is ~0.5 wt% at mantle geotherm, which is close to the maximum solubility of water in wadsleyite at this temperature (0.56 wt% at 1500–1600 °C) (Litasov and Ohtani, 2003a; Litasov et al., 2006a), the transition zone may accommodate ~1.5 ocean masses of water. The water storage potential of the lower mantle is poorly constrained. Ferropericlase and stishovite (in the case of the heterogeneous lower mantle) may be the most important water-bearing phases. If the lower mantle contains only 50 ppm water on average, the total amount of water will correspond to 0.2 ocean masses. If the lower mantle contains more water, which is a simple consequence of global volatile budget (e.g., expected high ^{3}He in the deep mantle; Clarke et al., 1969; Mamyrin et al., 1969; Lupton and Craig, 1975; Craig and Lupton, 1976), the state of this water cannot yet be clarified. The most plausible candidates should be a free fluid/melt and FeH_x. For example, 0.03 wt% FeH in the lower mantle can accommodate the amount of hydrogen corresponding to about one ocean mass.

CONCLUSIONS

Experimental studies of water-silicate systems indicate that many hydrous phases are stable in the mantle. Stability fields of most hydrous phases allow them to survive only in the cold part of the mantle related to subducting slabs. However, they may play the role of water carrier to the deep mantle, according to the temperature profiles of cold subduction. The most important water carriers to the mantle may be serpentine, chlorite, phase A, and superhydrous phase B in peridotite; zoisite, lawsonite, and phengite in basalt; and topaz-OH and phase Egg in sediments. Phase D, stable in peridotite composition, and δ-AlOOH phase, stable in sediment, may transport significant amounts of water at least to 1200–1500 km depth.

Studies of phase relations in hydrous peridotite show that the phase boundaries from olivine to wadsleyite and ringwoodite to Mg-perovskite + ferropericlase transitions shift to lower and higher pressures, respectively. Thus, elevation of the 410 km discontinuity and depression of the 650 km discontinuity in subduction zones may be partially affected by water. Hydration of slabs is supported by lack of evidence for metastable olivine wedges below 410 km, revealed from kinetic studies. Additionally, water may control separation of the basaltic layer of the slab near the 650 km discontinuity. Density crossover between peridotite and basalt near 650 km is absent under hydrous conditions due to a significant shift of the post-garnet transformation in basalt to lower pressures.

Water storage capacity of different mantle layers is greatly different. The transition zone may be a water reservoir in Earth's interior, which is supported by high water solubility in wadsleyite and ringwoodite. The transition zone may accommodate ~1.5 ocean masses of water along the average mantle geotherm. Recent geophysical and mineral physics data show that the transition zone at least locally may contain 0.2–1.5 wt% H_2O. In addition, the transition zone may serve as a water absorber for the water circulation system of the mantle. The upper mantle appears to be largely degassed through mid-ocean-ridge and hotspot volcanism and may accommodate 0.04–0.5 ocean masses of water stored in nominally anhydrous phases, whereas water storage potential of the lower mantle as well as hydrogen storage potential of the core are still grossly uncertain.

Water concentrations near major boundaries and discontinuities appear to be a major feature of water distribution in present-day Earth. On the spatial effect of water on the "boundaries," we can emphasize several potential dehydration sites in the mantle that may control water and material cycling after initiation of plate tectonics: (1) the mantle wedge above subducting slabs, (2) the zone above the 410 km discontinuity, (3) the top of the lower mantle, and possibly (4) the deep lower mantle.

ACKNOWLEDGMENTS

We greatly appreciate D. Frost and T. Inoue for the hard work of reviewing this manuscript and for many valuable comments and suggestions. We thank T. Kondo, A. Suzuki, T. Kubo, and H. Terasaki for collaborative discussions, and P. Gorjan for the technical corrections. Litasov thanks the Japan Society for Promotion of Sciences (JSPS) and 21st century Center of Excellence (COE) program of Tohoku University for research fellowships. This work was supported by the Grants-in-Aid for Scientific Researches from Ministry of Education, Culture, Sports, Science, and Technology, Japan (nos. 14102009 and 16075202) to E. Ohtani. This work was conducted as a part of the 21st century COE program "Advanced Science and Technology Center for the Dynamic Earth" at Tohoku University.

REFERENCES CITED

Abe, Y., Ohtani, E., Okuchi, T., Righter, K., and Drake, M., 2000, Water in the early Earth, *in* Canup, R., and Righter, K., eds., Origin of the Earth and Moon: Tucson, University of Arizona Press, p. 413–433.

Ackermann, L., Cemic, L., and Langer, K., 1983, Hydrogarnet substitution in pyrope: A possible location for "water" in the mantle: Earth and Planetary Science Letters, v. 62, p. 208–214, doi: 10.1016/0012-821X(83)90084-5.

Agee, C.B., 1998, Phase transformations and seismic structure in the upper mantle and transition zone: Mineralogical Society of America Reviews in Mineralogy, v. 37, p. 165–203.

Ai, Y., Zheng, T., Xu, W., He, Y., and Dong, D., 2003, A complex 660 km discontinuity beneath northeast China: Earth and Planetary Science Letters, v. 212, p. 63–71, doi: 10.1016/S0012-821X(03)00266-8.

Aines, R.D., and Rossman, G.R., 1984a, The water content of mantle garnets: Geology, v. 12, p. 720–723, doi: 10.1130/0091-7613(1984)12<720: WCOMG>2.0.CO;2.

Aines, R.D., and Rossman, G.R., 1984b, The hydrous component in garnets: Pyralspites: American Mineralogy v. 69, p. 1116–1126.

Aines, R.D., Kirby, S.H., and Rossman, G.R., 1984, Hydrogen speciation in synthetic quartz: Physics and Chemistry of Minerals, v. 11, p. 204–212, doi: 10.1007/BF00308135.

Akagi, T., and Matsuda, A., 1988, Isotopic and elemental evidence for a relationship between kimberlite and Zaire cubic diamonds: Nature, v. 336, p. 665–667, doi: 10.1038/336665a0.

Akaishi, M., Shaji Kumar, M.D., Kanda, H., and Yamaoka, S., 2000, Formation process of diamond from supercritical H_2O-CO_2 fluid under high pressure and high temperature conditions: Diamond and Related Materials, v. 9, p. 1945–1950, doi: 10.1016/S0925-9635(00)00366-6.

Akaogi, M., Ito, E., and Navrotsky, A., 1989, Olivine-modified spinel–spinel transition in the system Mg_2SiO_4-Fe_2SiO_4: calorimetric measurements, thermochemical calculation, and geophysical application: Journal of Geophysical Research, v. 94, p. 15,671–15,685.

Akimoto, S., and Akaogi, M., 1980, The system Mg_2SiO_4-MgO-H_2O at high pressures and temperatures—Possible hydrous magnesian silicates in the mantle transition zone: Physics of the Earth and Planetary Interiors, v. 23, p. 268–275, doi: 10.1016/0031-9201(80)90123-5.

Amthauer, G., and Rossman, G.R., 1998, The hydrous component in andradite garnet: The American Mineralogist, v. 83, p. 835–840.

Andersen, T., O'Reilly, S.Y., and Griffin, W.L., 1984, The trapped fluid phase in mantle xenoliths from Victoria, Australia: Implication for mantle metasomatism: Contributions to Mineralogy and Petrology, v. 88, p. 72–85, doi: 10.1007/BF00371413.

Anderson, D.L., and Bass, J.D., 1986, Transition region of the Earth's upper mantle: Nature, v. 320, p. 321–328, doi: 10.1038/320321a0.

Aoki, K., Fujino, K., and Akaogi, M., 1976, Titanchondrodite and titanclinohumite derived from the upper mantle in the Buell Park kimberlite, Arizona, USA: Contributions to Mineralogy and Petrology, v. 56, p. 243–253, doi: 10.1007/BF00466824.

Artioli, G., Fumagalli, P., and Poli, S., 1999, The crystal structure of $Mg_8(Mg_2Al_8Si_{12}(O,OH)_{56}$ pumpellyite and its relevance in ultramafic systems at high pressure: The American Mineralogist, v. 84, p. 1906–1914.

Asahara, Y., and Ohtani, E., 2001, Melting relations of the hydrous primitive mantle in the CMAS-H_2O system at high pressures and temperatures, and implications for generation of komatiites: Physics of the Earth and Planetary Interiors, v. 125, p. 31–44, doi: 10.1016/S0031-9201(01)00208-4.

Asahara, Y., Ohtani, E., Kondo, T., Kubo, T., Miyajima, N., Nagase, T., Fujino, K., Yagi, T., and Kikegawa, T., 2005, Formation of metastable cubic-perovskite in high pressure phase transformation of $Ca(Mg,Fe,Al)Si_2O_6$: The American Mineralogist, v. 90, p. 457–462, doi: 10.2138/am.2005.1649.

Aubaud C., Hauri, E.H., and Hirschmann, M.M., 2004, Hydrogen partition coefficients between nominally anhydrous minerals and basaltic melts: Geophysical Research Letters, v. 31, doi: 10.1029/2004GL021341.

Bai, Q., and Kohlstedt, D.L., 1992, Substantial hydrogen solubility in olivine and implications for water storage in the mantle: Nature, v. 357, p. 672–674, doi: 10.1038/357672a0.

Bai, Q., and Kohlstedt, D.L., 1993, Effect of chemical environment on solubility and incorporation mechanism of hydrogen in olivine: Physics and Chemistry of Minerals, v. 19, p. 460–471, doi: 10.1007/BF00203186.

Baikov, Y.M., and Shalkova, E.K., 1992, Hydrogen in perovskites: Journal of Solid State Chemistry, v. 97, p. 224–227, doi: 10.1016/0022-4596(92)90027-S.

Bebout, G.E., 1995, The impact of subduction-zone metamorphism on mantle-ocean chemical cycling: Chemical Geology, v. 126, p. 191–218, doi: 10.1016/0009-2541(95)00118-5.

Bebout, G.E., 1996, Volatile transfer and recycling at convergent margins: Mass-balance and insights from high-P/T metamorphic rocks zones in Bebout, G.E., Scholl, D.W., Kirby, S.H., and Platt, J.P., eds., Subduction Top to Bottom: American Geophysical Union Geophysical Monograph 96, p. 179–193.

Bell, D.R., and Rossman, G.R., 1992a, Water in Earth's mantle: The role of nominally anhydrous minerals: Science, v. 255, p. 1391–1397, doi: 10.1126/science.255.5050.1391.

Bell, D.R., and Rossman, G.R., 1992b, The distribution of hydroxyl in garnets from the sub-continental mantle of southern Africa: Contributions to Mineralogy and Petrology, v. 111, p. 161–178, doi: 10.1007/BF00348949.

Bell, D.R., Ihinger, P.D., and Rossman, G.R., 1995, Quantitative analysis of hydroxyl in garnet and pyroxene: The American Mineralogist, v. 80, p. 465–474.

Bell, D.R., Rossman, G.R., Maldener, J., Endisch, D., and Rauch, F., 2003, Hydroxide in olivine: A quantitative determination of the absolute amount and calibration of the IR spectrum: Journal of Geophysical Research, v. 108, doi: 10.1029/2001JB000679.

Bell, D.R., Rossman, G.R., and Moore, R.O., 2004a, Abundance and partitioning of OH in a high-pressure magmatic system: Megacrysts from the Monastery kimberlite, South Africa: Journal of Petrology, v. 45, p. 1539–1564, doi: 10.1093/petrology/egh015.

Bell, D.R., Rossman, G.R., Maldener, J., Endisch, D., and Rauch, F., 2004b, Hydroxide in kyanite: A quantitative determination of the absolute amount and calibration of the IR spectrum: The American Mineralogist, v. 89, p. 998–1003.

Benz, H.M., and Vidale, J.E., 1993, Sharpness of upper mantle discontinuities determined from high-frequency reflections: Nature, v. 365, p. 147–150, doi: 10.1038/365147a0.

Beran, A., 1991, Trace hydrogen in Vermeuil-grown corundum and its colour varieties, an IR spectroscopic study: European Journal of Mineralogy, v. 3, p. 971–975.

Beran, A., and Götzinger, M.A., 1987, The quantitative IR spectroscopic determination of structural OH groups in kyanites: Mineralogy and Petrology, v. 36, p. 41–49, doi: 10.1007/BF01164368.

Beran, A., and Putnis, A., 1983, A model of the OH positions in olivine, derived from infrared-spectroscopic investigations: Physics and Chemistry of Minerals, v. 9, p. 57–60, doi: 10.1007/BF00308148.

Beran, A., and Zemann, J., 1986, The pleochroism of a gem-quality enstatite in the region of the OH stretching frequency, with a stereochemical interpretation: Tschermaks Mineralogische und Petrographische Mitteilungen, v. 35, p. 19–25, doi: 10.1007/BF01081915.

Beran, A., Langer, K., and Andrut, M., 1993, Single crystal infrared spectra in the range of OH fundamentals of paragenetic garnet, omphacite and kyanite in an eclogitic mantle xenolith: Mineralogy and Petrology, v. 48, p. 257–268, doi: 10.1007/BF01163102.

Beran, A., Libowitzky, E., and Armbruster, T., 1996, A single-crystal infrared spectroscopic and X-ray diffraction study of untwinned San Benito perovskites containing OH groups: Canadian Mineralogist, v. 34, p. 803–809.

Bercovici, D., and Karato, S., 2003, Whole-mantle convection and the transition zone water filter: Nature, v. 425, p. 39–44, doi: 10.1038/nature01918.

Bina, C.R., and Helffrich, G.R., 1994, Phase transition Clapeyron slopes and transition zone seismic discontinuity topography: Journal of Geophysical Research, v. 99, p. 15,853–15,860, doi: 10.1029/94JB00462.

Bina, C.R., and Navrotsky, A., 2000, A possible presence of ice in cold subducting slabs: Nature, v. 408, p. 844–847, doi: 10.1038/35048555.

Bohlen, S.R., and Boettcher, A.L., 1982, The quartz-coesite transformation: A precise determination and the effects of other components: Journal of Geophysical Research, v. 87, p. 7073–7078.

Bolfan-Casanova, N., Keppler, H., and Rubie, D.C., 2000, Water partitioning between nominally anhydrous minerals in the MgO-SiO_2-H_2O system up to 24 GPa: Implications for the distribution of water in the Earth's mantle: Earth and Planetary Science Letters, v. 182, p. 209–221, doi: 10.1016/S0012-821X(00)00244-2.

Bolfan-Casanova, N., Keppler, H., and Rubie, D.C., 2002a, Hydroxyl in $MgSiO_3$ akimotoite: A polarized and high-pressure IR study: The American Mineralogist, v. 87, p. 603–608.

Bolfan-Casanova, N., Mackwell, S., Keppler, H., McCammon, C., and Rubie, D.C., 2002b, Pressure dependence of H solubility in magnesiowüstite up to 25 GPa: Implications for the storage of water in the Earth's lower mantle: Geophysical Research Letters, v. 29, doi: 10.1029/2001GL014457.

Bolfan-Casanova, N., Keppler, H., and Rubie, D.C., 2003, Water partitioning at 660 km depth and evidence for very low water solubility in magnesium silicate perovskites: Geophysical Research Letters, v. 30, doi: 10.1029/2003GL017182.

Bose, K., and Ganguly, J., 1995, Experimental and theoretical studies of the stabilities of talc, antigorite and phase A at high pressures with applications to subduction processes: Earth and Planetary Science Letters, v. 136, p. 109–121, doi: 10.1016/0012-821X(95)00188-I.

Bounama, C., Franck, S., and von Bloh, W., 2001, The fate of Earth's ocean: Hydrology and Earth System Sciences, v. 5, p. 569–575.

Brenan, J.M., Shaw, H.F., Ryerson, F.J., and Phinney, D.L., 1995, Mineral-aqueous fluid partitioning of trace elements at 900°C and 2.0 GPa: Constraints on the trace element chemistry of mantle and deep crust fluids: Geochimica et Cosmochimica Acta, v. 59, p. 3331–3350, doi: 10.1016/0016-7037(95)00215-L.

Bromiley, G.D., and Keppler, H., 2004, An experimental investigation of hydroxyl solubility in jadeite and Na-rich clinopyroxenes: Contributions to Mineralogy and Petrology, v. 147, p. 189–200, doi: 10.1007/s00410-003-0551-1.

Bromiley, G.D., and Pawley, A.R., 2002, The high-pressure stability of Mg-sursassite in a model hydrous peridotite: A possible mechanism for the deep subduction of significant volumes of H_2O: Contributions to Mineralogy and Petrology, v. 142, p. 714–723.

Bromiley, G.D., Keppler, H., McCammon, C., Bromiley, F.A., and Jacobsen, S.D., 2004, Hydrogen solubility and speciation in natural, gem-quality chromian diopside: The American Mineralogist, v. 89, p. 941–949.

Bureau, H., and Keppler, H., 1999, Complete miscibility between silicate melts and hydrous fluids in the upper mantle; experimental evidence and geochemical implications: Earth and Planetary Science Letters, v. 165, p. 187–196, doi: 10.1016/S0012-821X(98)00266-0.

Burgess, R., Layzelle, E., Turner, G., and Harris, J.W., 2002, Constraints on the age and halogen composition of mantle fluids in Siberian coated diamonds: Earth and Planetary Science Letters, v. 197, p. 193–203, doi: 10.1016/S0012-821X(02)00480-6.

Burnley, P.C., and Navrotsky, A., 1996, Synthesis of high-pressure hydrous magnesium silicates: Observations and analysis: The American Mineralogist, v. 81, p. 317–326.

Campbell, I.H., and Taylor, S.R., 1983, No water, no granites—No oceans, no continents: Geophysical Research Letters, v. 10, p. 1061–1064.

Carswell, D.A., 1975, Primary and secondary phlogopites and clinopyroxenes in garnet lherzolite xenoliths: Physics and Chemistry of the Earth, v. 9, p. 417–429, doi: 10.1016/0079-1946(75)90031-2.

Cartigny, P., Harris, J.W., Phillips, D., Girard, M., and Javoy, M., 1998, Subduction-related diamonds? The evidence for a mantle-derived origin from coupled δC^{13}–δN^{15} determinations: Chemical Geology, v. 147, p. 147–159, doi: 10.1016/S0009-2541(97)00178-2.

Chaussidon, M., and Jambon, A., 1994, Boron content and isotopic composition of oceanic basalts: Geochemical and cosmochemical implications: Earth and Planetary Science Letters, v. 121, p. 277–291, doi: 10.1016/0012-821X(94)90073-6.

Chen J., Inoue, T., Yurimoto, H., and Weidner, D.J., 2002, Effect of water on olivine-wadsleyite phase boundary in the $(Mg,Fe)_2SiO_4$ system: Geophysical Research Letters, v. 29, doi: 10.1029/2001GL014429.

Chinnery, N., Pawley, A.R., and Clark, S.M., 1999, In situ observation of the formation of 10Å phase from talc + H_2O at mantle pressures and temperatures: Science, v. 286, p. 940–942, doi: 10.1126/science.286.5441.940.

Cho, H., and Rossman, G.R., 1993, Single-crystal NMR studies of low-concentration hydrous species in minerals: Grossular garnet: The American Mineralogist, v. 78, p. 1149–1164.

Chrenko, R., McDonald, R., and Darrow, K., 1967, Infrared spectrum of diamond coat: Nature, v. 213, p. 474–476, doi: 10.1038/213474a0.

Christensen, N.I., 2004, Serpentinites, peridotites, and seismology: International Geology Review, v. 46, p. 795–816.

Chung, J.I., and Kagi, H., 2002, High concentration of water in stishovite in the MORB system: Geophysical Research Letters, v. 29, doi: 10.1029/2002GL015579.

Clarke, W.B., Beg, M.A., and Craig, H., 1969, Excess ^3He in the sea: Evidence for terrestrial primordial helium: Earth and Planetary Science Letters, v. 6, p. 213–220, doi: 10.1016/0012-821X(69)90093-4.

Closmann, C., and Williams, Q., 1995, In situ spectroscopic investigation of high-pressure hydrated $(Mg,Fe)SiO_3$ glasses: OH vibrations as a probe of glass structure: American Mineralogy, v. 80, p. 201–212.

Collier, J.D., Helffrich, G.R., and Wood, B.J., 2001, Seismic discontinuities and subduction zones: Physics of the Earth and Planetary Interiors, v. 127, p. 35–49, doi: 10.1016/S0031-9201(01)00220-5.

Condie, K.C., 2004, Supercontinents and superplume events: Distinguishing signals in the geologic record: Physics of the Earth and Planetary Interiors, v. 146, p. 319–332, doi: 10.1016/j.pepi.2003.04.002.

Constantin, M., 1999, Gabbroic intrusions and magmatic metasomatism in harzburgites from the Garret transform fault: Implications for the nature of the mantle-crust transition at fast spreading ridges: Contributions to Mineralogy and Petrology, v. 136, p. 111–130, doi: 10.1007/s004100050527.

Craig, H., and Lupton, J.E., 1976, Primordial neon, helium, and hydrogen in oceanic basalts: Earth and Planetary Science Letters, v. 31, p. 369–385, doi: 10.1016/0012-821X(76)90118-7.

Cynn, H., and Hofmeister, A.M., 1994, High pressure IR spectra of lattice modes and O-H vibration in Fe-bearing wadsleyite: Journal of Geophysical Research, v. 99, p. 17,717–17,727, doi: 10.1029/94JB01661.

Daniel, I., Fiquet, G., Gillet, P., Schmidt, M., and Hanfland, M., 2000, High-pressure behavior of lawsonite: A phase transition at 8.6 GPa: European Journal of Mineralogy, v. 12, p. 721–733.

Danyushevsky, L.V., Eggins, S.M., Falloon, T.J., and Christie, D.M., 2000, H_2O abundance in depleted to moderately enriched mid-ocean ridge magmas. Part 1: Incompatible behaviour, implications for mantle storage, and origin of regional variations: Journal of Petrology, v. 41, p. 1329–1364, doi: 10.1093/petrology/41.8.1329.

Davies, J.H., and Stevenson, D.J., 1992, Physical model of source region of subduction zone volcanics: Journal of Geophysical Research, v. 97, p. 2037–2070.

Deer, W.A., Howie, R.A., and Zussman, J., 1992, The rock-forming minerals (second ed.): Edinburgh, Pearson Education Ltd., 696 p.

Delaney, J.R., Muenow, D.W., and Graham, D.G., 1978, Abundance and distribution of water, carbon and sulfur in the glassy rims of submarine pillow basalts: Geochimica et Cosmochimica Acta, v. 42, p. 581–594, doi: 10.1016/0016-7037(78)90003-0.

Demouchy, S., Deloule, E., Frost, D.J., and Keppler, H., 2005, Pressure- and temperature-dependence of water solubility in Fe-free wadsleyite: The American Mineralogist, v. 90, p. 1084–1091, doi: 10.2138/am.2005.1751.

Deuss, A., and Woodhouse, J.H., 2001, Seismic observations of splitting of the mid transition zone discontinuity: Science, v. 294, p. 354–357, doi: 10.1126/science.1063524.

Dixon, J.E., and Clague, D.A., 2001, Volatiles in basaltic glasses from Loihi Seamount, Hawaii: Evidence for a relatively dry plume component: Journal of Petrology, v. 42, p. 627–654, doi: 10.1093/petrology/42.3.627.

Dixon, J.E., Stolper, E.M., and Delaney, J.R., 1988, Infrared spectroscopic measurements of CO_2 and H_2O in Juan de Fuca Ridge basaltic glasses: Earth and Planetary Science Letters, v. 90, p. 87–104, doi: 10.1016/0012-821X(88)90114-8.

Dixon, J.E., Leist, L., Langmuir, C., and Schilling, J.G., 2002, Recycled dehydrated lithosphere observed in plume-influenced mid-ocean-ridge basalt: Nature, v. 420, p. 385–389, doi: 10.1038/nature01215.

Domanik, K., and Holloway, J.R., 1996, The stability and composition of phengitic muscovite and associated phases from 5.5 to 11 GPa: Implications for deeply subducted sediments: Geochimica et Cosmochimica Acta, v. 60, p. 4133–4150, doi: 10.1016/S0016-7037(96)00241-4.

Domanik, K.J., and Holloway, J.R., 2000, Experimental synthesis and phase relations of phengitic muscovite from 6.5 to 11 GPa in a calcareous metapelite from the Dabie Mountains China: Lithos, v. 52, p. 51–77, doi: 10.1016/S0024-4937(99)00084-5.

Downs, J.W., 1989, Possible sites for protonation in β-Mg_2SiO_4 from an experimentally derived electrostatic potential: The American Mineralogist, v. 74, p. 1124–1129.

Duffy, T.S., Ahrens, T., and Lange, M.A., 1991, The shock wave equation of state of brucite $Mg(OH)_2$: Journal of Geophysical Research, v. 96, p. 14,319–14,330.

Duffy, T.S., Meade, C., Fei, Y., Mao, H.K., and Hemley, R.J., 1995, High-pressure phase transition in brucite, $Mg(OH)_2$: The American Mineralogist, v. 80, p. 222–230.

Dymek, R.F., Boak, J.L., and Brothers, S.C., 1988, Titanian chondrodite– and titanian clinohumite–bearing metadunite from the 3800 Ma Isua supracrustal belt, West Greenland: Chemistry, petrology, and origin: The American Mineralogist, v. 73, p. 547–558.

Eggleton, R.A., Boland, J.N., and Ringwood, A.E., 1978, High pressure synthesis of a new aluminum silicate: $Al_5Si_5O_{17}(OH)$: Geochemical Journal, v. 12, p. 191–194.

Evans, B.W., and Trommsdorff, V., 1983, Fluorine hydroxyl titanianclinohumite in Alpine recrystallized garnet peridotite: Compositional controls and petrologic significance: American Journal of Science, v. 283A, p. 355–369.

Feenstra, A., and Wunder, B., 2002, The dehydration of diasporite to corundite in nature and experiment: Geology, v. 30, p. 119–122, doi: 10.1130/0091-7613(2002)030<0119:DODTCI>2.0.CO;2.

Fei, Y., and Mao, H.K., 1993, Static compression of $Mg(OH)_2$ to 78 GPa at high temperature and constraints on the equation of state of fluid H_2O: Journal of Geophysical Research, v. 98, p. 11,875–11,884.

Fei, Y., Van Orman, J., Li, J., van Westrenen, W., Sanloup, C., Minarik, W., Hirose, K., Komabayashi, T., Walter, M., and Funakoshi, K., 2004, Experimentally determined postspinel transformation boundary in Mg_2SiO_4 using MgO as an internal pressure standard and its geophysical implications: Journal of Geophysical Research, v. 109, doi: 10.1029/2003JB002562.

Finger, L.W., Ko, J., Hazen, R.M., Gasparik, T., Hemley, R.J., Prewitt, C.T., and Weidner, D.J., 1989, Crystal chemistry of phase B and anhydrous analogue: Implications for water storage in the upper mantle: Nature, v. 341, p. 140–142, doi: 10.1038/341140a0.

Finger, L.W., Hazen, R.M., and Prewitt, C.T., 1991, Crystal structures of $Mg_{12}Si_4O_{19}(OH)_2$ (phase B) and $Mg_{14}Si_5O_{24}$ (phase AnhB): The American Mineralogist, v. 76, p. 1–7.

Flanagan, M.P., and Shearer, P.M., 1998, Global mapping of topography on transition zone velocity discontinuities by stacking SS precursors: Journal of Geophysical Research, v. 103, p. 2673–2692, doi: 10.1029/97JB03212.

Fockenberg, T., 1998, An experimental study of the pressure–temperature stability of MgMgAl-pumpellyite in the system $MgO-Al_2O_3-SiO_2-H_2O$: The American Mineralogist, v. 83, p. 220–227.

Frezzotti, M.L., 2001, Silicate-melt inclusions in magmatic rocks: Applications to petrology: Lithos, v. 55, p. 273–299, doi: 10.1016/S0024-4937(00)00048-7.

Frost, D.J., 1999, The stability of dense hydrous magnesium silicates in Earth's transition zone and lower mantle, in Fei, Y., Bertka, C.M., and Mysen B.O., eds., Mantle Petrology: Field Observations and High Pressure Experimentation: A Tribute to F.R. Boyd: Geochemical Society Special Publication, v. 6, p. 283–296.

Frost, D.J., 2003, The structure and sharpness of $(Mg,Fe)_2SiO_4$ phase transformations in the transition zone: Earth and Planetary Science Letters, v. 216, p. 313–328, doi: 10.1016/S0012-821X(03)00533-8.

Frost, D.J., and Fei, Y., 1998, Stability of phase D at high pressure and high temperature: Journal of Geophysical Research, v. 103, p. 7463–7474, doi: 10.1029/98JB00077.

Frost, D.J., Liebske, C., Langenhorst, F., McCammon, C.A., Trønnes, R.G, and Rubie, D.C., 2004, Experimental evidence for the existence of iron-rich metal in the Earth's lower mantle: Nature, v. 428, p. 409–412, doi: 10.1038/nature02413.

Fukao, Y., Widiyantoro, S., and Obayashi, M., 2001, Stagnant slabs in the upper and lower mantle transition region: Reviews of Geophysics, v. 39, p. 291–323, doi: 10.1029/1999RG000068.

Fukao, Y., Koyama, T., Obayashi, M., and Utada, H., 2004, Trans-Pacific temperature field in the mantle transition zone derived from seismic and electromagnetic tomography: Earth and Planetary Science Letters, v. 217, p. 425–434, doi: 10.1016/S0012-821X(03)00610-1.

Fumagalli, P., and Poli, S., 2005, Experimentally determined phase relations in hydrous peridotites to 6.5 GPa and their consequences on the dynamics of subduction zones: Journal of Petrology, v. 46, p. 555–578, doi: 10.1093/petrology/egh088.

Fumagalli, P., Stixrude, L., Poli, S., and Snyder, L., 2001, The 10 Å phase: A high pressure expandable sheet silicate during subduction of hydrated lithosphere: Earth and Planetary Science Letters, v. 186, p. 125–141, doi: 10.1016/S0012-821X(01)00238-2.

Funiciello, F., Morra, G., Regenauer-Lieb, K., and Giardini, D., 2003, Dynamics of retreating slabs: 1. Insights from two-dimensional numerical experiments: Journal of Geophysical Research, v. 108, doi: 10.1029/2001JB000898.

Fyfe, W.S., 1970, Lattice energies, phase transformations, and volatiles in the mantle: Physics of the Earth and Planetary Interiors, v. 3, p. 196–200, doi: 10.1016/0031-9201(70)90055-5.

Gallagher, K., and Hawkesworth, C., 1992, Dehydration melting and the generation of continental flood basalts: Nature, v. 358, p. 57–59, doi: 10.1038/358057a0.

Garnero, E.J., and Helmberger, D.V., 1995, On seismic resolution of lateral heterogeneity in the Earth's outermost core: Physics of the Earth and Planetary Interiors, v. 88, p. 117–130, doi: 10.1016/0031-9201(94)02976-I.

Gasparik, T., 1990, Phase relations in the transition zone: Journal of Geophysical Research, v. 95, p. 15,751–15,769.

Gasparik, T., 1993, The role of volatiles in the transition zone: Journal of Geophysical Research, v. 98, p. 4287–4299.

Gasparik, T., 1997, A model for the layered upper mantle: Physics of the Earth and Planetary Interiors, v. 100, p. 197–212, doi: 10.1016/S0031-9201(96)03240-2.

Gasparik, T., and Litvin, Y.A., 1997, Stability of $Na_2Mg_2Si_2O_7$ and melting relations on the forsterite-jadeite join at pressures up to 22 GPa: European Journal of Mineralogy, v. 9, p. 311–326.

Gasparik, T., Tripathi, A., and Parise, J.B., 2000, Structure of a new Al-rich phase, $[K,Na]_{0.9}[Mg,Fe][Mg,Fe,Al,Si]_6O_{12}$, synthesized at 24 GPa: The American Mineralogist, v. 85, p. 613–618.

Geiger, C.A., Langer, K., Bell, D.R., Rossman, G.R., and Winkler, B., 1991, The OH^- component in synthetic pyrope: The American Mineralogist, v. 76, p. 49–59.

Gerretsen, J., Paterson, M.S., and McLaren, A.C., 1989, The uptake and solubility of water in quartz at elevated pressure and temperature: Physics and Chemistry of Minerals, v. 16, p. 334–342, doi: 10.1007/BF00199553.

Grand, S.P., Van der Hilst, R.D., and Widiyantoro, S., 1997, Global seismic tomography: A snapshot of convection in the Earth: GSA Today, v. 7, no. 4, p. 1–7.

Green, D.H., and Falloon, T.J., 1998, Pyrolite: A Ringwood concept and its current expression, in Jackson, I., ed., The Earth's Mantle: Cambridge, Cambridge University Press, p. 311–378.

Griffin, W.L., Pearson, N.J., Belousova, E., Jackson, S.E., van Achterbergh, E., O'Reilly, S.Y., and Shee, S.R., 2000, The Hf isotope composition of cratonic mantle: LAM-MC-ICPMS analysis of zircon megacrysts in kimberlites: Geochimica et Cosmochimica Acta, v. 64, p. 133–147, doi: 10.1016/S0016-7037(99)00343-9.

Gu, Y.J., and Dziewonski, A.M., 2002, Global variability of transition zone thickness: Journal of Geophysical Research, v. 107, doi: 10/1029/2001JB000489.

Hacker, B.R., Aber, G.A., and Peacock, S.M., 2003a, Subduction factory: 1. Theoretical mineralogy, densities, seismic wave speeds, and H_2O contents: Journal of Geophysical Research, v. 108, doi: 10.1029/2001JB001127.

Hacker, B.R., Peacock, S.M., Aber, G.A., and Holloway, D., 2003b, Subduction factory: 2. Are intermediate-depth earthquakes in subducting slabs linked to metamorphic dehydration reactions?: Journal of Geophysical Research, v. 108, doi: 10.1029/2001JB001129.

Hallam, A., 1992, Phanerozoic Sea-Level Changes: New York, Columbia University Press, 278 p.

Hammer, V.M.F., and Beran, A., 1991, Variations in the OH concentration in rutile from different geological environments: Mineralogy and Petrology, v. 45, p. 1–9, doi: 10.1007/BF01164498.

Haq, B.U., Hardenbol, J., and Vail, P.R., 1987, Chronology of fluctuating sea levels since the Triassic (250 million years to present): Science, v. 235, p. 1156–1167, doi: 10.1126/science.235.4793.1156.

Hauri, E., 2002, SIMS analysis of volatiles in silicate glasses: 2. Isotopes and abundances in Hawaiian melt inclusions: Chemical Geology, v. 183, p. 115–141, doi: 10.1016/S0009-2541(01)00374-6.

Hawthorne, F.C., 1992, The role of OH and H_2O in oxide and oxysalt minerals: Zeitschrift für Kristallographie, v. 201, p. 183–206.

Hawthorne, F.C., 1994, Structural aspects of oxide and oxysalt crystals: Acta Crystallographica, v. B50, p. 481–510.

Helffrich, G.R., 1996, Subducted lithospheric slab velocity structure: Observations and mineralogical inferences, in Bebout, G.E., Scholl, D.W., Kirby, S.H., and Platt, J.P., eds., Subduction Top to Bottom: American Geophysical Union Geophysical Monograph 96, p. 215–222.

Helffrich, G.R., 2000, Topography of the transition zone seismic discontinuities: Reviews of Geophysics, v. 38, p. 141–158, doi: 10.1029/1999RG000060.

Helffrich, G.R., and Wood, B.J., 1996, 410-km discontinuity sharpness and the form of the olivine α-β phase diagram: Resolution of apparent seismic contradictions: Geophysical Journal International, v. 126, p. F7–F12.

Henderson, P., 1982, Inorganic Geochemistry: New York, Permagon Press, 353 p.

Hermann, J., 2002, Experimental constraints on phase relations in subducted continental crust: Contributions to Mineralogy and Petrology, v. 143, p. 219–235, doi: 10.1007/s00410-001-0336-3.

Herzberg, C., and Zhang, J., 1996, Melting experiment on anhydrous peridotite KLB-1: Composition of magmas in the upper mantle and transition zone: Journal of Geophysical Research, v. 101, p. 8271–8295, doi: 10.1029/96JB00170.

Herzberg, C., Raterron, P., and Zhang, J., 2000, New experimental observations on the anhydrous solidus for peridotite KLB-1: Geochemistry, Geophysics, Geosystems, v. 1, doi: 10.1029/2000GC000089.

Higo, Y., Inoue, T., Irifune, T., and Yurimoto, H., 2001, Effect of water on the spinel-postspinel transformation in Mg_2SiO_4: Geophysical Research Letters, v. 28, p. 3505–3508, doi: 10.1029/2001GL013194.

Hirose, K., 2002, Phase transition in pyrolitic mantle around 670-km depth: Implications for upwelling of plumes from the lower mantle: Journal of Geophysical Research, v. 107, doi: 10.1029/2001JB000597.

Hirose, K., and Fei, Y., 2002, Subsolidus and melting phase relations of basaltic composition in the uppermost lower mantle: Geochimica et Cosmochimica Acta, v. 66, p. 2099–2108, doi: 10.1016/S0016-7037(02)00847-5.

Hirose, K., and Kawamoto, T., 1995, Hydrous partial melting of lherzolite at 1 GPa: The effect of H_2O on the genesis of basaltic magmas: Earth and Planetary Science Letters, v. 133, p. 463–473, doi: 10.1016/0012-821X(95)00096-U.

Hirose, K., and Kushiro, I., 1993, Partial melting of dry peridotites at high pressures; determination of compositions of melts segregated from peridotite using aggregates of diamond: Earth and Planetary Science Letters, v. 114, p. 477–489, doi: 10.1016/0012-821X(93)90077-M.

Hirose, K., Fei, Y., Ma, Y., and Mao, H.K., 1999, The fate of the subducted basaltic crust in the Earth's lower mantle: Nature, v. 397, p. 53–56, doi: 10.1038/16225.

Hirschmann, M.M., Aubaud, C., and Withers, A.C., 2005, Storage capacity of H_2O in nominally anhydrous minerals in the upper mantle: Earth and Planetary Science Letters, v. 236, p. 167–181, doi: 10.1016/j.epsl.2005.04.022.

Hofmann, A.W., 1988, Chemical differentiation of the Earth: The relationship between mantle, continental crust, and oceanic crust: Earth and Planetary Science Letters, v. 90, p. 297–314, doi: 10.1016/0012-821X(88)90132-X.

Horiuchi, H., Morimoto, N., Yamamoto, K., and Akimoto, S., 1979, Crystal structure of $2Mg_2SiO_4 \cdot 3Mg(OH)_2$, a new high-pressure structure type: The American Mineralogist, v. 64, p. 593–598.

Ingrin, J., 1989, Water in diopside: An electron microscopy and infrared spectroscopy study: European Journal of Mineralogy, v. 1, p. 327–341.

Ingrin, J., and Skogby, H., 2000, Hydrogen in nominally anhydrous upper-mantle minerals: Concentration levels and implications: European Journal of Mineralogy, v. 12, p. 543–570.

Ingrin, J., Hercule, S., and Charton, T., 1995, Diffusion of hydrogen in diopside: Results of dehydration experiments: Journal of Geophysical Research, v. 100, p. 15,489–15,499, doi: 10.1029/95JB00754.

Inoue, T., 1994, Effect of water on melting phase relations and melt compositions in the system Mg_2SiO_4-$MgSiO_3$-H_2O up to 15 GPa: Physics of the Earth and Planetary Interiors, v. 85, p. 237–263, doi: 10.1016/0031-9201(94)90116-3.

Inoue, T., Yurimoto, H., and Kudoh, Y., 1995, Hydrous modified spinel, $Mg_{1.75}SiH_{0.5}O_4$: A new water reservoir in the mantle transition region: Geophysical Research Letters, v. 22, p. 117–120, doi: 10.1029/94GL02965.

Inoue, T., Irifune, T., Yurimoto, H., and Miyagi, I., 1998a, Decomposition of K-amphibole at high pressures and implications for subduction zone volcanism: Physics of the Earth and Planetary Interiors, v. 107, p. 221–231, doi: 10.1016/S0031-9201(97)00135-0.

Inoue, T., Weidner, D.J., Northrup, P.A., and Parise, J.B., 1998b, Elastic properties of hydrous ringwoodite (γ-phase) in Mg_2SiO_4: Earth and Planetary Science Letters, v. 160, p. 107–113, doi: 10.1016/S0012-821X(98)00077-6.

Ionov, D.A., Griffin, W.L., and O'Reilly, S.Y., 1997, Volatile-bearing minerals and lithosphere trace elements in the upper mantle: Chemical Geology, v. 141, p. 153–184, doi: 10.1016/S0009-2541(97)00061-2.

Irifune, T., and Isshiki, M., 1998, Iron partitioning in a pyrolite mantle and the nature of the 410-km seismic discontinuity: Nature, v. 392, p. 702–705, doi: 10.1038/33663.

Irifune, T., and Ringwood, A.E., 1993, Phase transformations in subducted oceanic crust and buoyancy relationships at depths of 600–800 km in the mantle: Earth and Planetary Science Letters, v. 117, p. 101–110, doi: 10.1016/0012-821X(93)90120-X.

Irifune, T., Ringwood, A.E., and Hibberson, W.O., 1994, Subduction of continental crust and terrigenous and pelagic sediments: An experimental study: Earth and Planetary Science Letters, v. 126, p. 351–368, doi: 10.1016/0012-821X(94)90117-1.

Irifune, T., Kubo, N., Isshiki, M., and Yamasaki, Y., 1998, Phase transformation in serpentine and transportation of water into the lower mantle: Geophysical Research Letters, v. 25, p. 203–206, doi: 10.1029/97GL03572.

Ito, E., and Takahashi, E., 1989, Post-spinel transformation in the system Mg_2SiO_4-Fe_2SiO_4 and some geophysical implications: Journal of Geophysical Research, v. 94, p. 10,637–10,646.

Ito, E.D., Harris, M., and Anderson, A.T., Jr., 1983, Alteration of oceanic crust and geologic cycling of chlorine and water: Geochimica et Cosmochimica Acta, v. 47, p. 1613–1624, doi: 10.1016/0016-7037(83)90188-6.

Iwamori, H., 1998, Transportation of H_2O and melting in subduction zones: Earth and Planetary Science Letters, v. 160, p. 65–80, doi: 10.1016/S0012-821X(98)00080-6.

Iwamori, H., 2004, Phase relations of peridotite under H_2O-saturated conditions and ability of subducting plate for transportation of H_2O: Earth and Planetary Science Letters, v. 227, p. 57–71, doi: 10.1016/j.epsl.2004.08.013.

Izraeli, E.S., Harris, J.W., and Navon, O., 2001, Brine inclusions in diamonds: A new upper mantle fluid: Earth and Planetary Science Letters, v. 187, p. 323–332, doi: 10.1016/S0012-821X(01)00291-6.

Jacobsen, S.D., Smyth, J.R., Spetzler, H., Holl, C.M., and Frost, D.J., 2004, Sound velocities and elastic constants of iron-bearing hydrous ringwoodite: Physics of the Earth and Planetary Interior, v. 143–144, p. 47–56.

Jagoutz, E., Palme, H., Baddenhausen, H., Blum, K., Cendales, M., Dreibus, G., Spettel, B., Lorenz, V., and Wanke, H., 1979, The abundances of major, minor and trace elements in the Earth's mantle as derived from primitive ultramafic nodules: Houston, Texas, Proceedings of the 10th Lunar and Planetary Scientific Conference, New York, Pergamon Press, v. 2, p. 2031–2050.

Jambon, A., 1994, Earth degassing and large scale geochemical cycling of volatile elements: Mineralogical Society of America Reviews in Mineralogy, v. 30, p. 479–517.

Jamtveit, B., Brooker, R., Brooks, K., Larsen, L.M., and Pedersen, T., 2001, The water content of olivines from the North Atlantic volcanic province: Earth and Planetary Science Letters, v. 186, p. 401–415, doi: 10.1016/S0012-821X(01)00256-4.

Jarrard, R.D., 2003, Subduction fluxes of water, carbon dioxide, chlorine, and potassium: Geochemistry, Geophysics, Geosystems, v. 4, doi: 10.1029/2002GC000392.

Johnson, L.H., Burgess, R., Turner, G., Milledge, H.J., and Harris, J.W., 2000, Noble gas and halogen geochemistry of mantle fluids: Comparison of African and Canadian diamonds: Geochimica et Cosmochimica Acta, v. 64, p. 717–732, doi: 10.1016/S0016-7037(99)00336-1.

Johnson, M.C., and Plank, T., 1999, Dehydration and melting experiments constrain the fate of subducted sediments: Geochemistry, Geophysics, Geosystems, v. 1, doi: 10.1029/1999GC000014.

Johnson, M.C., and Walker, D., 1993, Brucite [$Mg(OH)_2$] dehydration and the molar volume of H_2O to 15 GPa: The American Mineralogist, v. 78, p. 271–284.

Kagi, H., Lu, R., Davidson, P.M., Goncharov, A.F., Mao, H.K., and Hemley, R.J., 2000, Evidence for ice VI as an inclusion of cuboid diamonds from high P-T near infrared spectroscopy: Mineralogical Magazine, v. 64, p. 1089–1097, doi: 10.1180/002646100549904.

Kanzaki, M., 1991, Stability of hydrous magnesium silicates in the mantle transition zone: Physics of the Earth and Planetary Interiors, v. 66, p. 307–312, doi: 10.1016/0031-9201(91)90085-V.

Katayama, I., and Nakashima, S., 2003, Hydroxyl in clinopyroxene from the deep subducted crust: Evidence for H_2O transport into the mantle: The American Mineralogist, v. 88, p. 229–234.

Katayama, I., Hirose, K., Yurimoto, H., and Nakashima, S., 2003, Water solubility in majoritic garnet in subduction oceanic crust: Geophysical Research Letters, v. 30, doi: 10.1029/2003GL018127.

Kats, A., 1962, Hydrogen in alpha quartz: Phillips Research Report, v. 17, p. 133–195, 201–279.

Katsura, T., and Ito, E., 1989, The system Mg_2SiO_4-Fe_2SiO_4 at high pressures and temperatures: Precise determination of stabilities of olivine, modified spinel, and spinel: Journal of Geophysical Research, v. 94, p. 15,663–15,670.

Katsura, T., Yamada, H., Shinmei, T., Kubo, A., Ono, S., Kanzaki, M., Yoneda, A., Walter, M.J., Ito, E., Urakawa, S., Funakoshi, K., and Utsumi, W., 2003, Post-spinel transition in Mg_2SiO_4 determined by high P-T in situ X-ray diffraction: Physics of the Earth and Planetary Interiors, v. 136, p. 11–24, doi: 10.1016/S0031-9201(03)00019-0.

Kawamoto, T., 2004, Hydrous phase stability and partial melt chemistry in H_2O-saturated KLB-1 peridotite up to the uppermost lower mantle conditions: Physics of the Earth and Planetary Interiors, v. 143–144, p. 387–395, doi: 10.1016/j.pepi.2003.06.003.

Kawamoto, T., and Holloway, J.R., 1997, Melting temperature and partial melt chemistry of H_2O-saturated mantle peridotite to 11 gigapascals: Science, v. 276, p. 240–243, doi: 10.1126/science.276.5310.240.

Kawamoto, T., Leinenweber, K., Herving, R.L., and Holloway, J.R., 1995, Stability of hydrous minerals in H_2O-saturated KLB-1 peridotite up to 15 GPa, in Farley, K.A., ed., Volatiles in the Earth and Solar System: Melville, New York, American Institute of Physics Conference Proceeding, v. 341, p. 229–239.

Kawamoto, T., Hervig, R.L., and Holloway, J.R., 1996, Experimental evidence for a hydrous transition zone in the Earth's early mantle: Earth and Planetary Science Letters, v. 142, p. 587–592, doi: 10.1016/0012-821X(96)00113-6.

Kawamoto, T., Matsukage, K.N., Mibe, K., Isshiki, M., Nishimura, K., Ishimatsu, N., and Ono, S., 2004, Mg/Si ratios of aqueous fluids coexisting with forsterite and enstatite based on the phase relations in the Mg_2SiO_4-SiO_2-H_2O system: The American Mineralogist, v. 89, p. 1433–1437.

Kawamoto, T., Kanzaki, M., Mibe, K., Matsukage, K.N., and Ono, S., 2005, Direct observation of critical behaviors between aqueous fluids and an andesitic melt: Major element chemistry of supercritical fluids in mantle wedge [abs.]: Eos (Transactions, American Geophysical Union), v. 86, Fall Meeting Supplement, abstract V33C-04.

Kennedy, G.C., Wasseburg, G.J., Heard, H.C., and Newton, R.C., 1962, The upper three-phase region in the system SiO_2-H_2O: American Journal of Science, v. 260, p. 501–521.

Kepezhinskas, K.B., and Khlestov, V.V., 1977, The petrogenetic grid and subfacies for middle-temperature metapelites: Journal of Petrology, v. 18, p. 114–143.

Keppler, H., 1996, Constraints from partitioning experiments on the composition of subduction-zone fluids: Nature, v. 380, p. 237–240, doi: 10.1038/380237a0.

Kerr, A.C., 1998, Oceanic plateau formation: A cause of mass extinction and black shale deposition around the Cenomanian-Turonian boundary: Journal of the Geological Society of London, v. 155, p. 619–626.

Kerrick, D.M., and Connolly, J.A.D., 2001a, Metamorphic devolatilization of subducted marine sediments and the transport of volatiles into the Earth's mantle: Nature, v. 411, p. 293–296, doi: 10.1038/35077056.

Kerrick, D.M., and Connolly, J.A.D., 2001b, Metamorphic devolatilization of subducted oceanic metabasalts: Implications for seismicity, arc magmatism and volatile recycling: Earth and Planetary Science Letters, v. 189, p. 19–29, doi: 10.1016/S0012-821X(01)00347-8.

Kessel, R., Ulmer, P., Pettke, T., Schmidt, M.W., and Thompson, A.B., 2005, The water-basalt system at 4 to 6 GPa: Phase relations and second critical endpoint in a K-free eclogite at 700 to 1400°C: Earth and Planetary Science Letters, v. 237, p. 873–892, doi: 10.1016/j.epsl.2005.06.018.

Khisina, N.R., and Wirth, R., 2002, Hydrous olivine $(Mg_{1-y}Fe^{2+}_y)_{2-x}V_x$$SiO_4H_{2x}$—A new DHMS phase of variable composition observed as nanometer-sized precipitations in mantle olivine: Physics and Chemistry of Minerals, v. 29, p. 98–111, doi: 10.1007/s002690100205.

Khisina, N.R., and Wirth, R., 2004, First finding of high-pressure silicate $Mg_3Si_4(OH)nH_2O$ 10Å phase: Vestnik of the Department of the Earth Science RAS, v. 22, http://www.scgis.ru/russian/cp1251/h_dgggms/1-2004/informbul-1/term-9.pdf (in Russian).

Khisina, N.R., Wirth, R., Andrut, M., and Ukhanov, A.V., 2001, Extrinsic and intrinsic mode of hydrogen occurrence in natural olivines: FTIR and TEM investigation: Physics and Chemistry of Minerals, v. 28, p. 291–301, doi: 10.1007/s002690100162.

Khomenko, V.M., Langer, K., Beran, A., Koch-Müller, M., and Fehr, T., 1994, Titanium substitution and OH-bearing defects in hydrothermally grown pyrope crystals: Physics and Chemistry of Minerals, v. 20, p. 483–488, doi: 10.1007/BF00203218.

Kincaid, C., and Sacks, I.S., 1997, Thermal and dynamical evolution of the upper mantle in subduction zones: Journal of Geophysical Research, v. 102, p. 12,295–12,315, doi: 10.1029/96JB03553.

Kirby, S.H., Stein, S., Okal, E.A., and Rubie, D.C., 1996, Metastable mantle phase transformations and deep earthquakes in subducting oceanic lithosphere: Reviews of Geophysics, v. 34, p. 261–306, doi: 10.1029/96RG01050.

Kitamura, M., Kondoh, S., Morimoto, N., Miller, G.H., Rossman, G.R., and Putnis, A., 1987, Planar OH-bearing defects in mantle olivine: Nature, v. 328, p. 143–145, doi: 10.1038/328143a0.

Kleppe, A.K., Jephcoat, A.P., Smyth, J.R., and Frost, D.J., 2002, On protons, iron, and the high pressure behaviour of ringwoodite: Geophysical Research Letters, v. 29, doi: 10.1029/2002GL015276.

Knittle, E., and Jeanloz, R., 1991, Earth's core-mantle boundary: Results of experiments at high pressures and temperatures: Science, v. 251, p. 1438–1443, doi: 10.1126/science.251.5000.1438.

Koch-Müller, M., Fei, Y., Hauri, E., and Liu, Z., 2001, Location and quantitative analysis of OH in coesite: Physics and Chemistry of Minerals, v. 28, p. 693–705, doi: 10.1007/s002690100195.

Koch-Müller, M., Dera, P., Fei, Y., Reno, B.L., Sobolev, N.V., Hauri, E., and Wysoczanski, R., 2003, OH^- in synthetic and natural coesite: The American Mineralogist, v. 88, p. 1436–1445.

Koch-Müller, M., Matsyuk, S.S., and Wirth, R., 2004, Hydroxyl in omphacites and omphacitic clinopyroxenes of upper mantle to lower crustal origin beneath the Siberian platform: The American Mineralogist, v. 89, p. 921–931.

Koga, K., Hauri, E., Hirschmann, M., and Bell, D., 2003, Hydrogen concentration analyses using SIMS and FTIR: Comparison and calibration for nominally anhydrous minerals: Geochemistry, Geophysics, Geosystems, v. 4, doi: 10.1029/2002GC000378.

Kohlstedt, D.L., and Mackwell, S.J., 1998, Diffusion of hydrogen and intrinsic point defects in olivine: Zeitschrift für Physikalische Chemie, v. 207, p. 147–162.

Kohlstedt, D.L., Keppler, H., and Rubie, D.C., 1996, Solubility of water in the α, β, and γ phases of $(Mg,Fe)_2SiO_4$: Contributions to Mineralogy and Petrology, v. 123, p. 345–357, doi: 10.1007/s004100050161.

Kohn, S.C., 1996, Solubility of H_2O in nominally anhydrous mantle minerals using ^1H MAS NMR: The American Mineralogist, v. 81, p. 1523–1526.

Kohn, S.C., Brooker, R.A., Frost, D.J., Slesinger, A.E., and Wood, B.J., 2002, Ordering of hydroxyl defects in hydrous wadsleyite β-Mg_2SiO_4: The American Mineralogist, v. 87, p. 293–301.

Komabayashi, T., Omori, S., and Maruyama, S., 2004, Petrogenetic grid in the system MgO-SiO_2-H_2O up to 30 GPa, 1600°C: Applications to hydrous peridotite subducting into the Earth's deep interior: Journal of Geophysical Research, v. 109, p. B03206, doi: 10.1029/2003JB002651.

Konzett, J., and Fei, Y., 2000, Transport and storage of potassium in the Earth's upper mantle and transition zone: An experimental study to 23 GPa in simplified and natural bulk compositions: Journal of Petrology, v. 41, p. 583–603, doi: 10.1093/petrology/41.4.583.

Konzett, J., and Ulmer, P., 1999, The stability of hydrous potassic phases in lherzolitic mantle—An experimental study to 9.5 GPa in simplified and natural bulk compositions: Journal of Petrology, v. 40, p. 629–652, doi: 10.1093/petrology/40.4.629.

Konzett, J., Sweeney, R.J., Thompson, A.B., and Ulmer, P., 1997, Potassium amphibole stability in the upper mantle: An experimental study in a peralkaline KNCMASH system to 8.5 GPa: Journal of Petrology, v. 38, p. 537–568, doi: 10.1093/petrology/38.5.537.

Krawczynski, M.J., Fei, Y., and Hauri, E., 2004, Hydrous synthesis of aluminum bearing silicate perovskites: Implications for hydrogen storage in the lower mantle: Eos (Transactions, American Geophysical Union), v. 85, no. 47, p. F1731–F1732.

Kreuer, K.D., 1997, On the development of proton conducting materials for technological applications: Solid State Ionics, v. 97, p. 1–15, doi: 10.1016/S0167-2738(97)00082-9.

Kreuer, K.D., 1999, Aspects of the formation and mobility of protonic carriers and the stability of perovskite-type oxides: Solid State Ionics, v. 125, p. 285–302, doi: 10.1016/S0167-2738(99)00188-5.

Kronenberg, A.K., 1994, Hydrogen speciation and chemical weakening of quartz, in Heaney, P.J., Prewitt, C.T., and Gibbs, J.V., eds., Silica: Physical Behavior, Geochemistry, and Materials Applications: Mineralogical Society of America Reviews in Mineralogy, v. 29, p. 123–176.

Kronenberg, A.K., Kirby, S.H., Aines, R.D., and Rossman, G.R., 1986, Solubility and diffusional uptake of hydrogen in quartz at high water pressures: Implications for hydrolytic weakening: Journal of Geophysical Research, v. 91, p. 12,723–12,744.

Kudoh, Y., 2002, Predicted model for hydrous modified olivine (HyM-α): Physics and Chemistry of Minerals, v. 29, p. 387–395, doi: 10.1007/s00269-002-0245-7.

Kudoh, Y., and Inoue, T., 1999, Mg-vacant structural modules and dilution of the symmetry of hydrous wadsleyite, β-$Mg_{2-x}SiH_{2x}O_4$ with $0.00 = x = 0.25$: Physics and Chemistry of Minerals, v. 26, p. 382–388, doi: 10.1007/s002690050198.

Kudoh, Y., Finger, L.W., Hazen, R.M., Prewitt, C.T., Kanzaki, M., and Veblen, D.R., 1993, Phase E: A high-pressure hydrous silicate with unique crystal chemistry: Physics and Chemistry of Minerals, v. 19, p. 357–360, doi: 10.1007/BF00202972.

Kudoh, Y., Nagase, T., Mizohata, H., Ohtani, E., Sasaki, S., and Tanaka, M., 1997, Structure and crystal chemistry of phase G, a new hydrous magnesium silicate synthesized at 22 GPa and 1050°C: Geophysical Research Letters, v. 24, p. 1051–1054, doi: 10.1029/97GL00875.

Kudoh, Y., Kuribayashi, T., Mizobata, H., and Ohtani, E., 2000, Structure and cation disorder of hydrous ringwoodite, γ-$Mg_{1.89}Si_{0.98}H_{0.30}O_4$: Physics and Chemistry of Minerals, v. 27, p. 474–479, doi: 10.1007/s002690000091.

Kurosawa, M., Yurimoto, H., and Sueno, S., 1997, Patterns in the hydrogen and trace element compositions of mantle olivines: Physics and Chemistry of Minerals, v. 24, p. 385–395, doi: 10.1007/s002690050052.

Lager, G.A., and VonDreele, R.B., 1996, Neutron powder diffraction study of hydrogarnet to 9.0 GPa: The American Mineralogist, v. 81, p. 1097–1104.

Lager, G.A., Armbruster, T., Rotella, F.J., and Rossman, G.R., 1989, OH substitution in garnets—X-ray and neutron-diffraction, infrared, and geometric-modeling studies: The American Mineralogist, v. 74, p. 840–851.

Lanford, W.A., Trautvetter, H.P., Ziegler, J.F., and Keller, J., 1976, New precision technique for measuring the concentration versus depth of hydrogen in solid: Applied Physics Letters, v. 28, p. 566–570, doi: 10.1063/1.88826.

Langer, K., Robarick, E., Sobolev, N.V., Shatsky, V.S., and Wang, W., 1993, Single-crystal spectra of garnets from diamondiferous high-pressure metamorphic rocks from Kazakhstan: Indications for OH^-, H_2O and FeTi charge transfer: European Journal of Mineralogy, v. 5, p. 1091–1100.

Larson, R.L., 1991, Latest pulse of the Earth: Evidence for a mid-Cretaceous super plume: Geology, v. 19, p. 547–550, doi: 10.1130/0091-7613(1991)019<0547:LPOEEF>2.3.CO;2.

Li, W., Lu, R., Prewitt, C.T., and Fei, Y., 1997, Hydrogen in coesite crystals: Eos (Transactions, American Geophysical Union), v. 78, p. F736.

Li, Y.H., and Schoonmaker, J.E., 2003, Chemical composition and mineralogy of marine sediments, in Holland, H.D., and Turekian, K.K., eds., Treatise on Geochemistry, Volume 7: Oxford, Elsevier Ltd., p. 1–35.

Libowitzky, E., and Beran, A., 1995, OH defects in forsterite: Physics and Chemistry of Minerals, v. 22, p. 387–392, doi: 10.1007/BF00213336.

Libowitzky, E., and Rossman, G.R., 1997, An IR absorption calibration for water in minerals: The American Mineralogist, v. 82, p. 1111–1115.

Ligi, M., Bonatti, E., Cipriani, A., and Ottolini, L., 2005, Water-rich basalts at mid-ocean-ridge cold spots: Nature, v. 434, p. 66–69, doi: 10.1038/nature03264.

Liou, J.G., Zhang, R.Y., Ernst, W.G., Rumble, D., III, and Maruyama, S., 1998, High pressure minerals from deeply subducted metamorphic rocks: Mineralogical Society of America Reviews in Mineralogy, v. 37, p. 33–96.

Litasov, K.D., and Litasov, Yu.D., 1999, Biotite in megacryst assemblages of the alkaline basaltoids, Vitim Plateau: Geochemistry International, v. 37, p. 213–223.

Litasov, K.D., and Ohtani, E., 2002, Phase relations and melt compositions in CMAS pyrolite–H_2O system up to 25 GPa: Physics of the Earth and Planetary Interiors, v. 134, p. 105–127, doi: 10.1016/S0031-9201(02)00152-8.

Litasov, K.D., and Ohtani, E., 2003a, Stability of various hydrous phases in CMAS pyrolite–H_2O system up to 25 GPa: Physics and Chemistry of Minerals, v. 30, p. 147–156, doi: 10.1007/s00269-003-0301-y.

Litasov, K.D., and Ohtani, E., 2003b, Hydrous solidus of CMAS-pyrolite and melting of mantle plumes at the bottom of the upper mantle: Geophysical Research Letters, v. 30, doi: 10.1029/2003GL018318.

Litasov, K.D., and Ohtani, E., 2003c, Hydrous lower mantle: The water source for wet plumes?: Victoria, Canada, 8th International Kimberlite Conference, Long Abstracts, FLA030 (CD edition).

Litasov, K.D., and Ohtani, E., 2004, Relationship of the Al-bearing phases NAL and CF in the lower mantle: Russian Geology and Geophysics, v. 45, p. 1313–1325.

Litasov, K.D., and Ohtani, E., 2005, Phase relations in hydrous MORB at 18–28 GPa: Implications for heterogeneity of the lower mantle: Physics of the Earth and Planetary Interiors, v. 150, p. 239–263, doi: 10.1016/j.pepi.2004.10.010.

Litasov, K.D., Ohtani, E., and Taniguchi, H., 2001, Melting relations of hydrous pyrolite in CaO-MgO-Al_2O_3-SiO_2-H_2O system at the transition zone pressures: Geophysical Research Letters, v. 28, p. 1303–1306, doi: 10.1029/2000GL012291.

Litasov, K.D., Ohtani, E., Langenhorst, F., Yurimoto, H., Kubo, T., and Kondo, T., 2003, Water solubility in Mg-perovskites and water storage capacity in the lower mantle: Earth and Planetary Science Letters, v. 211, p. 189–203, doi: 10.1016/S0012-821X(03)00200-0.

Litasov, K.D., Ohtani, E., Suzuki, A., Kawazoe, T., and Funakoshi, K., 2004, Absence of density crossover between basalt and peridotite in the cold slabs passing through 660 km discontinuity: Geophysical Research Letters, v. 31, doi: 10.1029/2004GL021306.

Litasov, K.D., Ohtani, E., Sano, A., Suzuki, A., and Funakoshi, K., 2005a, In situ X-ray diffraction study of post-spinel transformation in a peridotite mantle: Implication to the 660-km discontinuity: Earth and Planetary Science Letters, v. 238, p. 311–328, doi: 10.1016/j.epsl.2005.08.001.

Litasov, K.D., Ohtani, E., Sano, A., Suzuki, A., and Funakoshi, K., 2005b, Wet subduction versus cold subduction: Geophysical Research Letters, v. 32, p. L13312, doi: 10.1029/2005GL022921.

Litasov, K.D., Ohtani, E., Kagi, H., and Ghosh, S., 2006a, Influence of water on olivine-wadsleyite phase transformation and water partitioning near 410-km seismic discontinuity, in Tohji, K., Tsuchiya, N., and Jeyadevan, B., eds., Proceedings of 3rd International Workshop on Water Dynamics: American Institute of Physics, v. 833, p. 150–155.

Litasov, K.D., Ohtani, E., and Sano, A., 2006b, Influence of water on major phase transitions in the Earth's mantle, in Jacobsen, S.D., and van der Lee, S., eds., Earth's Deep Water Cycle: Washington, D.C., American Geophysical Union Geophysical Monograph, v. 168, p. 95–111.

Litasov, K.D., Ohtani, E., Kagi, H., Lakshtanov, D.L., and Bass, J.D., 2006c, Hydrogen solubility in Al-rich stishovite and water transport to the lower mantle: Abstracts of the 16th annual V.M. Goldschmidt conference, Melbourne, Australia, Geochimica et Cosmochimica Acta, v.70, no.18, Supplement 1, p. A362

Liu, J., Bohen, S.R., and Ernst, W.G., 1996, Stability of hydrous phases in subduction oceanic crust: Earth and Planetary Science Letters, v. 143, p. 161–171, doi: 10.1016/0012-821X(96)00130-6.

Liu, L.G., 1987, Effects of H_2O on the phase behaviour of the forsterite-enstatite system at high pressures and temperatures and implications for the Earth: Physics of the Earth and Planetary Interiors, v. 49, p. 142–167, doi: 10.1016/0031-9201(87)90138-5.

Liu, L.G., 1993, Effect of H_2O on the phase behavior of the forsterite-enstatite system at high pressures and temperatures: Revisited: Physics of the Earth and Planetary Interiors, v. 76, p. 209–218, doi: 10.1016/0031-9201(93)90013-Y.

Liu, L.G., Lin, C.C., Mernagh, T.P., and Irifune, T., 1997a, Raman spectra of phase A at various pressures and temperatures: Journal of Physics and Chemistry of Solids, v. 58, p. 2023–2030, doi: 10.1016/S0022-3697(97)00121-2.

Liu, L.G., Mernagh, T.P., Lin, C.C., and Irifune, T., 1997b, Raman spectra of phase E at various pressures and temperatures with geophysical implications: Earth and Planetary Science Letters, v. 149, p. 57–65, doi: 10.1016/S0012-821X(97)00063-0.

Liu, L.G., Lin, C.C., Irifune, T., and Mernagh, T.P., 1998a, Raman study of phase D at various pressures and temperatures: Geophysical Research Letters, v. 25, p. 3453–3456, doi: 10.1029/98GL01415.

Liu, L.G., Lin, C.C., Mernagh, T.P., and Irifune, T., 1998b, Raman spectra of phase B at various pressures and temperatures: Journal of Physics and Chemistry of Solids, v. 59, p. 871–877, doi: 10.1016/S0022-3697(98)00029-8.

Lu, R., and Keppler, H., 1997, Water solubility in pyrope to 100 kbar: Contributions to Mineralogy and Petrology, v. 129, p. 35–42, doi: 10.1007/s004100050321.

Lu, R., Hofmeister, A.M., and Wang, Y., 1994, Thermodynamic properties of ferromagnesium silicate perovskites from vibrational spectroscopy: Journal of Geophysical Research, v. 99, p. 11,795–11,804, doi: 10.1029/94JB00501.

Luo, S.N., Mosenfelder, J.L., Asimow, P.D., and Ahrens, T.J., 2002, Stishovite and its implications in geophysics: New results from shock-wave experiments and theoretical modeling: Physics Uspekhi, v. 45, p. 435–439, doi: 10.1070/PU2002v045n04ABEH001155.

Lupton, J.E., and Craig, H., 1975, Excess ^3He in oceanic basalts: Evidence for terrestrial primordial helium: Earth and Planetary Science Letters, v. 26, p. 133–139, doi: 10.1016/0012-821X(75)90080-1.

Luth, R.W., 1995, Is phase A relevant to the Earth's mantle?: Geochimica et Cosmochimica Acta, v. 59, p. 679–682, doi: 10.1016/0016-7037(95)00319-U.

Luth, R.W., 1997, Experimental study of the system phlogopite-diopside from 3.5 to 17 GPa: The American Mineralogist, v. 82, p. 1198–1209.

Mackwell, S.J., and Kohlstedt, D.L., 1990, Diffusion of hydrogen in olivine: Implications for water in the mantle: Journal of Geophysical Research, v. 95, p. 5079–5088.

Mackwell, S.J., and Paterson, M.S., 1985, Water-related diffusion and deformation effect in quartz at pressures of 1500 and 300 MPa, in Schock, R., ed., Point defects in minerals: American Geophysical Union Geophysical Monograph 31, p. 141–150.

Maldener, J., Rauch, F., Gavranic, M., and Beran, A., 2001, OH absorption coefficients of rutile and cassiterite deduced from nuclear reaction analysis and FTIR spectroscopy: Mineralogy and Petrology, v. 71, p. 21–29, doi: 10.1007/s007100170043.

Mamyrin, B.A., Tolstikhin, I.N., Anufriev, G.S., and Kamensky, I.L., 1969, Anomalous isotopic composition of helium in volcanic gases: Transactions (Doklady): Akademiya Nauk SSSR, v. 184, p. 1197–1199.

Manning, C.E., 1994, The solubility of quartz in H_2O in the lower crust and upper mantle: Geochimica et Cosmochimica Acta, v. 58, p. 4831–4839, doi: 10.1016/0016-7037(94)90214-3.

Manning, C.E., 2004, The chemistry of subduction-zone fluids: Earth and Planetary Science Letters, v. 223, p. 1–16, doi: 10.1016/j.epsl.2004.04.030.

Martin, R.F., and Donnay, G., 1972, Hydrohyl in the mantle: The American Mineralogist, v. 57, p. 554–570.

Maruyama, S., 1999, Leaking water into the Earth: Eos (Transactions, American Geophysical Union), v. 80, Fall Meeting Supplement, abstract U31B-06.

Maruyama, S., Liou, J.G., and Terabayashi, M., 1996, Blueschists and eclogites of the world and their exhumation: International Geology Review, v. 38, p. 563–594.

Maruyama, S., Isozaki, Y., Nakashima, S., and Windley, B., 2001, History of the Earth and life, in Nakashima, S., et al., eds., Geochemistry and Origin of Life: Tokyo, Universal Academic Press, p. 285–325.

Matsukage, K.N., Jing, Z., and Karato, S., 2005, Density of hydrous silicate melt at the conditions of Earth's deep upper mantle: Nature, v. 438, p. 488–491, doi: 10.1038/nature04241.

Matsyuk, S.S., and Langer, K., 2004, Hydroxyl in olivines from mantle xenoliths in kimberlites of the Siberian platform: Contributions to Mineralogy and Petrology, v. 147, p. 413–437, doi: 10.1007/s00410-003-0541-3.

Matsyuk, S.S., Langer, K., and Hoesch, A., 1998, Hydroxyl defects in garnets from mantle xenoliths in kimberlites of the Siberian platform: Contributions to Mineralogy and Petrology, v. 132, p. 163–179, doi: 10.1007/s004100050414.

Matveev, S., O'Neil, H.St.C., Ballhaus, C., Taylor, W.R., and Green, D.H., 2001, Effect of silica activity on OH^- IR spectra of olivine: Implications for low $aSiO_2$ mantle metasomatism: Journal of Petrology, v. 42, p. 721–729, doi: 10.1093/petrology/42.4.721.

Matveev, S., Portnyagin, M., Ballhaus, C., Brooker, R., and Geiger, C.A., 2005, FTIR spectrum of phenocryst olivine as an indicator of silica saturation in magmas: Journal of Petrology, v. 46, p. 603–614, doi: 10.1093/petrology/egh090.

McCammon, C., 1997, Perovskite as a possible sink for ferric iron in the lower mantle: Nature, v. 387, p. 694–696, doi: 10.1038/42685.

McCammon, C.A., Lauterbach, S., Seifert, F., Langenhorst, F., and van Aken, P.A., 2004a, Iron oxidation state in the lower mantle mineral assemblages. I: Empirical relations derived from high-pressure experiments: Earth and Planetary Science Letters, v. 222, p. 435–449, doi: 10.1016/j.epsl.2004.03.018.

McCammon, C.A., Stachel, T., and Harris, J.W., 2004b, Iron oxidation state in the lower mantle mineral assemblages. II: Inclusions in diamonds from Kankan, Guinea: Earth and Planetary Science Letters, v. 222, p. 423–434, doi: 10.1016/j.epsl.2004.03.019.

McDonough, W.F., and Rudnick, R.L., 1998, Mineralogy and composition of the upper mantle: Mineralogical Society of America Reviews in Mineralogy, v. 37, p. 139–164.

McDonough, W.F., and Sun, S.-S., 1995, The composition of the Earth: Chemical Geology, v. 120, p. 223–254, doi: 10.1016/0009-2541(94)00140-4.

McGetchin, T.R., Silver, L.T., and Chodos, A.A., 1970, Titanclinohumite: A possible mineralogical site for water in the upper mantle: Journal of Geophysical Research, v. 75, p. 255–259.

McMillan, P.F., Akaogi, M., Sato, R., Poe, B., and Foley, J., 1991, Hydrohyl groups in β-Mg_2SiO_4: The American Mineralogist, v. 76, p. 354–360.

Meade, C., Reffner, J.A., and Ito, E., 1994, Synchrotron infrared absorbance measurements of hydrogen in $MgSiO_3$ perovskite: Science, v. 264, p. 1558–1560, doi: 10.1126/science.264.5165.1558.

Mengel, K., and Green, D.H., 1989, Stability of amphibole and phlogopite in metasomatized peridotite under water-saturated and water-undersaturated conditions: Proceedings of the 4th International Kimberlite Conference: Geological Society of Australia Special Publication, v. 14, no. 1, p. 571–581.

Mibe, K., Fujii, T., and Yasuda, A., 1998, Connectivity of aqueous fluid in the Earth's upper mantle: Geophysical Research Letters, v. 25, p. 1233–1236, doi: 10.1029/98GL00872.

Mibe, K., Fujii, T., and Yasuda, A., 1999, Control of the location of the volcanic front in island arcs by aqueous fluid connectivity in the mantle wedge: Nature, v. 401, p. 259–262, doi: 10.1038/45762.

Mibe, K., Fujii, T., and Yasuda, A., 2002, Composition of aqueous fluid coexisting with mantle minerals at high pressure and its bearing on the differentiation of the Earth's mantle: Geochimica et Cosmochimica Acta, v. 66, p. 2273–2285, doi: 10.1016/S0016-7037(02)00856-6.

Mibe, K., Yoshino, T., Ono, S., Yasuda, A., and Fujii, T., 2003, Connectivity of aqueous fluid in eclogite and its implications for fluid migration in the Earth's interior: Journal of Geophysical Research, v. 108, doi: 10.1029/2002JB001960.

Mibe, K., Fujii, T., and Yasuda, A., 2004a, Response to the comment by R. Stalder on "Composition of aqueous fluid coexisting with mantle minerals at high pressure and its bearing on the differentiation of the Earth's mantle": Geochimica et Cosmochimica Acta, v. 68, p. 929–930, doi: 10.1016/j.gca.2003.07.003.

Mibe, K., Kanzaki, M., Kawamoto, T., Matsukage, K.N., Fei, Y., and Ono, S., 2004b, Determination of the second critical end point in silicate-H_2O systems using high-pressure and high-temperature X-ray radiography: Geochimica et Cosmochimica Acta, v. 68, p. 5189–5195, doi: 10.1016/j.gca.2004.07.015.

Mibe, K., Kanzaki, M., Kawamoto, T., Matsukage, K.N., Fei, Y., and Ono, S., 2005, Second critical endpoint in the basalt-H_2O system: Eos (Transactions, American Geophysical Union), v. 86, Fall Meeting Supplement, abstract V33C-03.

Michael, P.J., 1988, The concentration, behavior and storage of water in the suboceanic upper mantle: Implications for mantle metasomatism: Geochimica et Cosmochimica Acta, v. 52, p. 555–566, doi: 10.1016/0016-7037(88)90110-X.

Michael, P.J., 1995, Regionally distinctive sources of depleted MORB: Evidence from trace element and H_2O: Earth and Planetary Science Letters, v. 131, p. 301–320, doi: 10.1016/0012-821X(95)00023-6.

Miller, G.H., Rossman, G.R., and Harlow, G.E., 1987, The natural occurrence of hydroxide in olivine: Physics and Chemistry of Minerals, v. 14, p. 461–472, doi: 10.1007/BF00628824.

Mitchell, R.H., 1986, Kimberlites: Mineralogy, Geochemistry and Petrology: New York, Plenum, 442 p.

Moore, J.C., and Vrolijk, P., 1992, Fluids in accretionary prisms: Reviews of Geophysics, v. 30, p. 113–135.

Morishima, H., Kato, T., Suto, M., Ohtani, E., Urakawa, S., Utsumi, W., Shimomura, O., and Kikegawa, T., 1994, The phase boundary between α- and β-Mg_2SiO_4 determined by in situ X-ray observation: Science, v. 265, p. 1202–1203, doi: 10.1126/science.265.5176.1202.

Mosenfelder, J.L., 2000, Pressure dependence of hydroxyl solubility in coesite: Physics and Chemistry of Minerals, v. 27, p. 610–617, doi: 10.1007/s002690000105.

Mosenfelder, J.L., Deligne, N.I., Asimov, P.D., and Rossman, G.R., 2006, Hydrogen incorporation in olivine from 2–12 GPa: The American Mineralogist, v. 91, p. 285–294, doi: 10.2138/am.2006.1943.

Mottana, A., Carswell, D.A., Chopin, C., and Oberhansli, R., 1990, Eclogite facies mineral paragenesis, in Carswell, D.A., ed., Eclogite Facies Rocks: New York, Blackie, p. 14–52.

Murakami, M., Hirose, K., Yurimoto, H., Nakashima, S., and Takafuji, N., 2002, Water in Earth's lower mantle: Science, v. 295, p. 1885–1887, doi: 10.1126/science.1065998.

Murakami, M., Hirose, K., Kawamura, K., Sata, N., and Ohishi, Y., 2004, Post-perovskite phase transition in $MgSiO_3$: Science, v. 304, p. 855–858, doi: 10.1126/science.1095932.

Mysen, B.O., Ulmer, P., Konzett, J., and Schmidt, M.W., 1998, The upper mantle near convergent plate boundaries: Mineralogical Society of America Reviews in Mineralogy, v. 37, p. 97–138.

Navon, O., Hutcheon, I.D., Rossman, G.R., and Wasseburg, G.L., 1988, Mantle-derived fluids in diamond micro-inclusions: Nature, v. 335, p. 784–789, doi: 10.1038/335784a0.

Navon, O., Izraeli, E.S., and Klein-BenDavid, O., 2003, Fluid inclusions in diamonds The carbonatitic connection: Victoria, Canada, 8th International Kimberlite Conference, Long Abstracts, FLA00107 (CD-edition).

Navrotsky, A., 1999, Mantle geochemistry: A lesson from ceramics: Science, v. 284, p. 1788–1789, doi: 10.1126/science.284.5421.1788.

Niida, K., and Green, D.H., 1999, Stability and chemical composition of pargasitic amphibole in MORB pyrolite under upper mantle conditions: Con-

tributions to Mineralogy and Petrology, v. 135, p. 18–40, doi: 10.1007/s004100050495.
Niu, F., and Kawakatsu, H., 1996, Complex structure of the mantle discontinuities at the tip of the subducting slab beneath northeast China: A preliminary investigation of broadband receiver functions: Journal of Physics of the Earth, v. 44, p. 701–711.
Norby, T., 1990, Proton conduction in oxides: Solid State Ionics, v. 40–41, p. 857–862, doi: 10.1016/0167-2738(90)90138-H.
Ohtani, E., 2002, Water transport by slab subduction and generation of wet plume in the mantle [abs.]: Tokyo, Superplume International Workshop, Tokyo Institute of Technology, p. 358–362.
Ohtani, E., 2005, Water in the mantle: Elements, v. 1, p. 25–30.
Ohtani, E., Shibata, T., Kubo, T., and Kato, T., 1995, Stability of hydrous phases in the transitional zone and the uppermost part of the lower mantle: Geophysical Research Letters, v. 22, p. 2553–2556, doi: 10.1029/95GL02338.
Ohtani, E., Mizobata, H., Kudoh, Y., Nagase, T., Arashi, H., Yurimoto, H., and Miyagi, I., 1997, A new hydrous silicate, a water reservoir, in the upper part of the lower mantle: Geophysical Research Letters, v. 24, p. 1047–1050, doi: 10.1029/97GL00874.
Ohtani, E., Mizobata, H., and Yurimoto, H., 2000, Stability of dense hydrous magnesium silicate phases in the system Mg_2SiO_4-H_2O and $MgSiO_3$-H_2O at pressures up to 27 GPa: Physics and Chemistry of Minerals, v. 27, p. 533–544, doi: 10.1007/s002690000097.
Ohtani, E., Litasov, K., Suzuki, A., and Kondo, T., 2001a, Stability field of new hydrous phase, δ-AlOOH, with implications for water transport into the deep mantle: Geophysical Research Letters, v. 28, p. 3991–3993, doi: 10.1029/2001GL013397.
Ohtani, E., Touma, M., Litasov, K., Kubo, T., and Suzuki, A., 2001b, Stability of hydrous phases and water storage capacity in the transitional zone and lower mantle: Physics of the Earth and Planetary Interiors, v. 124, p. 105–117, doi: 10.1016/S0031-9201(01)00192-3.
Ohtani, E., Toma, M., Kubo, T., Kondo, T., and Kikegawa, T., 2003, In situ X-ray observation of decomposition of superhydrous phase B at high pressure and temperature: Geophysical Research Letters, v. 30, doi: 10.1029/2002GL015549.
Ohtani, E., Litasov, K., Hosoya, T., Kubo, T., and Kondo, T., 2004, Water transport into the deep mantle and formation of a hydrous transition zone: Physics of the Earth and Planetary Interiors, v. 143–144, p. 255–269, doi: 10.1016/j.pepi.2003.09.015.
Ohtani, E., Hirao, N., Kondo, T., Ito, M., and Kikegawa, T., 2005, Iron-water reaction at high pressure and temperature, and hydrogen transport into the core: Physics and Chemistry of Minerals, v. 32, p. 77–82, doi: 10.1007/s00269-004-0443-6.
Okamoto, K., and Maruyama, S., 1999, The high pressure stability limits of lawsonite in the MORB + H_2O system: The American Mineralogist, v. 84, p. 362–373.
Okamoto, K., and Maruyama, S., 2004, The eclogite-garnetite transformation in the MORB + H_2O system: Physics of the Earth and Planetary Interiors, v. 146, p. 283–296, doi: 10.1016/j.pepi.2003.07.029.
Okamoto, K., Maruyama, S., and Poli, S., 2002, The eclogite-garnetite transformation in the MORB + H_2O system [abs]: Tokyo, Superplume International Workshop, Tokyo Institute of Technology, p. 342–346.
Ono, S., 1998, Stability limits of hydrous minerals in sediment and mid-ocean ridge basalt compositions: Implications for water transport in subduction zones: Journal of Geophysical Research, v. 103, p. 18,253–18,267, doi: 10.1029/98JB01351.
Ono, S., 1999, High temperature stability limit of phase Egg, $AlSiO_3(OH)$: Contributions to Mineralogy and Petrology, v. 137, p. 83–89, doi: 10.1007/s004100050583.
Ono, S., Mibe, K., and Yoshino, T., 2002, Aqueous fluid connectivity in pyrope aggregates: Water transport into the deep mantle by a subducted oceanic crust without any hydrous minerals: Earth and Planetary Science Letters, v. 203, p. 895–903.
Pacalo, R.E.G., and Parise, J.B., 1992, Crystal structure of superhydrous B, a hydrous magnesium silicate synthesized at 1400°C and 20 GPa: The American Mineralogist, v. 77, p. 681–684.
Palme, H., and Jones, A., 2003, Solar system abundances of the elements, *in* Holland, H.D., and Turekian, K.K., eds., Treatise on Geochemistry, Volume 1: Oxford, Elsevier Ltd., p. 41–61.
Pal'yanov, Yu.N., Sokol, A.G., Borzdov, Yu.M., Khokhryakov, A.F., and Sobolev, N.V., 1999, Diamond formation from mantle carbonate fluids: Nature, v. 400, p. 417–418, doi: 10.1038/22678.

Pal'yanov, Yu.N., Sokol, A.G., Borzdov, Yu.M., and Khokhryakov, A.F., 2002, Fluid-bearing alkaline-carbonate melts as the medium for the formation of diamonds in the Earth's mantle: An experimental study: Lithos, v. 60, p. 145–159, doi: 10.1016/S0024-4937(01)00079-2.
Panero, W.R., and Akber-Knutson, S., 2004, Solubility of hydrogen and aluminum in $CaSiO_3$-perovskite and the partitioning of water among silicates in the lower mantle: Eos (Transactions, American Geophysical Union), v. 85, no. 47, p. F1734.
Panero, W.R., and Stixrude, L.P., 2004, Hydrogen incorporation in stishovite at high pressure and symmetric bonding in δ-AlOOH: Earth and Planetary Science Letters, v. 221, p. 421–431, doi: 10.1016/S0012-821X(04)00100-1.
Panero, W.R., Benedetti, L.R., and Jeanloz, R., 2003, Transport of water into lower mantle: Role of stishovite: Journal of Geophysical Research, v. 108, doi: 10.1029/2002JB002053.
Parman, S.W., Dann, J.C., Grove, T.L., and de Wit, M.J., 1997, Emplacement conditions of komatiite magmas from the 3.49 Ga Komati Formation, Barberton greenstone belt, South Africa: Earth and Planetary Science Letters, v. 150, p. 303–323, doi: 10.1016/S0012-821X(97)00104-0.
Paterson, M.S., 1982, The determination of hydroxyl by infrared absorption in quartz, silicate glasses and similar materials: Bulletin de Mineralogie, v. 105, p. 20–29.
Paterson, M.S., 1986, The thermodynamics of water in quartz: Physics and Chemistry of Minerals, v. 13, p. 245–255, doi: 10.1007/BF00308276.
Pawley, A., 1994, The pressure and temperature stability limits of lawsonite: Implication for H_2O recycling in subduction zones: Contributions to Mineralogy and Petrology, v. 118, p. 99–108, doi: 10.1007/BF00310614.
Pawley, A.R., and Holloway, J.R., 1993, Water sources for subduction zone volcanism: New experimental constraints: Science, v. 260, p. 664–666, doi: 10.1126/science.260.5108.664.
Pawley, A.R., and Wood, B.J., 1996, The low-pressure stability of phase A, $Mg_7Si_2O_8(OH)_6$: Contributions to Mineralogy and Petrology, v. 124, p. 90–97, doi: 10.1007/s004100050176.
Pawley, A.R., McMillan, P.F., and Holloway, J.R., 1993, Hydrogen in stishovite, with implications for mantle water content: Science, v. 261, p. 1024–1026, doi: 10.1126/science.261.5124.1024.
Peacock, S.M., 1990, Fluid processes in subduction zones: Science, v. 248, p. 329–337, doi: 10.1126/science.248.4953.329.
Peacock, S.M., 1996, Thermal and petrological structure of subduction zones, *in* Bebout, G.E., Scholl, D.W., Kirby, S.H., and Platt, J.P., eds., Subduction Top to Bottom: American Geophysical Union Geophysical Monograph 96, p. 119–133.
Peacock, S.M., 2003, Thermal structure and metamorphic evolution of subducting slabs, *in* Eiler, J., ed., Inside the Subduction Factory: American Geophysical Union Geophysical Monograph 138, p. 7–22.
Peacock, S.M., and Wang, K., 1999, Seismic consequences of warm versus cool subduction metamorphism: Examples from southwest and northeast Japan: Science, v. 286, p. 937–939, doi: 10.1126/science.286.5441.937.
Pearson, G., Canil, D., and Shirey, S.B., 2003, Mantle samples included in volcanic rocks: Xenoliths and diamonds, *in* Holland, H.D., and Turekian, K.K., eds., Treatise on Geochemistry, Volume 2: Oxford, Elsevier Ltd., p. 171–275.
Peslier, A.H., Luhr, J.F., and Post, J., 2002, Low water contents in pyroxenes from spinel-peridotites of the oxidized, sub-arc mantle: Earth and Planetary Science Letters, v. 201, p. 69–86, doi: 10.1016/S0012-821X(02)00663-5.
Pineau, F., and Javoy, M., 1994, Strong degassing at ridge crests: The behaviour of dissolved carbon and water in basalt glasses at 14°N, Mid-Atlantic Ridge: Earth and Planetary Science Letters, v. 123, p. 179–198, doi: 10.1016/0012-821X(94)90266-6.
Plank, T., and Langmuir, C.H., 1998, The chemical composition of subducting sediment: Implications for the crust and mantle: Chemical Geology, v. 145, p. 325–394, doi: 10.1016/S0009-2541(97)00150-2.
Poli, S., 1993, The amphibolite-eclogite transformation—An experimental study on basalt: American Journal of Science, v. 293, p. 1061–1107.
Poli, S., and Fumagalli, P., 2003, Mineral assemblages in ultrahigh pressure metamorphism: A review of experimentally determined phase diagrams, *in* Carswell, D.A., and Compagnoli, R., eds., Ultrahigh Pressure Metamorphism: Budapest, Eotvos University Press, Eurpean Mineralogical Union Notes in Mineralogy, v. 5, p. 307–340.
Poli, S., and Schmidt, M.W., 1995, H_2O transport and release in subduction zones—Experimental constraints on basaltic and andesitic systems: Journal of Geophysical Research, v. 100, p. 22,299–22,314, doi: 10.1029/95JB01570.

Poli, S., and Schmidt, M.W., 1998, The high pressure stability of zoisite and phase relations of zoisite-bearing assemblages: Contributions to Mineralogy and Petrology, v. 130, p. 162–175, doi: 10.1007/s004100050357.

Poli, S., and Schmidt, M.W., 2002, Petrology of subducted slabs: Annual Review of Earth and Planetary Science, v. 30, p. 207–235, doi: 10.1146/annurev.earth.30.091201.140550.

Prewitt, C.T., and Parise, J.B., 2000, Hydrous phases and hydrogen bonding at high pressure, in Hazen, R.M., ed., Crystal Chemistry at High Pressures and Temperatures: Mineralogical Society of America Reviews in Mineralogy, v. 41, p. 309–334.

Ranero, C.R., Phipps Morgan, J., McIntosh, K.D., and Reichert, C., 2003, Bending, faulting, and mantle serpentinization at the Middle America Trench: Nature, v. 425, p. 367–373, doi: 10.1038/nature01961.

Rauch, M., and Keppler, H., 2002, Water solubility in orthopyroxene: Contributions to Mineralogy and Petrology, v. 143, p. 525–536.

Rea, D.K., and Ruff, L.J., 1996, Composition and mass flux of sediment entering the world's subduction zones: Implications for global sediment budgets, great earthquakes, and volcanism: Earth and Planetary Science Letters, v. 140, p. 1–12, doi: 10.1016/0012-821X(96)00036-2.

Revenaugh, J., and Sipkin, S.A., 1994, Seismic evidence for silicate melt atop the 410-km mantle discontinuity: Nature, v. 369, p. 474–476, doi: 10.1038/369474a0.

Richard, G., Monnereau, M., and Ingrin, J., 2002, Is the transition zone an empty water reservoir? Influence from numerical model of mantle dynamics: Earth and Planetary Science Letters, v. 205, p. 37–51, doi: 10.1016/S0012-821X(02)01012-9.

Ringwood, A.E., 1975, Composition and Petrology of the Earth's Mantle: New York, McGraw-Hill, 630 p.

Ringwood, A.E., 1994, Role of the transition zone and 660 km discontinuity in mantle dynamics: Physics of the Earth and Planetary Interiors, v. 86, p. 5–24, doi: 10.1016/0031-9201(94)05058-9.

Ringwood, A.E., and Major, A., 1967, High pressure reconnaissance investigation in the system $Mg_2SiO_4MgO-H_2O$: Earth and Planetary Science Letters, v. 2, p. 130–133, doi: 10.1016/0012-821X(67)90114-8.

Rossman, G.R., 1988, Vibrational spectroscopy of hydrous components, in Hawthorne, F.C., ed., Spectroscopic Methods in Mineralogy: Mineralogical Society of America Reviews in Mineralogy, v. 18, p. 193–206.

Rossman, G.R., 1996, Studies of OH in nominally anhydrous minerals: Physics and Chemistry of Minerals, v. 23, p. 299–304, doi: 10.1007/BF00207777.

Rossman, G.R., and Aines, R.D., 1991, The hydrous components in garnets: Grossular–hydrogrossular: The American Mineralogist, v. 76, p. 1153–1164.

Rossman, G.R., and Smyth, J.R., 1990, Hydroxyl contents of accessory minerals in mantle eclogites and related rocks: The American Mineralogist, v. 75, p. 765–780.

Rossman, G.R., Beran, A., Langer, K., and Chopin, Ch., 1989, The hydrous component in pyrope from the Dora Maira Massif: Western Alps: European Journal of Mineralogy, v. 1, p. 151–154.

Rovetta, M.R., Holloway, J.R., and Blacic, J.D., 1986, Solubility of hydroxyl in natural quartz annealed in water at 900°C and 1.5 GPa: Geophysical Research Letters, v. 13, p. 145–148.

Rovetta, M.R., Blacic, J.D., Hervig, R.L., and Holloway, J.R., 1989, An experimental study of hydrohyl in quartz using infrared spectroscopy and ion microprobe techniques: Journal of Geophysical Research, v. 94, p. 5840–5850.

Rubey, W.W., 1951, Geologic history of sea water: An attempt to state the problem: Geological Society of America Bulletin, v. 62, p. 1111–1147.

Rudnick, R.L., and Fountain, D.M., 1995, Nature and composition of the continental crust: A lower crustal perspective: Reviews of Geophysics, v. 33, p. 267–309, doi: 10.1029/95RG01302.

Rüpke, L.H., Phipps Morgan, J., Hort, M., and Connolly, J.A.D., 2004, Serpentine and the subduction zone water cycle: Earth and Planetary Science Letters, v. 223, p. 17–34, doi: 10.1016/j.epsl.2004.04.018.

Ryabchikov, I.D., Schreyer, W., and Abraham, K., 1982, Composition of aqueous fluid in equilibrium with pyroxenes and olivines at mantle pressures and temperatures: Contributions to Mineralogy and Petrology, v. 79, p. 80–84, doi: 10.1007/BF00376964.

Sakamaki, T., Suzuki, A., and Ohtani, E., 2006, Stability of hydrous melt at the base of the Earth's upper mantle: Nature, v. 439, p. 192–194, doi: 10.1038/nature04352.

Sanehira, T., Irifune, T., Inoue, T., and Nishiyama, N., 2002, Synthesis of hydrous aluminous perovskite, in 43rd High Pressure Conference Japan Abstracts: Review of High Pressure Science and Technology, v. 159, p. 118–126.

Sano, A., Ohtani, E., Kubo, T., and Funakoshi, K., 2004, In situ X-ray observation of decomposition of hydrous aluminum silicate $AlSiO_3OH$ and aluminum oxide hydroxide δ-AlOOH at high pressure and temperature: Journal of Physics and Chemistry of Solids, v. 65, p. 1547–1554, doi: 10.1016/j.jpcs.2003.12.015.

Sano, A., Litasov, K., Ohtani, E., Kubo, T., Hosoya, T., Funakoshi, K., and Kikegawa, T., 2006, Effect of water on garnet-perovskite transformation in MORB and implications for penetrating slab into the lower mantle: Physics of the Earth and Planetary Interiors (in press).

Scambelluri, M., and Philippot, P., 2001, Deep fluids in subduction zones: Lithos, v. 55, p. 213–227, doi: 10.1016/S0024-4937(00)00046-3.

Scambelluri, M., and Rampone, E., 1999, Mg-metasomatism of oceanic gabbros and its control on Ti-clinohumite formation during eclogitization: Contributions to Mineralogy and Petrology, v. 135, p. 1–17, doi: 10.1007/s004100050494.

Scambelluri, M., Piccardo, G.B., Philippot, P., Robbiano, A., and Negretti, L., 1997, High salinity fluid inclusions formed from recycled seawater in deeply subducted alpine serpentinite: Earth and Planetary Science Letters, v. 148, p. 485–500, doi: 10.1016/S0012-821X(97)00043-5.

Schiano, P., and Clocchiatti, R., 1994, Worldwide occurrence of silica-rich melts in sub-continental and sub-oceanic mantle minerals: Nature, v. 368, p. 621–624, doi: 10.1038/368621a0.

Schiffman, P., and Liou, J.G., 1980, Synthesis and stability relations of Mg-Al pumpellyite, $Ca_4Al_5MgSi_6O_{21}(OH)_7$: Journal of Petrology, v. 21, p. 441–474.

Schmidt, M.W., 1995, Lawsonite: Upper pressure stability and formation of higher density hydrous phases: The American Mineralogist, v. 80, p. 1286–1292.

Schmidt, M.W., 1996, Experimental constraints on recycling potassium from subducted oceanic crust: Science, v. 272, p. 1927–1930.

Schmidt, M.W., and Poli, S., 1998, Experimental base water budgets for dehydrating slabs and consequences for arc magma generation: Earth and Planetary Science Letters, v. 163, p. 361–379, doi: 10.1016/S0012-821X(98)00142-3.

Schmidt, M.W., and Poli, S., 2003, Generation of mobile components during subduction of oceanic crust, in Holland, H.D., and Turekian, K.K., eds., Treatise on Geochemistry, Volume 3: Oxford, Elsevier Ltd., p. 567–591.

Schmidt, M.W., Finger, L.W., Angel, R.J., and Dinnbier, R.E., 1998, Synthesis, crystal structure, and phase relations of $AlSiO_3OH$, a high-pressure hydrous phase: The American Mineralogist, v. 83, p. 881–888.

Schneider, M.E., and Eggler, D.H., 1986, Fluids in equilibrium with peridotite minerals: Implications for mantle metasomatism: Geochimica et Cosmochimica Acta, v. 50, p. 711–724, doi: 10.1016/0016-7037(86)90347-9.

Schrauder, M., and Navon, O., 1994, Hydrous and carbonatitic mantle fluids in fibrous diamonds from Jwaneng, Botswana: Geochimica et Cosmochimica Acta, v. 58, p. 761–771, doi: 10.1016/0016-7037(94)90504-5.

Schrauder, M., Koeberl, C., and Navon, O., 1996, Trace element analyses of fluid-bearing diamonds from Jwaneng, Botswana: Geochimica et Cosmochimica Acta, v. 60, p. 4711–4724, doi: 10.1016/S0016-7037(96)00274-8.

Schreyer, W., Maresch, W.V., Medenbach, O., and Baller, T., 1986, Calcium-free pumpellyite, a new synthetic hydrous Mg-Al silicate formed at high pressures: Nature, v. 321, p. 510–511, doi: 10.1038/321510a0.

Sciurto, P.F., and Ottonello, G., 1995, Water-rock interaction on Zabargad Island, Red Sea—A case study: I. Application of the concept of local equilibrium: Geochimica et Cosmochimica Acta, v. 59, p. 2187–2206, doi: 10.1016/0016-7037(95)00100-E.

Sclar, C.B., 1970, High-pressure studies in the system $MgO-SiO_2-H_2O$: Physics of the Earth and Planetary Interiors, v. 3, p. 333, doi: 10.1016/0031-9201(70)90072-5.

Sclar, C.B., Carrison, L.C., and Schwartz, C.M., 1965, High pressure synthesis and stability of a new hydronium bearing layer silicate in the system $MgO-SiO_2-H_2O$: Eos (Transactions, American Geophysical Union), v. 46, p. 184.

Scott, H.P., and Williams, Q., 1999, An infrared spectroscopic study of lawsonite to 20 GPa: Physics and Chemistry of Minerals, v. 26, p. 437–445, doi: 10.1007/s002690050206.

Shen, A.H., and Keppler, H., 1997, Direct observation of complete miscibility in the albite-H_2O system: Nature, v. 385, p. 710–712, doi: 10.1038/385710a0.

Shieh, S.R., Mao, H.K., Hemley, R.J., and Ming, L.C., 1998, Decomposition of phase D in lower mantle and the fate of dense hydrous silicates in subducting slabs: Earth and Planetary Science Letters, v. 159, p. 13–23, doi: 10.1016/S0012-821X(98)00062-4.

Shieh, S.R., Mao, H.K., Hemley, R.J., and Ming, L.C., 2000, In situ X-ray diffraction studies of dense hydrous magnesium silicates at mantle conditions: Earth and Planetary Science Letters, v. 177, p. 69–80, doi: 10.1016/S0012-821X(00)00033-9.

Shimizu, K., Komiya, T., Hirose, K., Shimizu, N., and Maruyama, S., 2001, Cr-spinel, an excellent micro-container for retaining primitive melts—Implications for hydrous plume origin for komatiite: Earth and Planetary Science Letters, v. 189, p. 177–188, doi: 10.1016/S0012-821X(01)00359-4.

Shmulovich, K., Graham, C., and Yardley, B., 2001, Quartz, albite and diopside solubilities in H_2O–NaCl and H_2O–CO_2 fluids at 0.5–0.9 GPa: Contributions to Mineralogy and Petrology, v. 141, p. 95–108.

Simmons, N.A., and Gurrola, H., 2000, Multiple seismic discontinuities near the base of the transition zone in the Earth's mantle: Nature, v. 405, p. 559–562, doi: 10.1038/35014589.

Simons, K., Dixon, J., Schilling, J.-G., Kingsley, R., and Poreda, R., 2002, Volatiles in basaltic glasses from the Easter-Salas y Gomez Seamount Chain and Easter microplate: Implications for geochemical cycling of volatile elements: Geochemistry, Geophysics, Geosystems, v. 3, doi: 10.1029/2001GC000173.

Skogby, H., 1994, OH incorporation in synthetic clinopyroxene: The American Mineralogist, v. 79, p. 240–249.

Skogby, H., and Rossman, G.R., 1989, OH in pyroxenes: An experimental study of incorporation mechanisms and stability: The American Mineralogist, v. 74, p. 1059–1069.

Skogby, H., Bell, D., and Rossman, G., 1990, Hydroxide in pyroxene: Variations in the natural environment: The American Mineralogist, v. 75, p. 764–774.

Smith, D., 1979, Hydrous minerals and carbonates in peridotite inclusions from the Green Knobs and Buell Park kimberlitic diatremes on the Colorado Plateau, in Boyd, F.R., and Meyer, H.O.A., eds., The Mantle Sample: Inclusions in Kimberlites and Other Volcanics: Washington, D.C., American Geophysical Union, p. 345–356.

Smith, D., 1995, Chlorite-rich ultramafic reaction zones in Colorado Plateau xenoliths: Recorders of sub-Moho hydration: Contributions to Mineralogy and Petrology, v. 121, p. 185–200, doi: 10.1007/s004100050098.

Smith, J.V., and Dawson, J.B., 1981, Storage of F and Cl in the upper mantle: Geochemical implications: Lithos, v. 14, p. 133–147, doi: 10.1016/0024-4937(81)90050-5.

Smyth, J.R., 1987, Beta-Mg_2SiO_4: A potential host for water in the mantle?: The American Mineralogist, v. 72, p. 1051–1055.

Smyth, J.R., 1989, Electrostatic characterization of oxygen sites in minerals: Geochimica et Cosmochimica Acta, v. 53, p. 1101–1110, doi: 10.1016/0016-7037(89)90215-9.

Smyth, J.R., 1994, A crystallographic model for hydrous wadsleyite: An ocean the Earth's interior?: The American Mineralogist, v. 79, p. 1021–1025.

Smyth, J.R., and Frost, D.J., 2002, The effect of water on the 410-km discontinuity: An experimental study: Geophysical Research Letters, v. 29, doi: 10.1029/2001GL014418.

Smyth, J.R., and Hatton, C.J., 1977, A coesite-sanidine grospydite from the Roberts-Victor kimberlite: Earth and Planetary Science Letters, v. 34, p. 284–290, doi: 10.1016/0012-821X(77)90012-7.

Smyth, J.R., and Kawamoto, T., 1997, Wadsleyite II: A new high pressure hydrous phase in the peridotite-H_2O system: Earth and Planetary Science Letters, v. 146, p. E9–E16, doi: 10.1016/S0012-821X(96)00230-0.

Smyth, J.R., and McCormick, T.C., 1995, Crystallographic data for minerals, in Ahrens, T.J., ed., Mineral Physics and Crystallography, Volume 2: A Handbook of Physical Constants: Washington, D.C., American Geophysical Union, p. 1–17.

Smyth, J.R., Bell, D., and Rossman, G., 1991, Incorporation of hydroxyl in upper-mantle clinopyroxenes: Nature, v. 351, p. 732–735, doi: 10.1038/351732a0.

Smyth, J.R., Swope, R.J., and Pawley, A.J., 1995, H in rutile-type compounds: II. Crystal chemistry of Al substitution in H-bearing stishovite: The American Mineralogist, v. 80, p. 454–456.

Smyth, J.R., Kawamoto, T., Jacobsen, S.D., Swope, R.J., Hervig, R.L., and Holloway, J.R., 1997, Crystal structure of monoclinic hydrous wadsleyite: The American Mineralogist, v. 82, p. 270–275.

Smyth, J.R., Holl, C.M., Frost, D.J., Jacobsen, S.D., Langenhorst, F., and McCammon, C.A., 2003, Structural systematics of hydrous ringwoodite and water in the Earth's interior: The American Mineralogist, v. 88, p. 1402–1407.

Smyth, J.R., Holl, C.M., Langenhorst, F., Laustsen, H.M.S., Rossman, G.R., Kleppe, A., McCammon, C.A., Kawamoto, T., and van Aken, P.A., 2005, Crystal chemistry of wadsleyite II and water in the Earth's interior: Physics and Chemistry of Minerals, v. 31, p. 691–705, doi: 10.1007/s00269-004-0431-x.

Smyth, J.R., Frost, D.J., Nestola, F., Holl, C.M., and Bromiley, G., 2006, Olivine hydration in the deep upper mantle: effect of temperature and silica activity: Geophysical Research Letters, v. 33, L15301, doi:10.1029/2006GL026194.

Sobolev, A.V., and Chaussidon, M., 1996, H_2O concentrations in primary melts from supra-subduction zones and mid-ocean ridges: Implications for H_2O storage and recycling in the mantle: Earth and Planetary Science Letters, v. 137, p. 45–55, doi: 10.1016/0012-821X(95)00203-O.

Sobolev, N.V., and Shatsky, V.S., 1990, Diamond inclusions in garnets from metamorphic rocks—A new environment for diamond formation: Nature, v. 343, p. 742–746, doi: 10.1038/343742a0.

Sokol, A.G., Pal'yanov, Yu.N., Pal'yanova, G.A., Khokhryakov, A.F., and Borzdov, Yu.M., 2001, Diamond and graphite crystallization from COH fluids under high pressure and high temperature conditions: Diamond and Related Material, v. 10, p. 2131–2136, doi: 10.1016/S0925-9635(01)00491-5.

Song, T.R., Helmberger, D.V., and Grand, S.P., 2004, Low-velocity zone atop the 410-km seismic discontinuity in the northwestern United States: Nature, v. 427, p. 530–533, doi: 10.1038/nature02231.

Stalder, R., 2004, Comment on K. Mibe, T. Fujii, and A. Yasuda (2002), "Composition of aqueous fluid coexisting with mantle minerals at high pressure and its bearing on the differentiation of the Earth's mantle" (Geochimica et Cosmochimica Acta, v. 66, p. 2273–2285): Geochimica et Cosmochimica Acta, v. 68, p. 927–928, doi: 10.1016/j.gca.2003.01.001.

Stalder, R., and Ulmer, P., 2001, Phase relations of a serpentine composition between 5 and 14 GPa: Significance of clinohumite and phase E as water carriers into the transition zone: Contributions to Mineralogy and Petrology, v. 140, p. 670–679.

Stalder, R., Ulmer, P., Thompson, A.B., and Günther, D., 2000, Experimental approach to constrain second critical end points in fluid/silicate systems: Near-solidus fluids and melts in the system albite-H_2O: The American Mineralogist, v. 85, p. 68–77.

Stalder, R., Ulmer, P., Thompson, A.B., and Gunther, D., 2001, High pressure fluids in the system MgO-SiO_2-H_2O under upper mantle conditions: Contributions to Mineralogy and Petrology, v. 140, p. 607–618.

Staudigel, J., Plank, T., White, B., and Schmincke, H.U., 1996, Geochemical fluxes during seafloor alteration of the basaltic upper oceanic crust: DSDP Sites 417 and 418, in Bebout, G.E., Scholl, D.W., Kirby, S.H., and Platt, J.P., eds., Subduction Top to Bottom: American Geophysical Union Geophysical Monograph 96, p. 19–38.

Stern, R.J., 2002, Subduction zones: Reviews of Geophysics, v. 40, p. 3-1–3-38.

Stevenson, D.J., 2004, Inside history in depth: Nature, v. 428, p. 476–477, doi: 10.1038/428476a.

Stone, W.E., Deloule, E., Larson, M.S., and Lesher, C.M., 1997, Evidence for hydrous high-Mg melts in the Precambrian: Geology, v. 25, p. 143–146, doi: 10.1130/0091-7613(1997)025<0143:EFHHMM>2.3.CO;2.

Stracke, A., Hofmann, A.W., and Hart, S.R., 2005, FOZO, HIMU, and the rest of the mantle zoo: Geochemistry, Geophysics, Geosystems, v.6, Q05007, doi: 10.1029/2004GC000824.

Sudo, A., and Tatsumi, Y., 1990, Phlogopite and K-amphibole in the upper mantle: Implication for magma genesis in subduction zones: Geophysical Research Letters, v. 17, p. 29–32.

Suzuki, A., Ohtani, E., and Kamada, T., 2000a, A new hydrous phase δ-AlOOH synthesized at 21 GPa and 1000°C: Physics and Chemistry of Minerals, v. 27, p. 689–693, doi: 10.1007/s002690000120.

Suzuki, A., Ohtani, E., Morishima, H., Kubo, T., Kanbe, Y., Kondo, T., Okada, T., Terasaki, H., Kato, T., and Kikegawa, T., 2000b, In situ determination of the phase boundary between wadsleyite and ringwoodite in Mg_2SiO_4: Geophysical Research Letters, v. 27, p. 803–806, doi: 10.1029/1999GL008425.

Sweeney, R., 1997, The role of hydrogen in geological processes in the Earth's interior: Solid State Ionics, v. 97, p. 393–397, doi: 10.1016/S0167-2738(97)00040-4.

Swope, R.J., Smyth, J.R., and Larson, A.C., 1995, H in rutile-type compounds: I. Single-crystal neutron and X-ray diffraction study of H in rutile: The American Mineralogist, v. 80, p. 448–453.

Takahashi, E., 1986, Melting of a dry peridotite KLB-1 up to 14 GPa: Implications on the origin of peridotitic upper mantle: Journal of Geophysical Research, v. 91, p. 9367–9382.

Tatsumi, Y., 1989, Migration of fluid phase and genesis of basalt magmas in subduction zones: Journal of Geophysical Research, v. 94, p. 4697–4707.

Taylor, S.R., and McLennan, S.M., 1985, The Continental Crust: Its Composition and Evolution: Oxford, Blackwell Science Publication, 312 p.

Thompson, A.B., 1992, Water in the Earth's mantle: Nature, v. 358, p. 295–302, doi: 10.1038/358295a0.

Thompson, J.B., Laird, J., and Thompson, A.B., 1982, Reactions in amphibolite, greenschist and blueschist: Journal of Petrology, v. 23, p. 1–17.

Trønnes, R.G., 2002, Stability range and decomposition of potassic richterite and phlogopite end members at 5–15 GPa: Mineralogy and Petrology, v. 74, p. 129–148, doi: 10.1007/s007100200001.

Turner, G., Burgess, R., and Bannon, M., 1990, Volatile-rich mantle fluids inferred from inclusions in diamond and mantle xenoliths: Nature, v. 344, p. 653–655, doi: 10.1038/344653a0.

Turner, S., and Hawkesworth, C., 1995, The nature of the subcontinental mantle: Constraints from the major-element composition of continental flood basalts: Chemical Geology, v. 120, p. 295–314, doi: 10.1016/0009-2541(94)00143-V.

Ulmer, P., and Trommsdorff, V., 1995, Serpentine stability to mantle depths and subduction related magmatism: Science, v. 268, p. 858–861.

Ulmer, P., and Trommsdorff, V., 1999, Phase relations of hydrous mantle subducting to 300 km, in Fei, Y., Bertka, C.M., and Mysen B.O., eds., Mantle Petrology: Field Observations and High Pressure Experimentation: A Tribute to F.R. Boyd: The Geochemical Society Special Publication, v. 6, p. 259–281.

Van der Hilst, R.D., Widiyantoro, S., Creager, K.C., and McSweeney, T., 1998, Deep subduction and aspherical variations in P-wave speed at the base of Earth's mantle, in Girnis, M., Wysselton, M.E., Knittle, E., and Buffett, B.A., eds., Observational and Theoretical Constraints on the Core Mantle Boundary Region: American Geophysical Union Geodynamics Series 28, p. 5–20.

Vielzeuf, D., and Schmidt, M.W., 2001, Melting relations in hydrous systems revisited: Applications to pelites, greywackes and basalts: Contributions to Mineralogy and Petrology, v. 141, p. 251–267.

Vlassopoulos, D., Rossman, G.R., and Haggerty, S.E., 1993, Coupled substitution of H and minor elements in rutile and implications of high OH contents in Nb-Cr rutile from the upper mantle: The American Mineralogist, v. 78, p. 1181–1191.

Von Huene, R., and Scholl, D.W., 1991, Observations at convergent margins concerning sediment subduction, subduction erosion, and the growth of continental crust: Reviews of Geophysics, v. 29, p. 279–316.

Walker, J.A., Roggensack, K., Patino, L.C., Cameron, B.I., and Otoniel, M., 2003, The water and trace element contents of melt inclusions across an active subduction zone: Contributions to Mineralogy and Petrology, v. 146, p. 62–77, doi: 10.1007/s00410-003-0482-x.

Wallace, M.E., and Green, D.H., 1991, The effect of bulk rock composition on the stability of amphibole in the upper mantle: Implications for solidus positions and mantle metasomatism: Mineralogy and Petrology, v. 44, p. 1–19, doi: 10.1007/BF01167097.

Wang, L., Zhang, Y., and Essene, E.J., 1996, Diffusion of the hydrous component in pyrope: The American Mineralogist, v. 81, p. 706–718.

Watson, E.B., and Lupulescu, A.A., 1993, Aqueous fluids connectivity and chemical transport in clinopyroxene-rich rocks: Earth and Planetary Science Letters, v. 117, p. 279–294, doi: 10.1016/0012-821X(93)90133-T.

Watson, E.B., Brenan, J.M., and Baker, D.R., 1990, Distribution of fluids in the continental mantle, in Menzies, M., ed., Continental Mantle: New York, Oxford University Press, p. 111–125.

Wiens, D.A., and Smith, G.P., 2003, Seismological constraints on structure and flow patterns within the mantle wedge, in Eiler, J., ed., Inside the Subduction Factory: American Geophysical Union Geophysical Monograph 138, p. 59–82.

Wilkins, R.W., and Sabine, W., 1973, Water content of some nominally anhydrous silicates: American Mineralogy, v. 58, p. 508–516.

Williams, Q., and Hemley, R.J., 2001, Hydrogen in the deep Earth: Annual Review of Earth and Planetary Science, v. 29, p. 365–418, doi: 10.1146/annurev.earth.29.1.365.

Wilson, A.H., Shirey, S.B., and Carlson, R.W., 2003, Archaean ultra-depleted komatiites formed by hydrous melting of cratonic mantle: Nature, v. 423, p. 858–861, doi: 10.1038/nature01701.

Withers, A., Wood, B., and Carroll, M., 1998, The OH content of pyrope at high pressure: Chemical Geology, v. 147, p. 161–171, doi: 10.1016/S0009-2541(97)00179-4.

Wood, B.J., 1995, The effect of H_2O on the 410-kilometer seismic discontinuity: Science, v. 268, p. 74–76.

Wood, B.J., 2000, Phase transformations and partitioning relations in peridotite under lower mantle conditions: Earth and Planetary Science Letters, v. 174, p. 341–354, doi: 10.1016/S0012-821X(99)00273-3.

Woodhead, J.A., Rossman, G.R., and Thomas, A.P., 1991, Hydrous species in zircon: The American Mineralogist, v. 76, p. 1533–1546.

Wright, K., and Catlow, C.R.A., 1994, A computer simulation study of OH defects in olivine: Physics and Chemistry of Minerals, v. 20, p. 515–518.

Wunder, B., 1998, Equilibrium experiments in the system $MgO-SiO_2-H_2O$ (MSH): Stability fields of clinohumite-OH [$Mg_9Si_4O_{16}(OH)_2$], chondrodite-OH [$Mg_5Si_2O_8(OH)_2$] and phase A [$Mg_7Si_2O_8(OH)_6$]: Contributions to Mineralogy and Petrology, v. 132, p. 111–120, doi: 10.1007/s004100050410.

Wunder, B., and Schreyer, W., 1997, Antigorite: High-pressure stability in the system $MgO-SiO_2-H_2O$ (MSH): Lithos, v. 41, p. 213–227, doi: 10.1016/S0024-4937(97)82013-0.

Wunder, B., Medenbach, O., Krause, W., and Schreyer, W., 1993a, Synthesis, properties and stability of $Al_3Si_2O_7(OH)_3$ (phase Pi), a hydrous high-pressure phase in the system $Al_2O_3-SiO_2-H_2O$ (ASH): European Journal of Mineralogy, v. 5, p. 637–649.

Wunder, B., Rubie, D.C., Ross, C., II, Medenbach, O., Seifert, F., and Schreyer, W., 1993b, Synthesis, stability, and properties of $Al_2SiO_4(OH)_2$: A fully hydrated analogue of topaz: The American Mineralogist, v. 78, p. 285–297.

Wunder, B., Medenbach, O., Daniels, P., and Schreyer, W., 1995, First synthesis of the hydroxyl end-member of humite, $Mg_7Si_3O_{12}(OH)_2$: The American Mineralogist, v. 80, p. 638–640.

Wunder, B., Andrut, M., and Wirth, R., 1999, High-pressure synthesis and properties of OH-rich topaz: European Journal of Mineralogy, v. 11, p. 803–813.

Wyllie, P.J., and Ryabchikov, I.D., 2000, Volatile components, magmas, and critical fluids in upwelling mantle: Journal of Petrology, v. 41, p. 1195–1206, doi: 10.1093/petrology/41.7.1195.

Xia, Q.K., Sheng, Y.M., Yang, X.Z., and Yu, H.M., 2005, Heterogeneity of water in garnet from UHP eclogites, eastern Dabieshan, China: Chemical Geology, v. 224, p. 237–246, doi: 10.1016/j.chemgeo.2005.08.003.

Yagi, A., Suzuki, T., and Akaogi, M., 1994, High pressure transitions in the system $KAlSi_3O_8-NaAlSi_3O_8$: Physics and Chemistry of Minerals, v. 21, p. 12–17, doi: 10.1007/BF00205210.

Yamada, A., Inoue, T., and Irifune, T., 2004, Melting of enstatite from 13 to 18 GPa under hydrous conditions: Physics of the Earth and Planetary Interiors, v. 147, p. 45–56, doi: 10.1016/j.pepi.2004.05.005.

Yamamoto, K., and Akimoto, S., 1977, The system $MgO-SiO_2-H_2O$ at high pressures and temperatures: Stability field of hydroxyl-chondrodite, hydroxyl-clinohumite and 10 Å-phase: American Journal of Science, v. 277, p. 288–312.

Yang, H., Prewitt, C.T., and Frost, D.J., 1997, Crystal structure of the dense hydrous magnesium silicate, phase D: The American Mineralogist, v. 82, p. 651–654.

Yang, H., Konzett, J., and Prewitt, C.T., 2001, Crystal structure of phase X, a high pressure alkali-rich hydrous silicate and its anhydrous equivalent: The American Mineralogist, v. 86, p. 1483–1488.

Yasuda, A., Fujii, T., and Kurita, K., 1994, Melting phase relations of anhydrous mid-oceanic ridge basalts from 3 to 20 GPa: Implications for the behavior of subducted oceanic crust in the mantle: Journal of Geophysical Research, v. 99, p. 9401–9414, doi: 10.1029/93JB03205.

Yesinowski, J.P., Eckert, H., and Rossman, G.R., 1988, Characterization of hydrous species by high speed ^1H MAS NMR: Journal of the American Chemical Society, v. 110, p. 1367–1375, doi: 10.1021/ja00213a007.

Yoder, H.S., 1952, The $MgO-Al_2O_3-SiO_2-H_2O$ system and the related metamorphic facies: American Journal of Science, v. 250A, p. 569–627.

Young, T.E., Green, H.W., II, Hofmeister, A.M., and Walker, D., 1993, Infrared spectroscopic investigation of hydrohyl in $\beta-(Mg,Fe)_2SiO_4$ and coexisting olivine: Implications for mantle evolution and dynamics: Physics and Chemistry of Minerals, v. 19, p. 409–422, doi: 10.1007/BF00202978.

Yurimoto, H., Kurosawa, M., and Sueno, S., 1989, Hydrogen analysis in quartz crystals and quartz glasses by secondary ion mass spectrometry: Geochimica et Cosmochimica Acta, v. 53, p. 751–755, doi: 10.1016/0016-7037(89)90018-5.

Zhang, J., and Herzberg, C., 1994, Melting experiments on anhydrous peridotite KLB-1 from 5.0 to 22.5 GPa: Journal of Geophysical Research, v. 99, p. 17,729–17,742, doi: 10.1029/94JB01406.

Zhang, J., Li, B., Utsumi, W., and Liebermann, R.C., 1996, In situ X-ray observations of the coesite-stishovite transition: Reversed phase boundary and kinetics: Physics and Chemistry of Minerals, v. 23, p. 1–10, doi: 10.1007/BF00202987.

Zhang, R.Y., Liou, J.G., and Gong, B.L., 1995, Ultrahigh-pressure metamorphosed talc-, magnesite-, and Ti-clinohumite-bearing mafic-ultramafic complex, Dabie Mountains, east-central China: Journal of Petrology, v. 36, p. 1011–1037.

Zhao, D., 2001, Seismological structure of subduction zones and its implications for arc magmatism and dynamics: Physics of the Earth and Planetary Interiors, v. 127, p. 197–214, doi: 10.1016/S0031-9201(01)00228-X.

Zhao, D., 2004, Global tomographic images of mantle plumes and subducting slabs: Insight into deep Earth dynamics: Physics of the Earth and Planetary Interiors, v. 146, p. 3–34, doi: 10.1016/j.pepi.2003.07.032.

Zhao, D., Xu, Y., Wiens, D., Dorman, L., Hildebrand, J., and Webb, S., 1997, Depth extent of the Lau back-arc spreading center and its relation to subduction processes: Science, v. 278, p. 254–257, doi: 10.1126/science.278.5336.254.

Zhao, Y.H., Ginsberg, S.G., and Kohlstedt, D.L., 2004, Solubility of hydrogen in olivine: Dependence of temperature and Fe content: Contributions to Mineralogy and Petrology, v. 147, p. 155–161, doi: 10.1007/s00410-003-0524-4.

MANUSCRIPT ACCEPTED BY THE SOCIETY 14 AUGUST 2006

Printed in the USA

Geophysical applications of nuclear resonant spectroscopy

Wolfgang Sturhahn
Jennifer M. Jackson[†]
Advanced Photon Source, Argonne National Laboratory, 9700 South Cass Avenue, Argonne, Illinois 60439, USA

ABSTRACT

We summarize recent developments of nuclear resonant spectroscopy methods, such as nuclear resonant inelastic X-ray scattering and synchrotron Mössbauer spectroscopy, and their uses for the geophysical sciences. The inelastic method provides specific vibrational information, for example, the phonon density of states, and, in combination with compression data, it permits the determination of sound velocities and Grüneisen parameters under high pressure and high temperature. The Mössbauer method provides hyperfine interactions between the resonant nucleus and electronic environment, such as isomer shifts, quadrupole splittings, and magnetic fields, which provide important information on valence, spin state, and magnetic ordering. Both methods use a nuclear resonant isotope as a probe and can be applied under high pressure and high temperature. The physical mechanism of nuclear resonant scattering and the specifics in applications to Earth materials are presented with reference to several high-pressure studies on iron-bearing compounds.

Keywords: phonon density of states, sound velocities, Grüneisen parameter, magnetism, nuclear resonant scattering.

INTRODUCTION

The chemical composition, the elastic and transport properties, and the thermodynamic parameters of materials identified or expected in Earth's deep interior are of general importance to geochemical modeling (Tolstikhin and Hofmann, 2005; Coltice and Ricard, 1999), geodynamic simulation (Nakagawa and Tackley, 2005), and interpretation of seismic-wave observations (Birch, 1952; Kellogg et al., 1999; van der Hilst and Kàrason, 1999; Ishii and Tromp, 1999; Ni et al., 2002; Trampert et al., 2004; Mattern et al., 2005). Inferences about the mineralogy and chemical composition of Earth's deep interior must depend on comparisons of accurate, laboratory-derived properties of candidate phases with geophysical observations because our ability to directly sample deep Earth is severely limited. Typically, models of deep Earth have been made by extrapolating mineral properties to appropriate pressure-temperature (P-T) conditions and comparing them with seismologically determined properties, such as sound velocities and density. Many extrapolations use either low P-T data, infer that the chemically complex minerals behave the same as their Mg end-members, use analogue materials, and/or neglect the behavior of the shear properties (due in part to lack of experimental data). Although useful, these assumptions may not accurately reflect the behavior or chemistries of the actual components deep in Earth. Recent advances in experimental methods and theoretical calculations show that elements such as iron, aluminum,

---

[†]Present address: Seismological Laboratory, Division of Geological and Planetary Sciences, California Institute of Technology, Pasadena, California, USA.

Sturhahn, W., and Jackson, J.M., 2007, Geophysical applications of nuclear resonant spectroscopy, *in* Ohtani, E., ed., Advances in High-Pressure Mineralogy: Geological Society of America Special Paper 421, p. 157–174, doi: 10.1130/2007.2421(09). For permission to copy, contact editing@geosociety.org. ©2007 Geological Society of America. All rights reserved.

and calcium have appreciable effects on the shear properties of lower-mantle silicates and oxides (Jacobsen et al., 2002; Kung et al., 2002; Karki and Crain, 1998; Kiefer et al., 2002; Jackson et al., 2004; Jacobsen et al., 2004; Jackson et al., 2005b) and therefore alter our interpretations of geophysical observations. The addition of light elements such as silicon, oxygen, sulfur, and hydrogen to iron also has significant effects on the shear properties (Machova and Kadeckova, 1977; Struzhkin et al., 2001; Lin et al., 2003, 2004a; Mao et al., 2004b; Jacobsen et al., 2004). Several experiments have been performed to understand the crystal chemistry of these complex systems (Irifune, 1994; Kesson et al., 1995; Wood and Rubie, 1996; Hirose et al., 1999; Frost and Langenhorst, 2002), and they have shown that even though most of the iron in mantle materials is divalent (Fe^{2+}), a significant amount of trivalent iron (Fe^{3+}) was determined for aluminum-bearing $(Mg,Fe)SiO_3$-perovskite (McCammon, 1997). Related to these effects are the possible occurrences of high-spin to low-spin crossovers in the iron component of lower-mantle phases (Badro et al., 2003, 2004; Li et al., 2004; Jackson et al., 2005a; Lin et al., 2005a). Theoretical predictions indicate that material properties influenced by the spin state of iron change smoothly along the geotherm (Sturhahn et al., 2005), but, owing to the possible significance of these electronic changes to our interpretation of geophysical observations, more experiments need to be done to clarify the nature of these changes at typical P-T conditions. Even less is known about the recently discovered post-perovskite phase (e.g., Murakami et al., 2004). The effect of iron (Mao et al., 2004a) and aluminum (Akber-Knutson et al., 2005; Caracas and Cohen, 2005) on the perovskite to post-perovskite transition pressure has been studied experimentally and theoretically, respectively, but still little is known about the physical and chemical properties of post-perovskite. It is necessary to study the elastic parameters (especially the shear properties) of deep Earth materials, the valence state of iron in these materials, and the elemental partitioning under appropriate P-T conditions to constrain the chemistry and composition of this region. Many of these unknowns could be determined using nuclear resonant scattering methods under simultaneous high-pressure and high-temperature conditions.

The environmental conditions of planetary interiors often require challenging studies of materials under pressures exceeding 100 GPa and high temperatures. The combination of synchrotron radiation techniques with high-pressure studies has been very successful over the last decades to address these challenges. In particular, the brightness of third-generation synchrotron radiation facilities has permitted scientists to reach ever higher pressures with increasingly small samples. Focusing optics have been refined to concentrate a significant portion of the X-ray beam into an area of less than 10×10 μm^2 size, and, in combination with efficient high-resolution monochromators, the application of inelastic X-ray scattering and nuclear resonant scattering methods to high-pressure problems has become feasible (see the overview by Hemley et al., 2005). Among nuclear resonant scattering techniques, practical importance has been achieved by nuclear resonant inelastic X-ray scattering (NRIXS) for the study of lattice dynamics and by synchrotron Mössbauer spectroscopy (SMS) for the study of magnetism and valence states. In brief, SMS includes recoilless scattering processes, i.e., without participation of lattice vibrations, whereas NRIXS uses the possibility of simultaneous excitation of nuclear resonance and lattice vibrations. A comprehensive overview of the field has been given in a collection of review articles (Gerdau and de Waard, 1999), and also in more recent reviews on nuclear resonant spectroscopy (Alp et al., 2001; Sturhahn, 2004; Scheidt et al., 2005). The NRIXS and SMS methods can be applied to materials that contain a nuclear resonant isotope, such as ^{57}Fe for geophysical studies.

In this chapter, we will first briefly explain the basics of nuclear resonances, as well as some experimental aspects of NRIXS and SMS measurements. Following the experimental section, we will discuss the relevant geophysical parameters that can be determined using nuclear resonant scattering, such as sound velocities, Grüneisen parameters, temperature, magnetism, valence, and spin state.

BASICS ON NUCLEAR RESONANCES

Standard text books on experimental techniques with synchrotron radiation discuss mechanisms that are based on the scattering of X-rays by electronic charge and spin. An argument from classical electrodynamics usually provides the justification to ignore the nuclear charge. The Thomson-scattering cross section for electromagnetic radiation by a point charge q with mass m is given by

$$\sigma_T = \frac{8\pi}{3}\left(\frac{q^2}{mc^2}\right)^2. \qquad (1)$$

The scattering strength of the nucleus compared to the electron shell is then reduced by a factor $(Z m/M)^2 \approx 10^{-7}$, where m and M are the masses of the electron and nucleus, respectively, and Z is the atomic number. In this long-wavelength limit, which is justified for X-rays, the nuclear charge scattering is reduced by orders of magnitude because of the large nuclear mass, and it seems completely appropriate to ignore scattering contributions from the nucleus. However, this argument fails if the time scale of internal nuclear dynamics matches the energy of the X-ray photon, i.e., the nucleus experiences a resonant excitation. The nuclear resonant cross section is then calculated as

$$\sigma_N = \frac{\lambda^2}{2\pi}\frac{1}{1+\alpha}\frac{2I'+1}{2I+1}, \qquad (2)$$

where λ is the wavelength of the resonant X-rays, α is the internal conversion coefficient, and I and I' are the spins of the nuclear ground and excited state, respectively. The probability that an excited nucleus directly transfers its excess energy to the electron shell followed by expulsion of an inner electron is given by $\alpha/(1+\alpha)$. Values of α for relevant nuclear transitions range from 1 to 1000. The nuclear resonant cross section can become very large, e.g., using values of $\lambda = 86$ pm (1 pm = 10^{-12}m), $\alpha = 8.6$, $I = 1/2$, and $I' = 3/2$, the 14.4 keV nuclear transition of ^{57}Fe gives a value of $\sigma_N = 2.56$ Mbarn (1 barn = 10^{-28}m^2) and a ratio $\sigma_N/\sigma_T \approx 5700$. It should be noted that the photoelectric cross section σ_{pe} often exceeds σ_T, but we still observe $\sigma_N/\sigma_{pe} \approx 450$. Even though the nuclear resonant cross section is very large, the energy width of such resonances is very narrow. The weakness of the nuclear coupling to its electronic surroundings and to the electromagnetic field results in a weakly damped resonance of high quality, e.g., the energy width of the ^{57}Fe resonance is only $\Gamma = 4.66$ neV. Such extraordinarily narrow resonances escape traditional X-ray scattering methods because the best energy resolutions of X-ray optics are in the 100 μeV ($\approx$24 GHz) regime (Toellner et al., 2001; Yabashi et al., 2001). A resonant enhancement over a neV scale remains unresolved and unnoticed because even an experiment with bandwidth $\delta E \approx 100$ μeV would find $\sigma_N \Gamma \ll \sigma_{pe} \delta E$.

Nuclear resonant scattering techniques are nevertheless possible in light of the inverse relationship between lifetime and level width of the nuclear state, $\tau \Gamma = \hbar$. The value of τ determines the time scale on which a sample containing resonant nuclei would respond to an excitation by a synchrotron radiation pulse. Whereas values for τ are in the nanosecond to microsecond range, the duration of a synchrotron radiation pulse is typically less than 100 ps, and the electronic scattering of X-rays occurs typically on the time scale of femtoseconds, virtually immediately compared to nuclear resonant contributions. Detectors with time resolutions of ns or better can exploit this mismatch with a "time-discrimination trick" as illustrated in Figure 1, and we understand how nuclear resonant signals are cleanly separated from other scattering contributions. In Figure 2, we show the collection of nuclear isotopes that possess resonances below 150 keV. The higher transition energies are less favorable because the intensity of synchrotron radiation sources typically decreases with increasing X-ray energy, and X-ray optics, as well as detectors, become less efficient at higher energies. The time resolution of X-ray detectors (presently around 1 ns) limits the feasibility of short-lived isotopes, whereas very long lifetimes lead to low signal rates and to difficulties with the storage ring operations.

EXPERIMENTAL METHODS

The pioneering experimental work on nuclear resonant scattering with synchrotron radiation (Gerdau et al., 1985) strongly suggested the utilization of new time-resolved techniques instead of the energy-resolved measurements of conventional Mössbauer spectroscopy. This novel approach resulted from the particular property of synchrotron radiation, which is emitted as a sequence of very short X-ray pulses of typically less than 100 ps duration. The time-decay pattern of X-rays scattered off or transmitted through samples containing a suitable nuclear-resonant iso-

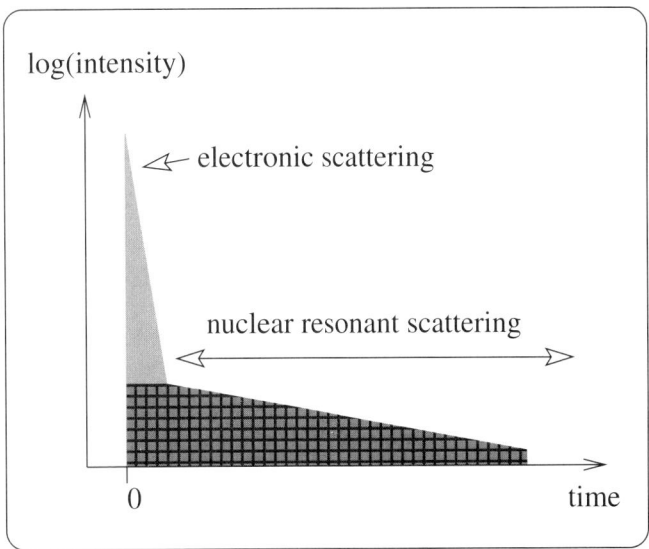

Figure 1. Scattering intensity versus time. At zero time, a synchrotron radiation pulse excites a sample containing a nuclear resonant isotope. Electronic scattering is prompt, whereas the response of the resonant nuclei is delayed. Time discrimination is the key to distinguishing nuclear and electronic scattering.

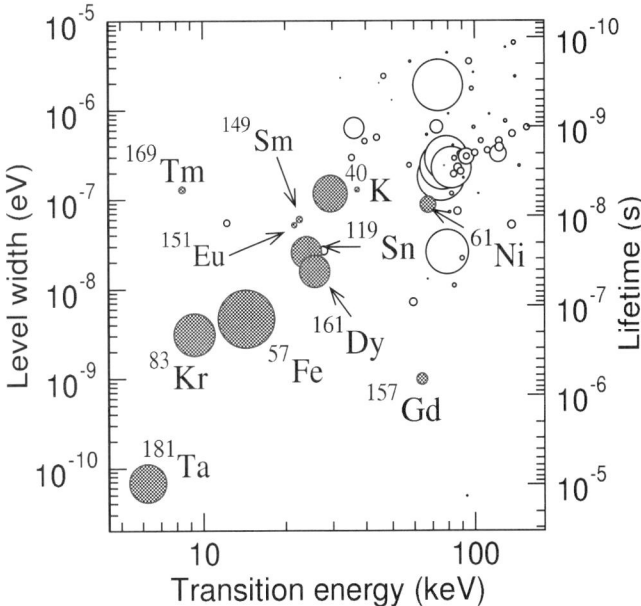

Figure 2. Level widths and lifetimes of nuclear resonances below 150 keV. The symbol size is proportional to the nuclear resonant cross section. Isotopes that have been used in nuclear resonant scattering experiments are identified.

tope is equivalent to an energy spectroscopy in the μeV to neV (GHz to MHz) range. The measurement and analysis of such time spectra constitute the main part of the SMS method. During the last decade, we saw the refinement of the SMS technique and its application in the area of high-pressure research (Nasu, 1994, 1998; Zhang et al., 1999; Pleines et al., 1999; Lübbers et al., 1999b; Lin et al., 2004a; Barla et al., 2004a, 2004b, 2005a, 2005b; Jackson et al., 2005a).

Conventional Mössbauer spectroscopy as well as SMS are based on the often appreciable probability that the resonant nuclei absorb X-rays without participation of lattice vibrations, i.e., recoilless. The exclusion of phonons from the absorption process leads to the extremely high resolving power of these methods. However, theoretical calculations also predicted the exciting opportunity to observe the phonon density of states (DOS) via nuclear resonant excitation (Visscher, 1960; Singwi and Sjölander, 1960). Demonstration experiments using strong radioactive sources that were tuned by large Doppler shifts (Weiss and Langhoff, 1979; Endres et al., 1981) were limited in practice by the notoriously small inelastic absorption cross section and overwhelmingly large background signals. Only recently was the idea of using nuclear resonances to observe phonon excitations revived in synchrotron radiation experiments (Seto et al., 1995; Sturhahn et al., 1995; Chumakov et al., 1995), and the extraction of the phonon DOS was demonstrated (Sturhahn et al., 1995). The use of pulsed synchrotron radiation and a time-discrimination technique circumvented the background problem, and the new NRIXS method started to produce unique results.

The phonon DOS is an important quantity to describe the low-energy collective states of solids and is used to calculate thermodynamic properties related to lattice vibrations. Experimentally, the phonon DOS is often obtained indirectly, e.g., by calculation from the phonon dispersion relations measured on single crystals using inelastic neutron scattering. The NRIXS technique provides direct access to the partial and projected phonon DOS of the resonant isotope only (Kohn et al., 1998; Sturhahn and Kohn, 1999). The apparent disadvantage of results restricted to vibrational information from a few nuclear resonant isotopes has in turn led to several unique applications. The most suitable nuclear resonant isotope is ^{57}Fe (Fig. 2), and iron-containing materials, molecules, and proteins are of tremendous interest in geophysics, thin-film research, and biophysics. NRIXS signals originate from particular resonant nuclei only, and this complete isotope selectivity is truly unique among techniques for the study of lattice vibrations. For example, materials surrounding the sample that do not contain resonant nuclei produce no unwanted background, and this feature now permits experiments under extreme pressure-temperature conditions that were impossible before (Lübbers et al., 2000a; Mao et al., 2001, 2004b; Struzhkin et al., 2001; Lin et al., 2003, 2004a, 2004b, 2005b; Shen et al., 2004; Papandrew et al., 2004; Kobayashi et al., 2004; Zhao et al., 2004). Details on the scattering mechanisms and methodology for NRIXS and SMS have been published elsewhere (Chumakov and Sturhahn, 1999; Sturhahn, 2004).

A schematic of the typical setup for nuclear resonant scattering experiments found at third-generation synchrotron radiation facilities is shown in Figure 3. The X-ray source consists of electron bunches that are orbiting in the storage ring and periodically pass through an undulator. The X-rays are monochromatized in two steps using a premonochromator and a high-resolution monochromator to an energy bandwidth of ~1 meV or 0.24 THz (Toellner, 2000).

NRIXS

For NRIXS measurements, the energy bandwidth of the incident X-rays determines the resolution of the phonon spectra of the samples. The high-resolution monochromator is tuned around the nuclear transition energy, and the X-rays excite the resonant nuclei in the sample. The incoherently re-emitted radiation is observed with an avalanche photodiode detector that is placed as close as possible to the sample but away from any strong coherent scattering directions (Fig. 3). The integrated delayed counting rate is recorded. The NRIXS method directly provides the Fourier-transformed self-intermediate scattering function

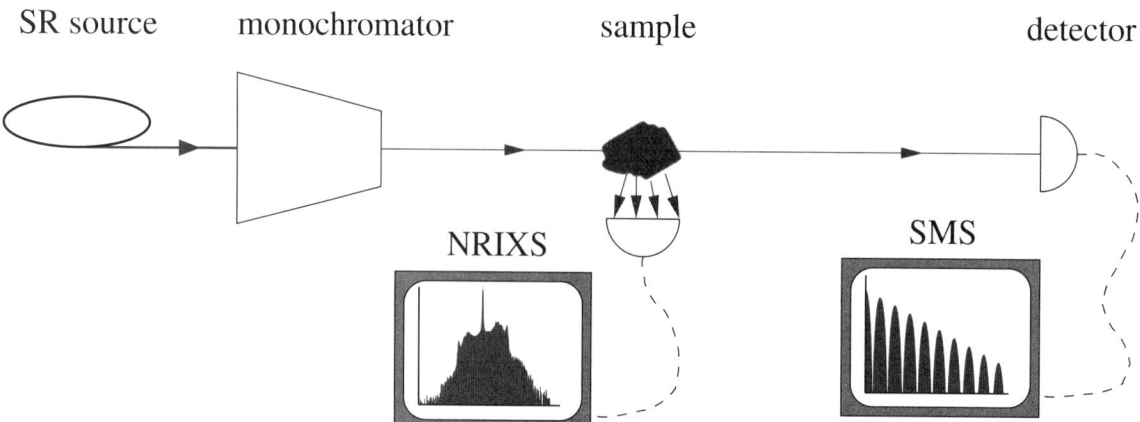

Figure 3. Experimental setup for NRIXS (nuclear resonant inelastic x-ray scattering; incoherent) and SMS (synchrotron Mössbauer spectroscopy; elastic coherent forward-directed scattering). For high-pressure applications, the sample is mounted inside a diamond anvil cell and focusing mirrors are installed after the monochromator.

$$S(\mathbf{k}, E) = \frac{1}{2\pi\hbar} \int \langle e^{i\mathbf{k}\mathbf{r}(t)} e^{-i\mathbf{k}\mathbf{r}(0)} \rangle e^{iEt/\hbar} dt, \qquad (3)$$

where $2\pi\hbar$ is Planck's constant, $\mathbf{k}$ is the wave vector of the X-rays incident on the sample, and $\mathbf{r}(t)$ is the displacement operator of the resonant nucleus (Sturhahn and Kohn, 1999; Sturhahn, 2004). The quasi-harmonic model of lattice vibrations is then used to extract the partial (due to information about motions of the resonant nuclei only) and projected (due to a potential angular dependence on $\mathbf{k}$) phonon DOS from $S(\mathbf{k},E)$ (Sturhahn et al., 1995; Kohn et al., 1998; Hu et al., 1999; Chumakov and Sturhahn, 1999; Sturhahn, 2000, 2004). Typical acquisition times for a NRIXS spectrum range between one hour for iron-rich samples under ambient conditions and days for dilute samples under very high pressures. The evaluation of the measured NRIXS spectra can be performed using the PHOENIX software (Sturhahn, 2000).

The Quasi-Harmonic Model

The vibrations in a solid are determined by the interatomic potential, which is harmonic in the lowest order of the displacement of the atoms from their equilibrium positions. If higher-order terms in the interatomic potential are neglected, a solid with N atoms shows $3N-6$ independent collective vibration modes, also known as phonons. The equations of motion can be solved exactly, and the self-intermediate scattering function can be calculated for the thermalized ensemble (Sturhahn and Kohn, 1999) with the result

$$\begin{aligned} S(\mathbf{k}, E) &= f(\mathbf{k})\delta(E) + \sum_{n=1}^{\infty} S_n(\mathbf{k}, E) \\ S_1(\mathbf{k}, E) &= f(\mathbf{k}) \frac{E_R}{E(1 - \exp[-\beta E])} g(\mathbf{k}, |E|) \\ S_n(\mathbf{k}, E) &= \frac{1}{nf(\mathbf{k})} \int S_{n-1}(\mathbf{k}, E') S_1(\mathbf{k}, E - E') dE' \\ f(\mathbf{k}) &= \exp\left[-\int \frac{E_R}{E} \coth\left(\frac{\beta E}{2}\right) g(\mathbf{k}, E) dE \right] \end{aligned} \qquad (4)$$

In this expression, $f(\mathbf{k})$ is the probability for recoilless absorption of the X-rays, also known as the Lamb-Mössbauer factor, $E_R = E_0^2/(\tilde{m}c^2)$ is the recoil energy (with nuclear transition energy E_0, mass of the nuclear resonant isotope $\tilde{m}$, and speed of light c), $\beta = 1/(k_B T)$ is the inverse temperature (with Boltzmann's constant k_B), and $g(\mathbf{k},E)$ is the partial and projected phonon DOS. The expansion has a clear physical interpretation: the value of $S_n(\mathbf{k},E)dE$ gives the probability for the creation/annihilation of n phonons with a total energy between E and $E + dE$. The extraction of the phonon DOS from measured data based on Equation 4 using the Fourier-log inversion method has been described previously (Kohn et al., 1998; Hu et al., 1999; Sturhahn, 2000).

Beyond the very powerful harmonic model, anharmonic interatomic potentials can arise either from very special atomic arrangements (Tse et al., 2005) or from increasing atomic separation caused by temperature (Chumakov et al., 1996). Most often, anharmonic effects are described by perturbations of the harmonic approximation using the concepts of "phonon lifetime" and "phonon scattering length." In the analysis of the self-intermediate scattering function, we distinguish situations of weak and strong anharmonicity. The former is characterized by renormalization of phonon energies (mode softening) and finite phonon lifetime but still assumes the existence of phonons as such, i.e., the coupling between the renormalized vibrational modes is negligible. So, the use of Equation 4 is still justified and will provide the renormalized phonon DOS. Situations of strong anharmonicity defy the phonon interpretation because the collective excitations are very different from phonons and renormalization is not sufficient, e.g., collective excitations in liquids and fast atomic diffusion in solids fall into this category. The quasi-harmonic model assumes the validity of Equation 4 and gives reliable results for cases of weak anharmonicity, as explained already.

If a NRIXS spectrum has been measured on a solid at certain P-T conditions, the validity of the weak anharmonicity assumption may be of concern, particularly at temperatures close to melting. Several tests based on the internal consistency of Equation 4 have been suggested (Chumakov and Sturhahn, 1999). The Lamb-Mössbauer factor, the kinetic energy per resonant atom, and the average force constant of the resonant atom can be calculated either as moments of the measured $S(\mathbf{k},E)$ or from the phonon DOS (Lipkin, 1962; Chumakov and Sturhahn, 1999; Sturhahn and Chumakov, 1999). In addition, proper normalization and the positiveness of the phonon DOS can be tested.

Directional Dependence

The dependence of $S(\mathbf{k},E)$ on the direction of the incident X-rays is implicitly contained in Equation 3 and is expressed via the directional dependence of the phonon DOS. The potential anisotropy of the phonon DOS must not be confused with the elastic anisotropy: the description of the former is given by a symmetric second-rank tensor (Sturhahn and Kohn, 1999), whereas the latter requires a symmetric fourth-rank tensor. Therefore, in crystals with cubic symmetry, such as ambient body-centered cubic (bcc) iron, the phonon DOS is isotropic, even though the elastic anisotropy is large ($C_{11} = 230$ GPa, $C_{12} = 135$ GPa, $C_{44} = 117$ GPa, anisotropy A = $[2C_{44} + C_{12} - C_{11}]/C_{11}$ of ~60%). Anisotropic behavior of the phonon DOS has been observed in $FeBO_3$ (Chumakov et al., 1997; Kohn et al., 1998) and hexagonal close-packed (hcp) iron (Giefers et al., 2000), but the anisotropy of the Lamb-Mössbauer factor was reported to be below the detection limits of 3.5% and 0.1%, respectively. Even single crystals of planar organic molecules show only a variation of ~2.5% in the Lamb-Mössbauer factor (Rai et al., 2002). The NRIXS spectrum of polycrystalline materials provides an average value $\langle S(\mathbf{k},E) \rangle$, and we write, neglecting the small variations of the Lamb-Mössbauer factor,

$$\langle S(\mathbf{k},E)\rangle = f\delta(E) + \sum_{n=1}^{\infty}\langle S_n(\mathbf{k},E)\rangle$$

$$\langle S_1(\mathbf{k},E)\rangle = \frac{fE_R}{E(1-\exp[-\beta E])}\langle g(\mathbf{k},|E|)\rangle$$

$$\langle S_n(\mathbf{k},E)\rangle = \frac{1}{nf}\int\langle S_{n-1}(\mathbf{k},E')S_1(\mathbf{k},E-E')\rangle dE' \quad (5)$$

$$f = \exp\left[-\int \frac{E_R}{E}\coth\left(\frac{\beta E}{2}\right)\langle g(\mathbf{k},E)\rangle dE\right]$$

Unfortunately the averaged higher-order terms cannot be derived from the averaged one-phonon term, and the formal inversion of Equation 5 to obtain <g(**k**,E)>, e.g., with the Fourier-log method, requires further justification. If we use the Fourier-log method nevertheless, we would obtain a result for the averaged one-phonon term that would deviate somewhat from the correct expression. This deviation is approximately given by $-1/2f\int$ <$\delta_1(\mathbf{k},E')\delta_1(\mathbf{k},E-E')$> dE' with $\delta_1(\mathbf{k},E) = S_1(\mathbf{k},E) - <S_1(\mathbf{k},E)>$.

We note that this correction is a two-phonon term and of second order in the integrated anisotropy functions δ_1. Reported cases of anisotropy (Chumakov et al., 1997; Giefers et al., 2000) lead to estimates of much less than 1% for this term. Direct inversion of the measured NRIXS spectrum therefore provides a reliable value of the averaged phonon DOS, unless the material exhibits a small Lamb-Mössbauer factor and a large anisotropy (which has yet to be observed).

SMS

For SMS measurements, the energy bandwidth should be as small as practicably achievable with reasonable efficiency. The high-resolution monochromator is tuned to the nuclear transition energy and kept as stable as possible. X-rays that are transmitted through the sample excite the resonant nuclei coherently and are observed with an avalanche photodiode detector that is placed far enough away from the sample to avoid contamination from incoherent scattering. The delayed events are mapped as a function of elapsed time between arrival of a synchrotron radiation pulse and detection of transmitted X-ray photons—this constitutes the time spectrum of the nuclei in the sample.

The delayed transmitted intensity can be expressed in terms of the nuclear contribution to the index of refraction of the sample (Sturhahn, 2004)

$$\delta_N(E) = \frac{\lambda}{4\pi}\rho\sigma Nf\sum_{mm'}\frac{W_{mm'}}{z_{mm'}(E)-i}, \quad (6)$$

where λ is the X-ray wavelength, ρ is the volume density of resonant nuclei, σ_N is the nuclear resonant cross section, and f is the Lamb-Mössbauer factor. The sum is over all sublevels of nuclear ground and excited states. The function $z_{mm'} = 2(E_{mm'}-E)/\Gamma$ depends on the energy difference between excited and ground states $E_{mm'}$ and the nuclear-level width Γ. The weight of each resonance at $E_{mm'}$ is given by the second-rank tensor $W_{mm'}$. The weights are normalized by $\sum_{mm'}W_{mm'} = 1$. The index of refraction has sharp maxima at energies $E = E_{mm'}$ corresponding to the positions of the nuclear transitions, which are determined by the electronic environment of the nucleus. The time spectrum is given by

$$\frac{dI}{dt} = \frac{I_0}{(2\pi\hbar)^2}\left|\int(\exp[ikD\delta n]-1)e^{-iEt/\hbar}dE\right|^2$$

$$= \frac{I_0}{(2\pi\hbar)^2}\left|\int\left(\exp\left[i\frac{\eta}{2}\sum_{mm'}\frac{W_{mm'}}{z_{mm'}(E)-i}\right]-1\right)e^{-iEt/\hbar}dE\right|^2, \quad (7)$$

where I_0 is the incident intensity corrected by electronic absorption, $k = E/(\hbar c)$ is the wave number, and D is the physical thickness of the sample. The effective thickness, $\eta = \rho\sigma_N fD$, is a useful parameter to describe total intensity and the influence of sample thickness on the shape of the time spectra. An effective thickness between 10 and 50 usually provides a good compromise between a distortion of the time spectra for larger values of η and a small total counting rate for smaller values of η. Typical acquisition times for an SMS time spectrum range between minutes for iron-rich samples under ambient conditions and a few hours for dilute samples under high pressures. Measured time spectra can be evaluated with the CONUSS software (Sturhahn, 2000). It should be noted that the index of refraction cannot be calculated by a simple Fourier transformation of the time spectrum. This situation has been described as the phase problem of SMS (Sturhahn, 2001), and experimental schemes to circumvent this problem have been proposed (Sturhahn et al., 2004). At present, such schemes would lead to a large increase in data collection time and are not routinely applied in high-pressure experiments.

SOUND VELOCITIES

The starting point for sound velocity measurements is the phonon DOS that is extracted from the NRIXS data. The connection between the phonon DOS and sound velocities may not be immediately obvious. In solids, sound waves and acoustic phonons of wavelengths much larger than interatomic distances describe the same physical phenomenon. The "phonon-picture" emphasizes microscopic properties, like interatomic force constants, whereas the macroscopic descriptors, such as elastic moduli and density, dominate understanding of the "sound-wave picture." The energy of an acoustic phonon of mode s with a small wave number q (long wavelength) that propagates in the direction **q** is given by $E_s = \hbar qv_s(\mathbf{q})$, where $2\pi\hbar$ is Planck's constant and $v_s(\mathbf{q})$ is the sound velocity. The number of phonon states in momentum space is then $dN_s = V\int k_s^2 dk_s d\Omega_q$, where $k_s = E_s/(\hbar v_s)$, V is a normalization volume, and the integration is performed over all directions **q**. The linear phonon dispersion leads to a Debye-like phonon DOS

$$D(E) = \sum_s \frac{dN_s}{dE} = \frac{VE^2}{\hbar^3} \sum_s \int \frac{1}{v_s^3(\boldsymbol{q})} d\Omega_q. \quad (8)$$

This relationship is exact for small energies (long phonon wavelengths) and has been experimentally shown to hold even in the case of a partial phonon DOS of a particular type of atom in a compound (Hu et al., 2003). Here, the previous equation is given in a form suitable for quantitative analysis

$$D(E) = \frac{\tilde{m}}{2\pi^2 \hbar^3 \rho} \frac{1}{v_D^3} E^2 \text{ with } \frac{1}{v_D^3} = \frac{1}{3} \sum_s \int \frac{1}{v_s^3(\boldsymbol{q})} \frac{d\Omega_q}{4\pi}, \quad (9)$$

where v_D is the Debye sound velocity, ρ is the density of the material, and $\tilde{m}$ is the mass of the nuclear resonant isotope.

The derivation of the Debye sound velocity from a NRIXS spectrum relies on the validity of several assumptions that may be violated in practice, leading to systematic errors. In the determination of the phonon DOS, we rely on the quasi-harmonic model (see section titled "The-Quasi-Harmonic Model") and a weak anisotropy of the phonon DOS (see section "Directional-Dependence"). The extraction of the curvature of the phonon DOS can lead to errors related to the limited energy range in which Equation 9 is valid; this is discussed in detail in the following section (Debye Sound Velocity). In special cases (discussed in Compound Mixtures section), small impurities can affect the low-energy region of the phonon DOS strongly and lead to large systematic errors in the Debye sound velocity of the majority phase. The calculation of compression- and shear-wave velocities requires additional knowledge about the static compression behavior of the material and this is discussed in the Compression- and Shear-Wave Velocities section. Systematic errors from this procedure are very small for the shear-wave velocities but are potentially more important for the compression-wave velocities. Identification of the sources for systematic errors in a particular experimental situation is a complex process, but under appropriate conditions cumulative systematic errors can be less than 1%.

Debye Sound Velocity

The quantitative description of the low-energy region of the phonon DOS via Equation 9 provides the Debye sound velocity. In Figure 4, the low-energy region of the phonon DOS is shown for iron metal under different pressures and temperatures. Fits to the data provide the curvature of the parabola defined by Equation 9, and with the known density, one can derive Debye sound velocities of 3.49(5) km/s for ambient conditions, 4.54(6) km/s for 50 GPa and 300 K, and 3.98(1) km/s for 55 GPa and 1500 K, where values in paranthesis represent standard errors. In a similar way, Debye sound velocities have been obtained for iron metal (Mao et al., 2001), Fe-Ni and Fe-Si alloys (Lin et al., 2003), $Fe_{0.9x}O$ (Struzhkin et al., 2001), Fe_3S (Lin et al., 2004a), FeH_x (Mao et al., 2004b), and FeS (Kobayashi et al., 2004) under high pressure. Very recently, the same method was applied to measure

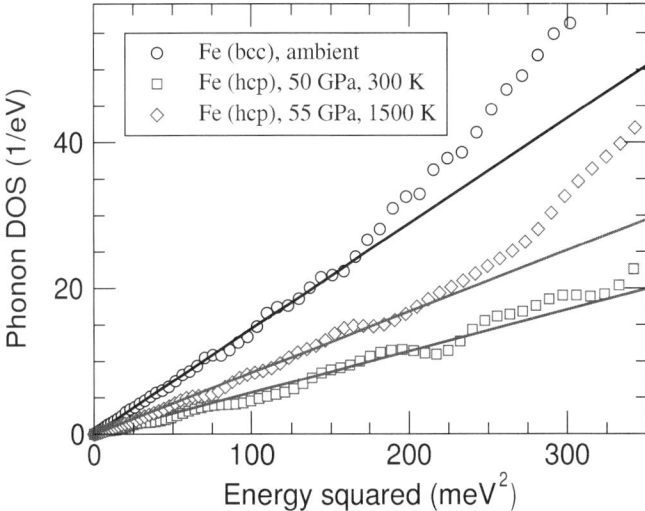

Figure 4. Phonon density of states (DOS) versus energy squared. The straight lines are fits to the data for energy below ~15 meV (3.6 THz). In this energy region, Debye-like behavior is observed. The data were taken from Sturhahn (2004, ambient) and Lin et al. (2005b, high pressure).

the temperature dependence of sound velocities in compressed iron metal (Lin et al., 2005b). In Table 1, we compare Debye sound velocities from NRIXS measurements with calculated Debye sound velocities from reported values for the elastic constants. We numerically determined sound velocities from the Christoffel equation (Musgrave, 1970) for all crystal directions and applied the averaging procedure of Equation 9 to obtain the Debye sound velocity.

The derivation of Debye sound velocities from the phonon DOS relies on a linear phonon dispersion that will only be accurate in a limited energy range as seen from the deviations apparent in Figure 4 for energies above ~15 meV (3.6 THz). The energy resolution of the NRIXS method depends on the bandwidth of the X-rays incident on the sample and reaches ~1 meV (0.24 THz) with present monochromator technology. In measured data, we therefore expect the energy region below ~2–3 meV (0.48–0.72 THz) to be obscured by elastic scattering and to be less reliable for sound velocity determination. Even though this does not seem to be a problem for the analysis of the data on pure iron that is shown in Figure 4, other materials may be more problematic. We will estimate the systematic errors resulting from the use of Equation 9 by an improved, empirical relation for the dispersion of the acoustic phonons. Assuming that the phonon energies reach a maximum value of E_{cs} at the Brillouin-zone boundary and that we can describe the phonon energies by $E_s = E_{cs} \sin(\hbar q v_s / E_{cs})$, an approximate relationship used before to extract compressional sound velocities using momentum-resolved inelastic X-ray scattering (Krisch et al., 1997; Fiquet et al., 2001; Antonangeli et al., 2005), we obtain a relationship similar to Equation 9 but with an energy-dependent Debye velocity

TABLE 1. MEASURED DEBYE SOUND VELOCITIES OF EARTH-RELEVANT MATERIALS FROM NRIXS (NOTED WITH *) COMPARED WITH CALCULATED DEBYE SOUND VELOCITIES FROM AVAILABLE SINGLE-CRYSTAL ELASTICITY INFORMATION

Material	Density (g/cm³)	v_D (m/s)	Reference
α-Fe	7.86†	3521 ± 50*	Sturhahn (2004)
α-Fe	7.86†	3519 ± 48*	Hu et al. (2003)
α-Fe	7.874	3538	Hearmon (1984)
$Fe_{0.85}Si_{0.15}$	7.42 ± 0.01	3281 ± 10*	Lin et al. (2003)
$Fe_{0.91}Si_{0.09}$	7.601	3541	Machova and Kadeckova (1977)
$Fe_{0.94}Si_{0.06}$	7.675	3556	Machova and Kadeckova (1977)
$Fe_{0.92}Ni_{0.08}$	8.40 ± 0.02	3530 ± 6*	P = 7.5 GPa (Lin et al., 2003)
Fe_3S	7.05	2902 ± 6*	Lin et al. (2004)
FeS	4.6	2831 ± 10*	P = 1.5 GPa, from Kobayashi et al. (2004)
$Fe_{0.947}O$	5.708	2801*	P = 0.9 GPa, from Struzhkin et al. (2001)
$Fe_{0.943}O$	5.708	3210	Jackson and Khanna (1990)
$Fe_{0.946}O$	5.721 ± 0.006	3223 ± 7	Jacobsen et al. (2002)
Fe_2O_3	5.254	4597 ± 100*	This study
Fe_2O_3	5.254	4654 ± 100	Liebermann and Schreiber (1968)§

Note: Errors for values from nuclear resonant inelastic X-ray scattering (NRIXS) include statistical contributions only. All values are for ambient conditions, unless mentioned otherwise.

*Debye sound velocity from NRIXS data and corrected for natural Fe-enrichment using $Fe_{(1-x)}R_x$ and $v_{natural} = v_{enriched} \sqrt{\frac{(1-x)57 + xA_R}{(1-x)56 + xA_R}}$, where A_R represents the atomic mass of the nonenriched portion. Note that all other Debye sound velocities were calculated from available single-crystal elasticity data (unless mentioned otherwise) using the Christoffel equation and the averaging method outlined in Equation 9.

†Weast (1984).

§Polycrystalline sample.

$$\frac{1}{v_D^3(E)} = \frac{1}{3} \sum_s \frac{\arcsin^2 \eta_s}{\eta_s^2 \sqrt{1-\eta_s^2}} \int \frac{1}{v_s^3(q)} \frac{d\Omega_q}{4\pi}, \quad (10)$$

where $\eta_s = E/E_{cs}$. If we use the energy interval $[E_1, E_2]$ for analysis with a parabolic fit as indicated in Figure 4, the derived Debye sound velocity is approximately given by

$$\bar{v}_D = v_D(0) \left\{ 1 - \frac{5}{63} \left(\frac{E_2}{E_c}\right)^2 \frac{1-(E_1/E_2)^7}{1-(E_1/E_2)^5} \right\}$$
$$\approx v_D(0) \left\{ 1 - \frac{5}{63} \left(\frac{E_2}{E_c}\right)^2 \right\}, \quad (11)$$

where $v_D(0)$ is the true Debye sound velocity. The reasonable assumption $E_2 > 2E_1$ allows us to neglect the fraction. The correction term in this expression is always negative, and analysis of the measured data by a parabolic fit provides a reduced Debye sound velocity depending on the values of E_1, E_2, and E_c. For the ambient iron data shown in Figure 4, parameters of E_c = 22 meV, E_1 = 3 meV, and E_2 = 12 meV lead to ~2.4% reduction, but a value of E_2 < 8 meV brings this effect below 1%. In Figure 5,

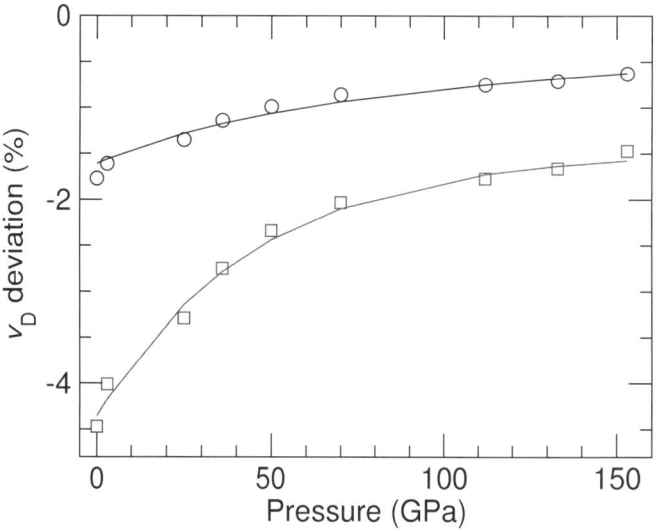

Figure 5. Systematic error of the Debye sound velocity from a parabolic fit to the phonon density of states (DOS) of iron metal versus pressure. Values of E_c between 22 meV (ambient) and 36 meV (153 GPa) were obtained from previously published data (Mao et al., 2001), and systematic errors were estimated using Equation 11 with E_2 = 10 meV (circles) and E_2 = 15 meV (squares). The solid lines are guides to the eye.

we show systematic errors from this deviation versus pressure for iron metal. At higher pressures, the phonon spectrum shifts to higher energies, the parabolic region in the phonon DOS is extended, and the derived Debye sound velocity becomes more accurate. The errors can be reduced further by decreasing E_2, even though at present, the statistical quality of high-pressure NRIXS data would probably require $E_2 \geq 10$ meV.

Compression- and Shear-Wave Velocities

Seismic data of Earth's interior distinguish between compressional waves and shear waves corresponding to longitudinally and transversely polarized phonons, respectively. The polarization vectors $\mathbf{e}_s(\mathbf{q})$ of the phonons can be used to define average compression- and shear-wave velocities, v_P and v_S, by

$$\frac{1}{v_P^3} = \sum_s \int \frac{[\mathbf{q} \cdot \mathbf{e}_s(\mathbf{q})]^2}{v_s^3(\mathbf{q})} \frac{d\Omega_q}{4\pi} \text{ and}$$
$$\frac{1}{v_S^3} = \frac{1}{2}\sum_s \int \frac{[\mathbf{q} \cdot \mathbf{e}_s(\mathbf{q})]^2}{v_s^3(\mathbf{q})} \frac{d\Omega_q}{4\pi}. \quad (12)$$

The NRIXS method provides the Debye sound velocity, which, according to Equations 9 and 12, can now be expressed as

$$\frac{3}{v_D^3} = \frac{1}{v_P^3} + \frac{2}{v_S^3}. \quad (13)$$

For isotropic media, v_P and v_S are independent of direction and follow the additional relationship

$$\frac{K_s}{\rho} = v_\phi^2 = v_P^2 - \frac{4}{3}v_S^2, \quad (14)$$

where K_s and ρ are the adiabatic bulk modulus and density, respectively. Equations 13 and 14 have widely been used to derive compression- and shear-wave velocities from NRIXS data with additional knowledge of adiabatic bulk modulus and density. These equations have general solutions

$$\begin{aligned} v_S &= 0.952\, v_D - 0.041\, v_\phi \\ v_P &= 0.908 \\ v_\phi &+ 0.297\, v_D - 0.243\, v_D^2/v_\phi \end{aligned}, \quad (15)$$

which are accurate to better than 0.1%. A variation of v_ϕ has only a minor effect on the shear-wave velocity, i.e., $\delta v_S/\delta v_\phi = -0.041$, whereas $\delta v_P/\delta v_\phi = 0.908 - 0.243\, v_D^2/v_\phi^2$ reflects the strong influence of v_ϕ on the value for the compression-wave velocity. These findings are material independent, and the weak effect of v_ϕ on

the shear-wave velocity determination of FeH$_x$ has been pointed out previously (Mao et al., 2004b). As a result, the NRIXS method is particularly suitable to provide accurate values for the average shear-wave velocity, where the averaging mechanism is defined by Equation 12. In Figure 6, we show examples of sound velocities of iron and iron alloys at room temperature. The straight lines are fits to the NRIXS results and suggest the validity of Birch's law (Birch, 1952) for v_P and v_S, i.e., $v_{P,S} \propto \rho$. The same linear dependence of v_P and v_S was also found independently in inelastic X-ray scattering studies (Antonangeli et al., 2004). In addition, recent NRIXS experiments on iron under high pressure and high temperature have demonstrated an explicit temperature dependence of the sound velocities at constant density (Lin et al., 2005b).

It is important to realize that Equation 14 only holds approximately for elastically anisotropic materials. For example, a calculation using elastic constants of iron metal at ambient conditions ($C_{11} = 230$ GPa, $C_{12} = 135$ GPa, $C_{44} = 117$ GPa, anisotropy $A = [2C_{44} + C_{12} - C_{11}]/C_{11}$ of ~60%) shows that the right side of Equation 14 will fall short of its correct value by ~4% when using the average velocities according to Equation 12, and potentially more for materials with larger elastic anisotropy. Nevertheless average shear-wave velocities can be determined by NRIXS with great precision and high accuracy. Consider solutions (v_P, v_S) of Equations 13 and 14 where v_D and v_ϕ serve as input parameters. A variation of v_ϕ can then simulate the effects of elastic anisotropy

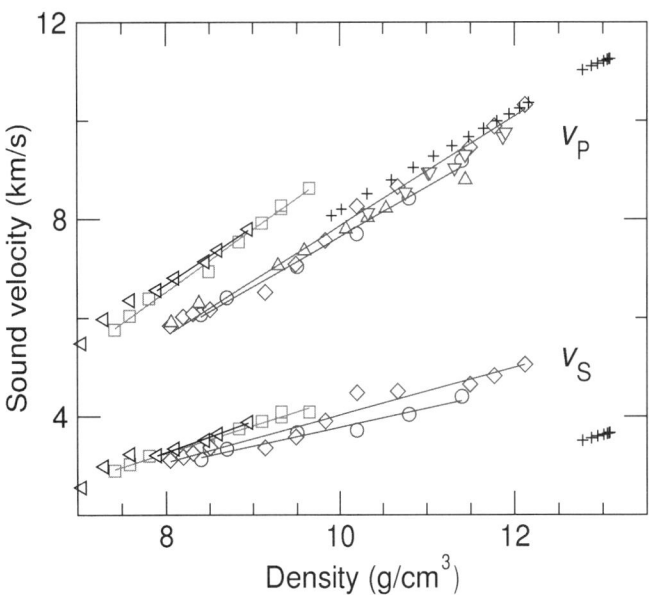

Figure 6. Compression-wave and shear-wave velocities versus density. Results from NRIXS (nuclear resonant inelastic x-ray scattering) data: diamonds—Fe (Mao et al., 2001); circles—Fe$_{0.92}$Ni$_{0.08}$; rectangles—Fe$_{0.85}$Si$_{0.15}$ (Lin et al., 2003); left triangles—Fe$_3$S (Lin et al., 2004a). Other results: triangles up—Fe, IXS (Fiquet et al., 2001); triangles down—Fe, shock-wave (Brown and McQueen, 1986); crosses—PREM (preliminary reference Earth model) (Dziewonski and Anderson, 1981). The straight lines are fits to the NRIXS results and suggest the validity of Birch's law.

and uncertainty in the bulk modulus. According to Equation 15, uncertainty about the elastic anisotropy will then mainly influence v_P, whereas v_s remains almost unchanged.

Compound Mixtures

The previous discussion implicitly assumed that the material under investigation was homogeneous. Even though most experiments may try to prepare and study pure samples, the effect of a mixture of compounds with different elastic properties on the sound velocity determination is important. For simplicity, we will assume a mixture of only two compounds with concentrations α_1 and $\alpha_2 = 1 - \alpha_1$. The NRIXS spectrum will then be given by $S(E) = \alpha_1 S_1(E) + \alpha_2 S_2(E)$, where $S_j(E)$ are the self-intermediate scattering functions according to Equation 3 for each compound. The extracted phonon DOS is nonlinearly related to $S(E)$ and is in general not the linear superposition of the phonon DOS of the two compounds (see the discussion in Sturhahn and Chumakov, 1999). If we still assume that the extracted phonon DOS is a linear superposition, we will make an error of the order $(f - 1 - f \ln f)/(f \ln f)$, which can be approximated by $3(1-f)^2/2$ for Lamb-Mössbauer factors close to unity. For example, hcp-iron at 300 K gives $f > 0.83$, and the error would be only a few percent. Under these conditions, it is justified to assume a linear superposition of the phonon DOS, which leads to the following addition rule for the Debye sound velocities:

$$f \rho v_D^{-3} = f_1 \alpha_1 \rho_1 v_{D1}^{-3} + f_2 \alpha_2 \rho_2 v_{D2}^{-3}, \qquad (16)$$

where f_j, ρ_j, and v_{Dj} are the Lamb-Mössbauer factors, densities, and Debye sound velocities of the two compounds, respectively, and $f = \alpha_1 f_1 + \alpha_2 f_2$. The value on the left side of Equation 16 is obtained from the measured NRIXS spectrum, but the individual values for the two compounds are not known. An important application of Equation 16 is the assessment of effects caused by a contamination of the compound to be studied. In this case, we may (erroneously) assume that $\rho = \rho_1$ and rewrite Equation 16 as follows

$$\frac{v_D}{v_{D1}} = \left[\frac{\alpha_1 + \alpha_2 \zeta}{\alpha_1 + \alpha_2 \zeta \eta} \right]^{1/3}, \qquad (17)$$

where $\xi = f_1/f_2$ and $\eta = (\rho_2 v_{D1}^3)/(\rho_1 v_{D2}^3)$. In Figure 7, we show the ratio of measured and actual Debye sound velocities v_D/v_{D1} for various values of η. The influence of ξ on the results is small, and we chose $\xi = 1$, which is reasonable for most iron-bearing materials under high pressure. The approximation of Equation 17 for small α_2 reads $v_D/v_{D1} \approx 1 + \alpha_2 \xi (1-\eta)/3$, which suggests a strong effect for large η. Even small amounts of a dense contaminant with small Debye sound velocity added to a light material with high Debye sound velocity (large values of η) can lead to very

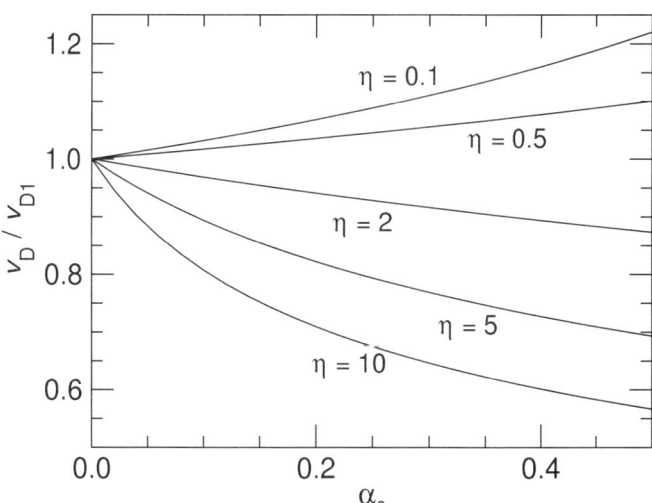

Figure 7. Correction of the measured Debye sound velocity in mixed compounds versus the concentration of the contaminant. Equation 17 was used with the specified values for $\eta = (\rho_2 v_{D1}^3)/(\rho_1 v_{D2}^3)$. Large η describes the admixture of a contaminant with high density and/or low Debye sound velocity.

different results for the measured value of v_D. For example, a mixture of 0.5% FeO ($\rho_2 = 5.721$ g/cm^3, $v_{D2} = 3.221$ km/s) and 99.5% perovskite $Mg_{0.9}Fe_{0.1}SiO_3$ ($\rho_1 = 4.106$ g/cm^3, $v_{D1} = 7.21$ km/s; from Sinogeikin et al. [2004] for iron-free $MgSiO_3$) gives $\eta = 15.6$ and $\alpha_2 = 0.16$, which would lead to a measured Debye sound velocity of only $v_D = 4.24$ km/s, a 41% reduction. The method is much more robust in the opposite scenario with small values of η. For example, a mixture of 10% FeO ($\rho_2 = 5.721$ g/cm^3, $v_{D2} = 3.221$ km/s) and 90% iron metal ($\rho_1 = 7.86$ g/cm^3, $v_{D1} = 3.52$ km/s) gives $\eta = 0.95$ and $\alpha_2 = 0.06$, which leads to a change in the measured Debye sound velocity of only 0.1%.

GRÜNEISEN PARAMETERS

The volume and temperature dependence of vibrational modes contains important information about the thermodynamic behavior of condensed matter. A microscopic picture was developed by Grüneisen (Grüneisen, 1926) that introduced the isothermal change of the energy ω_l of the vibrational mode l with volume as a characteristic parameter. These mode-specific isothermal Grüneisen parameters are defined by the equation

$$\gamma_l = -\frac{V}{\omega_l} \left(\frac{\partial \omega_l}{\partial V} \right)_T. \qquad (18)$$

In principle, there are $3N$ possible different values for γ_l if N is the number of atoms in the material, and the independent determination of all these parameters is experimentally not feasible. However, the volume dependence of the phonon DOS can be determined by NRIXS.

Several types of Grüneisen parameters have been introduced and are used in the literature (see, e.g., Poirier, 2000). In particular, the Debye-Grüneisen parameter, given by

$$\gamma_D = \frac{1}{3} + \frac{\rho}{v_D}\left(\frac{\partial v_D}{\partial \rho}\right)_T, \quad (19)$$

can be extracted from the measured Debye sound velocities v_D. An ansatz of the type $\gamma_D = \gamma_{D0}(\rho_0/\rho)^q$ (Anderson, 1979) permits us to integrate the previous equation and leads to the following functional description:

$$v_D = v_{D0}\zeta^{1/3}\exp\left(-\frac{\gamma_{D0}}{q}(\zeta^q - 1)\right), \quad (20)$$

where $\zeta = \rho_0/\rho$. We used Equation 20 to derive values for γ_{D0} and q with density-dependent Debye sound velocities obtained from published NRIXS data on bcc-Fe and hcp-Fe (Mao et al., 2001) as well as $Fe_{0.92}Ni_{0.08}$ and $Fe_{0.85}Si_{0.15}$ (Lin et al., 2003). The best agreement was achieved with $\gamma_{D0} = 1.9$ and $q = 1.71$ for bcc-iron, $\gamma_{D0} = 2.36$ and $q = 1.67$ for hcp-iron, $\gamma_{D0} = 1.67$ and $q = 1.1$ for $Fe_{0.92}Ni_{0.08}$, and $\gamma_{D0} = 2.01$ and $q = 1.13$ for $Fe_{0.85}Si_{0.15}$. These Debye-Grüneisen parameters and previously published experimental values on hcp-Fe versus the reduced density ρ/ρ_0 are displayed in Figure 8. Previous NRIXS studies provided $\gamma_{D0} = 2.0$ (bcc-Fe) and $\gamma_{D0} = 1.8$ (hcp-Fe in the pressure range below 40 GPa) but assumed $q = 0$ (Giefers et al., 2000).

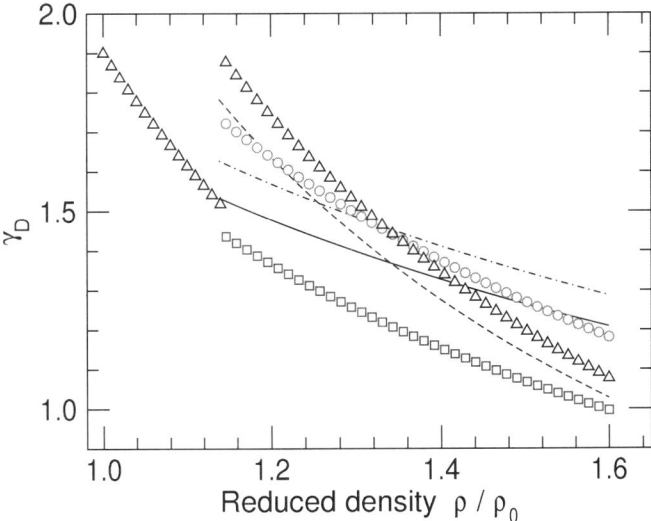

Figure 8. Debye-Grüneisen parameter versus reduced density. Results derived from NRIXS (nuclear resonant inelastic x-ray scattering) data: triangles—Fe (Mao et al., 2001); rectangles—$Fe_{0.92}Ni_{0.08}$; circles—$Fe_{0.85}Si_{0.15}$ (Lin et al., 2003). For comparison, hexagonal close-packed (hcp) iron: solid line—Raman spectroscopy (Merkel et al., 2000); dashed line—shock compression (Jeanloz, 1979); dashed-dotted line—X-ray diffraction (Dubrovinsky et al., 2000a, 2000b).

TEMPERATURE

The NRIXS raw data have a very fundamental property that is independent of the material under investigation: the spectra follow a detailed balance principle (Sturhahn and Kohn, 1999). If the NRIXS data are given by $I(E)$ with $E = 0$ as the exact nuclear transition energy, we can write

$$I(-E) = e^{-\beta E}I(E), \quad (21)$$

where $\beta = 1/(k_B T)$ is the inverse temperature, and k_B is Boltzmann's constant. This relation permits us to determine the temperature of the sample from the spectral intensity ratios of phonon creation ($E > 0$) and annihilation ($E < 0$) parts. Recently, this method has been applied to determine the temperature of heated samples of iron under pressures up to 29 GPa (Shen et al., 2004). A comparison of the temperature from the NRIXS data with a temperature obtained by a fit of the thermal radiation spectra to the Planck radiation function up to 1700 K and 58 GPa confirmed independently the validity of temperatures determined from the spectroradiometric method in laser-heated diamond cell experiments (Lin et al., 2004b). In Figure 9, we illustrate the effect of temperature on NRIXS data by introducing a thermal asymmetry function defined by

$$A = \frac{I(E) - I(-E)}{I(E) + I(-E)}. \quad (22)$$

The detailed balance principle predicts the thermal asymmetry to behave as $A = \tanh(\beta E/2)$. This function was fitted to the data in the region between 5 meV and 50 meV with only temperature as an adjustable parameter. The central part was excluded to avoid the influence of the elastic peak. At large energies, the statistical accuracy of the data points decreases rapidly. A detailed analysis of error sources related to temperature determination using this method has been published previously (Shen et al., 2004). In Figure 10, we show temperatures determined with the spectroradiometric method and from NRIXS spectra on iron in a diamond anvil cell (Lin et al., 2004b). The spectroradiometric temperatures have been averaged over the collection time of the corresponding NRIXS spectra of typically 8 h.

MAGNETISM

Magnetic ordering in a material causes a characteristic Zeeman splitting of the nuclear levels of the resonant isotope. According to Equations 6 and 7, the SMS time spectrum carries the signature of such a magnetic splitting. The SMS method has been used to investigate magnetism under high pressure using the

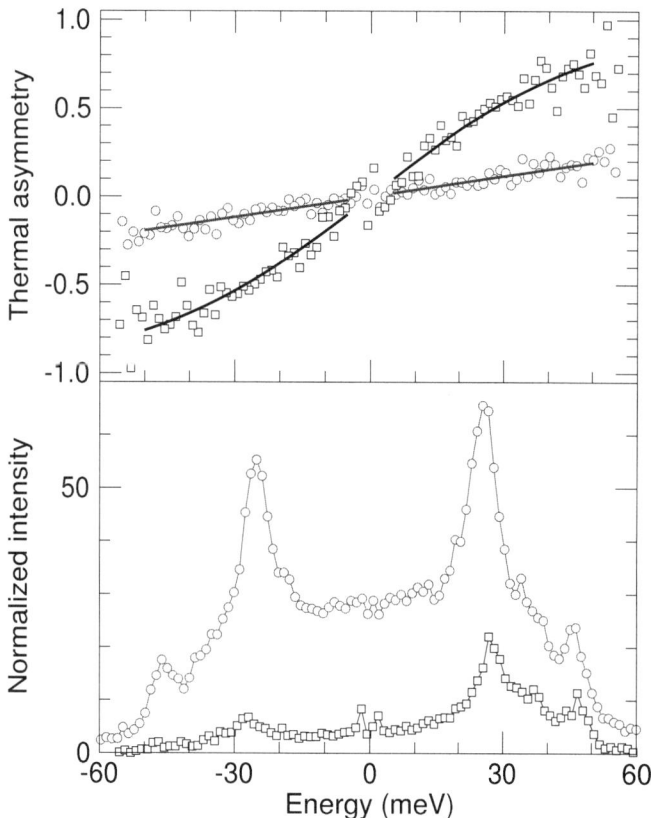

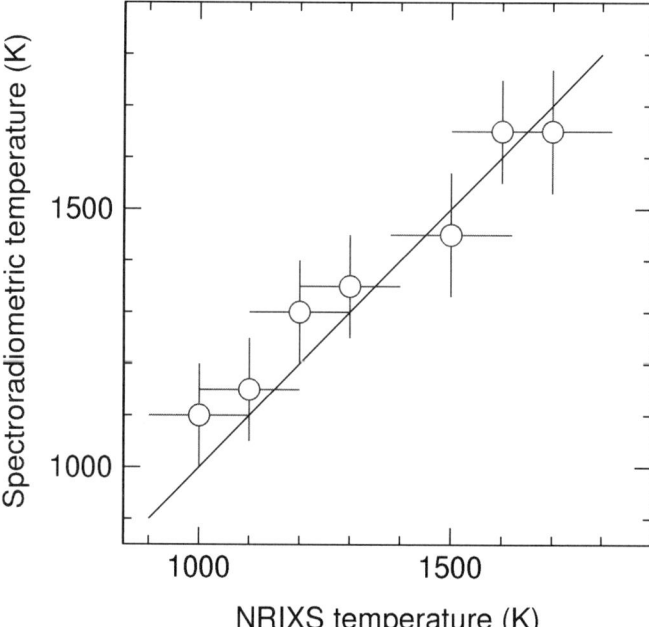

Figure 10. Temperatures determined by spectroradiometric method and from NRIXS (nuclear resonant inelastic X-ray scattering) data. The iron sample was mounted inside a diamond anvil cell and pressurized to ~58 GPa (Lin et al., 2004b). The red symbol at 1500 K indicates the temperature determined from the high-temperature spectrum in Figure 9. The straight line indicates identical temperatures.

Figure 9. The effect of the detailed balance rule on NRIXS (nuclear resonant inelastic x-ray scattering) data. Bottom graph: normalized intensity for hexagonal close-packed (hcp) iron at 50 GPa and ambient temperature (squares), and at 55 GPa and 1500 K (circles) (Lin et al., 2004b). The elastic peak has been subtracted. Top graph: thermal asymmetry calculated according to Equation 22 with solid lines corresponding to fits with temperatures of 293(15) K and 1500(100) K.

^{57}Fe isotope (Nasu, 1998; Lübbers et al., 1999a, 1999b, 2000b; Rupprecht et al., 2000; Wortmann et al., 2002; Lin et al., 2004a; Mao et al., 2004b), the ^{151}Eu isotope (Lübbers et al., 1999a; Lengsdorf et al., 2004), the ^{119}Sn isotope (Barla et al., 2005a), and the ^{149}Sm isotope (Barla et al., 2004a, 2004b, 2005b). Geophysical applications of the SMS method examined the pressure dependence of magnetism in materials like Fe_3S (Lin et al., 2004a) or FeH_x (Mao et al., 2004b). Fe_3S is the most iron-rich sulfide known to date and may be an important component in the iron-sulfur system at high pressures (Fei et al., 2000; Martin et al., 2004). In Figure 11, we show representative time spectra and the magnetic hyperfine fields derived from the time spectra of the SMS method together with normalized sound velocities (Lin et al., 2004a). The low-pressure magnetic phase contains two sites with different magnetic hyperfine fields. A collapse of the magnetic order occurs around 21 GPa, and simultaneous SMS and NRIXS experiments have shown that the magnetic to nonmagnetic transition significantly affects the elastic, thermodynamic, and vibrational properties of Fe_3S (Lin et al., 2004a).

VALENCE AND SPIN STATE

Most of the minerals and polymorphs expected in Earth's interior are believed to incorporate low concentrations of Fe^{2+} and/or Fe^{3+} of ~10 atom% or less. They are not expected to be magnetically ordered in Earth's lower mantle because of the low Fe content and the elevated temperatures. However, valence and spin state of iron in minerals may still be relevant with respect to density, iron partitioning, partial melting, radiative thermal conductivity, and compositional layering (Shannon and Prewitt, 1969; Gaffney and Anderson, 1973; Sherman, 1988, 1991; Sherman and Jansen, 1995; Badro et al., 2003, 2004; Li et al., 2004). The SMS method provides quadrupole splittings and isomer shifts similar to traditional Mössbauer spectroscopy, but the high brilliance of the synchrotron radiation reduces the data collection times tremendously, allows easier access for high-pressure studies, and reduces pressure gradients in the observed data (Sturhahn et al., 1998; Lübbers et al., 1999b; Sturhahn, 2004). The assignment of a set of parameter values to valence and spin state is usually based on a fingerprinting scheme (Bancroft et al., 1967).

Traditional Mössbauer spectroscopy had been used previously to study ferro-magnesium silicate perovskite (hereafter referred to as Pv) and Al-bearing Pv under ambient pressure (McCammon, 1997). Recently, the SMS method was applied to Pv using compositions $Mg_{1-y}Fe_ySiO_3$ with $y = 0.05$ and $y = 0.1$ up to 120 GPa at room temperature (Jackson et al., 2005a). In

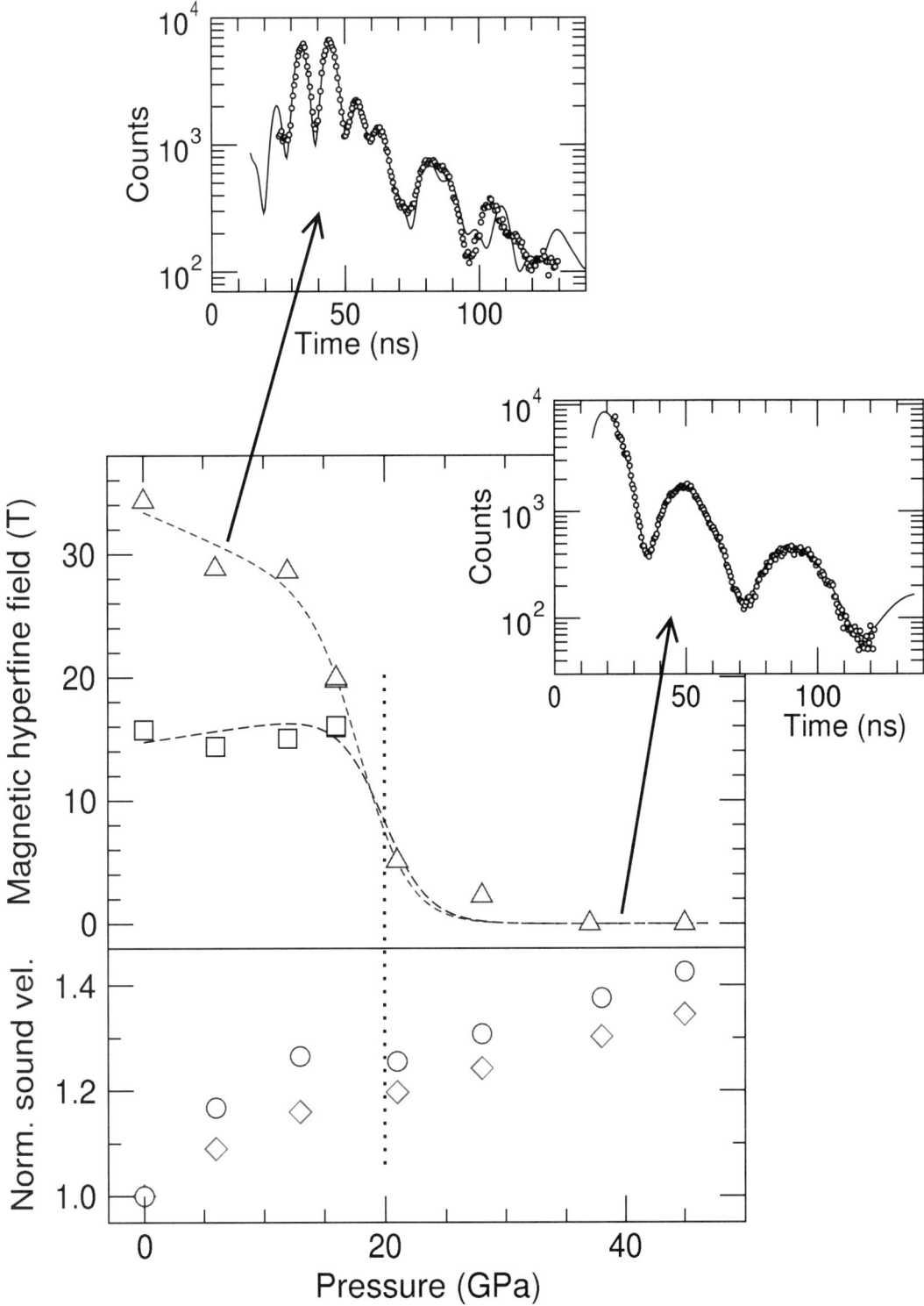

Figure 11. Magnetic hyperfine fields and normalized sound velocities of Fe_3S with increasing pressure and time spectra recorded at 6 GPa and at 45 GPa. Below ~20 GPa, the low-pressure magnetic phase displays two magnetic field sites. The higher magnetic field (triangles) decreases with increasing pressure, whereas the lower magnetic field (squares) remains almost constant. A magnetic collapse occurs at ~21 GPa. The dashed lines are guides to the eye. The magnetic collapse is accompanied by a change in the pressure dependence of the normalized compressional- (diamonds) and shear-wave (circles) velocities as shown in the lower panel. At low pressure, the fast oscillations in the time spectrum clearly indicate the magnetic nuclear level splitting. At high pressure, the oscillations are significantly slower and result from the remaining quadrupole splitting and the rather large thickness of the sample. The time spectra were evaluated with the CONUSS programs (Sturhahn, 2000). Black circles—experimental time spectra; solid lines—best calculated time spectra. The data were taken from Lin et al. (2004a).

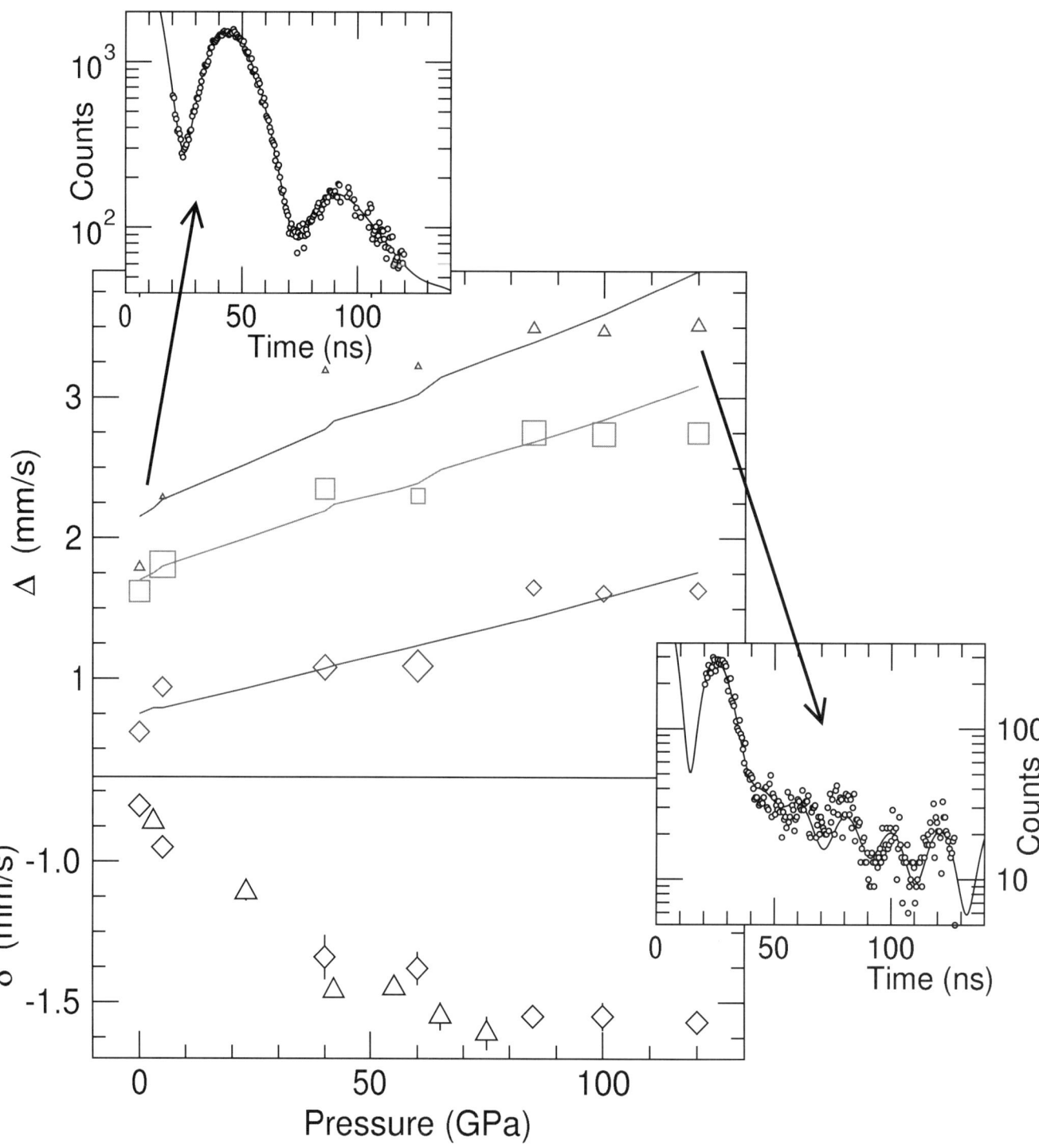

Figure 12. Quadrupole splittings (Δ) of the three Fe sites in $Mg_{0.95}Fe_{0.05}SiO_3$ versus pressure, isomer shifts (δ) between Fe^{2+} and Fe^{3+} sites, and time spectra recorded at ambient pressure and at 120 GPa. The solid lines were calculated using an isomorphic volume reduction of the unit cell (Jackson et al., 2005a). In the top panel, triangles and rectangles—sites associated with Fe^{2+}; diamonds—Fe^{3+}. The symbol size is proportional to the weight of the corresponding Fe site. In the lower panel, triangles—$Mg_{0.9}Fe_{0.1}SiO_3$; diamonds—$Mg_{0.95}Fe_{0.05}SiO_3$. Above ~70 GPa, the isomer shift between Fe^{2+} and Fe^{3+} sites changes little. At low pressure, the slower oscillations in the time spectrum clearly indicate smaller nuclear level splittings for all sites. The time spectra were evaluated with the CONUSS program to provide quadrupole splittings, weights, and isomer shifts of each Fe site (Sturhahn, 2000). Circles—experimental time spectra; solid lines—best calculated time spectra. The data were taken from Jackson et al. (2005a).

Figure 12, we show the pressure dependence of the quadrupole splittings of the three Fe sites in Pv. The increase in the splitting was explained by the compression of the perovskite lattice. A change in the pressure dependence of the isomer shift between Fe^{2+} and Fe^{3+} around 70 GPa possibly indicates a change in the Fe^{3+} spin state (Jackson et al., 2005a). X-ray emission spectroscopy has also been applied to probe the spin state of Fe in Pv (Badro et al., 2004; Li et al., 2004), Al-bearing Pv (Li et al., 2004), and (Mg,Fe)O ferropericlase (Badro et al., 2003; Lin et al., 2005a). In contrast to X-ray emission spectroscopy, SMS can distinguish valences in addition to the spin states of Fe. Specifically, SMS data provide us with different Fe sites characterized by isomer shift and quadrupole splitting.

CONCLUSION

We introduced two particular nuclear resonant scattering techniques that are applied to a variety of problems in condensed-matter physics, material science, geophysics, biophysics, and chemistry. In this contribution, we focused on the geophysical applications. The NRIXS and SMS methods have prospered with the commissioning of third-generation synchrotron radiation sources. We expect even greater potential when the brilliance of these sources is further increased by optimization of undulator technology or operating parameters of the storage ring. If fourth-generation sources (X-ray lasers) become a reality at sub-Ångström wavelengths, orders of magnitude increases in spectral flux density could lead to tremendous opportunities for the "flux-hungry" nuclear resonant and inelastic X-ray scattering techniques.

The study of planetary interiors encompasses a wide area of research activity. At present, the amount of available facts is far too limited to arrive at definite conclusions about the status quo and the evolutionary history of our planet. Experimental tools that provide information about candidate materials under high-pressure and high-temperature conditions are invaluable in progressing with this task. In this chapter, we have demonstrated how nuclear resonant scattering methods can contribute to the study of Earth materials like iron and iron-bearing compounds. The assembly of a database of sound velocities, Grüneisen parameters, and electronic properties related to valence and spin state of Fe has just begun. Over the next decade, we anticipate a further increase in available X-ray intensity, which will permit more accurate measurements. In addition, the synergy of nuclear resonant scattering methods with other X-ray techniques, e.g., X-ray diffraction for in situ density determination, will be another crucial step toward improved data reliability in the high-pressure, high-temperature sector.

ACKNOWLEDGMENTS

This work was supported by the U.S. Department of Energy, Office of Science, Office of Basic Energy Sciences, under contract no. DE-AC02-06CH11357.

REFERENCES CITED

Akber-Knutson, S., Steinle-Neumann, G., and Asimow, P.D., 2005, Effect of Al on the sharp-ness of the $MgSiO_3$ perovskite to post-perovskite phase transition: Geophysical Research Letters, v. 32, L14303, doi: 10.1029/2005GL023192.

Alp, E.E., Sturhahn, W., and Toellner, T.S., 2001, Lattice dynamics and inelastic nuclear resonance X-ray scattering: Journal of Physical Chemistry–Condensed Matter, v. 13, p. 7645, doi: 10.1088/0953-8984/13/34/311.

Anderson, O.L., 1979, Evidence supporting the approximation $\rho\gamma$ = const for the Grüneisen parameter of the Earth's lower mantle: Journal of Geophysical Research, v. 81, p. 3537–3542.

Antonangeli, D., Occelli, F., Requardt, H., Badro, J., Fiquet, G., and Krisch, M., 2004, Elastic anisotropy in textured hcp-iron to 112 GPa from sound wave propagation measurements: Earth and Planetary Science Letters, v. 225, p. 243–251, doi: 10.1016/j.epsl.2004.06.004.

Antonangeli, D., Krisch, M., Fiquet, G., Farber, D.L., Bossak, A., and Merkel, S., 2005, Aggregate and single-crystalline elasticity of hcp cobalt at high pressure: Physical Review B, v. 72, 134303, doi: 10.1103/PhysRevB.72.134303.

Badro, J., Struzhkin, V.V., Shu, J., Hemley, R.J., Mao, H.K., Rueff, J.P., and Kao, C.C., 2003, Iron partitioning in Earth's mantle: Toward a deep lower mantle discontinuity: Science, v. 300, p. 789–791, doi: 10.1126/science.1081311.

Badro, J., Rueff, J.P., Vanko, G., Monaco, G., Fiquet, G., and Guyot, F., 2004, Electronic transitions in perovskite: Possible nonconvecting layers in the lower mantle: Science, v. 305, p. 383–386, doi: 10.1126/science.1098840.

Bancroft, G.M., Maddock, A.G., and Burns, R., 1967, Applications of the Mössbauer effect to silicate mineralogy: 1. Iron silicates of known crystal structure: Geochimica et Cosmochimica Acta, v. 31, p. 2219–2246, doi: 10.1016/0016-7037(67)90062-2.

Barla, A., Sanchez, J.P., Malaman, B., Doyle, B.P., and Ruffer, R., 2004a, Sm magnetism in the layered compound $SmMn_2Ge_2$: Physical Review B, v. 69, 220405, doi: 10.1103/PhysRevB.69.220405.

Barla, A., Sanchez, J.P., Haga, Y., Lapertot, G., Doyle, P., Leupold, O., Rüffer, R., Abd-Elmeguid, M.M., Lengsdorf, R., and Flouquet, J., 2004b, Pressure-induced magnetic order in golden SmS: Physical Review Letters, v. 92, 066401, doi: 10.1103/PhysRevLett.92.066401.

Barla, A., Sanchez, J.-P., Aksungur, A., Lengsdorf, R., Plessel, J., Doyle, B.P., Ruffer, R., and Abd-Elmeguid, M.M., 2005a, Delocalization of the U 5f magnetic moments in $U(In_{0.6}Sn_{0.4})_3$ and UNiSn under high pressure: Journal of Physics: Condensed Matter, v. 17, p. S859–S870, doi: 10.1088/0953-8984/17/11/015.

Barla, A., Sanchez, J.-P., Derr, J., Salce, B., Lapertot, G., Flouquet, J., Doyle, B.P., Leupold, O., Rüffer, R., Abd-Elmeguid, M.M., and Lendsdorf, R., 2005b, Valence and magnetic instabilities in Sm compounds at high pressures: Journal of Physics: Condensed Matter, v. 17, p. S837–S848, doi: 10.1088/0953-8984/17/11/013.

Birch, F., 1952, Elasticity and composition of the Earth's interior: Journal of Geophysical Research, v. 57, p. 227–286.

Brown, J.M., and McQueen, R.G., 1986, Phase-transitions; Grüneisen parameter; and elasticity for shocked iron between 77 GPa and 400 GPa: Journal of Geophysical Research, v. 91, p. 7485–7494.

Caracas, R., and Cohen, R.E., 2005, Effect of chemistry on the stability and elasticity of the perovskite and post-perovskite phases in the $MgSiO_3$-$FeSiO_3$-Al_2O_3 system and implications for the lowermost mantle: Geophysical Research Letters, v. 32, L16310, doi: 10.1029/2005GL023164.

Chumakov, A.I., and Sturhahn, W., 1999, Experimental aspects of inelastic nuclear resonant scattering: Hyperfine Interactions, v. 781, p. 133–124.

Chumakov, A.I., Rüffer, R., Grünsteudel, H., Grünsteudel, H.F., Grubel, G., Metge, J., and Goodwill, H.A., 1995, Energy dependence of nuclear recoil measured with incoherent nuclear scattering of synchrotron radiation: Europhysics Letters, v. 30, p. 427.

Chumakov, A.I., Rüffer, R., Baron, A.Q.R., Grünsteudel, H., and Grünsteudel, H.F., 1996, Temperature dependence of nuclear inelastic absorption of synchrotron radiation in α-^{57}Fe: Journal of Physics: Condensed, v. 54, p. R9596.

Chumakov, A.I., Rüffer, R., Baron, A.Q.R., Grünsteudel, H., Grünsteudel, H.F., and Kohn, V.G., 1997, Anisotropic inelastic nuclear absorption: Physical Review B: Condensed Matter and Materials Physics, v. 56, p. 10,758.

Coltice, N., and Ricard, Y., 1999, Geochemical observations and one layer mantle convection: Earth and Planetary Science Letters, v. 174, p. 125–137, doi: 10.1016/S0012-821X(99)00258-7.

Dubrovinsky, L.S., Saxena, S.K., Dubrovinskaia, N.A., Rekhi, S., and Bihan, T.L., 2000a, Grüneisen parameter of ε-iron up to 300 GPa from in-situ X-ray study: The American Mineralogist, v. 85, p. 386–389.

Dubrovinsky, L.S., Saxena, S.K., Tutti, F., Rekhi, S., and Bihan, T.L., 2000b, In situ X-ray study of thermal expansion and phase transition of iron at multimegabar pressure: Physical Review Letters, v. 84, p. 1720–1723, doi: 10.1103/PhysRevLett.84.1720.

Dziewonski, A., and Anderson, D.L., 1981, Preliminary reference Earth model: Physics of the Earth and Planetary Interiors, v. 25, p. 297–356, doi: 10.1016/0031-9201(81)90046-7.

Endres, G., Strokendl, F., Langhoff, H., and Gmelin, F., 1981, Observation of localized modes in TbO_y and $TbAl_2$ by resonance absorption of γ-quanta: Zeitschrift für Physik B, v. 44, p. 253, doi: 10.1007/BF01294160.

Fei, Y.W., Li, J., Bertka, C.M., and Prewitt, C.T., 2000, Structure type and bulk modulus of Fe_3S; a new iron-sulfur compound: The American Mineralogist, v. 85, p. 1830–1833.

Fiquet, G., Badro, J., Guyot, F., Requardt, H., and Krisch, M., 2001, Sound velocities in iron to 110 gigapascals: Science, v. 291, p. 468–471, doi: 10.1126/science.291.5503.468.

Frost, D.J., and Langenhorst, F., 2002, The effect of Al_2O_3 on the Fe-Mg partitioning between magnesiowüstite and magnesium silicate perovskite: Earth and Planetary Science Letters, v. 199, p. 227–241, doi: 10.1016/S0012-821X(02)00558-7.

Gaffney, E.S., and Anderson, D.L., 1973, Effect of low-spin Fe^{2+} on the composition of the lower mantle: Journal of Geophysical Research, v. 78, p. 7005–7014.

Gerdau, E., and de Waard, H., eds., 1999, Nuclear Resonant Scattering of Synchrotron Radiation (Parts A&B): Hyperfine Interactions: Amsterdam, Baltzer Science Publishers, v. 123–125.

Gerdau, E., Rüffer, R., Winkler, H., Tolksdorf, W., Klages, C.P., and Hannon, J.P., 1985, Nuclear Bragg diffraction of synchrotron radiation in yttrium iron garnet: Physical Review Letters, v. 54, p. 835–838.

Giefers, H., Lübbers, R., Rupprecht, K., Wortmann, G., Alfè, D., and Chumakov, A.I., 2000, Phonon spectroscopy of oriented hcp iron: High Pressure Research, v. 22, p. 501–506, doi: 10.1080/08957950212817.

Grüneisen, E., 1926, The state of a solid body, in Handbuch der Physik, Volume 1: Berlin, Springer-Verlag, 145 p.

Hearmon, R.F.S., 1984, The elastic constants of crystals and other anisotropic materials: Springer-Verlag, 559 p.

Hemley, R.J., Mao, H.-K., and Struzhkin, V.V., 2005, Synchrotron radiation and high pressure: New light on materials under extreme conditions: Journal of Synchrotron Radiation, v. 12, p. 135–154, doi: 10.1107/S0909049504034417.

Hirose, K., Fei, Y., Ma, Y., and Mao, H.K., 1999, The fate of subducted basaltic crust in the Earth's lower mantle: Nature, v. 397, p. 53–56, doi: 10.1038/16225.

Hu, M.Y., Sturhahn, W., Toellner, T.S., Hession, P., Sutter, J., and Alp, E.E., 1999, Data analysis for inelastic nuclear resonant absorption experiments with synchrotron radiation: Nuclear Instruments & Methods in Physics Research, Section A—Accelerators, Spectrometers, Detectors and Associated Equipment, v. 428, p. 551, doi: 10.1016/S0168-9002(99)00134-5.

Hu, M.Y., Sturhahn, W., Toellner, T.S., Mannheim, P.D., Brown, D.E., Zhao, J., and Alp, E.E., 2003, Measuring velocity of sound with nuclear resonant inelastic X-ray scattering: Physical Review B, v. 67, 094304, doi: 10.1103/PhysRevB.67.094304.

Irifune, T., 1994, Absence of an aluminum phase in the upper part of the Earth's lower mantle: Nature, v. 370, p. 131–133, doi: 10.1038/370131a0.

Ishii, M., and Tromp, J., 1999, Normal-mode and free-air gravity constraints on lateral variations in velocity and density of the Earth's mantle: Science, v. 285, p. 1231–1236, doi: 10.1126/science.285.5431.1231.

Jackson, I., and Khanna, S., 1990, Shear-mode softening and high pressure polymorphism of wüstite ($Fe_{1-x}O$): Journal of Geophysical Research, v. 95, p. 21,671–21,685.

Jackson, J.M., Zhang, J., and Bass, J.D., 2004, Sound velocities and elasticity of aluminous $MgSiO_3$ perovskite: Implications for aluminum heterogeneity in Earth's lower mantle: Geophysical Research Letters, v. 31, L10614, doi: 10.1029/2004GL019918.

Jackson, J.M., Sturhahn, W., Shen, G., Zhao, J., Hu, M.Y., Errandonea, D., Bass, J.D., and Fei, Y., 2005a, A synchrotron Mössbauer spectroscopy study of (Mg,Fe)SiO_3 perovskite up to 120 GPa: The American Mineralogist, v. 90, no. 1, p. 199–205, doi: 10.2138/am.2005.1633.

Jackson, J.M., Zhang, J., Shu, J., Sinogeikin, S., and Bass, J.D., 2005b, High-pressure sound velocities and elasticity of aluminous $MgSiO_3$ perovskite to 45 GPa: Implications for lateral heterogeneity in Earth's lower mantle: Geophysical Research Letters, v. 32, L21305, doi: 10.1029/2005GL023522.

Jacobsen, S.D., Reichmann, H.J., Spetzler, H.A., Mackwell, S.J., Smyth, J.R., Angel, R.J., and McCammon, C.A., 2002, Structure and elasticity of single-crystal (Mg,Fe)O and a new method of generating shear waves for gigahertz interferometry: Journal of Geophysical Research, v. 107, no. B2, 2037, doi: 10.1029/2001JB000490.

Jacobsen, S.D., Spetzler, H., Reichmann, H.J., and Smyth, J.R., 2004, Shear waves in the diamond-anvil cell reveal pressure-induced instability in (Mg,Fe)O: Proceedings of the National Academy of Sciences of the United States of America, v. 101, p. 5867–5871, doi: 10.1073/pnas.0401564101.

Jeanloz, R., 1979, Properties of iron at high pressure and the state of the core: Journal of Geophysical Research, v. 84, p. 6059–6069.

Karki, B.B., and Crain, J., 1998, First-principles determination of elastic properties of $CaSiO_3$ perovskite at lower mantle pressures: Geophysical Research Letters, v. 25, p. 2741–2744, doi: 10.1029/98GL51952.

Kellogg, L.H., Hager, B.H., and van der Hilst, R.D., 1999, Compositional stratification in the deep mantle: Science, v. 283, p. 1881–1884.

Kesson, S.E., Gerald, J.D.F., Shelley, J.M.G., and Withers, R.L., 1995, Phase relations, structure and crystal chemistry of some aluminous silicate perovskites: Earth and Planetary Science Letters, v. 134, p. 187–201, doi: 10.1016/0012-821X(95)00112-P.

Kiefer, B., Stixrude, L., and Wentzcovitch, R., 2002, Elasticity of (Mg,Fe)SiO_3-perovskite at high-pressures: Geophysical Research Letters, v. 29, no. 11, 1539, doi: 10.1029/2002GL014683.

Kobayashi, H., Kamimura, T., Alfè, D., Sturhahn, W., Zhao, J., and Alp, E.E., 2004, Phonon density of states and compression behavior in iron sulfide under pressure: Physical Review Letters, v. 93, 195503, doi: 10.1103/PhysRevLett.93.195503.

Kohn, V.G., Chumakov, A.I., and Rüffer, R., 1998, Nuclear resonant inelastic absorption of synchrotron radiation in an anisotropic single crystal: Physical Review B, v. 58, p. 8437.

Krisch, M.H., Mermet, A., Miguel, A.S., Sette, F., Masciovecchio, C., Ruocco, G., and Verbeni, R., 1997, Acoustic-phonon dispersion in CdTe at 7.5 GPa: Physical Review B, v. 56, p. 8691, doi: 10.1103/PhysRevB.56.8691.

Kung, J., Li, B., Weidner, D.J., Zhang, J., and Liebermann, R.J., 2002, Elasticity of $(Mg_{0.83}Fe_{0.17})O$ ferropericlase at high pressure: Ultrasonic measurements in conjunction with X-radiation techniques: Earth and Planetary Science Letters, v. 203, p. 557–566, doi: 10.1016/S0012-821X(02)00838-5.

Lengsdorf, R., Baria, A., Alonso, J.A., Martinez-Lope, M.J., Micklitz, H., and Abd-Elmeguid, M.M., 2004, The observation of the insulator metal transition in $EuNiO_3$ under high pressure: Journal of Physics: Condensed Matter, v. 16, p. 3355–3360, doi: 10.1088/0953-8984/16/20/006.

Li, J., Struzhkin, V.V., Mao, H.-K., Shu, J., Hemley, R.J., Fei, Y., Mysen, B., Dera, P., Prakapenka, V., and Shen, G., 2004, Electronic spin state of iron in lower mantle perovskite: Proceedings of the National Academy of Sciences of the United States of America, v. 101, no. 39, p. 14,027–14,030, doi: 10.1073/pnas.0405804101.

Liebermann, R.C., and Schreiber, E., 1968, Elastic constants of polycrystalline hematite as a function of pressure to 3 kilobars: Journal of Geophysical Research, v. 73, p. 6585–6590.

Lin, J.-F., Struzhkin, V.V., Sturhahn, W., Huang, E., Zhao, J., Hu, M.Y., Alp, E.E., Mao, H.-K., Boctor, N., and Hemley, R.J., 2003, Sound velocities of iron-nickel and iron-silicon alloys at high pressures: Geophysical Research Letters, v. 30, 2112, doi: 10.1029/2003GL018405.

Lin, J.-F., Fei, Y., Sturhahn, W., Zhao, J., Mao, H.-K., and Hemley, R.J., 2004a, Magnetic transition and sound velocities of Fe_3S at high pressure: Implications for Earth and planetary cores: Earth and Planetary Science Letters, v. 226, p. 33–40, doi: 10.1016/j.epsl.2004.07.018.

Lin, J.-F., Sturhahn, W., Zhao, J., Shen, G., Mao, H.-K., and Hemley, R.J., 2004b, Absolute temperature measurement in a laser-heated diamond anvil cell: Geophysical Research Letters, v. 31, L14611, doi: 10.1029/2003GL020599.

Lin, J.F., Struzhkin, V.V., Jacobsen, S.D., Hu, M.Y., Chow, P., Kung, J., Liu, H., Mao, H.K., and Hemley, R.J., 2005a, Spin transition of iron in magnesiowüstite in the Earth's lower mantle: Nature, v. 436, p. 377–380, doi: 10.1038/nature03825.

Lin, J.-F., Sturhahn, W., Zhao, J., Shen, G., Mao, H.-K., and Hemley, R.J., 2005b, Sound velocities of hot dense iron: Birch's law revisited: Science,

v. 308, p. 1892–1894, doi: 10.1126/science.1111724.

Lipkin, H.J., 1962, Some simple features of the Mössbauer effect. II. Sum rules and the moments of the energy spectrum: Annals of Physics, v. 18, p. 182–197, doi: 10.1016/0003-4916(62)90066-0.

Lübbers, R., Pleines, M., Hesse, H.-J., Wortmann, G., Grünsteudel, H.F., Rüffer, R., Leupold, O., and Zukrowski, J., 1999a, Magnetism under high pressure studied by ^{57}Fe and ^{151}Eu nuclear scattering of synchrotron radiation: Hyperfine Interactions, v. 120/121, p. 49–58, doi: 10.1023/A:1017017827058.

Lübbers, R., Wortmann, G., and Grünsteudel, H.F., 1999b, High-pressure studies with nuclear scattering of synchrotron radiation: Hyperfine Interactions, v. 123/124, p. 529, doi: 10.1023/A:1017032125551.

Lübbers, R., Grünsteudel, H.F., Chumakov, A.I., and Wortmann, G., 2000a, Density of phonon states in iron at high pressure: Science, v. 287, p. 1250–1253.

Lübbers, R., Rupprecht, K., and Wortmann, G., 2000b, High-pressure Mössbauer studies of magnetism in RFe$_2$ Laves phases and Eu-chalcogenides: Hyperfine Interactions, v. 128, p. 115–135, doi: 10.1023/A:1012675330403.

Machova, A., and Kadeckova, S., 1977, Elastic constants of iron-silicon alloy single crystals: Czech. Journal of Physics B, v. 27, p. 555–563.

Mao, H.K., Xu, J., Struzhkin, V.V., Shu, J., Hemley, R.J., Sturhahn, W., Hu, M.Y., Alp, E.E., Vocadlo, L., Alfè, D., Price, G.D., Gillan, M.J., Schwoerer-Böhning, M., Häusermann, D., Eng, P., Shen, G., Giefers, H., Lübbers, R., Wortmann, G., 2001, Phonon density of states of iron up to 153 gigapascals: Science, v. 292, p. 914, doi: 10.1126/science.1057670.

Mao, W.L., Shen, G., Prakapenka, V.B., Meng, Y., Campbell, A.J., Heinz, D.L., Shu, J., Hemley, R.J., and Mao, H.K., 2004a, Ferromagnesium postperovskite silicates in the D″ layer of the Earth: Proceedings of the National Academy of Sciences of the United States of America, v. 101, p. 15,866–15,869.

Mao, W.L., Sturhahn, W., Heinz, D.L., Mao, H.-K., Shu, J., and Hemley, R.J., 2004b, Nuclear resonant X-ray scattering of iron hydride at high pressure: Geophysical Research Letters, v. 31, L15618, doi: 10.1029/2004GL020541.

Martin, P., Vocadio, L., Alfè, D., and Price, G.D., 2004, An ab initio study of the relative stabilities and equations of state of Fe$_3$S polymorphs: Mineralogical Magazine, v. 68, p. 813–817.

Mattern, E., Matas, J., Ricard, Y., and Bass, J., 2005, Lower mantle composition and temperature from mineral physics and thermodynamic modelling: Geophysical Journal International, v. 160, p. 973–990, doi: 10.1111/j.1365-246X.2004.02549.x.

McCammon, C.A., 1997, Perovskite as a possible sink for ferric iron in the lower mantle: Nature, v. 387, p. 694–696, doi: 10.1038/42685.

Merkel, S., Goncharov, A.F., Mao, H.-K., Gillet, P., and Hemley, R.J., 2000, Raman spectroscopy of iron to 152 gigapascals: Implications for Earth's inner core: Science, v. 288, p. 1626–1629, doi: 10.1126/science.288.5471.1626.

Murakami, M., Hirose, K., Kawamura, K., Sata, N., and Onishi, Y., 2004, Post-perovskite phase transition in MgSiO$_3$: Science, v. 304, p. 855–858, doi: 10.1126/science.1095932.

Musgrave, M.J.P., 1970, Crystal Acoustics: Introduction to the Study of Elastic Waves and Vibrations in Crystals: San Francisco, California, Holden-Day, Inc., 72 p.

Nakagawa, T., and Tackley, P.J., 2005, The interaction between the post-perovskite phase change and a thermo-chemical boundary layer near the core-mantle boundary: Earth and Planetary Science Letters, v. 238, p. 204–216, doi: 10.1016/j.epsl.2005.06.048.

Nasu, S., 1994, High pressure Mössbauer spectroscopy using a diamond anvil cell: Hyperfine Interactions, v. 90, p. 59, doi: 10.1007/BF02069118.

Nasu, S., 1998, High pressure experiments with synchrotron radiation: Hyperfine Interactions, v. 113, p. 97, doi: 10.1023/A:1012663330441.

Ni, S., Tan, E., Gurnis, M., and Helmberger, D., 2002, Sharp sides to the African superplume: Science, v. 296, p. 1850–1852, doi: 10.1126/science.1070698.

Papandrew, A.B., Yue, A.F., Fultz, B., Halevy, I., Sturhahn, W., Toellner, T.S., Alp, E.E., and Mao, H.-K., 2004, Vibrational modes in nanocrystalline iron under high pressure: Physical Review B, v. 69, 144301, doi: 10.1103/PhysRevB.69.144301.

Pleines, M., Lübbers, R., Strecker, M., Wortmann, G., Leupold, O., Shvyd'ko, Y.V., Gerdau, E., and Metge, J., 1999, Pressure-induced valence transition in EuNi$_2$Ge$_2$ studied by ^{151}Eu nuclear forward scattering of synchrotron radiation: Hyperfine Interactions, v. 120/121, p. 181, doi: 10.1023/A:1017002818398.

Poirier, J.-P., 2000, Introduction to the Physics of the Earth's Interior (2nd edition): Cambridge, Cambridge University Press, 46 p.

Rai, B.K., Durbin, S.M., Prohofsky, E.W., Sage, J.T., Ellison, M.K., Scheidt, W.R., Sturhahn, W., and Alp, E.E., 2002, Iron normal mode dynamics in a porphyrin-imidazole model for deoxyheme proteins: Physical Review E, v. 66, p. 051,904, doi: 10.1103/PhysRevE.66.051904.

Rupprecht, K., Friedmann, T., Giefers, H., Wortmann, G., Doyle, B., and Zukrowski, J., 2000, High-pressure/high-temperature NFS study of magnetism in LuFe$_2$ and ScFe$_2$: High Pressure Research, v. 22, p. 189–194, doi: 10.1080/08957950211356.

Scheidt, W.R., Durbin, S.M., and Sage, J.T., 2005, Nuclear resonance vibrational spectroscopy (NRVS): Journal of Inorganic Biochemistry, v. 99, p. 60–71, doi: 10.1016/j.jinorgbio.2004.11.004.

Seto, M., Yoda, Y., Kikuta, S., Zhang, X., and Ando, M., 1995, Observation of nuclear resonant scattering accompanied by phonon excitation using synchrotron radiation: Physical Review Letters, v. 74, p. 3828, doi: 10.1103/PhysRevLett.74.3828.

Shannon, R.D., and Prewitt, C.T., 1969, Effective ionic radii in oxides and fluorides: Acta Crystallographica, v. B25, p. 925–946, doi: 10.1107/S0567740869003220.

Shen, G., Sturhahn, W., Alp, E.E., Zhao, J., Toellner, T.S., Prakapenka, V.B., Meng, Y., and Mao, H.-K., 2004, Phonon density of states in iron at high pressures and high temperatures: Physics and Chemistry of Minerals, v. 31, p. 353, doi: 10.1007/s00269-004-0403-1.

Sherman, D.M., 1988, High-spin to low-spin transition of iron(II) oxides at high pressures: Possible effects on the physics and chemistry of the lower mantle, in Ghose, S., Coey, J.M.D., and Salje, E., eds., Structural and Magnetic Phase Transitions in Minerals: New York, Springer Verlag, p. 113–128.

Sherman, D.M., 1991, The high-pressure electronic structure of magnesiowüstite (Mg,Fe)O: Applications to the physics and chemistry of the lower mantle: Journal of Geophysical Research, v. 96, p. 14,299–14,312.

Sherman, D.M., and Jansen, H.J.F., 1995, First-principle prediction of the high-pressure phase transition and electronic structure of FeO: Implications for the chemistry of the lower mantle and core: Geophysical Research Letters, v. 22, p. 1001–1004, doi: 10.1029/94GL03010.

Singwi, K.S., and Sjölander, A., 1960, Resonance absorption of nuclear gamma rays and the dynamics of atomic motions: Physical Review, v. 120, p. 1093, doi: 10.1103/PhysRev.120.1093.

Sinogeikin, S.V., Zhang, J., and Bass, J.D., 2004, Elasticity of single crystal and polycrystalline MgSiO$_3$ perovskite by Brillouin spectroscopy: Geophysical Research Letters, v. 31, L06620, doi: 10.1029/2004GL019559.

Struzhkin, V.V., Mao, H.-k., Hu, J., Schwoerer-Böhning, M., Shu, J., Hemley, R.J., Sturhahn, W., Hu, M.Y., Alp, E.E., Eng, P., and Shen, G., 2001, Nuclear inelastic X-ray scattering of FeO to 48 GPa: Physical Review Letters, v. 87, 255501, doi: 10.1103/PhysRevLett.87.255501.

Sturhahn, W., 2000, CONUSS and PHOENIX: Evaluation of nuclear resonant scattering data: Hyperfine Interactions, v. 125, p. 149, doi: 10.1023/A:1012681503686.

Sturhahn, W., 2001, Phase problem in synchrotron Mössbauer spectroscopy: Physical Review B, v. 63, 094105, doi: 10.1103/PhysRevB.63.094105.

Sturhahn, W., 2004, Nuclear resonant spectroscopy: Journal of Physics: Condensed, v. 16, p. S497–S530, doi: 10.1088/0953-8984/16/5/009.

Sturhahn, W., and Chumakov, A.I., 1999, Lamb-Mössbauer factor and second-order Doppler shift from inelastic nuclear resonant absorption: Hyperfine Interactions, v. 123/124, p. 809, doi: 10.1023/A:1017060931911.

Sturhahn, W., and Kohn, V., 1999, Theoretical aspects of inelastic nuclear resonant scattering: Hyperfine Interactions, v. 123/124, p. 367, doi: 10.1023/A:1017071806895.

Sturhahn W., Toellner, T.S., Alp, E.E., Zhang, X., Ando, M., Yoda, Y., Kikuta, S., Seto, M., Kimball, C.W., and Dabrowski, B., 1995, Phonon density of states measured by inelastic nuclear resonant scattering: Physical Review Letters, v. 74, p. 3832, doi: 10.1103/PhysRevLett.74.3832.

Sturhahn, W., Alp, E.E., Toellner, T.S., Hession, P., Hu, M., and Sutter, J., 1998, Introduction to nuclear resonant scattering with synchrotron radiation: Hyperfine Interactions, v. 113, p. 47, doi: 10.1023/A:1012607212694.

Sturhahn, W., L'abbé, C., and Toellner, T.S., 2004, Exo-interferometric phase determination in nuclear resonant scattering: Europhysics Letters, v. 66, p. 506, doi: 10.1209/epl/i2003-10235-7.

Sturhahn, W., Jackson, J.M., and Lin, J.-F., 2005, The spin state of iron in minerals of Earth's lower mantle: Geophysical Research Letters, v. 32, L12307, doi: 10.1029/2005GL022802.

Toellner, T.S., 2000, Monochromatization of synchrotron radiation for nuclear resonant scattering experiments: Hyperfine Interactions, v. 125, p. 3–28, doi: 10.1023/A:1012621317798.

Toellner, T.S., Hu, M.Y., Sturhahn, W., Bortel, G., Alp, E.E., and Zhao, J., 2001, Crystal monochromator with a resolution beyond 10^8: Journal of Synchrotron Radiation, v. 8, p. 1082.

Tolstikhin, I., and Hofmann, A.W., 2005, Early crust on top of the Earth's core: Physics of the Earth and Planetary Interiors, v. 148, p. 109–130, doi: 10.1016/j.pepi.2004.05.011.

Trampert, J., Deschamps, F., Resovski, J., and Yuen, D., 2004, Probabilistic tomography maps chemical heterogeneities throughout the lower mantle: Science, v. 306, p. 853–856, doi: 10.1126/science.1101996.

Tse, J., Klug, D., Zhao, J., Sturhahn, W., Alp, E., Baumert, J., Gutt, C., and Johnson, M., 2005, Anharmonic motions of Kr in the clathrate hydrate: Nature Materials, doi: 10.1038/nmat1525.

van der Hilst, R.D., and Kàrason, H., 1999, Compositional heterogeneity in the bottom 1000 kilometers of Earth's mantle: Toward a hybrid convection model: Science, v. 283, p. 1885–1888, doi: 10.1126/science.283.5409.1885.

Visscher, W.M., 1960, Study of lattice vibrations by resonance absorption of nuclear gamma rays: Annals of Physics, v. 9, p. 194–210, doi: 10.1016/0003-4916(60)90028-2.

Weast, R.C., 1984, CRC Handbook of Chemistry and Physics (65th edition): Boca Raton, Florida, CRC Press, 204 p.

Weiss, H., and Langhoff, H., 1979, Observation of localized modes in TbO_2 using the Mössbauer effect: Zeitschrift für Physik B, v. 33, p. 365, doi: 10.1007/BF01319926.

Wood, B.J., and Rubie, D.C., 1996, The effect of alumina on phase transformations at the 660-kilometer discontinuity from Fe-Mg partitioning experiments: Science, v. 273, p. 1522–1524, doi: 10.1126/science.273.5281.1522.

Wortmann, G., Rupprecht, K., and Giefers, H., 2002, High-pressure studies of magnetism and lattice dynamics by nuclear resonant scattering of synchrotron radiation: Hyperfine Interactions, v. 144/145, p. 103–117, doi: 10.1023/A:1025493303182.

Yabashi, M., Tamasaku, K., Kikuta, S., and Ishikawa, T., 2001, X-ray monochromator with an energy resolution of 8×10^{-9} at 14.41 keV: The Review of Scientific Instruments, v. 72, p. 4080–4083, doi: 10.1063/1.1406925.

Zhang, L., Stanek, J., Hafner, S.S., Ahsbahs, H., Grünsteudel, H.F., Metge, J., and Ruffer, R., 1999, ^{57}Fe nuclear forward scattering of synchrotron radiation in hedenbergite $CaFeSi_2O_6$ at hydrostatic pressures up to 68 GPa: The American Mineralogist, v. 84, p. 447.

Zhao, J., Sturhahn, W., Lin, J.-F., Shen, G., Alp, E.E., and Mao, H.-K., 2004, Nuclear resonant scattering at high pressure and high temperature: High Pressure Research, v. 24, p. 447–457, doi: 10.1080/08957950412331331727.

MANUSCRIPT ACCEPTED BY THE SOCIETY 14 AUGUST 2006

Single-crystal structure and electron-density analyses of Earth's interior under high-pressure and high-temperature conditions using synchrotron radiation

Takamitsu Yamanaka[†]

Department of Earth and Space Science, Graduate School of Science, Osaka University, Machikaneyama Toyonaka, Osaka 560-0043, Japan

ABSTRACT

Crystal structures together with physical properties under nonambient conditions are significant subjects in the effort to understand geophysical phenomena or solid-state physics. Miniature diamond anvil pressure cell (DAC) and multianvil high-pressure apparatus have become effective tools for the observation of pressure effects on crystalline materials, not only for X-ray diffraction measurements but also for physical property studies such as electrical conductivity and magnetism. These high-pressure studies have been made at high temperatures by electric resistance heater or laser and at low temperatures by cryostat.

For the last twenty years, synchrotron radiation facilities have accelerated the study of high-pressure crystallography because of their great advantages for diffraction studies at nonambient conditions. Application of synchrotron radiation enhances structure analyses as a function of pressure. Pressure dependence of electron-density distributions around atoms is elucidated by single-crystal diffraction study using deformation electron-density analysis. In this study, compression mechanisms were investigated through structure analyses. The maximum entropy method (MEM) based on the observed structure $F_{obs}(hkl)$ of reflection hkl was applied to reveal an electron-density map, and the results were compared with difference Fourier synthesis based on $F_{obs}(hkl) - F_{calc}(hkl)$. Radial electron distribution revealed the localization or delocalization of electrons around atomic positions together with bonding electron densities. The diffraction intensity measurements of $FeTiO_3$ ilmenite and γ-SiO_2 stishovite single crystals were made at high pressures. In both cases, the valence electrons became more localized around the cations with increasing pressure. This is consistent with molecular orbital calculations—both methods show that the bonding electron density becomes smaller with pressure. The thermal displacement parameters of both samples were reduced with increasing pressure.

Keywords: electron-density distribution, high-pressure transition, maximum entropy method, single-crystal structure analysis, Earth's interiors.

[†]E-mail: t.yamanaka@kce.biglobe.ne.jp

INTRODUCTION

A half-century of developments in high-pressure apparatus have enhanced in situ investigations of materials constituting Earth's interior under extreme conditions. Because many minerals in the crust and mantle undergo structural transformations, solid-solid reaction, exsolution, decomposition, and melting under high-pressure conditions, in situ high-pressure experiments are mandatory for a proper understanding of Earth's interior.

Understanding of crystal structures together with physical properties of Earth materials under nonambient conditions is an important step in the effort to understand geophysical phenomena. Since miniature diamond anvil high-pressure cell (DAC) and multianvil high-pressure apparatus were developed and improved over the last three decades, numerous studies of crystal structure and equations of state (EOS) have been repeatedly carried out on materials of geophysical interest. These high-pressure apparatuses have become effective tools for observations of pressure effects on crystalline materials, not only by X-ray diffraction measurements but also through the study of physical properties such as electric conductivity and magnetism. Optical investigations by Raman and infrared spectroscopy also take advantage of DAC methods. These studies have been executed using a single crystal or polycrystalline samples under compression at high temperatures generated by electric resistive heaters or lasers and at low temperatures using cryostats.

During the last twenty years, synchrotron radiation facilities have been used to develop high-pressure crystallography because of their great advantages for diffraction studies at nonambient conditions. Application of synchrotron radiation enhances structural analyses as a function of pressure, and many reports on crystal structures using DAC have been published.

VARIOUS PRESSURE-INDUCED PHASE TRANSFORMATIONS

Dynamic processes of structural change are significant topics in high-pressure crystallography. Structure changes under compression, such as phase transformation, lattice deformation, cation ordering change, decomposition, and amorphization, and phase transitions under compression due to lattice instability, electronic state change, and magnetic spin ordering can be elucidated by X-ray diffraction, absorption, and resonance phenomena using DAC together with other X-ray inelastic measurements. Developments of various types of detectors and the intense X-ray from synchrotron radiation have accelerated the studies on kinetics and dynamics of structure changes and phase transitions. Many different types of phase transformations have been reported under static high-pressure conditions during the last twenty years, as shown in Table 1.

EFFECT OF PRESSURE ON CRYSTAL STRUCTURE

A macroscopic system under pressure can be quantitatively expressed by statistical thermodynamics. Pressure (P), volume (V), and temperature (T) are interconnected by an equation of state. The virial equation of state in a power series is

$$P - \frac{Nk_BT}{V}\left[1 + \frac{NC_2(T)}{V} + \frac{N^2C_3(T)}{V^2} + \ldots\right] = 0, \quad (1)$$

where $C_2(T)$ and $C_3(T)$ are virial coefficients. N denotes a constant, k_B indicates the Boltzmann constant.

Virial theorem gives a view of the compression mechanism of substances under pressure.

TABLE 1. VARIOUS TYPES OF THE PRESSURE-INDUCED STRUCTURE TRANSFORMATIONS

Lattice change	Electronic state	Spin-lattice interaction
Displacive		
No coordination change	Electronic charge	Jahn-Teller transition
Lattice type change	$2Fe_{4+} \Leftrightarrow Fe_{3+} + Fe_{5+}$	CuO, $NaNiO_2$
Mn_2O_3	$CaFeO_3$	
Molecular dissociation	$2Au_{2+} \Leftrightarrow Au_{1+} + Au_{3+}$	
Br_2, I_2, SnI_4	$CsAuI_3$	
	$2Mn^{3+} \Leftrightarrow Mn_{2+} + Mn_{4+}$	
	Mn_2O_3	
Order-disorder	Density of state	High-low spin transition
$(K, Na)(Al, S)_4O_8$	$s \Leftrightarrow d$ orbital: Cs, Rb, K, Ca	FeS_2, FeO, CoO,
$(Ca, Sr)TiO_3$	$f \Leftrightarrow d$ orbital: Ce	Co_2O_3, $\alpha\text{-}Fe_2O_3$
Reconstructive	Orbital rearrange	
Polyhedral joint change	K_2CuF_3 (ferro? antiferro)	Spin-Peierls transition
Corner $\Leftrightarrow$ edge joint		$CuGeO_3$
$MgSiO_3$, Al_2SiO_5	Band overlap	
Edge $\Leftrightarrow$ face joint	O_2 I_2 H_2	
Mg_2SiO_4, $MgSiO_3$	Mott transition	
Coordination change	NiO, FeO, O_2	
$3 \rightarrow 4$ fold C, BN		
$4 \rightarrow 6$ fold SiO_2, $MgSiO_3$, $ZnSiO_3$		
$6 \rightarrow 7$ fold Sr_2PbO_4, Mn_2GeO_4, Mn_2SnO_4		
$6 \rightarrow 8$ fold SrO, CaO, SnO_2, PbO_2		
$7 \rightarrow 8 \rightarrow 9$ fold ZrO_2, HfO_2		

Pressure provided by the external system P_{ext} can be expressed by the following equation:

$$P_{ext} = \frac{Nk_B T}{V} - \frac{1}{3V}\left[\sum_{j=1}^{N}\sum_{j>1}^{N}\left(-\frac{\partial \psi_{ij}}{\partial r_{ij}}\right)r_{ij}\right]$$
$$= \frac{Nk_B T}{V} - \frac{1}{3V}\left[\sum_{j=1}^{N}\sum_{j>1}^{N} F_{ij} \cdot r_{ij}\right], \quad (2)$$

where V is the cell volume, N is the number of particles, ψ_{ij} is an internal potential between atom i and j, r_{ij} is the interatomic distance between atom i and j,

$$\text{and } F_{ij}\left(=-\frac{\partial \psi_{ij}}{\partial r_{ij}}\right)$$

is the interatomic force that comes out from atom j to atom i.

This equation implies that the external pressure can be represented by

$$\sum_{j=1}^{N}\sum_{j>1}^{N} F_{ij} \cdot r_{ij}$$

in the equilibrium state with pressure of the compressed substance. Stress is represented by the following tensors:

$$P = \begin{pmatrix} Pxx & Pxy & Pxz \\ & Pyy & Pyz \\ & & Pzz \end{pmatrix} \quad (3)$$

One stress component is

$$P_{\alpha\beta} = \frac{1}{V}\left(Nk_B T_{\alpha\beta} - \left\langle \sum\sum \frac{|F_{ij}|}{|r_{ij}|} r_{ij\alpha} r_{ij\beta}\right\rangle\right) \quad (4)$$

and

$$T_{\alpha\beta} = \frac{m_i V_{i\alpha} V_{i\beta}}{Nk_B}, \quad (5)$$

where $v_{i\alpha}$ and $v_{i\beta}$ indicate velocity of particle i. m_i represents the mass of i-th component. α and β are the position in the structure.

Structure analysis under high pressure evaluates r_{ij} as a function of pressure. Positional information of atoms can be observed by X-ray or neutron diffraction intensity measurement. F_{ij} can be elucidated by lattice dynamical experiments, such as Raman and infrared spectroscopy under high pressure. Electron spectroscopy for chemical analysis and photoelectron spectroscopy are also candidates for providing interatomic force. Pressure dependence of volume V is induced from the energy change $F_{ij} \cdot r_{ij}$ with pressure. Consequently, the energy change is reflected by volume compression, which is expressed by the equation of state.

DIAMOND ANVIL CELL FOR SINGLE-CRYSTAL STRUCTURE ANALYSIS

X-ray diffraction study under high pressure using diamond anvil cell (DAC) was originally executed by camera method (Weir et al., 1965). The first DAC suitable for single-crystal X-ray diffraction experiments was developed by Merrill and Bassett (1975). The design of the cells has been improved over the last three decades by Keller and Holzapfel (1977), Yamaoka et al. (1979), Mao et al. (1988), Schiferl (1977), Ahsbahs (1984), and Koepke et al. (1985). Some cells are specialized for powder diffraction study at high pressures over megabar regions or measurements of physical properties. Powder diffraction has been achieved by DAC at pressures over 300 GPa (Mao et al., 1978). Combined with either laser or internal electric resistive heater, high-temperature and high-pressure X-ray diffraction measurements have been carried out by Ming and Bassett (1974), Hazen and Finger (1981a, 1981b), Alaska et al. (1983), Arashi and Ishigami (1982), Boehler et al. (1986), Mao et al. (1987), Jeanloz and Heinz (1984), and Heinz and Jeanloz (1987).

Finger and King (1978) and Denner et al. (1978) established the basic techniques of diffraction intensity measurement using DAC for the single-crystal structure and data correction such as absorption correction of beryllium backing plate and diamond anvil. Single-crystal structure analyses under pressure often encounter the following difficulties in intensity measurements: (1) hydrostaticity, (2) a large blind region due to the pressure cell, (3) X-ray absorption, (4) small sample space, and (5) limitation of compression. The DAC used in this study (Fig. 1) solves some of these problems and reveals electron-density distributions under pressures up to 30 GPa (Yamanaka et al., 2001). High-pressure single-crystal X-ray diffraction studies have been reported previously on ruby and pyrope to 31 GPa and 33 GPa by Kim-Zajonz et al. (1998) and Zhang et al. (1998), respectively.

Generally, because of their very low X-ray absorption, beryllium hemispheres or plates have been used as backing plates for single-crystal diffraction measurement. However, the backing plates give many broad and fairly strong spotty powder rings. These rings often overlap with the diffraction peaks of the sample and cause ambiguity in peak profiles and peak intensities. Large single-crystal diamond backing plates overcome these problems because of the absence of powder rings and relatively small X-ray absorption (Yamanaka et al., 2001). In addition, the absorption correction for diffraction intensities is much simpler because both backing plates and anvils are made of diamond. Beryllium plates or disk windows do not permit diffraction studies at pressures over 20 GPa because of their softness. However, use of the diamond backing plates enables diffraction studies over 50 GPa and a wide window with maximum diffraction angle of 80°. Use of DAC with the backing plates combined with short wavelength enables observation of a large number of reflections within the value of Q < 5.1303 (d > 0.44150 Å). These advantages can clarify electron densities and atomic thermal displacement as a function of pressure. Since single-crystal diamond plates are transparent

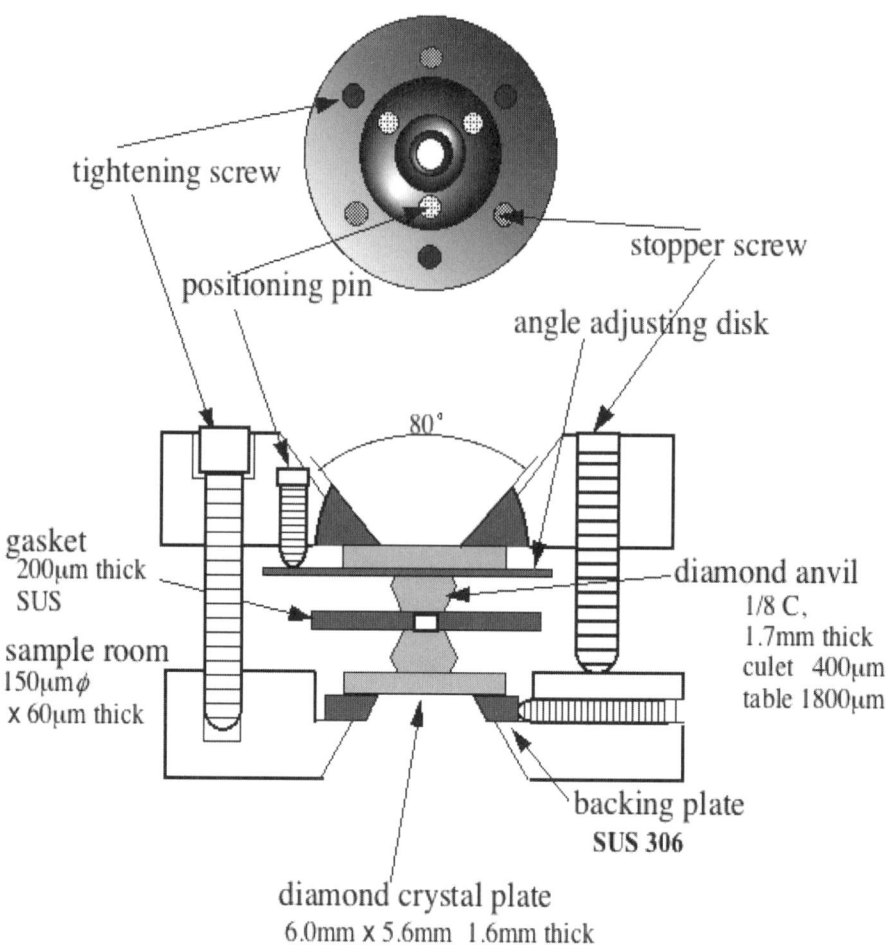

Figure 1. Diamond anvil cell for X-ray single-crystal diffraction study. Large single-crystal diamonds were used as the windows. Anvils of 1/8-carat brilliant cut diamonds were applied. Both diamond plates and anvils were directly fixed on (100) plane of both crystals.

widows, these plates are very convenient for alignment of opposed diamond anvil and setting sample in DAC apparatus.

HIGH-PRESSURE STRUCTURE STUDY USING SYNCROTRON RADIATION

Since the first synchrotron radiation studies, structure studies as well as phase stability studies have been more intensively made under high pressure using DAC or multianvil apparatus. For example in SPring-8 (Super Phone ring 8Gev) in Japan, seven beam lines are designed for high-pressure studies by means of X-ray Absorption Fine Structure, seven-axes diffractometer, powder diffractometer with laser heating system or low-temperature cooling system, inelastic scattering spectroscopy, infrared spectroscopy, and Mössbauer systems. High-pressure studies have often encountered difficulties with in situ observation. Significant progress with in situ diffraction of materials constituting Earth's interior under extreme conditions has been made using synchrotron radiation facilities.

Synchrotron radiation has excellent beam characteristics for high-pressure diffraction studies using DAC. The brilliance of the synchrotron radiation emitted from the bending magnet is $\sim 10^{12}$ (photons s^{-1} mm^{-2} $mrad^{-2}$, 0.1% band width), which is $\sim 10^4$ times larger than that of the conventional rotating anode X-ray generator (10^8 photons $s^{-1\,mm-2}$ $mrad^{-2}$, 0.1% band width). Insertion devices such as wigglers or undulators further enhance the source intensity.

The high source intensity of synchrotron radiation has the following advantages: intensive transmittance for anvil or window in the X-ray path, high signal-to-noise (S/N) ratio of the diffraction intensity, detection of weak diffraction peaks, narrow peak profile, and the short time interval of measurement.

Because of white or continuous X-ray provided by synchrotron radiation, the wavelength can be tuned and optimized to provide suitable wavelengths for high-pressure diffraction studies. Since the small aperture angle of DAC or multianvil high-pressure apparatuses gives a limited diffraction angle, a short wavelength has the advantage of providing access to a large area of reciprocal space with a large Q value (= $1/d^2$). In the present experiment, a synchrotron radiation beam with 8 GeV and 100 mA of beam current at SPring-8 provided a critical wavelength of ~30 keV ($\lambda = 0.413$ Å), where the maximum photon count is expected in continuous energy distribution. The short wavelength enabled observations of higher-order reflections, and the number of observed reflections was four times more than those from the laboratory source.

Synchrotron radiation beam convergence by Pt-coated mirror was produced as an incident beam through the evacuated guide pipe, which enormously reduced the background intensity. A collimator of 100 μm in diameter was generally adopted, because the gasket hole was 200 μm and the sample size was several ten of microns across. A receiving slit of 1° angle was used. An evacuated collimator and guide pipe were located near the DAC.

STRUCTURE TRANSITIONS UNDER NONHYDROSTATIC AND HYDROSTATIC CONDITIONS

A mixture of ethanol and methanol was used as a pressure-transmitting medium to preserve hydrostatic condition for diffraction intensity measurement at pressures up to 15 GPa. Argon gas was also applied under pressures over 10 GPa, but the inert gas at that point solidifies and then does not guarantee hydrostaticity. Helium gas was also used as a medium. However, owing to the great compressibility of helium, the shrinkage of the gasket hole under those pressures could not hold enough space for the sample under pressures >10 GPa.

Nonhydrostaticity at high pressure affects phase transformation behavior because shear stress in lattice does not necessarily lead to a thermodynamically stable phase. Nonhydrostatic compression may result in metastable states or intermediate states, because they are formed in the dynamical process. Besides the stress-strain relation, crystallite size is an effective parameter of a phase transition under compression. Metastable or unstable phases have been formed by the Ostwald's step rule.

Phase transformation under compression is not necessary to produce a thermodynamically stable phase. Virial theorem shown in Equation 2 proposes the relation of structure stability under equilibrium conditions between summations of interatomic potential and surrounding pressure P_{ext} around the crystal, from the viewpoint of a statistical process, macroscopic time, and space average of the structure transition under pressure.

Ostwald's step rule suggests that multiple transformation pathways are possible. Metastable phases having small energy barriers, i.e., activation energies lower than that of the stable phase, can form during the transformation process. Anisotropic stress easily accelerates several steps of metastable phases. The sheer stress P_{ij} (Eqn. 4) induced from the nonhydrostatic compression brings about lattice deformation and gliding of the slabs, such as that which occurs in the martensitic transitions in random stacking. The phase transition is possible by the nondiffusion martensitic transformation mechanism under a stress field.

The kinetics of a structural transition under a stress field such as the martensitic transition urge a metastable phase. One example of nonhydrostatic structure changes is $FeGeO_3$. X-ray diffraction experiments of $FeGeO_3$ at high pressures under nonhydrostatic condition show a sequential high-pressure transformation: high-pressure–type clinopyroxene $FeGeO_3$(I) ($C2/c$, $Z = 4$) transforms to $FeGeO_3$(II) pseudo-ilmenite ($P\bar{1}$, $Z = 4$) to ilmenite type ($R\bar{3}$, $Z = 6$). $FeGeO_3$(II) is a metastable phase that occurs by a slight deformation of the oxygen arrangement and short-range movement of cations (Fig. 2). The volume changes of $FeGeO_3$ (I), $FeGeO_3$ (II), and ilmenite-type $FeGeO_3$ are shown in Figure 3. These transitions belong to a reconstructive-type transformation; however, they often have a topotaxic structure relation. $FeGeO_3$ (I) transforms to ilmenite structure through a

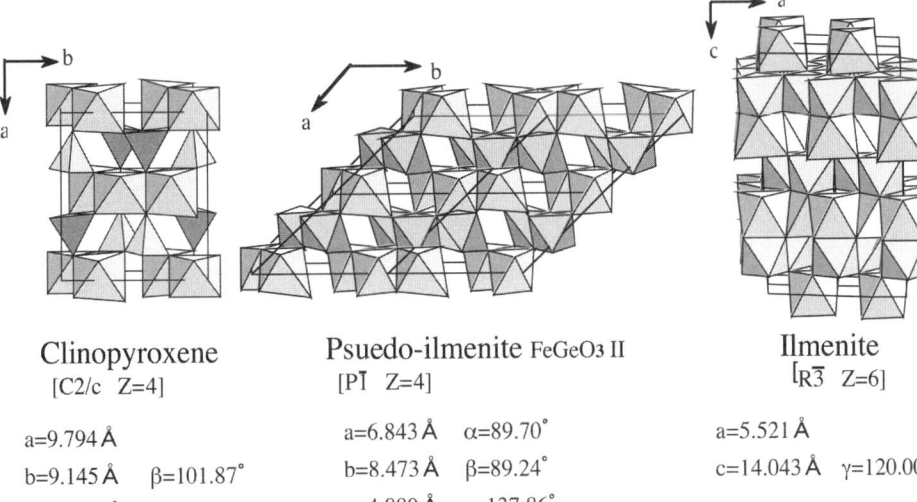

Figure 2. $FeGeO_3$ high-pressure polymorphs under nonhydrostatic conditions. Clinopyroxene (C2/c, Z = 4) transforms to a metastable form of $FeGeO_3$ II (P$\bar{1}$, Z = 4) (pseudo-ilmenite) under nonhydrostatic stress conditions. Under higher compression, it shows a high-pressure stable form of ilmenite-type structure (R$\bar{3}$, Z = 6).

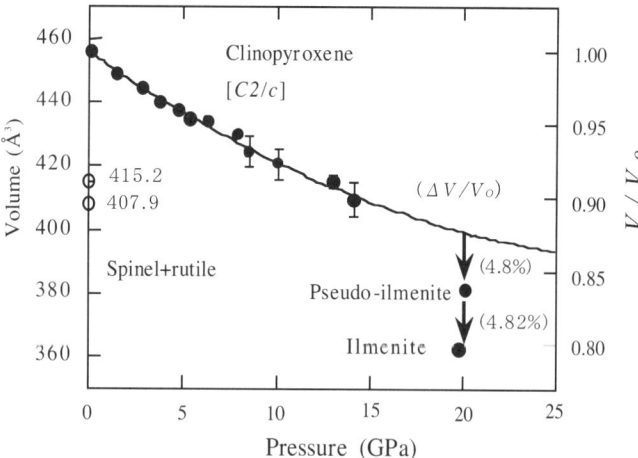

Figure 3. Volume change of $FeGeO_3$ high-pressure polymorphs (clinopyroxene–pseudo-ilmenite–ilmenite) with pressure. The numbers in parenthesis are volume changes ($\Delta V/V_0$) at structure transitions.

low-activation-energy path under the thermodynamically low-temperature conditions (Hattori et al., 2001).

The phase transition is induced by the nondiffusion martensitic transformation mechanism under a stress field. Molecular dynamics (MD) calculation reproduces the nonhydrostatic transformations and provides an anisotropic property in the structure. MD calculation suggests a mechanism of pressure-induced modification that is a precursor to atomic movement during the transition process. High-pressure transformation should be controlled by the elastic stability, which is derived by Born criteria. Consequently, a nonhydrostatic condition often reduces a critical transition pressure and produces metastable phases. First-order phase transitions show a hysteresis against pressure.

In order to conduct kinetic studies of pressure-induced transformations, the current devices can perform in situ and time-dependent measurements of diffraction intensities with high accuracy. The transition rate can be determined by a profile-fitting method (Yamanaka et al., 1992). Time-resolved diffraction study of the phase transition becomes possible by combination of a strong source with a quick photon counting system, because the time-dependent structure analysis requires repetition of the diffraction measurement within a short time interval. Time-resolved and in situ diffraction studies using synchrotron radiation illuminate the dynamical observation of pressure-induced structure changes. Also, the present system offers more reliable structure information under pressure.

SINGLE-CRYSTAL STRUCTURE REFINEMENT AT HIGH PRESSURE

A pre-indented gasket hole holds a single-crystal sample of several ten of microns in diameter and thickness, ruby chip pressure maker, and pressure-transmitting media. The pressure measurement is conducted by the ruby fluorescence technique (Mao et al., 1986). Lattice constants of the sample under pressure were determined by the least-squares calculation with 2θ values giving ~25 independent reflections. The volume change of V/V_0 with pressure provides the isothermal bulk modulus K and its pressure derivative dK/dP using the Birch-Murghanan equation of sate.

Diffraction intensity measurement applies 2θ scanning with fixed ϕ-rotation axis. The observed structure factor $F_{obs}(hkl)$ of reflection hkl is obtained from the observed diffraction intensity of $I_{obs}(hkl)$.

Reflections with $|F_{obs}| > 3\sigma|F_{obs}|$ are used for the structure refinement.

The structure factor $F_{calc}(h)$ of reflection hkl is expressed by

$$F_{calc}(\vec{h}) = K \sum_j a_j f_j(\vec{h}) \sum T_{js}(\vec{h}) \exp\{2\pi i(hx_{js} + ky_{js} + lz_{js})\}, \quad (6)$$

where K is a scale factor, f_j represents the atomic scattering factor of atom j, a_j is a constant and

$$f(\vec{h}) = \int_{atom} \rho(\vec{r}) \exp(2\pi i \vec{h} \cdot \vec{r}) dv. \quad (7)$$

An inverse Fourier transform of the structure factor leads the electron density in real space by

$$\rho(x,y,z) = \frac{1}{V} \sum_h \sum_k \sum_l F(hkl) \exp\{-2\pi i(hx + ky + lz)\}. \quad (8)$$

The structure refinement using each data set previously has been conducted using the full matrix least-squares program (Sasaki and Tsukimura, 1987). Atomic scattering factors for neutral or ionized atoms are taken from International Tables for X-Ray Crystallography (1974). Scale factor, atomic coordinates, isotropic temperature factors, and isotropic extinction parameter are chosen as the variable parameters. Initial structure parameters are inferred from the previously reported structure based on the data take at ambient condition.

The reliability factor (R) of the least-squares refinement is defined as

$$\begin{aligned}
R &= \Sigma(||F_{obs}| - |F_{calc}||)/\Sigma|F_{obs}| \\
R &= \Sigma w(||F_{obs}| - |F_{calc}||)/\Sigma w|F_{obs}| \\
&\text{or} \\
R &= [\Sigma(||F_{obs}| - |F_{calc}||^2)/\Sigma|F_{obs}|^2]^{1/2} \\
R &= [w\Sigma(||F_{obs}| - |F_{calc}||^2)/\Sigma w|F_{obs}|^2]^{1/2},
\end{aligned} \quad (9)$$

where $w = 1/\sigma^2(|F_{obs}|)$.

After the above-structure refinement, anharmonic thermal parameters can be refined on the basis of the Gram-Charier expansion or cumulant expansion in consideration of site symmetry restrictions on thermal-motion tensor coefficients given in International Tables for X-Ray Crystallography (1974).

The temperature factor $T(h)$ can be explained by the Fourier transform of the probability density function $P(u)$ of atomic thermal displacement (u) parameters (Willis and Pryor, 1975; Zucker and Schulz, 1982; Kontio and Stevens, 1982). The anharmonic potential can generate the true atomic thermal displacement, resulting in the thermal expansion of cell volume (Yamanaka et al., 1984; Yamanaka and Morimoto, 1996).

According to the pseudo-potential model, inner-core electrons are frozen with bonding effects. But valence electron clouds are breezing by neighboring coordination and thermal atomic vibration due to the weak bond to the nuclei. Their interactions with the core electrons are relatively weak. Accordingly, they are more sensitive to the interatomic potential affected by the coordination of the adjacent atoms. Since the deformation electron densities are assumed to be very small because of spherical electron orbits except excitation of d-electron, a monopole refinement is applied instead of the multipole deformation density. The κ parameter (Coppens et al., 1979; van der Wal and Stewart, 1984) was applied in the atomic scattering factor, which is an indicator of the radial distributions of electrons. The atomic scattering factor $f(s)$ was modified from a Hartree-Fock approximation based on the isolated atom model. Perturbed valence electron density was

$$f(s/2) = \Sigma[p_{j,\text{core}} f_{j,\text{core}}(s/2) + p_{j,\text{valence}} f_{j,\text{valence}}(\kappa_j, s/2) + f'_j + if''_j]. \quad (10)$$

The valence scattering of the perturbed atom at $s/2 (= \sin\theta/2\lambda)$ is given by

$$f_{\text{M-core}}(\kappa_j, s/2) = f_{j,\text{M-core(free atom)}} (\sin\theta/\lambda \cdot 1/\kappa_j). \quad (11)$$

A localized electron distribution such as indicated by $\kappa = 1.0$ implies more ionic character in the bonding nature. The detailed formalization is discussed in Yamanaka et al. (2000). The pressure change in the electron-density distribution can be disclosed by deformation electron density $\Delta\rho(xyz)$, which can be derived from difference Fourier transformation of $F_{\text{obs}}(hkl) - F_{\text{calc}}(hkl)$:

$$\Delta\rho(x,y,z) = \frac{1}{V} \sum_h \sum_k \sum_l \left\{ F_{\text{obs}}(hkl) - F_{\text{calc}}(hkl) \right\} \exp\{i\psi_{\text{calc}}(hkl)\} \exp\{-2\pi i(hx + ky + lz)\} \quad (12)$$

Stishovite Structure under High Pressure

We applied the new DAC design with diamond backing plates described here to the structure analysis of rutile-type IVb oxides MO_2 (M = Si, Ge, Sn) ($P4_2/mnm$, Z = 2). Sinclair and Ringwood (1978) were the first to undertake a single-crystal structure analysis of stishovite. Electron-density distribution in stishovite has been investigated by X-ray diffraction study (Hill et al., 1983; Spackman et al., 1987; Yamanaka et al., 2000). The crystal structure analyses of this phase have been carried out up to a pressure of 6 GPa (Sugiyama et al., 1987) and 16 GPa (Ross et al., 1990).

Molecular orbital calculation allows a precise discussion of electronic properties of bond nature. The bonding electron observed from the X-ray diffraction study is interpreted by the optimized pair potential and molecular orbital calculation (Svane and Antoncik, 1987; Mimaki et al., 2000). The electron orbital overlap and the bonding energy render the deformation of MO_6 octahedra of rutile-type phases from the bond character (Simunek et al., 1993; Camargo et al., 1996; Gibbs et al., 1997, 1998). The structure change with pressure is characterized by the molecular orbital overlap with metal-metal interaction. However, the electron-density distribution has not been studied yet under pressure due to the poor resolution of Fourier analysis.

Single-crystal structure analysis was carried out at 29.1 GPa. The lattice constants, atomic coordinates, and interatomic distances under pressure are presented in Table 2. The unit-cell volume at 29.1 GPa was reduced by as much as 9%. Isothermal bulk modulus K and dK/dP were determined by the volume change using the Birch-Murghanan equation of state (Table 3). The large K value indicates that stishovite is a noticeably hard crystal.

Structure refinements from two sets of diffraction intensities obtained at ambient pressure and 29.1 GPa provide valence electron densities in the unit cell. In order to estimate the valence electron distributions from the κ parameter, the reliable factor R was minimized with optimization by changing κ parameter and population parameter (P) in Equation 10. The value of the κ parameter for oxygen atom was 0.94 at ambient pressure and 1.11 at 29.1 GPa.

Fourier transform of $f(s)$, including the population parameter (P) of the valence electrons, defines electron density, and the spatial integration of $\rho(r)$ gives the effective charge of oxygen atoms. The κ parameters of Si and oxygen are presented in Table 4. The electron distributions become more localized with increasing pressure. A smaller κ parameter indicates more bonding electrons and intensifies more covalent-bond nature at high pressure. Our study of rutile-type MO_2 (M = Si, Ge, and Sn) indicates that the κ parameter of stishovite (SiO_2) has a relatively strong covalent bond in comparison with the other two compounds of GeO_2 and SnO_2 (Yamanaka et al., 2000).

After the refinement with the spherical-atom model, the deformations of electron distributions of stishovite at ambient pressure and 29.1 GPa are disclosed by difference Fourier map on the plane (110) as shown in Figure 4. The map of stishovite at ambient pressure is very similar to that of Hill et al. (1983) and Spackman et al. (1987). The nonspherical residual electron density around the Si site is caused by overlapping orbitals of d-electron of Si and p-electron of oxygen, resulting in d-p-π bond. Hence, the noticeable residual electron density on the Si-O bond indicates a bonding electron. A large negative den-

TABLE 2. RESULTS OF STRUCTURE REFINEMENTS OF STISHOVITE AT HIGH PRESSURES					
Pressure	1 atm	5.23 GPa	9.26 GPa	12.3 GPa	29.1 GPa
Diffractometer	Rigaku AFC5	Rigaku AFC6R			Huber (512.1)
wavelength	MoKα (0.7107Å)	AgKα (0.5608 Å)			SR (0.4077 Å)
Energy	150 kV, 50 mA	150 kV, 50 mA			8 GeV, 100 mA
Monochrometer	Graphite (002)	Graphite (002)			Si(111) double Crystal
Gasket	–	Spring steel			Spring steel
Pressure media	–	M:E:W = 16:3:1 Ar			
Scan mode	ω–2θ	φ-fix ω-scan			φ-fix ω-scan
Crystal size (μm)	50 × 60 × 80	40 × 40 × 60			20 × 20 × 40
2θ angle	120	53	47	54	49
Sin θ/λ	1.219	0.794	0.714	0.801	1.011
2θ angle/MoKα	120	69	61	70	92
Ref.(observed)	210	81	79	82	147
Ref. (independent)	126	25	25	26 57	
a (Å)	4.1812(1)	4.152(1)	4.134(1)	4.118(2)	4.044(6)
c (Å)	2.6662(3)	2.6590(8)	2.6540(7)	2.649(1)	2.619(20)
c/a	0.6377	0.6404	0.6420	0.6433	0.6476
V (Å^3)	46.61	45.84	45.36	44.92	42.83
No. ref.	126	25	25	26	36
R(F)	0.0253	0.0440	0.0312	0.0345	0.0330
wR(F)	0.0243	0.0234	0.0104	0.0227	0.0282
Si (000)	–	–	–	–	–
β11	0.0045(1)	0.0126(27)	0.0055(21)	0.0088(20)	0.0035(11)
β33	0.0037(5)	0.0261(18)	0.0142(12)	0.0103(13)	0.0131(13)
β12	0.0002(2)	0.0004(21)	0.0021(14)	0.0009 (12)	0.0019(15)
O (xx0)	0.3063(1)	0.3063(20)	0.3056(9)	0.3058(19)	0.3039(7)
β11	0.0051(2)	0.0075(37)	0.0051(29)	0.0104(21)	0.0095(13)
β33	0.0036(3)	0.0031(17)	0.0090(28)	0.0054(18)	0.0074(17)
β12	–0.0009(3)	0.0005(35)	0.0009(29)	0.0007(35)	0.0004(16)
Si-O(eq) (Å)	1.7559(9)	1.750(11)	1.748(8)	1.742(13)	1.724(3)
Si-O(ap) (Å)	1.8111(9)	1.798(4)	1.784(2)	1.781(4)	1.738(2)
ap/eq	1.0314	1.0274	1.0205	1.0223	1.0081
Vol (SiO$_6$) (Å^3)	7.374	7.266	7.178	7.134	6.806
O1-O2 (sh) (Å)	2.2906(10)	2.277(5)	2.275(3)	2.262(5)	2.242(3)
O1-O1(unsh) (Å)	2.6662(3)	2.6590(8)	2.6540(7)	2.649(1)	2.619(2)
O1-O3 (Å)	2.5226(4)	2.509(10)	2.498(10)	2.482(17)	2.448(2)
sh/unsh	0.8591	0.8563	0.8572	0.8539	0.8567

Notes: R = $\Sigma(||F_{obs}|-|F_{calc}||)/\Sigma w|F_{obs}|$ and Rw = $[\Sigma w(||F_{obs}|^2 - |F_{calc}|^2|)/\Sigma w|F_{obs}|^2]^{1/2}$, where $w = 1/\sigma^2(|F_{obs}|)$. Abbreviations of equatorial and apical bond are indicated by eq and ap and those of shared and unshared edge are indicated by sh and unsh, respectively.

TABLE 3. ISOTHERMAL BULK MODULUS OF STISHOVITE					
	K_T (GPa)	dK/dP	P_{max} (GPa)	Data	Sample and remark
Present data	292(13)	6(fixed)	29.1	5	Single crystal
Andrault et al. (1998)	291	4.29	53.2	17	Powder[†]
Ross et al. (1990)	302(5)	2.60(0.8)	16	6	Single crystal
Ross et al. (1990)[‡]	287(2)	6(fixed)			
Sugiyama et al. (1987)	313(4)	6(fixed)	6	9	Single crystal
Weidner et al. (1982)	306(4)				Brillion scattering

[†] The cell volume data are at pressures of 0.0001–15 GPa from Ross et al. (1990), 24.6–49.4 GPa from Hemley et al. (1994), and 48.1–53.2 GPa from Andrault et al. (1998).
[‡] When dK_T/dP is fixed to 6.

TABLE 4. κ PARAMETER, EFFECTIVE CHARGE, AND DIPOLE MOMENT OF Si IN STISHOVITE		
	Ambient condition	29.1 GPa
κ parameter of oxygen	0.94	1.11
Residual electron peak (position from Si)	0.86A	0.77A
Effective charge	+2.12(8)	+2.26(15)
Dipole moment	2.44	2.94

sity in the bridging O-O bond plays an important role for hindrance of cation repulsion. The bonding electron distribution at 29.1 GPa is less remarkable compared with that at 1 atm. This feature is caused by the observed κ parameter and effective charge. These data indicate that stishovite becomes more ionic with increasing pressure.

Interatomic distances (r_{Si-O}) represented by Equation 2 can be determined directly by the diffraction study as a function of pressure. The pressure effect on interatomic distances results in the compression of the lattice constants of a bulk crystal. X-ray diffraction study gives electron-density distributions that include valence electrons and bonding electrons. The charge density analysis based on the diffraction intensities provides the view of effective charge of ions and dipole moment. On the other hand, the term of interatomic force (F_{Si-O}) in Equation 2 can be determined by force constants from lattice dynamical experiments.

The charge distribution reveals a significant admixture of covalency in the chemical bonds of stishovite, and the appropriate charge state of the cations turns out to be far from a formal charge of Si^{4+} electron configuration. Our results are consistent with the energy band calculation (Svane and Antoncik, 1987). The significant *d*-electron population indicates some degree of nonsphericity of valence electron distribution around the cation. The difference Fourier map of stishovite reveals apparently nonspherical electron distribution around Si, as shown in Figure 4.

The bonding electrons are related to the overlapped orbits of *s* and *p* electrons with *d* electrons. In order to investigate the bond character of rutile-type phases SiO_2, GeO_2, and SnO_2, we carried out the molecular orbital calculation. The deformation

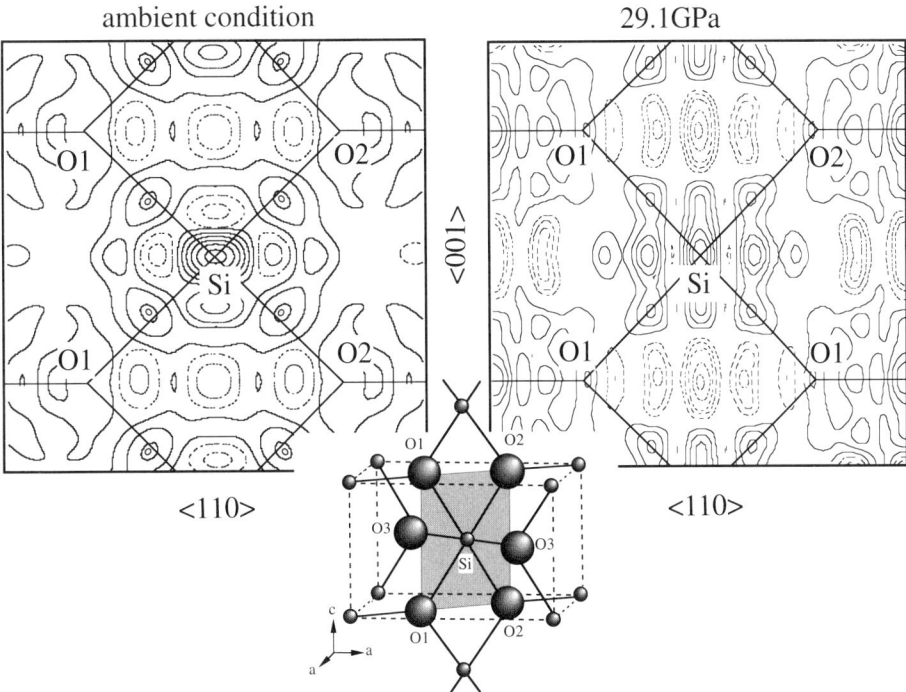

Figure 4. Difference Fourier map projected on (110) plane of stishovite SiO_2 at ambient pressure and 29.1 GPa. The cross-hatched plane is (110) plane, which is composed of equatorial oxygen atoms of O1 and O2 in rutile structure. Contours are at intervals of 0.2 eÅ^{-3}, and positive and negative contours are expressed by solid and broken lines, respectively. Residual valence electron density shows the distribution around cation and bonding electron distribution on Si-O bond.

electron-density map at 30 GPa obtained from molecular orbital calculation is shown in Figure 5. The detailed procedure of the calculation has been described in our previous paper (Mimaki et al., 2000).

Apparent dipole moment (μ) may be experimentally determined by the summation of product of charge (q) and interatomic distance (r):

$$\mu_{obs} = \sum_{atom} q_i r_i, \qquad (13)$$

where q_i is the same value as the population P_i obtained from the present κ refinement.

An apparent relative ionicity of rutile-type oxides can be expressed by μ_{obs}/μ_{ideal}, where μ_{ideal} is determined by the formal charge and interatomic distance. The result of the effective charge is presented in Table 4. The deformation of octahedral coordination SiO_6 of the rutile-type structures can be deduced from the degree of covalent bond character. The density of state (DOS) of Si and the band gap at 30 GPa and 0 K (Mimaki et al., 2000) are presented in Figure 6. A considerably large amount of d-electron density is found. The d-electrons of cations increase the degree of d-p-π bond in Si-O. The ratio between shared and unshared edge distance of O-O has a strong relation with the interatomic repulsive force between two cations M-M and the degree of π bond of Si-O.

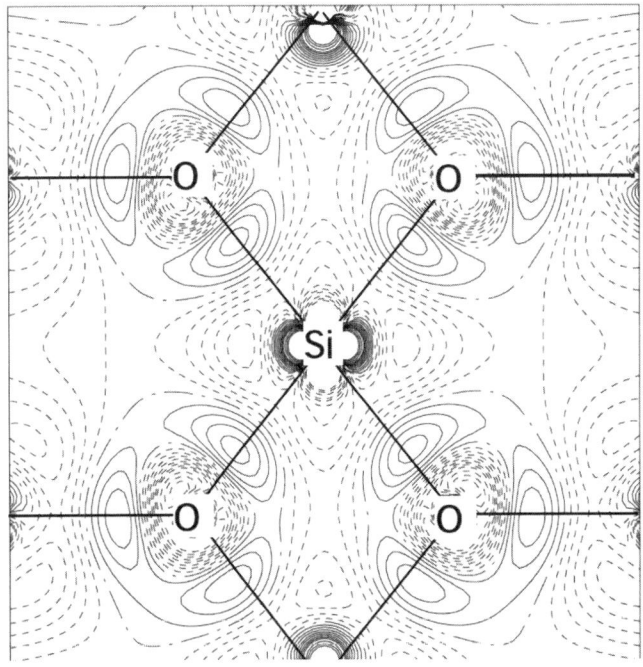

Figure 5. Deformation electron-density map of stishovite SiO_2 at 30 GPa and 0 K obtained from the molecular orbital calculation (Mimaki et al., 2000).

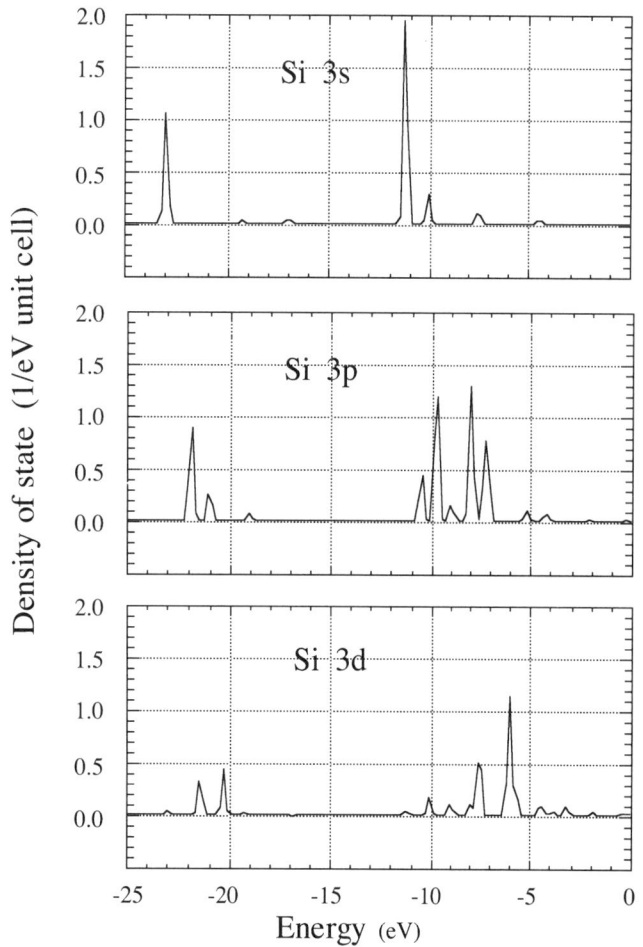

Figure 6. Density of state (DOS) of Si and the band gap of stishovite SiO_2 at 30 GPa and 0 K (Mimaki et al., 2000).

PRESSURE DEPENDENCY OF ELECTRON-DENSITY DISTRIBUTION BY THE MAXIMUM ENTROPY METHOD

Much attention has been paid to structure transformations under high-pressure conditions of transition element–bearing oxides in Earth's interior and those related to ferroelectrics, ferromagnetics, and solid ionics. For the structural studies at high pressure, the conventional difference Fourier analysis (Eqn. 12) cannot provide accurate electron-density distribution because of the termination effect of the Fourier series. This is because the diffraction angle using DAC limits the number of $F(\vec{h})$ in the reciprocal space, which lowers the quality of structure refinement. Electron-density distribution analysis by the maximum-entropy method (MEM) overcomes the difficulty and presents much more reliable electron distributions. Without any initial structure model, maximum-entropy method statistically estimates the most reliable electron-density distribution.

The ideal entropy (S) of maximum-entropy method was introduced by Jaynes (1968), and the theory of the method was applied for structure analysis by Collins (1982), Sakata and Sato (1990), Sakata et al. (1992), and Yamamoto et al. (1996). S is represented by

$$S = -\sum_j \rho'(\vec{r}_j) \ln \frac{\rho'(\vec{r}_j)}{\tau'(\vec{r}_j)}, \quad (14)$$

where electron density $\rho'(\vec{r}_j)$ and preliminary electron density $\tau'(\vec{r}_j)$ are

$$\rho'(\vec{r}_j) = \frac{\rho(\vec{r}_j)}{\sum_j \rho(\vec{r}_j)} \text{ and } \tau'(\vec{r}_j) = \frac{\rho(\vec{r}_j)}{\sum_j \rho(\vec{r}_j)}, \quad (15)$$

and $\tau'(\vec{r}_j)$ indicates an electron density one cycle before the iteration of least-squares calculation of $\rho'(\vec{r}_j)$. Fourier transform of the estimated $\tau'(\vec{r}_j)$ provides $F_{\text{calc}}(\vec{h})$:

$$F_{\text{calc}}(\vec{h}) = V \sum_j \tau(\vec{r}_j) \exp(-2\pi i \vec{h} \cdot \vec{r}_j), \quad (16)$$

where V is the unit cell volume and $\rho'(\vec{r}_j)$ is defined by

$$\rho(\vec{r}_j) = \tau(\vec{r}_j) \exp\left[\frac{\lambda F_0}{N} \sum_{\vec{h}} \frac{1}{\sigma^2(\vec{h})} \{F_{\text{obs}}(\vec{h}) - F_{\text{calc}}(\vec{h})\} \exp(-2\pi i \vec{h} \cdot \vec{r}_j)\right], \quad (17)$$

where N is the number of the observed $F_{\text{calc}}\vec{h}$; λ is a Lagrange's undetermined coefficient; F_0 is a total number of electrons. The term $\sigma(h)$ in Equation 17 is defined by

$$\sigma(\vec{h}) = a \times \frac{\sin\theta}{\lambda} \times b, \quad (18)$$

where a and b are empirical constants.

The calculated $\rho(\vec{r}_j)$ replaces $\tau(\vec{r}_j)$ in the next cycle and calculates $F_{\text{calc}}(\vec{h})$ again by Equation 15. The least-squares calculation is repeated up to the cycle where $\rho(r_j)$ satisfies the condition of $\rho(\vec{r}_j) \cong \tau(\vec{r}_j)$. Finally, the calculation results in the most reliable $F_{MEM}(\vec{h})$ and $\rho_{MEM}(\vec{r}_j)$ values. The detailed procedure of the maximum-entropy method calculation based on $F_{\text{obs}}(\vec{h})$ under high-pressure conditions is presented in Yamanaka (2005).

An example of the electron-density distribution analysis of $FeTiO_3$ ilmenite ($R\bar{3}$, Z = 6) using maximum-entropy method analysis by X-ray single-crystal diffraction study at high pressures up to 8.2 GPa is presented next. Electrical resistivity measurements at high pressures using a single crystal have proved the anisotropy between the a-axis and the c-axis directions.

Maximum-entropy method calculation using $F_{MEM}(\vec{h}) - F_{calc}(\vec{h})$ provides much more precise electron distributions that indicate the deformation electron density. Fe and Ti cations across the shared face are alternatively located along the c axis. FeO_6 and TiO_6 octahedra are alternatively located in a plane parallel to (001) and linked with adjacent octahedra with shared edge, as shown in Figure 7.

The diffraction intensity measurements of $FeTiO_3$ ilmenite at high pressures of 3.6, 5.3, and 8.2 GPa were made using wavelength $\lambda = 0.61907$ Å of synchrotron radiation at BL-10A, KEK with the aforementioned DAC. The structure parameters, including the isotropic thermal parameters of $FeTiO_3$ at various pressures, are presented in Table 5. Bond distances of Fe-O, Ti-O, O-O octahedral edges, selected bond angles, and FeO_6 and TiO_6 volumes are also presented in Table 5. The cation positions move in the direction of the c axis and approach the center of FeO_6 and TiO_6 with increasing pressure. The regularity of TiO_6 octahedra is enhanced at higher pressure. All oxygen atoms O1 to O9 (indicated in Fig. 7) occupy a crystallographically equivalent position. These octahedra have a pair of three equivalent bond distances, shared face M-O (sh face) and unshared face M-O (unsh face) (M: Fe and Ti). M-O (sh face) and M-O (unsh face) bond lengths are presented as a function of pressure.

The three M-O bonds (sh face) of the FeO_6 and TiO_6 octahedra are longer than those of M-O (unsh face). The longer M-O (sh face) bonds become more shortened under pressure than shorter M-O (unsh face). Therefore, the cation shifts toward the center of the octahedron. The shared edge of the octahedron, O-O (sh edge), is less influenced by compression than the unshared edge O-O (unsh edge). This tendency is more remarkable in FeO_6 octahedra compared to TiO_6.

The best conditions for the least squares of maximum-entropy method using each data set at various pressures are summarized in Table 6. The maximum-entropy method analysis reveals the valence electron densities of Fe and Ti. Figure 8 and Figure 9 show the electron distribution on the (010) and (110) plane at 0.0001, 3.6, 5.3, and 8.2 GPa. The electron-density distributions around Fe^{2+} ($3d^6$) and Ti^{4+} ($3d^0$) cations are not spherical but are elongated along the direction of the c axis due to the d-electron orbital in the octahedral site. The compression affects the Coulomb potential in M-O (M = Fe, Ti) bonds, and the repulsive force is enhanced under high-pressure conditions.

TABLE 5. RESULTS OF STRUCTURE REFINEMENT OF $FeTiO_3$ ILMENITE AT HIGH PRESSURES

Pressure		1 atm	3.6 GPa	5.3 GPa	8.2 GPa
$2\theta_{Max}$		80	80	80	80
Sin θ/λ (Å^{-1})		0.91	0.90	0.90	0.90
a (Å)		5.08810(4)	5.0678(11)	5.0567(13)	5.0398(10)
c (Å)		14.0910(10)	13.9956(9)	13.8892(10)	13.7968(12)
c/a		2.769	2.762	2.747	2.738
Vol (Å^3)		315.93(5)	310.40(13)	307.57(16)	303.49(12)
No. ref. (used)		408	215	214	200
R (%)		1.81	3.29	3.56	4.19
wR (%)		2.99	4.40	4.71	5.78
Atomic coordinates					
Fe (00z)	z	0.355430(9)	0.35570(5)	0.35568(5)	0.35611(7)
	B_{eq}(Å^2)	0.457(1)	0.390(9)	0.36(1)	0.22(1)
Ti (00z)	z	0.146429(9)	0.14641(5)	0.14641(6)	0.14695(8)
	B_{eq}(Å^2)	0.352(1)	0.31(1)	0.28(1)	0.29(1)
O (xyz)	x	0.31717(9)	0.3169(5)	0.3169(5)	0.3185(8)
	y	0.02351(9)	0.0233(5)	0.0232(5)	0.0233(7)
	z	0.24498(3)	0.24538(9)	0.24533(10)	0.24600(15)
	B_{eq}(Å^2)	0.48(1)	0.33(11)	0.30(11)	0.35(15)
	$G_{iso}(\times 10^4)$	0.180	0.198	0.197	0.196
Interatomic distances					
Fe-O(sh)	×3	2.2017(6)	2.185(2)	2.178(3)	2.170(4)
Fe-O(unsh)	×3	2.0795(6)	2.064(3)	2.059(3)	2.041(4)
<Fe-O>		2.1406	2.125	2.119	2.106
Ti-O(sh)	×3	2.0867(6)	2.077(3)	2.069(3)	2.066(5)
Ti-O(unsh)	×3	1.8745(6)	1.868(3)	1.863(3)	1.859(4)
<Ti-O>		1.9806	1.973	1.967	1.963
Fe-Ti		2.9450(3)	2.9210(10)	2.9065(11)	2.8858(15)
Fe-Fe $_{(1)}$†		3.0029(2)	2.9918(7)	2.9850(8)	2.9769(8)
Fe-Fe $_{(2)}$		4.0743(3)	4.0275(9)	4.0091(10)	3.9704(14)
Ti-Ti $_{(1)}$		2.9925(2)	2.9800(7)	2.9732(8)	2.9602(8)
Ti-Ti $_{(2)}$		4.1267(4)	4.0860(10)	4.0671(12)	4.0548(16)

†Metal-metal distances are indicated as follows: 1—across shared edge between adjacent metal sites; 2—across vacant octahedral position, along <001>.

TABLE 6. PARAMETERS USED FOR THE MAXIMUM-ENTROPY METHOD (MEM) CALCULATION OF ILMENITE $FeTiO_3$ AT VARIOUS PRESSURES

Pressure	1 atm†	1 atm	3.6 GPa	5.3 GPa	8.2 GPa
No. of reflections	933	408	215	214	200
Maximum 2θ	120	80	80	80	80
Maximum sin θ/λ	1.22	0.91	0.90	0.90	0.90
a‡	1.0	1.0	1.2	1.2	1.5
b‡	1.5	1.5	3.5	4.0	4.5
λ ($\times 10^5$)	0.2	0.5	0.3	0.35	0.35
No. of MEM cycles	863	285	801	779	879
R factor (%)	1.03	2.12	2.52	2.78	3.26

†Data without using diamond anvil cell.
‡$\sigma(h) = a(\sin\theta/\lambda) + b$.

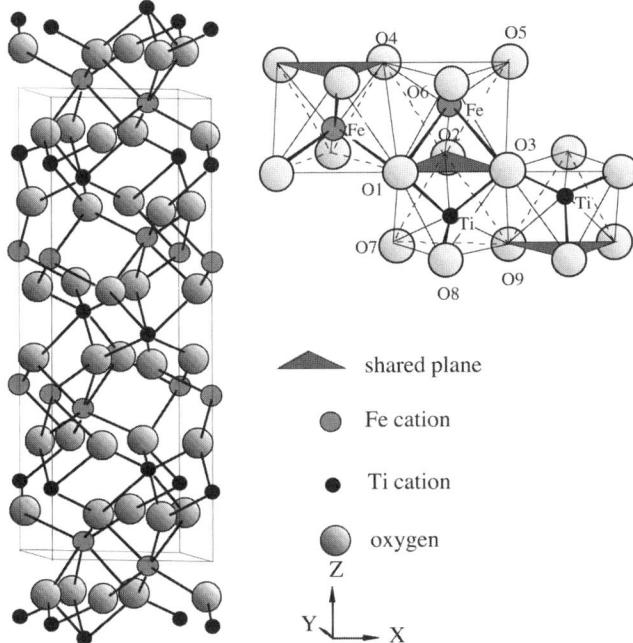

Figure 7. Structure of $FeTiO_3$ and configuration of FeO_6 and TiO_6 octahedra.

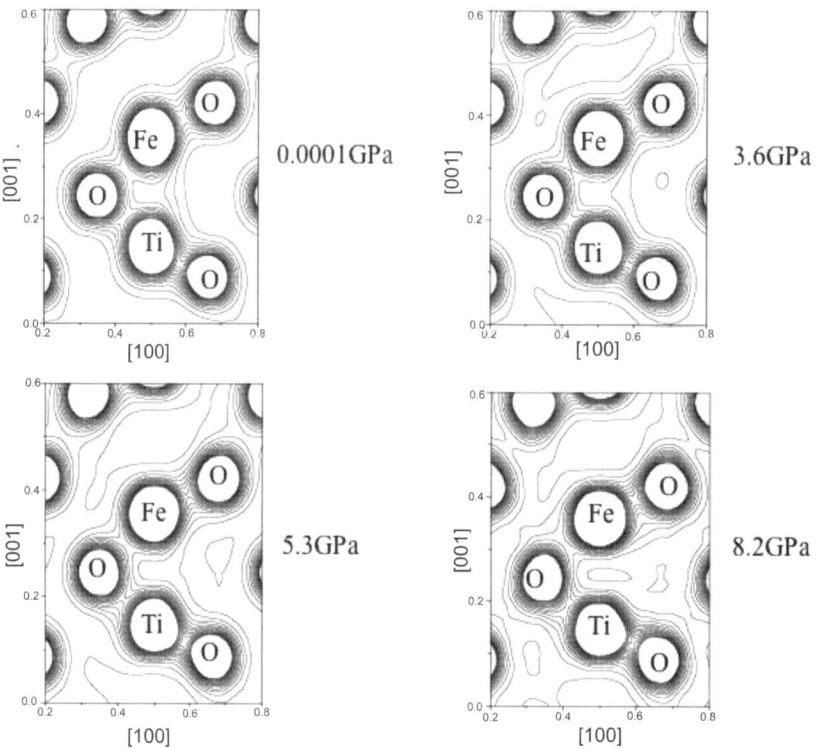

Figure 8. Electron-density change with pressure on the (010) plane of FeTiO$_3$ calculated by the maximum-entropy method (MEM). The data sets of sin θ/λ < 0.90 are used for all calculations. The contour lines are drawn from 0.2 to 4.0 eÅ^{-3} with 0.2 eÅ^{-3} intervals.

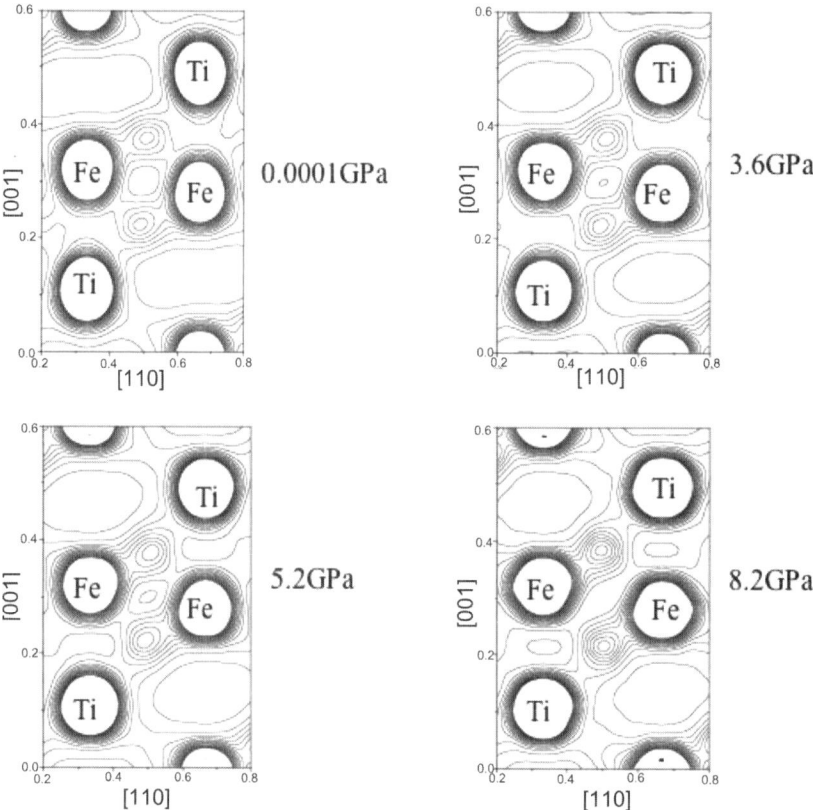

Figure 9. Electron-density change with pressure on the (110) plane of FeTiO$_3$ by MEM. The MEM map-setting conditions are same as Figure 8.

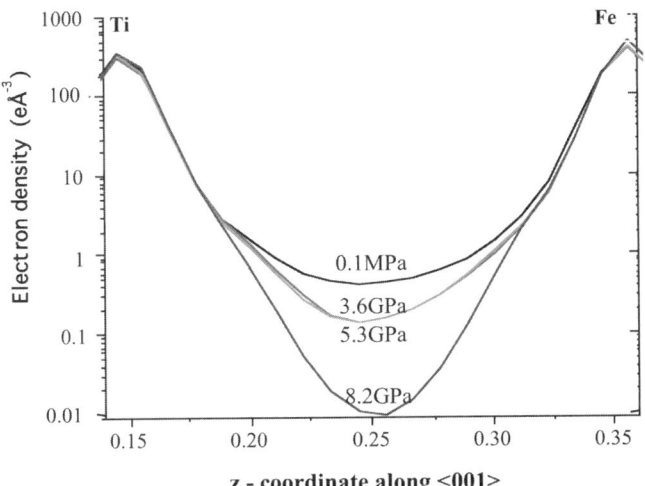

Figure 10. Normalized radial distribution of electron density between Fe and Ti at various pressures. The abscissa indicates the z coordination of Fe and Ti position along <001>.

The pressure dependence of the electron radial distribution obviously shows more localization of electrons around cations with increasing pressure, as shown in Figure 10. The radial distribution indicates that the electron density between Fe and Ti cations across the shared face becomes smaller with increasing pressure. This is because of the repulsion of *d*-electrons between two cations, which is enlarged with external compression. The electron distribution between Fe and Ti along <001> becomes lower with pressure. Therefore, neither charge transfer nor electron hopping is possible in $FeTO_3$ under pressure.

SUMMARY

Earth's materials have many high-pressure polymorphs. In this paper, their *P-T-X* phase diagrams and equation of states have been studied using X-ray diffraction techniques. Pressure-induced structure transitions have been also give much attention because of their importance in Earth's mantle. These structure changes are mainly caused by the dynamical lattice deformation or electron-phonon interaction under pressure.

The single-crystal X-ray diffraction studies as a function of pressure and temperature elucidate the electron-density distributions and aid our understanding of their transition mechanisms. Maximum-entropy method studies based on the diffraction intensities under high-pressure conditions provide the pressure change in the effective charge of ions, dipole moment, and bonding electrons.

REFERENCES CITED

Ahsbahs, H., 1984, Diamond-anvil high-pressure cell for improved single-crystal X-ray diffraction measurements: The Review of Scientific Instruments, v. 55, p. 99–102, doi: 10.1063/1.1137568.

Alaska, T.H., Kikegawa, T., and Iwasaki, H., 1983, X-ray diamond-anvil press for use at high temperatures: The Review of Scientific Instruments, v. 54, p. 1023–1025, doi: 10.1063/1.1137521.

Andrault, D., Fiquet, G., Guyot, F., and Hanfland, M., 1998, Pressure-induced Landau-type transition in stishovite: Science, v. 282, p. 720–724, doi: 10.1126/science.282.5389.720.

Arashi, H., and Ishigami, M., 1982, Diamond anvil pressure cell and pressure sensor for high-temperature use: Japanese Journal of Applied Physics, v. 21, p. 1647–1649, doi: 10.1143/JJAP.21.1647.

Boehler, R., Nicol, M., Zha, C.S., and Johnson, M.L., 1986, Resistance heating of Fe and W in diamond-anvil cells: Physica B: Condensed Matter (Amsterdam), v. 139–140, p. 916–918.

Camargo, A.C., Iguarada, J.A., Beltrain, A., Lhustar, R., Lango, E., and Anders, J., 1996, An ab initio perturbed ion study of structural properties of TiO_2, SnO_2 and GeO_2 rutile lattice: Chemical Physics, v. 212, p. 381–391, doi: 10.1016/S0301-0104(96)00228-5.

Collins, D.M., 1982, Electron density images from imperfect data by iterative entropy maximization: Nature, v. 298, p. 49–51, doi: 10.1038/298049a0.

Coppens, P., Guru Row, T.N., Leung, P., Stevens, E.D., Becker, P.J., and Yang, W., 1979, Net atom charges and molecular dipole moments from spherical-atom X-ray refinements, and the relation between atomic charge and shape: Acta Crystallographica, v. A35, p. 63–72.

Denner, W., Schulz, H., and D'Amour, H., 1978, A new measuring procedure for data collection with a high-pressure cell on an X-ray four-circle diffractometer: Journal of Applied Crystallography, v. 11, p. 260–264, doi: 10.1107/S002188987801328X.

Finger, L.W., and King, H., 1978, A revised method of operating of the single-crystal diamond cell and refinement of the structure of NaCl at 32 kbar: American Mineralogy, v. 63, p. 337–342.

Gibbs, G.V., Hill, F.C., and Boisen, M.B., Jr., 1997, The SiO bond and electron density distributions: Physics and Chemistry of Minerals, v. 24, p. 167–178, doi: 10.1007/s002690050030.

Gibbs, G.V., Boisen, M.B., Hill, F.C., Tamada, O., and Downs, R.T., 1998, SiO and GeO bonded interactions as inferred from the bond critical point properties of electron density distributions: Physics and Chemistry of Minerals, v. 25, p. 574–584, doi: 10.1007/s002690050150.

Hattori, T., Tsuchiya, T., Nagai, T., and Yamanaka, T., 2001, Sequential high-pressure transformations of $FeGeO_3$ high-P clinopyroxene (C2/c) at temperatures up to 365°: Physics and Chemistry of Minerals, v. 28, p. 377–387, doi: 10.1007/s002690100174.

Hazen, R., and Finger, L.W., 1981a, High-temperature diamond-anvil pressure cell for single crystal studies: The Review of Scientific Instruments, v. 52, p. 75–79, doi: 10.1063/1.1136450.

Hazen, R., and Finger, L.W., 1981b, Calcium fluoride as an internal pressure standard in high-pressure/high-temperature crystallography: Journal of Applied Crystallography, v. 14, p. 234–236, doi: 10.1107/S0021889881009266.

Heinz, D.L., and Jeanloz, R., 1987, Temperature measurement in a laser-heated diamond cell, *in* Manghanani, M.H., and Syono, Y., eds., High-Pressure Research in Mineral Physics: Washington, D.C., American Geophysical Union, p. 113–127.

Hemley, R.J., Prewitt, C.T., and Kingma, K.J., 1994, High-pressure behavior of silica, High-pressure behavior of silica, *in* Heaney, P., Gibbs, G.V., and Prewitt, C.T., eds., Reviews of Mineralogy: Washington, D.C., Mineralogical Society of America, p. 41–81.

Hill, R.J., Newton, M.D., and Gibbs, G.V., 1983, A crystal chemical study of stishovite: Journal of Solid State Chemistry, v. 47, p. 185–200, doi: 10.1016/0022-4596(83)90007-5.

Jaynes, E.T., 1968, Prior probability: IEEE Transactions on Systems Science and Cybernetics SSC4, 227 p.

Jeanloz, R., and Heinz, D.L., 1984, Experiments at high temperature and pressure: Laser heating through the diamond cell: Journal of Physics, Part E: Scientific Instruments, v. C8, p. 83–92.

Keller, R., and Holzapfel, W.B., 1977, Diamond anvil device for X-ray diffraction on single crystals under pressures up to 100 kilobar: The Review of Scientific Instruments, v. 48, p. 517–523, doi: 10.1063/1.1135065.

Kim-Zajonz, J., Werner, S., and Schulz, H., 1998, High-pressure single crystal X-ray diffraction study on ruby up to 31 GPa: Zeitschrift für Kristallographie, v. 214, p. 331–336.

Koepke, J., Dieterich, W., Glinnemann, J., and Schruz, H., 1985, Improved diamond anvil high-pressure cell for single-crystal work: Review of Scientific Instruments, v. 56, p. 2119–2122, doi: 10.1063/1.1138380.

Kontio, A., and Stevens, E.D., 1982, Determination of the one-particle potential for an atom with highly anharmonic thermal motion: Acta Crystallographica, v. A38, p. 623–629.

Mao, H.K., Bell, P.M., Shaner, J.W., and Steinberg, D.J., 1978, Specific volume measurements of Mo, Pd, and Ag and calibration of ruby R1 fluorescence pressure gauge from 0.06 to 1Mbar: Journal of Applied Physics, v. 49, p. 3276–3283, doi: 10.1063/1.325277.

Mao, H.K., Xu, J., and Bell, P.M., 1986, Calibration of the ruby pressure gauge to 800-kbar under quasi-hydrostatic conditions: Journal of Geophysical Research, v. 91, p. 4673–4676.

Mao, H.K., Bell, P.M., and Hadidiacos, C., 1987, Experimental phase relations of iron to 360kbar 1400°C, determined in an internally heated diamond anvil apparatus, in Manghanani, M.H., and Syono, Y., eds., High-Pressure Research in Mineral Physics: TERRAPUB/AGU, p. 135–138.

Mao, H.K., Hemley, R.J., Wu, Y., Jephcoat, A.P., Finger, L.W., Zha, C.S., and Basset, W.A., 1988, High-pressure phase diagram and equation of state of solid helium from single-crystal X-ray diffraction to 23.3 GPa: Physical Review Letters, v. 60, p. 2649–2652, doi: 10.1103/PhysRevLett.60.2649.

Merrill, L., and Bassett, A., 1974, Miniature diamond anvil pressure cell for single crystal X-ray diffraction studies: The Review of Scientific Instruments, v. 45, p. 290–294, doi: 10.1063/1.1686607.

Mimaki, J., Tsuchiya, T., and Yamanaka, T., 2000, The bond character of rutile type SiO_2, GeO_2 and SnO_2 investigated by molecular orbital calculation: Zeitschrift für Kristallographie, v. 215, p. 419–423, doi: 10.1524/zkri.2000.215.7.419.

Ming, L., and Bassett, A., 1974, Laser heating in the diamond anvil press up to 2000C sustained and 3000C pulsed at pressure up to 260 kbar: The Review of Scientific Instruments, v. 45, p. 1115–1118, doi: 10.1063/1.1686822.

Ross, N.L., Shu, J.F., Hazen, R.M., and Gasparik, T., 1990, High-pressure crystal chemistry of stishovite: American Mineralogy, v. 75, p. 739–747.

Sakata, M., and Sato, M., 1990, Accurate structure analysis by the maximum-entropy method: Acta Crystallographica, v. A46, p. 263–270.

Sakata, M., Uno, T., Takata, M., and Mori, R., 1992, Electron density in rutile (TiO_2) by the maximum-entropy method: Acta Crystallographica, v. B48, p. 591–598.

Sasaki, S., and Tsukimura, K., 1987, Atomic positions of K-shell electrons in crystals: Journal of the Physical Society of Japan, v. 56, p. 437–440, doi: 10.1143/JPSJ.56.437.

Schiferl, D., 1977, 50-kilobar gasketed diamond anvil cell for single-crystal X-ray diffractometer use with the crystal structure of Sb up to 26 kilobar as a test problem: The Review of Scientific Instruments, v. 48, p. 24–30, doi: 10.1063/1.1134861.

Simunek, A., Vackar, J., and Wiech, G., 1993, Local s, p and d charge distributions, and X-ray emission bands of SiO_2: α-quartz and stishovite: Journal of Physics Condensed Matter, v. 5, p. 867–874, doi: 10.1088/0953-8984/5/7/013.

Sinclair, W., and Ringwood, A.E., 1978, Single crystal analysis of the structure of stishovite: Nature, v. 272, p. 714–715, doi: 10.1038/272714a0.

Spackman, M.A., Hill, R.J., and Gibbs, G.V., 1987, Exploration of structure and bonding in stishovite with Fourier and pseudoatom refinement methods using single crystal and powder X-ray diffraction data: Physics and Chemistry of Minerals, v. 14, p. 139–150, doi: 10.1007/BF00308217.

Sugiyama, M., Endo, S., and Koto, K., 1987, The crystal structure of stishovite under pressure up to 6 GPa: Mineralogical Journal, v. 13, p. 455–466.

Svane, A., and Antoncik, E., 1987, Electronic structure of rutile SnO_2, GeO_2 and TeO_2: Journal of Physics and Chemistry of Solids, v. 48, p. 171–180, doi: 10.1016/0022-3697(87)90081-3.

van der Wal, R.J., and Stewart, R.F., 1984, Shell population and κ-refinements with canonical and density—Localized scattering factors in analytical form: Acta Crystallographica, v. A40, p. 587–593.

Weidner, D.J., Bass, J.D., Ringwood, A.E., and Sinclair, W., 1982, The single-crystal elastic-moduli of stishovite: Journal of Geophysical Research, v. 87, p. 4740–4746.

Weir, C.E., Block, S., and Piermarine, G.L., 1965, Single-crystal X-ray diffraction at high pressures: Journal of Research of the National Bureau of Standards, v. 69C, p. 275–281.

Willis, B.T.M., and Pryor, A.W., 1975, Thermal Vibration in Crystallography: Cambridge, Cambridge University Press.

Yamamoto, K., Takahashi, Y., Ohshima, K., Okamura, F.P., and Yukino, K., 1996, MEM analysis of electron-density distributions for silicon and diamond using short-wavelength X-rays (WKα): Acta Crystallographica, v. A52, p. 606–613.

Yamanaka, T., 2005, Structure changes induced by lattice-electron interactions: Journal of Synchrotron Radiation, v. 12, p. 566–576, doi: 10.1107/S0909049505022016.

Yamanaka, T., and Morimoto, S., 1996, Isotope effect on anharmonic thermal atomic vibration and κ-parameter of ^{12}C and ^{13}C diamond: Acta Crystallographica, v. B52, p. 232–238.

Yamanaka, T., Takeuchi, Y., and Tokonami, M., 1984, Anharmonic thermal vibrations of atoms in $MgAl_2O_4$ spinel at temperatures up to 1933 K: Acta Crystallographica, v. B40, p. 96–102.

Yamanaka, T., Sugiyama, K., and Ogata, K., 1992, Kinetic study of the GeO_2 transition under high pressures using synchrotron X-radiation: Journal of Applied Crystallography, v. 25, p. 11–15, doi: 10.1107/S0021889891008750.

Yamanaka, T., Kurashima, R., and Mimki, J., 2000, X-ray diffraction study of bond character of rutile-type SiO_2, GeO_2 and SnO_2: Zeitschrift für Kristallographie, v. 215, p. 424–428, doi: 10.1524/zkri.2000.215.7.424.

Yamanaka, T., Fukuda, T., Hattori, T., and Sumiya, H., 2001, New diamond anvil cell for single-crystal analysis: The Review of Scientific Instruments, v. 72, p. 1458–1462, doi: 10.1063/1.1336821.

Yamaoka, S., Fukunaga, O., Shimomura, O., and Nakazawa, H., 1979, Versatile type miniature diamond anvil high-pressure cell: The Review of Scientific Instruments, v. 50, p. 1163–1164, doi: 10.1063/1.1136017.

Zhang, L., Ahsbahs, H., and Kutoglu, A., 1998, Hydrostatic compression and crystal structure of pyrope to 33 GPa: Physics and Chemistry of Minerals, v. 25, p. 301–307, doi: 10.1007/s002690050118.

Zucker, U.H., and Schulz, H., 1982, Statistical approaches for the treatment of anharmonic motion in crystals. 2: Anharmonic thermal vibration and effective atomic potentials in the fast ionic conductor lithium nitride (Li_3N): Acta Crystallographica, v. A38, p. 568–576.

MANUSCRIPT ACCEPTED BY THE SOCIETY 14 AUGUST 2006

Phase-relation studies of mantle minerals by in situ X-ray diffraction using multianvil apparatus

Tomoo Katsura[†]

Institute for Study of the Earth's Interior, Okayama University, Misasa, Tottori-ken 682-0193, Japan

ABSTRACT

High-pressure and high-temperature in situ X-ray diffraction in a multianvil apparatus is the most reliable method for determination of phase boundaries of high-pressure transitions. Several basic experimental techniques are reviewed in this article. Recommendations include the equations of state for NaCl, MgO, and Au given by Brown (1999), Matsui et al. (2000), and Fei et al. (2004a), respectively, for practical use, as well as phase boundaries given by the following studies: coesite-stishovite transition in SiO_2 by Zhang et al. (1995), olivine-ringwoodite transition in Fe_2SiO_4 by Yagi et al. (1997), and olivine-wadsleyite transition in Mg_2SiO_4 by Morishima et al. (1994). Although the akimotoite-perovskite transition in $MgSiO_3$ and dissociation of ringwoodite to perovskite + periclase in Mg_2SiO_4 have been studied repeatedly, no reasonable agreements have been established to date. Although the wadsleyite-ringwoodite transition in Mg_2SiO_4 and dissociation of garnet to perovskite + corundum have also been studied, the results are still preliminary. The present state-of-the-art experimental techniques have several problems in accurately determining positions of phase boundaries, namely: (1) uncertainty of temperature distribution in a high-pressure cell, (2) an unknown pressure effect on electromotive force (EMF), (3) low equation of state reliability for pressure determination at relatively low pressures and high temperatures, (4) pressure intensification and reduction of pressure standards that are mixed with other materials to prevent grain growth, and (5) sluggish kinetics.

Keywords: mantle minerals, phase relation, high pressure, high temperature, in situ X-ray diffraction.

INTRODUCTION

The fine structures of the upper part of Earth's mantle have primarily been a field of study for seismologists. Global topographies of the 410 and 660 km discontinuities have been determined with a resolution of 1–5 km (cf. Flanagan and Shearer, 1998). Global seismic features of the mantle are usually attributed to phase transitions and reactions of mantle minerals. Comparison of seismological features with detailed phase relations could make it possible to obtain information on the chemical composition, mineralogical constitution, temperature distribution, dynamic motion, and evolution of the mantle. For example,

[†]E-mail: tkatsura@misasa.okayama-u.ac.jp.

temperatures at the seismic discontinuities could be estimated by comparing the transition pressures with the depths of the discontinuities. For this purpose, the phase relations of mantle minerals have to be determined with accuracy better than 0.2 GPa.

There are two popular types of high-pressure apparatuses used to study phase transitions and reactions at high pressures and temperatures. One is the multianvil apparatus, and the other is the diamond anvil cell. Although recent developments of experimental techniques have made it possible to determine phase relations of mantle minerals using the diamond anvil cell (cf. Chudinovskikh and Boehler, 2004), the experimental results obtained by diamond anvils cells are not precise enough to discuss detailed seismic structures (cf. Shim et al., 2001). Phase relations are determined more precisely by using a multianvil apparatus, because of the large sample volume and relatively uniform pressure and temperature distributions.

Traditionally, determination of phase relations using the multi-anvil experiments has been conducted in the following way. A worker loads a sample in a furnace assembly, compresses the sample to a certain press load, heats it to a certain temperature using a resistance heater, quenches it by cutting the electric power to the heater, and recovers it to ambient conditions. Phase relations are determined by analyzing run products with optical microscopy, powder X-ray diffraction, and electron microprobe analysis. This procedure is the so-called quench method.

It has been realized, however, that the quench method has some serious problems in obtaining reliable results on phase relations. A run product can be analyzed only after recovery. We do not know what kind of phase transitions and chemical reactions the run product underwent at the pressure-temperature (P-T) conditions of interest. In order to obtain a definitive answer to the stability between two phases, we have to observe normal and reversed reactions between them. The sample temperature is set to increase at a certain rate, and therefore, the sample could react at temperatures lower than the final temperature. Such a complex reaction history causes an ambiguity in the interpretation of observations.

Another problem is the poor assessment of generated pressures. Generated pressures are usually estimated by calibration against press load by detecting several known phase transitions and reactions. However, pressures are not only a function of press load but also of temperature. Pressure usually drops during heating, and the degree of pressure drop depends on the cell design and heating history. For example, although ~10% of total pressure is lost by heating in many runs, sometimes 5 GPa is lost at 20 GPa if hexagonal BN (boron nitride) is used as a pressure cell component because of the transition from hexagonal BN to wurtzite or cubic BN. Therefore, it is very difficult to estimate generated pressures at high temperatures in a conventional multi-anvil apparatus.

In situ X-ray diffraction is a powerful method that can solve these problems. One of the most important advantages of diamond anvil cells is the transparency of diamond. Many kinds of measurements can be conducted using the wide energy range of electromagnetic waves in a diamond anvil cell. Unfortunately, anvils and gaskets of multianvil apparatuses are opaque to the electromagnetic waves with energies lower than soft X-rays. However, a sample in a multianvil apparatus can be observed using hard X-rays. By means of in situ X-ray diffraction with hard X-rays, phase transitions and reactions of samples can be observed in real time. Sample pressure can be estimated by measuring cell volume of a pressure standard with a reasonably known equation of state.

In the early days, workers used a laboratory X-ray source. However, their X-ray fluxes were insufficient, and their beam divergences were large. Because of the development of synchrotron radiation facilities, we can now obtain hard X-rays with sufficiently high flux and low beam divergence to obtain high-quality diffraction patterns of samples in a multianvil apparatus. This technical development enables us to study phase relations of the mantle minerals with high reliability.

This paper presents a review of phase-relation studies of important minerals in the upper part of the mantle by means of in situ X-ray diffraction in a multianvil apparatus.

EXPERIMENTAL TECHNIQUES

There are nine multianvil apparatuses operating at synchrotron radiation facilities around the world: MAX-80, MAX-90, and MAX-III at KEK (High Energy Accelerator Research Organization) (Shimomura et al., 1985, 1992), SPEED-1500 and SPEED-Mk.II at SPring-8 (Utsumi et al., 1998; Katsura et al., 2004a, 2004b), SAM85 at NSLS (National Synchrotron Light Source) (Weidner et al., 1992), 2.5 and 10 MN LVP's (Large Volume Press) at APS (Advanced Photon Source) (Wang et al., 2000), MAX80 at HASYLAB (Hamburger Synchrontronstrahlungslabor) (Mueller et al., 2002), and these are summarized in Table 1. These apparatuses are divided into two groups: single-stage cubic anvil apparatuses (MAX-80, MAX-90, SAM85, and MAX80), and double-stage KAWAI-type apparatuses (cf. Katsura et al., 2003). KAWAI-type apparatuses can be further divided into two groups. In one group, the KAWAI-type assembly is compressed by a DIA-type guide block system (MAX-90, MAX-III, SPEED-1500, and SPEED-Mk.II). In the other group, the assembly is compressed by a T-cup–type guide block (2.5 and 10 MN LVP's and SAM85). In this section, SPEED-Mk.II is used as an example, due to the author's familiarity with SPEED-Mk.II. Another reason is that it is one of the most advanced, large-volume, high P-T, in situ X-ray diffraction systems operating today.

The KAWAI-type assembly is compressed by a uniaxial hydraulic press with DIA-type guide blocks. The maximum press load is 15 MN. The maximum pressures generated so far are 42 and 63 GPa using tungsten carbide and sintered diamond anvils, respectively. Practically, phase relation studies can be conducted at pressures up to 30 and 50 GPa using these two kinds of anvils. The configuration of in situ X-ray diffraction is shown in Figure 1. The KAWAI-type assembly is placed so that

TABLE 1. MULTIANVIL APPARATUSES IN THE SYNCHROTRON RADIATION FACILITIES

Apparatus (reference)	Capacity (MN)	Compression method	Facility	Country
MAX-80 (1)	5	Single-stage cubic anvils with DIA-type guide blocks	Photon Factory	Japan
MAX-90† (2)	3.5	KAWAI-type assembly with DIA-type guide blocks		
MAX-III	7	KAWAI-type assembly with DIA-type guide blocks		
SPEED-1500 (3)	15	KAWAI-type with DIA-type guide blocks	SPring-8	
SPEED-Mk.II (4)	15	KAWAI-type with DIA-type guide blocks		
SAM85 (5)	2.5	Single-stage cubic anvils with DIA-type guide blocks KAWAI-type with T-cup guide blocks	NSLS	USA
2.5 MN LVP (6)	2.5	Single-stage cubic anvils with DIA-type guide blocks KAWAI-type with T-cup guide blocks	APS	
10 MN LVP (6)	10	Single-stage cubic anvils with DIA-type guide blocks KAWAI-type with T-cup guide blocks		
MAX80 (7)	2.5	Single-stage cubic anvils with DIA-type guide blocks	HASYLAB	Germany

Note: References: 1—Shimomura et al. (1985), 2—Shimomura et al. (1992), 3—Utsumi et al. (1998), 4—Katsura et al. (2004a), 5—Weidner et al. (1992), 6—Wang et al. (2000), 7—Mueller et al. (2002).
†Out of use for in situ X-ray diffraction experiments.

the (100) and (110) planes of the anvils are faced in the vertical direction and in the direction of X-ray incidence, respectively. The incident X-rays pass between the anvils and through the gasket and pressure medium. The diffracted X-rays from the sample in a pressure medium are directed through the gasket and pressure medium and between the anvils.

SPEED-Mk.II is installed at the white X-ray beam line BL04B1 in SPring-8. The energy-dispersive X-ray diffraction is conducted using the white X-rays. In this method, diffracted X-rays are detected with a fixed diffraction angle (2θ) as a function of energy, using a Ge solid-state detector and 4096-channel multichannel analyzer. The energy range is up to 120 keV. This technique is useful for diffraction patterns with a relatively high number of counts for a limited collection time, which is typically a couple of hundred seconds. Energy width of a single channel of the multichannel analyzer is ~0.03 keV. The d-values are calculated as:

$$d = \frac{ch}{2\varepsilon \sin\theta}, \quad (1)$$

where d is the d-value, c is the speed of light, h is the Plank constant, ε is the X-ray energy, and θ is the diffraction angle.

The X-ray beam is cut into squares, which are several hundred micrometers in width and height, before entering the experimental hutch in order to decrease noise by scattered

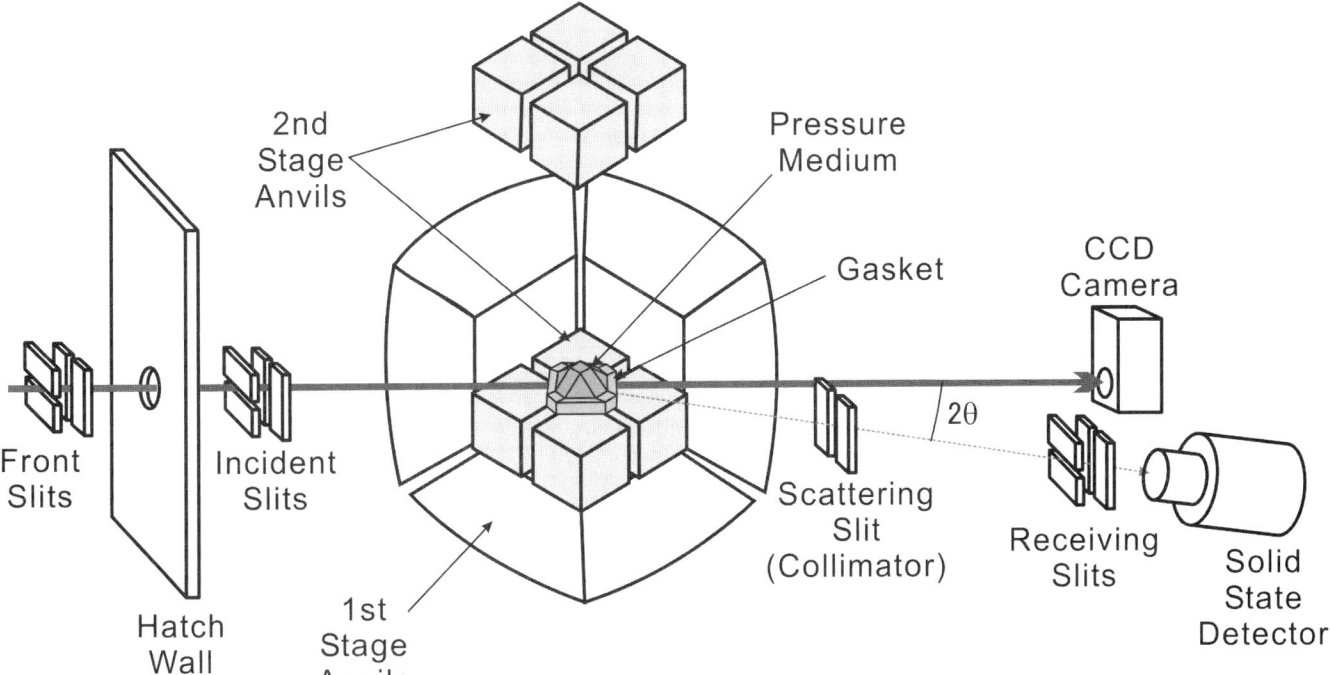

Figure 1. Schematic drawing of the in situ X-ray diffraction experiments with a Kawai-type multianvil apparatus. The incident X-rays (gray solid line) penetrates the gaskets, pressure medium, and sample through the anvil gap. CCD—charge coupled device.

X-rays. The incident X-rays are collimated by incident slits, typically 200 and 50 µm in vertical and horizontal directions, respectively. The diffracted X-rays are collimated by a horizontal scattering slit (usually called a collimator) and horizontal and vertical receiving slits, the widths of which are typically 50 µm, 200 µm, and 2 mm, respectively. The scattering slit is located ~150 mm away from the sample. The distance between the scattering and receiving slits is ~1 m. The fixed diffraction angle (2θ) is usually between 5 and 8°. The typical length of the diffraction area in the direction of X-ray incidence is 1.5 mm. One of advantages of the energy-dispersive X-ray diffraction is that the position and size of the diffraction area do not change because of the fixed diffraction angle. In the case of the angle-dispersive X-ray diffraction, the length of the diffraction area becomes shorter with increasing diffraction angle. It is also possible for the position of the diffraction area to shift with changing diffraction angles. SPEED-Mk.II is specially equipped with an oscillation system so that high-quality diffraction patterns can be obtained against grain growth of samples at high temperatures.

The energies of X-rays collected by the SSD (solid state detector) and multichannel analyzer are calibrated by the characteristics X-rays of several elements, usually Cu, Ag, Mo, Ta, Pt, Au, and Pb by assuming that the energy of each channel is a linear function of the channel number. The 2θ value is calibrated by taking the diffraction patterns of the standard material. The d-value of each peak is calculated using Equation 1. Otherwise, the d-value at each channel is calibrated using diffraction peaks of a standard material, and the d-value of the peak is directly calculated from the peak position.

Figure 2 shows one example of furnace assemblies. This assembly is compressed using tungsten carbide anvils with a 2.5 mm truncated edge length, with which 2500 K and 30 GPa can be generated simultaneously. The cylindrical heater is positioned with its axis almost parallel to the incident X-ray. Both ends of the heater are pinched out by the compression. Therefore, the temperature gradient along the furnace axis is small. At a certain temperature, the beginning of phase transition can be observed. However, the phase transition is not complete at that temperature, which seems to be because of temperature inhomogeneity in the sample. In many cases, the phase transition is completed by increasing the temperature by 50 K. Therefore, temperature inhomogeneity in the sample would be 50 K. Temperature variation is not observed in the horizontal direction, judging by unit cell volumes of a pressure standard.

The half disks of the sample and pressure standard, which is MgO in this case, with 0.6 mm thickness are then put in a cylindrical Re heater. The $W_{97}Re_3$-$W_{75}Re_{25}$ thermocouple is inserted into the heater through holes in the vertical direction. The junction is sandwiched by the sample and pressure marker. Diffractions of the sample and pressure marker are obtained separately by shifting the high-pressure apparatus horizontally. The position of the sample that is irradiated by X-rays can be observed by monitoring the transmitted X-rays using a CCD (charge coupled device) camera (Fig. 1). Rods of amorphous boron, bonded by epoxy (boron/epoxy), are placed behind and in front of the sample and pressure marker. The purpose of the boron/epoxy rods is to prevent diffractions from the low-temperature parts of the sample and pressure marker, because the diffraction area (1.5 mm) is fairly long in comparison with the length of the heater (4 mm).

The sample pressure is determined using the unit cell volume of MgO determined by X-ray diffraction and temperature measured by the thermocouple. In the best case, seven diffraction lines of MgO, that is (111), (200), (220), (311), (400), (222), and (420) are observed at 2θ = 6° so that the unit cell volume can be determined with a precision of 0.02%. The precision in pressure using MgO is better than 0.1 GPa.

In a usual procedure, the sample is compressed to a desired press load at ambient temperatures. It is then heated to a desired temperature. Pressures attained at cold compression are usually reduced by 10%–20% due to the heating, after which, if the temperature is decreased, pressure decreases with a nearly constant rate. If the temperature is increased again after cooling, pressure recovers almost completely. An example of such P-T paths is shown in Figure 3.

In the case of a univariant reaction, stability of phases can be judged by detecting the growth and reduction of their diffraction lines. One phase (referred to as phase I) should be more stable than the other (referred to as phase II) at given P-T conditions if the diffraction lines of phase I and II either appear or disappear. Examples of changes in diffraction patterns associated with a phase transition are shown in Figure 4. However, the intensity

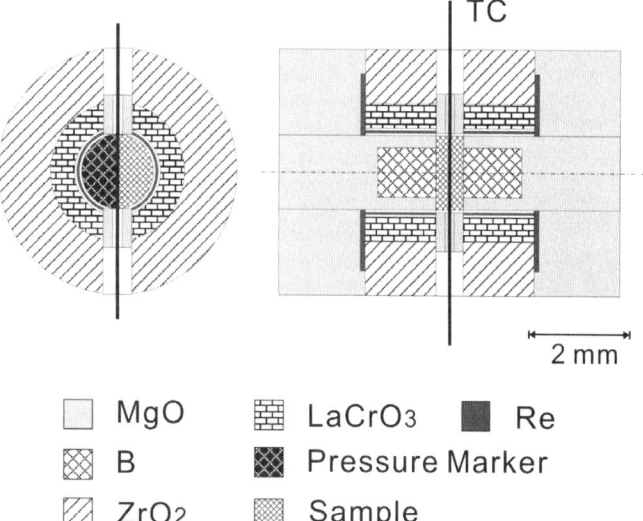

Figure 2. Schematic drawing of an example of the furnace assembly. Left: a cross section normal to the X-ray incidence; right: a cross section parallel to the X-ray incidence. The furnace system has a cylindrical symmetry. The incident X-rays go through the sample in the direction of the cylindrical furnace axis. The diffraction patterns of the sample and pressure marker are taken separately by shifting the press horizontally. TC—thermocouple.

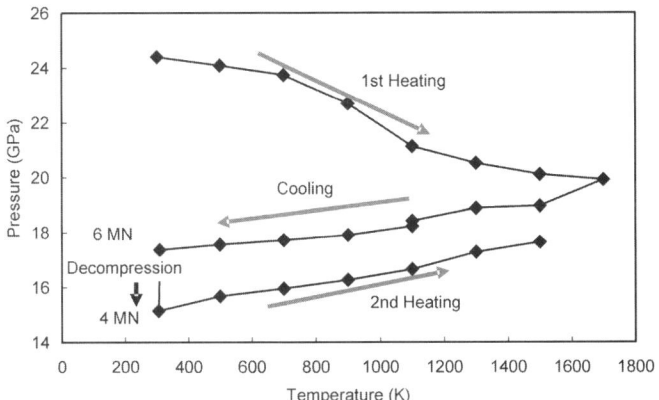

Figure 3. Example of a pressure-temperature (*P-T*) path of the multi-anvil experiment. Pressure gradually drops with increasing temperature in the first heating cycle. Pressure almost decreases linearly with decreasing temperature at a rate of 1.8 MN/K. In the second heating cycle, pressure increases almost linearly with increasing temperature at a rate of 1.7 MN/K. Although the sample was decompressed from 6 to 4 MN between the cooling and second heating cycle, the rates in the cooling and second heating cycles are nearly the same.

changes do not necessarily occur as we expect because of the sluggish kinetics. In this case, we encounter serious difficulties in determining the phase relations.

In a case with more than two reaction freedoms, we have to determine compositions of the coexisting phases. In this case, the sample is compressed to a desired press load and heated to the desired temperature, which are kept constant for a preferred duration, and pressures are measured and phases are determined by taking the diffraction patterns of the pressure standard and sample. After that, the sample is quenched by cutting the electric power, and it is decompressed to ambient pressure. The composition of the coexisting phases in the recovered sample are then examined by powder X-ray diffraction, electron microprobe, and analytical transmission-electron microscopy. The chemical equilibrium of the coexisting phases has to be confirmed in separate runs.

PRESSURE STANDARD

There are three popular pressure standards used for in situ X-ray diffraction in a multianvil press: sodium chloride (NaCl), gold (Au), and magnesium oxide (MgO). Thermoelastic properties of these materials have been studied extensively. Hence, relatively reliable equations of state have been constructed. Next is a summary of the features for the pressure standard of these materials.

NaCl is the most conventional pressure standard. It has a very low bulk modulus: ~25 GPa at ambient conditions. Therefore, pressure can be estimated with high precision. For example, if the volume compression is determined with precision of 0.1%, then pressure can be estimated with a nominal precision of 0.03 and 0.14 GPa at around 0 and 20 GPa, respectively. The equation of state proposed by Decker (1971) (Decker scale) has been used to calculate pressure for the past 30 yr. Recently, Brown (1999) revised the equation of state using experimental data available after Decker (1971) (Brown scale). These two scales give similar pressure values. However, the Brown scale gives somewhat lower pressures than the Decker scale. For example, the Brown scale gives 0.2 GPa lower pressure values than the Decker scale at around 1100 K and 20 GPa.

One of the advantages of NaCl as a pressure standard is its low yield strength. Effects of deviatoric stresses seem to be released by heating to 1000 K. On the other hand, NaCl has three disadvantages. It has very low melting points: the melting points are 1127 K at ambient temperature, and ~2100 K at 20 GPa (Boehler et al., 1997). Therefore, zero-pressure thermoelastic properties can be obtained in only a limited temperature range. Extrapolation of the equation of state to mantle temperatures may cause significant uncertainties. The second disadvantage is rapid grain growth, which is related to the low melting temperatures. It is becoming difficult to determine pressures above 700 K by loosing the diffraction peaks. If it is mixed with other materials, pressure could be determined at higher temperatures. In this case, however, pressure applied to NaCl could be different from the bulk pressure. The third disadvantage is its limited stability field; NaCl transforms from a B1 to B2 structure at 22 GPa and 1600 K and at 20 GPa and 2000 K, respectively (Nishiyama et al., 2003). This pressure scale cannot be used to study phase relations at pressures higher than the mantle transition zone.

MgO has very high melting points. The melting point at ambient pressure is ~3100 K. Therefore, its high-temperature thermoelastic properties are well studied over a wide temperature range at ambient pressure (Isaak et al., 1989; Dubrovinsky and Saxena, 1997). The high melting temperatures lead to low grain growth rate in comparison with NaCl. Usually, MgO can be used up to 1700 K. Thus, it is a very useful pressure standard for high-temperature experiments. In addition, no phase transition is known for this material, and therefore, it can be used to very high-pressure conditions. For these reasons, MgO is the best material as a pressure standard.

Matsui et al. (2000) proposed an equation of state for MgO that models experimental data such as compression at ambient temperature, thermal expansion at ambient pressure, and temperature derivative of elastic constants at ambient pressure, by means of molecular dynamics simulation with the breathing model (Matsui scale). Speziale et al. (2001) proposed another equation of state for MgO (Speziale scale). They constructed an equation of state by combining static compression data and thermal pressure from shock-wave data with the Mie-Grüneisen formula. Although these two studies constructed the equations of state by adopting different approaches, pressures given by them are very close, usually within 0.2 GPa. However, the Speziale scale does not reproduce thermal expansion at ambient pressure (Dubrovinsky and Saxena,

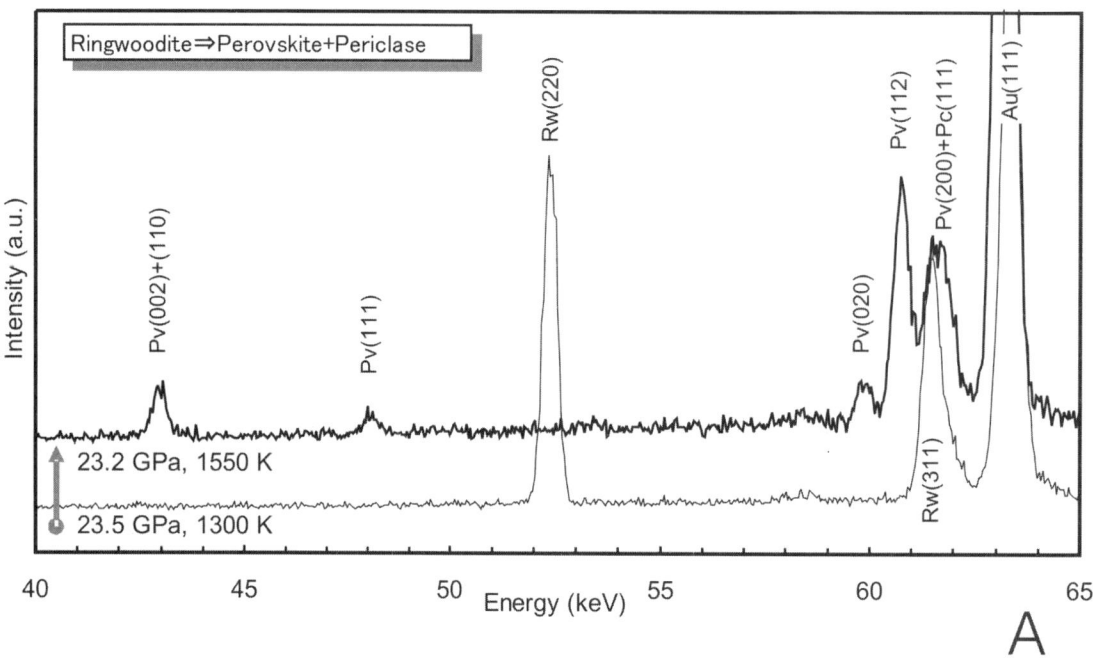

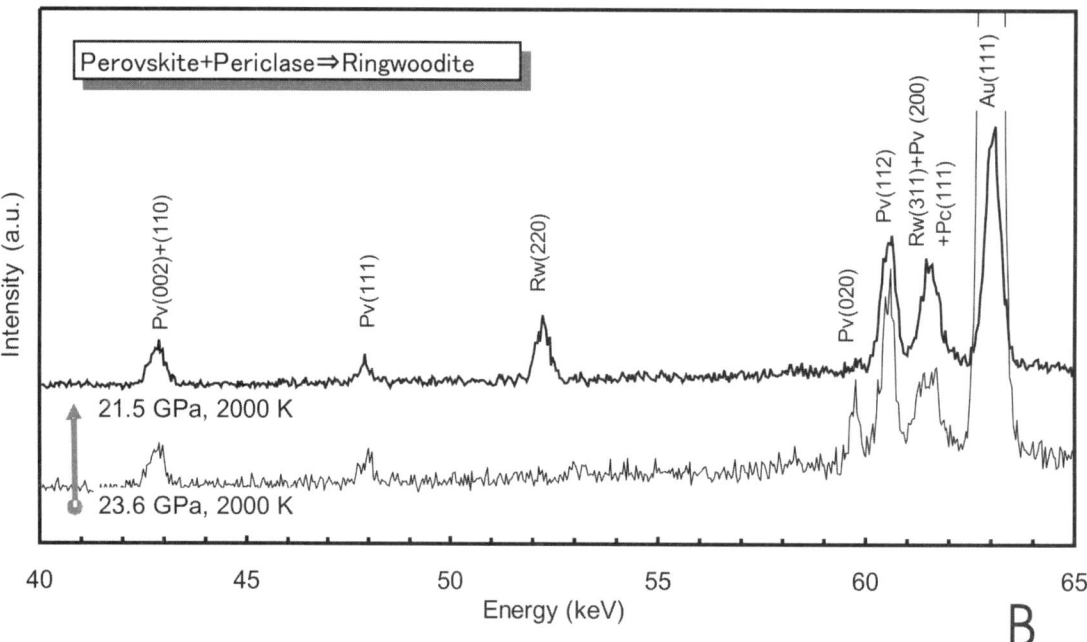

Figure 4. Examples of change of diffraction patterns associated with a phase transition. (A) Dissociation of ringwoodite to perovskite (Pv) + periclase (Pc) in Mg_2SiO_4 from 23.5 GPa and 1300 K to 23.2 GPa and 1550 K. The perovskite (002) + (110), (111), (020), (112), and (200) peaks appear, and the ringwoodite (220) peak disappears. The (111) peak of Au, which is mixed to determine pressure, shifts to the lower energy because of a pressure decrease and temperature increase. The dissociation reaction has proceeded completely. (B) Formation of ringwoodite (Rw) from perovskite + periclase in Mg_2SiO_4 by a pressure decrease from 23.6 to 21.5 GPa at 2000 K. The ringwoodite (220) peak appears, and the perovskite (020) and (112) peaks become weaker. However, the intensity changes of the perovskite (002) + (110) and (111) peaks remain unclear. The formation of ringwoodite was not completed, probably due to the sluggish kinetics. a.u.—arbitrary unit.

1997). Therefore, the Matsui scale should be more reliable than Speziale scale.

MgO has two disadvantages as a pressure standard. The stresses caused by cold compression can be released only by heating above 1500 K (Weidner et al., 1994). Therefore, the sample has to be heated above this temperature in order to obtain reliable pressure values. Another disadvantage is its relatively high bulk modulus, which is 164 GPa (Isaak et al., 1989). Hence, pressures obtained by MgO have much larger errors than those by NaCl. Very good sample conditions and careful calibration are required in order to determine pressure with a precision better than 0.1 GPa.

Au has a much larger atomic number than the elements that comprise mantle minerals. By mixing a small amount of Au in a sample (<10%), diffraction lines strong enough to determine pressure can be obtained. Therefore, this is considered to be a very useful pressure standard when the sample volume is very limited. Au grains are dispersed among sample grains, and therefore, grain growth does not occur. However, this advantage also can be a disadvantage. Pressures applied on Au grains may not be equal to the bulk pressure if surrounding materials have high creep strengths (Sato et al., 1973). Au has a relatively low melting point at ambient pressure, that is, 1336 K. Its zero-pressure thermoelastic properties can be obtained in only a limited temperature range like NaCl. The melting point is above 2200 K at 20 GPa (Katsura, own data), and there is no phase transition in Au. Therefore, it can be used to study the phase relation of mantle minerals.

The equation of state of Au given by Anderson et al. (1989) has been widely used to calculate pressure (Anderson scale). Shim et al. (2001) revised the equation of state of Au by using a pressure derivative of the bulk modulus from compression data (Takemura, 2001) (Shim scale). However, Shim et al. (2001) were not able to obtain a logarithmic volume dependence of Grüneisen parameter, q. Although they recommended $q = 1$, it contains a large uncertainty.

Fei et al. (2004a) proposed a modification of the Shim scale (Fei scale). They suggested a $q = 0.7$, so that pressures obtained by Au are more consistent with those obtained by the MgO standard using the Speziale scale. The volume data of Au and MgO used in this modification were obtained in similar experiments at pressures of 20–24 GPa and temperatures of 1470–2170 K (Fei et al., 2004a). It should be noted that their data suggests that pressures obtained using this Au pressure scale on average still give values lower by 0.2 GPa than those given by the Speziale scale. Since the Speziale scale gives ~0.2 GPa higher pressures than the Matsui scale, the Fei scale could be consistent with the Matsui scale.

Next, reviews are presented for the boundaries of important phase transitions and reactions of mantle minerals in the literature. Previous studies have used various equations of state to calculate pressures. In this review, pressures are recalculated using the following equations of state: for NaCl, the Brown scale; for MgO, the Matsui scale; and for Au, the Fei scale.

STUDIED PHASE RELATIONS

Coesite-Stishovite Transition

Silica polymorphs are expected to be present in the basaltic layer of subducted slabs. Knowledge of the phase transitions in SiO_2 is important to understand subduction processes. The coesite-stishovite transition was studied in the pioneering study by Yagi and Akimoto (1976) using a cubic anvil apparatus with a laboratory X-ray source. Later, it was reinvestigated by Zhang et al. (1996) using SAM-85. Pressures were measured using NaCl, which was mixed with the SiO_2 sample in Yagi and Akimoto (1976), and with BN in Zhang et al. (1996). Zhang et al. (1996) achieved two improvements in experimental technique over Yagi and Akimoto's (1976) study in order to obtain more reliable results. First, Yagi and Akimoto (1976) conducted runs in the temperature range of 770–1370 K, whereas Zhang et al. (1996) conducted runs in a much wider range (770–1800 K). Second, Yagi and Akimoto (1976) observed only the forward reaction from coesite to stishovite, whereas Zhang et al. (1996) conducted both the normal and reversal runs.

Zhang et al. (1996) successfully observed both normal and reversed transitions at temperatures of 1370–1570 K. Below 1270 K, they did not observe the transition near the P-T conditions predicted by back extrapolation of the phase boundary determined between 1370 and 1570 K. Zhang et al. (1996) concluded that the equilibrium phase boundary of the coesite-stishovite transition cannot be determined below 1270 K because of the sluggish kinetics. The phase boundary given by Zhang et al. (1996) was more than twice as steep as that given by Yagi and Akimoto (1976). This is because of the large overpressure necessary to drive the phase transition at low temperatures in Yagi and Akimoto's (1976) study. Therefore, it has been clearly demonstrated that a pair of both normal and reversal runs has to be conducted in order to determine the phase boundary accurately. From a practical viewpoint, however, it is difficult to conduct both normal and reversal runs, as we will see later.

The phase boundaries given by Yagi and Akimoto (1976) and Zhang et al. (1996) are shown in Figure 5, as recalculated using the Brown scale. The phase boundaries obtained by the quench method (Suito et al., 1977) and thermochemical calculation (Akaogi and Navrotsky, 1984; 1995) are also shown in Figure 5. The proposed phase boundaries given in various studies are summarized in Table 2.

Fayalite-Ringwoodite Transition

The upper mantle is considered to be mainly composed of $(Mg,Fe)_2SiO_4$ compounds. The fayalite-ringwoodite transition was studied in the early stages of high-pressure research (Akimoto et al., 1965). This phase transition was determined at temperatures of 1073–1473 K by Yagi et al. (1987) using MAX-80. They determined stability of fayalite and ringwoodite by observing increases or decreases in the diffraction intensities of these

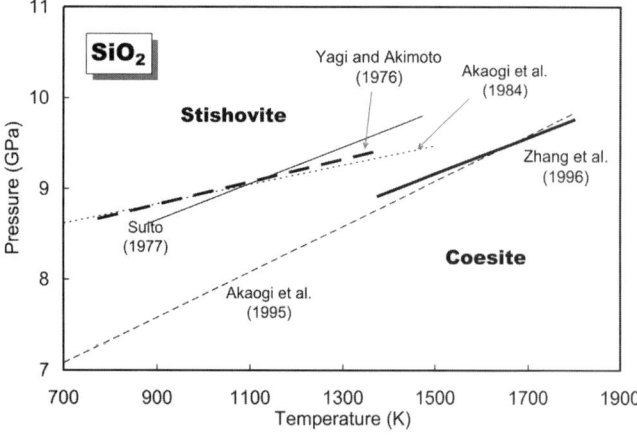
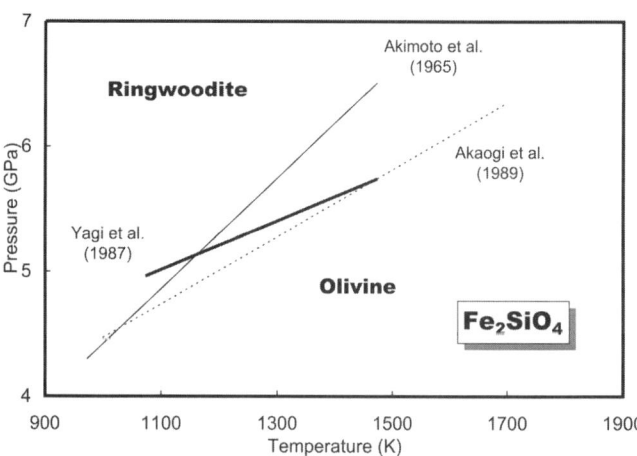

Figure 5. Phase boundaries of coesite-stishovite transition in SiO_2 determined by in situ X-ray diffraction (Yagi and Akimoto, 1976; Zhang et al., 1996), quench method (Suito, 1977), and thermochemical calculation (Akaogi and Navrotsky, 1984; Akaogi et al., 1995). The phase boundaries by Yagi and Akimoto (1976) and Zhang et al. (1996) were recalculated using Brown's (1999) scale.

Figure 6. Phase boundaries of olivine-ringwoodite transition in Fe_2SiO_4 determined by in situ X-ray diffraction (Yagi et al., 1987), quench method (Akimoto, 1987), and thermochemical calculation (Akaogi et al., 1989). The phase boundary by Yagi et al. (1987) has been recalculated using the Brown's (1999) scale.

TABLE 2. REPORTED PHASE BOUNDARIES OF COESITE-STISHOVITE TRANSITION IN SiO_2

Study	Slope of phase boundary (MPa/K)	Intercept of phase boundary (GPa)	Temperature range (K)	Method
Yagi and Akimoto (1976)	1.2[†]	7.7[†]	770–1370	IXD, NaCl
Suito (1977)	2.0	6.8	870–1470	Q
Akaogi and Navrotsky (1984)	1.1	7.9	–	TC
Akaogi et al. (1995)	2.5	5.3	–	TC
Zhang et al. (1993)	2.4	6.3	2270–3070	Q
Serghiou et al. (1995)	1	7.8	2570–2900	LHD, ruby
Zhang et al. (1996)	2.0[†]	6.2[†]	1370–1800	IXD, NaCl

Note: Abbreviations: IXD—in situ X-ray diffraction, Q—quench method, TC—thermochemical calculation, LHD—laser-heated diamond anvil cell.
[†]Recalculated using the Brown scale.

TABLE 3. REPORTED PHASE BOUNDARIES OF OLIVINE-RINGWOODITE TRANSITION IN Fe_2SiO_4

Study	Slope of phase boundary (MPa/K)	Intercept of phase boundary (GPa)	Temperature range (K)	Method
Akimoto et al. (1965)	4.4	0.0	1473–1873	Q
Yagi et al. (1987)	1.9[†]	2.9[†]	1073–1473	IXD, NaCl
Akaogi et al. (1989)	2.7	1.8	-	TC

Note: Abbreviations: Q—quench method, IXD—in situ X-ray diffraction, TC—thermochemical calculation.
[†]Recalculated using the Brown scale.

two phases. They used NaCl as a pressure maker, and they calculated pressures using the Decker scale. The phase boundary as recalculated using the Brown scale, is shown in Figure 6 with those obtained by the quench technique (Akimoto et al., 1965) and thermochemical calculation (Akaogi et al., 1989) (summarized in Table 3).

Woodland and Angel (2000) demonstrated that Fe_2SiO_4 ringwoodite contains a Fe_3O_4 component. The Fe_3O_4 component may affect the phase relations. The use of B-epoxy pressure media in Yagi et al.'s (1987) study, however, would be expected to cause highly reduced conditions in the sample environment, which should have minimized formation of ferric iron. Actually, they observed a reduction of ferric iron above 1473 K.

Although Yagi et al. (1987) is the oldest work to study high P-T phase relations of mantle minerals in a multianvil press with in situ X-ray diffraction, it is still reasonably reliable.

Olivine-Wadsleyite Transition in Mg_2SiO_4 and $(Mg,Fe)_2SiO_4$

The olivine-wadsleyite transition in $(Mg,Fe)_2SiO_4$ is believed to be responsible for the 410 km seismic discontinuity. Hence, determination of the phase relation in this transition is of primary importance in geophysics.

The phase relations of the olivine-wadsleyite transition in Mg_2SiO_4 were determined by Morishima et al. (1994) using MAX-90 with sintered diamond anvils. They used a mixture of olivine and wadsleyite as starting material. They judged the stability of these two phases by growth and extinction of these phases. They used NaCl as a pressure standard. The phase boundary, as recalculated using the Brown scale, is shown with the quench method (Katsura and Ito, 1989) and thermochemical calculation (Akaogi et al., 1989) in Figure 7. The experimental results are summarized in Table 4. The recalculated slope (3.2 MPa/K) by Morishima et al. (1994) is steeper than that given by the quench method (Katsura and Ito, 1989) (2.5 MPa/K) and the thermochemical calculation (Akaogi et al., 1989) (1.5 MPa/K).

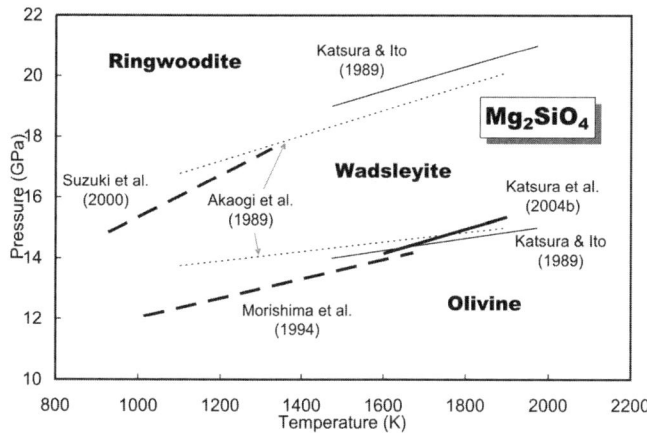

Figure 7. Phase boundaries of olivine-wadsleyite-ringwoodite transition in Mg_2SiO_4 determined by in situ X-ray diffraction (Morishima et al., 1994; Suzuki et al., 2000; Katsura et al., 2004b), quench method (Katsura and Ito, 1989), and thermochemical calculation (Akaogi et al., 1989). The phase boundary by Morishima et al. (1994) has been recalculated using Brown's (1999) scale.

The binary phase relations of the olivine-wadsleyite transition in $(Mg,Fe)_2SiO_4$ were studied by Katsura et al. (2004b). They loaded olivine solid solutions in a sample capsule with a pressure standard of MgO mixed with BN. They kept the samples at high-pressure and high-temperature conditions, while taking diffractions of MgO to monitor pressures. After that, they quenched and recovered the sample and determined the composition of coexisting olivine and wadsleyite. Using such a procedure, they constructed the phase relations at 1600 K and 1900 K. They suggested that the pressure interval of the olivine-wadsleyite transition at the conditions of the 410 km discontinuity would be 0.5 GPa. Although they did not conduct reversed experiments in this study, Katsura and Ito (1989) confirmed the chemical equilibrium at 1870 K by pairing mixing and unmixing experiments. However, there are no experimental data for the chemical equilibrium at temperatures around 1600 K. Consequently, Katsura et al. (2004b) could have overestimated the binary loop at 1600 K, because some traces of the low-temperature reaction could remain in the samples.

Katsura et al. (2004b) estimated the phase boundary in the end-member Mg_2SiO_4 by extrapolation of the binary loop. Morishima et al. (1994) and Katsura et al. (2004b) adopted different experimental procedures and pressure scales. Nevertheless, the phase boundaries given by these two studies are in good agreement. This fact may suggest a certain reliability of these two studies.

Wadsleyite-Ringwoodite Transition in Mg_2SiO_4

The wadsleyite-ringwoodite transition in $(Mg,Fe)_2SiO_4$ may be responsible for the 520 km seismic discontinuity. The phase relations of the wadsleyite-ringwoodite transition in Mg_2SiO_4 were determined by Suzuki et al. (2000) using MAX-80 and MAX-90. They loaded a fine-powder mixture of olivine, wadsleyite, and ringwoodite as starting material and observed a relative intensity change between wadsleyite and ringwoodite. Pressures were determined by NaCl with the Brown scale. They observed the phase transition from ringwoodite to wadsleyite at temperatures of 1020–1770 K. However, they observed the reversal transition at only one P-T condition, which was far away from the phase boundary. Instead of observing the direct transition from wadsleyite to ringwoodite, they observed growth of ringwoodite from the olivine starting material.

TABLE 4. REPORTED PHASE BOUNDARIES OF OLIVINE-WADSLEYITE TRANSITION IN Mg_2SiO_4

Study	Slope of phase boundary (MPa/K)	Intercept of phase boundary (GPa)	Temperature range (K)	Method
Katsura and Ito (1989)	2	11.1	1473–1873	Q
Akaogi et al. (1989)	1.5	12.0	–	TC
Morishima et al. (1994)	3.2†	8.9†	1013–1673	IXD, NaCl
Katsura et al. (2004b)	4.0	7.8	1600–1900	IXD, MgO

Note: Abbreviations: Q—quench method, TC—thermochemical calculation, IXD—in situ X-ray diffraction.
†Recalculated using the Brown scale.

Figure 7 shows the phase boundary obtained by Suzuki et al. (2000) along with those obtained by the quench experiment (Katsura and Ito, 1989) and thermochemical calculation (Akaogi et al., 1989). The reported phase boundaries are also summarized in Table 5. In addition to the olivine-wadsleyite transition, the phase boundary of the wadsleyite-ringwoodite transition given by the in situ X-ray diffraction has a steeper slope (6.9 MPa/K) than those obtained by the quench method (4 MPa/K) and thermochemical calculation (4.1 MPa/K). Because of insufficient reversed experiments, the phase boundary by Suzuki et al. (2000) may contain errors, similar to those of Yagi and Akimoto's (1976) coesite-stishovite boundary.

Akimotoite-Perovskite Transition in MgSiO$_3$

Mg-rich pyroxenes are the secondmost important mantle minerals next to olivine. MgSiO$_3$ clinopyroxene decomposes into wadsleyite + stishovite and ringwoodite + stishovite, then transforms to the ilmenite structure known as akimotoite. Akimotoite transforms into the perovskite structure. Therefore, knowledge about the akimotoite-perovskite transition in MgSiO$_3$ is of primary importance in understanding mantle mineralogy.

Kato et al. (1995) studied this phase transition using MAX-80 and MAX-90 with double-stage compression. They loaded MgSiO$_3$ enstatite as a starting material, compressed it, and heated it to conditions between 17 and 28 GPa and 920 and 1570 K. They used NaCl as a pressure standard with the Decker scale. They also used the Au pressure standard, the pressures of which were internally calibrated by NaCl in their study. The temperature fluctuations in their study were substantial, that is, between 50 and 200 K. Such large temperature fluctuations caused serious errors in pressure estimations of up to 2.5 GPa. They did not identify the phase present in situ, and identification was conducted after recovery to ambient conditions. Thus, they did not conduct either normal nor reversed experiments in a strict sense.

Kuroda et al. (2000) tried to determine the phase boundary using SPEED-1500. However, they were not able to because of serious grain growth of akimotoite and sluggish kinetics. Instead of conducting in situ X-ray diffraction experiments directly, they presented a phase boundary that was determined by the quench method with a pressure calibration based on the reaction boundary of ringwoodite to perovskite + periclase in Mg$_2$SiO$_4$ determined by Irifune et al. (1998). However, the samples of Irifune et al. (1998) and Kuroda et al. (2000) experienced different P-T paths. As discussed later, dissociation of ringwoodite to perovskite + periclase also has a serious kinetic problem. Hence, the phase boundary given by Kuroda et al. (2000) contains significant errors.

Ono et al. (2001) conducted both normal and reversal runs for this transition at temperatures of 1300–1700 K, with an Au pressure standard using SPEED-1500. The starting material of enstatite was mixed with forsterite in order to prevent grain growth of akimotoite and obtain high-quality diffraction patterns. Nevertheless, Ono et al. (2001) still found very sluggish kinetics for this transition. This group often observed that no transition occurred, even though the experimental conditions were supposed to cross the phase boundary. Ono et al. (2001) judged the location of the phase boundary by the relative intensity change between akimotoite and perovskite. However, the intensity changes associated with the phase transition were not always clear. In spite of these problems, Ono et al. (2001) proposed a phase boundary of this transition, as is shown in Figure 8, together with those obtained by other studies (Ito and Takahashi, 1989; Akaogi and Ito, 1993; Kato et al., 1995; Kuroda et al., 2000; Hirose et al., 2001b; Chudinovskikh and Boehler, 2004).

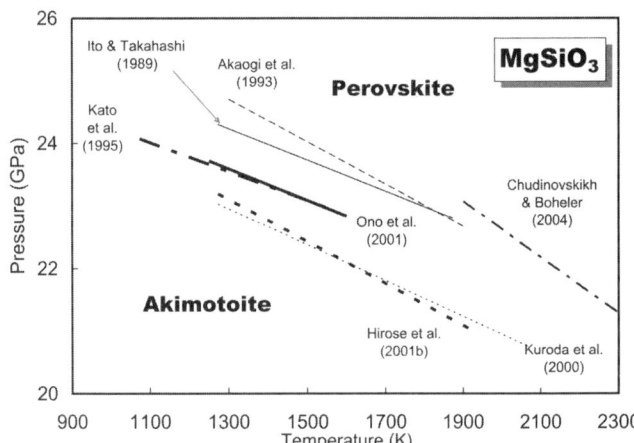

Figure 8. Phase boundaries of akimotoite-perovskite transition in MgSiO$_3$ determined by in situ X-ray diffraction (Kato et al., 1995; Ono et al., 2001; Hirose et al., 2001b), quench experiment (Ito and Takahashi, 1989; Kuroda et al., 2000), and thermochemical calculation (Akaogi and Ito, 1993). The phase boundaries by Ono et al. (2001) and Hirose et al. (2001b) have been recalculated using Fei's scale (Fei et al., 2004a). The phase boundary by quench experiment by Kuroda et al. (2000), which was calibrated using the reaction boundary of dissociation of ringwoodite to perovskite + periclase by Irifune et al. (1998), has also been recalculated using Fei's scale (Fei et al., 2004a).

TABLE 5. REPORTED PHASE BOUNDARIES OF OLIVINE-RINGWOODITE TRANSITION IN Mg$_2$SiO$_4$

Study	Slope of phase boundary (MPa/K)	Intercept of phase boundary (GPa)	Temperature range (K)	Method
Katsura and Ito (1989)	4	13.1	773–1373	Q
Akaogi et al. (1989)	4.1	12.2	–	TC
Suzuki et al. (2000)	6.9	8.4	927–1327	IXD, NaCl

Note: Abbreviations: Q—quench method, TC—thermochemical calculation, IXD—in situ X-ray diffraction.

Hirose et al. (2001a) conducted a series of quench experiments with pressure measurement by in situ X-ray diffraction using Au and Pt standards. They analyzed phases present after recovery to ambient conditions. They mentioned their experimental pressures dropped by 0.2–0.5 GPa while maintaining a constant temperature. The pressure drop and sluggish transition kinetics may have caused an underestimation of transition pressures. The phase boundary proposed by Hirose et al. (2001a), recalculated using the Fei scale, is shown in Figure 8. The proposed phase boundaries are summarized in Table 6.

The phase boundaries proposed by these studies have similar slopes (−2.5–3.4 MPa/K). However, the location of the phase boundaries varies by 2 GPa. Ono et al.'s (2001) study adopted the most proper way to determine the phase boundary. However, all the phase boundaries, including Ono et al.'s (2001), probably have significant errors due to the sluggish kinetics. At present, it is not possible to propose a reliable phase boundary from these studies.

Dissociation of Ringwoodite to Perovskite + Periclase in Mg_2SiO_4

The dissociation of ringwoodite to perovskite + ferropericlase in $(Mg,Fe)_2SiO_4$ is considered to be responsible for the 660 km seismic discontinuity. Irifune et al. (1998) studied the reaction boundary of ringwoodite to perovskite + periclase in Mg_2SiO_4 using SPEED-1500 at SPring-8. They observed the formation of ringwoodite from perovskite + periclase at three P-T conditions between 1670 and 2070 K, with a decreasing pressure with temperature decrease at a constant press load. They also described the dissociation of ringwoodite at 1920 K where an increasing pressure results in a temperature increase at a constant press load. Although they indicated a number of data points, where the intensity ratio of ringwoodite/perovskite + periclase increased or decreased, they also stated difficulties in judging an increase or decrease of the ratio. They placed the phase boundary near 21 GPa at 1900 K based on the Anderson scale, which was more than 2 GPa lower than that expected from the depth of the 660 km discontinuity. After that, several arguments were raised about the reliability of the pressure determination using in situ X-ray diffraction in multianvil apparatuses (cf. Shim et al., 2002).

Katsura et al. (2003) reinvestigated this reaction boundary using a similar technique as that used by Irifune et al. (1998). They found that dissociation of ringwoodite to perovskite + periclase was difficult to initiate even at high temperatures (~2000 K). In order to overcome the sluggish kinetics of dissociation of ringwoodite to perovskite + periclase, they conducted rapid continuous heating or cooling at a constant press load. They believed that this procedure created defects in ringwoodite crystals due to thermal stress. It appeared that formation of ringwoodite from perovskite + periclase was easy to initiate by decompression at high temperatures, but it was difficult to specify the beginning of the reaction. Although they made efforts to overcome the kinetic problems, they were unable to constrain the phase boundary tightly. Nevertheless, their data clearly demonstrated that the slope of this reaction boundary is, at steepest, −2.0 MPa/K, which is significantly less steep than previously considered.

Fei et al. (2004b) also studied this reaction. They compressed a sample to a desired press load and heated it to a desired temperature. They kept the press load and temperature constant for longer than two hours while continuously monitoring the sample pressure given by MgO and Au. They quenched the sample, recovered it to ambient conditions, and identified phases present in the sample by means of micro-Raman spectroscopy. They claimed that their method should provide a definitive answer regarding the phase relations. However, their data may suffer from kinetic problems, as discussed in the section on the coesite-stishovite transition.

Figure 9 shows the phase boundaries given by these three studies together with those obtained by quench experiments (Ito and Takahashi, 1989) and thermochemical calculation (Akaogi and Ito, 1993) (summarized in Table 7). Katsura et al. (2004b) presented a feasible range for the slope; the suggested boundaries of the middle and most negative slopes are shown. Even if we recalculated the phase boundaries using the most reliable pressure scale at present, all of the phase boundaries obtained by in situ X-ray diffraction are located at pressure values 0.5–1.5 GPa lower than those corresponding to the 660 km discontinuity (23.4 GPa). At present, it is still difficult to explain the origin of the 660 km discontinuity by dissociation of ringwoodite to perovskite + periclase in $(Mg,Fe)_2SiO_4$.

TABLE 6. REPORTED PHASE BOUNDARIES OF AKIMOTOITE-PEROVSKITE TRANSITION IN $MgSiO_3$

Study	Slope of phase boundary (MPa/K)	Intercept of phase boundary (GPa)	Temperature range (K)	Method
Ito and Takahashi (1989)	−2.5	27.5	1273–1873	Q
Akaogi and Ito (1993)	−3.4	29.1	–	TC
Kato et al. (1995)	−2.3[†]	26.6[†]	1073–1573	IXD, NaCl
Kuroda et al. (2000)	−2.9	26.7	1273–2073	Q
Ono et al. (2001)	−2.5[‡]	26.9[‡]	1250–1600	IXD, Au
Hirose et al. (2001b)	−3.3[‡]	27.5[‡]	1273–1973	IXD, Au

Note: Abbreviations: Q—quench method, TC—thermochemical calculation, IXD—in situ X-ray diffraction.
[†]Recalculated using the Brown scale.
[‡]Recalculated using the Fei scale.

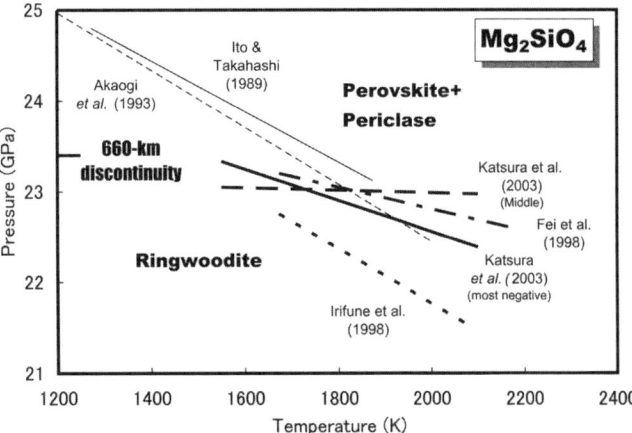

Figure 9. Phase boundaries of post-spinel in Mg_2SiO_4 determined by in situ X-ray diffraction (Irifune et al., 1998; Katsura et al., 2003; Fei et al., 2004b), quench experiment (Ito and Takahashi, 1989), and thermochemical calculation (Akaogi et al., 1995). The phase boundaries by Irifune et al. (1998) and Katsura et al. (2003) have been recalculated using Fei's scale (Fei et al., 2004a).

The recent two studies, Katsura et al. (2003) and Fei et al. (2004b), suggested very shallow gradients for the boundary. If the 660 km discontinuity is attributed to the dissociation of ringwoodite to perovskite + periclase, the depression of the 660 km discontinuity under the subduction zone could be explained by the negative slope of the reaction boundary combined with the low temperature of the subducted slab. Seismic observations suggest that the 660 km discontinuity is depressed by 20 km under the western Pacific subduction zones (Flanagan and Shearer, 1998). If we accept the slopes given by Katsura et al. (2003) and Fei et al. (2004b), the depression of the 660 km discontinuity implies that temperatures in these regions are lower by at least 400 K than the surrounding mantle.

The good agreement of the results of Katsura et al. (2003) and Fei et al. (2004b), who adopted different experimental techniques to address the kinetic problem, might suggest a certain reliability. At present, however, the results cannot provide the most probable boundary location of the dissociation of ringwoodite to perovskite + periclase because of the kinetic problem.

Dissociation of Garnet to Perovskite + Corundum in $Mg_3Al_2Si_3O_{12}$

The dissociation of garnet to perovskite + corundum could be responsible for the high velocity gradient at the top of the lower mantle. The dissociation reaction in $Mg_3Al_2Si_3O_{12}$ was studied by Hirose et al. (2001a). They adopted Au as a pressure marker. They used a glass sample with $Mg_3Al_2Si_3O_{12}$ as starting material. Their approach was essentially the same as Hirose et al. (2001b) and Fei et al. (2004b). As has already been repeatedly mentioned, although such a procedure is useful to obtain preliminary results, the results may contain significant errors due to the kinetic problem. Figure 10 shows their phase boundary together with that obtained by quench experiment (Kubo and Akaogi, 2000) and thermochemical calculation (Akaogi et al., 2002) (summarized in Table 8).

DISCUSSION

As mentioned in the introduction, it is expected that the phase relations of important mantle minerals in the upper part of

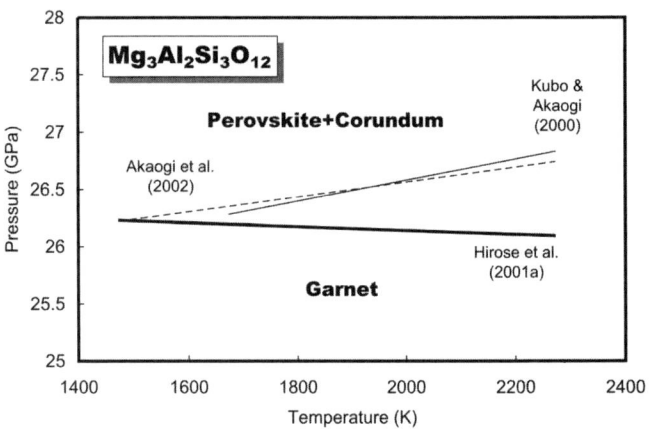

Figure 10. Reaction boundaries of dissociation of garnet to perovskite + corundum determined by in situ X-ray diffraction (Hirose et al., 2001a), quench experiment (Kubo and Akaogi, 2000), and thermochemical calculation (Akaogi et al., 2002). The phase boundary by Hirose et al. (2001a) has been recalculated using Fei's scale (Fei et al., 2004a).

TABLE 7. REPORTED REACTION BOUNDARIES BETWEEN RINGWOODITE AND PEROVSKITE + PERICLASE IN Mg_2SiO_4

Study	Slope of phase boundary (MPa/K)	Intercept of phase boundary (GPa)	Temperature range (K)	Method
Ito and Takahashi (1989)	−2.8	28.4	1273–1873	Q
Akaogi and Ito (1993)	−3.2	28.8	–	TC
Irifune et al. (1998)	−3.0[†]	27.8[†]	1673–2073	IXD, Au
Katsura et al. (2003)	−1.7[†]	26.0[†]	1550–2100	IXD, Au
	0.1[†]	23.3[†]		
Fei et al. (2004b)	−1.2[‡]	25.2[‡]	1673–2173	IXD, MgO

Note: Abbreviations: Q—quench method, TC—thermochemical calculation, IXD—in situ X-ray diffraction.
[†]Recalculated using the Fei scale.
[‡]Recalculated using the Matsui scale.

TABLE 8. REPORTED REACTION BOUNDARIES BETWEEN GARNET AND PEROVSKITE + CORUNDUM IN $Mg_3Al_2Si_3O_{12}$

Study	Slope of phase boundary (MPa/K)	Intercept of phase boundary (GPa)	Temperature range (K)	Method
Kubu and Akaogi (2000)	0.9	24.7	1773–2273	Q
Akaogi et al. (2002)	0.7	25.3	–	TC
Hirose et al. (2001a)	–0.2*	26.5*	1473–2273	IXD, Au

Note: Abbreviations: Q—quench method, TC—thermochemical calculation, IXD—in situ X-ray diffraction.
*Recalculated using the Fei scale

the mantle should be determined with an accuracy of 0.1 GPa. High-pressure workers have expected that in situ X-ray diffraction in a multianvil apparatus should reach such a requirement. However, there are still a number of technical difficulties in the effort to determine the phase boundaries with high accuracy. In this section, the technical difficulties that should be solved in order to determine phase relations accurately will be discussed.

Temperature Measurement

In order to determine the pressure from an equation of state of a pressure standard, accurate measurements of temperature are indispensable. For example, a temperature error of 100 K results in pressure errors of 0.3, 0.7, and 0.6 GPa at 1800 K, and 20 GPa for NaCl, Au, and MgO scales, respectively.

There is a temperature gradient in a sample because of the limited cell volume, especially in the case of experiments above 20 GPa. As mentioned before, the diffraction volume is usually elongated to more than 1 mm in the direction of the incident X-rays, whereas the length of the heater is 2–4 mm in many cases. The diffraction volume could include much lower-temperature portions than indicated by a thermocouple. In order to avoid this uncertainty, workers tend to load a sample that is smaller than the diffraction length. However, the typical thickness of the samples is still several hundred micrometers. The temperature inhomogeneity in the sample should be still significant.

The temperature distribution in the sample could be estimated by inserting two independent thermocouples. However, the temperature distribution would be largely affected by the insertion of the second thermocouple because of the small heater size. The temperature distribution could also be measured by measuring the variation of element partitioning of coexisting phases (Takahashi et al., 1982). However, appropriate reference data on the partition coefficients as a function of temperature and pressure above 20 GPa do not exist. In addition, it is difficult to achieve a chemical equilibrium.

Another problem with temperature measurement is the pressure dependence of the electromotive force. Some workers have compared temperatures indicated by different types of thermocouples at high pressures. For example, Li et al. (2003) compared the $W_{95}Re_5$-$W_{74}Re_{26}$ and $Pt_{90}Rh_{10}$-Pt thermocouples. They found up to 35 K difference between these two thermocouples at 15 GPa and temperatures up to 2070 K. Although such comparisons have been made repeatedly, temperature deviation indicated by each thermocouple at high pressures from the actual temperature have rarely been studied (Getting and Kennedy, 1970). In order to determine the location of the phase boundary with an accuracy of 0.1 GPa, calibration of the electromotive force of thermocouples at high pressures is required. Measurement of Johnson noise is one way to calibrate thermocouples, which is sill under development (I. Getting, personal commun.).

Pressure Measurement

The reliability of the equation of state of the pressure standard is a key issue in phase relation studies. An equation of state is usually constructed based on room-pressure compression and thermal pressure estimated from the shock Hugoniot. Along the shock Hugoniot, however, temperature increases to only a few hundred K up to 50 GPa in cases of Au and MgO (Jamieson et al., 1982). Hence, the equations of state constructed in this way could contain considerable errors in thermal pressure. For example, Shim et al. (2001) found a large increase of volume dependence of the Grüneisen parameter, q, from –6 to 1 with compression up to $V/V_0 = 0.75$, when constructing their equation of state for Au. They recommended $q = 1$, but this value does not have a sufficient reliability. Although an equation of state primarily constructed on the base of the shock Hugoniot may provide a fairly accurate pressure value under relatively high-pressure and low-temperature conditions, it does not at relatively low-pressure and high-temperature conditions, which is the case for the phase-relations studies in a multianvil apparatus. An equation of state should be constructed using volume data at high temperatures and low pressures, as was implemented by Matsui et al. (2000).

One of the promising ways to establish a reliable equation of state of a pressure standard is simultaneous combination of a volume measurement by in situ X-ray diffraction and velocity measurement (cf. Mueller et al., 2002). The velocity measurement gives us a pressure derivative of volume, rendering it possible to estimate pressures with successive compression. If this method can be conducted at high temperatures corresponding to the upper mantle, we will obtain a reliable thermal equation of state for the phase-relation studies.

In addition to the reliability of the equations of state, pressure intensification or reduction in composite materials could also be a problem (Sato et al., 1973). The pressure standard is usually mixed with other materials in order to avoid grain growth at high temperatures. In addition, the high X-ray absorbance requires that Au grains be inlaid in a host material composed of light elements. The pressure standard and mixed materials have different thermoelastic properties. If they do not creep sufficiently, they compress or expand at different rates, and as a result, pressure applied to them would be different. The pressure applied to each of them would be also different from the bulk pressure.

Figure 11 shows one example of pressure reduction in a composite material. In this experiment, $MgSiO_3$ perovskite and a pressure marker of MgO are compressed in one high-pressure cell, the cross sections of which are shown in Figure 2. After annealing at 1700 K, the volumes of $MgSiO_3$ perovskite and MgO were measured at high pressures and ambient temperature by in situ X-ray diffraction. After the high-pressure measurements, the volumes of perovskite and MgO were carefully measured at ambient conditions to determine their V_0. Pressures were calculated from the V/V_0 of MgO using the Matsui scale. In one run, the pressure marker of MgO was mixed with 20 wt% diamond, whereas in another run, it was not mixed with anything. Data from these two runs were fitted to the third-order Birch-Murnaghan equation of state. With fixed K_T of 262 GPa, we obtained a reasonable K′ value of 4.2(4) from the data using the pure MgO pressure marker. On the other hand, we obtained an anomalously low K′ value of 2.0(2) from the data set using the MgO pressure marker mixed with 20 wt% diamond. With the volume of perovskite as a standard, pressures applied on the MgO mixed with diamond should be 10% lower than those of pure MgO. The framework support of diamond should reduce pressures applied to the MgO grains.

Thus, the pressure standard should not be mixed with other materials. However, it is practically impossible to use pure Au and NaCl as a pressure standard, as explained earlier. Hence, use of MgO is essentially important in order to determine pressure accurately.

Kinetics

The sluggish kinetics are also a serious problem in determining the phase relations. The starting material is relatively easy to initiate and transform to a high-pressure phase at relatively low temperatures, such as 900–1100 K (cf. Yagi and Akimoto, 1976; Irifune et al., 1998; Suzuki et al., 2000; Katsura et al., 2003). However, once the first phase transition occurs, it is much more difficult to initiate transition to another phase in the newly formed phase (Kuroda et al., 2000; Ono et al., 2001; Katsura et al., 2003). Even though the sample is heated to a higher temperature, the further phase transition tends to remain difficult or become more difficult to initiate.

Such kinetic features could be understood in view of the defect density. A powdered starting material contains tens of percents porosity. Cold compression causes strong local deviatoric stresses on the grains to fill gaps among them. High-density dislocations and stacking faults would be formed by local deviatoric stresses. This phenomenon is inferred from the broadening of diffraction peaks during cold compression (Utsumi et al., 1998). The phase transition should be easy to start at the defects. Once the phase transition occurs, the local deviatoric stresses should be released. The newly formed high-pressure phase should contain only a limited number of dislocations and stacking faults. This fact is inferred from the sharp diffraction peaks of the newly formed phase. The further phase transition is difficult to initiate because the energy barrier for the phase transition should be very high in a perfect crystal.

The grain boundary is a planar defect, where the energy barrier for the phase transition should be lower than within the grain. The phase transition could be initiated on the grain boundary. At low temperatures, the grain sizes of the newly formed high-pressure phase are small. With increasing temperature, the grain sizes increase because of the quasi-hydrostatic conditions that result from softening of the pressure medium at high temperatures (Karato et al., 1980). In addition, the quasi-hydrostatic conditions should also decrease the densities of dislocations and stacking faults. These could account for the increased difficulty of initiating the further phase transition from the newly formed high-pressure phase at higher temperatures.

The inertness appears to become more severe for the reaction at higher pressures. In the case of coesite-stishovite transition in SiO_2, Zhang et al. (1996) succeeded in reversing the phase transition from stishovite to coesite, in which stishovite was synthesized during the in situ X-ray diffraction experiment. In the case of an olivine-wadsleyite transition in Mg_2SiO_4, Morishima et al. (1994) were able to conduct normal and reversal runs by using the mixture of olivine and wadsleyite. However,

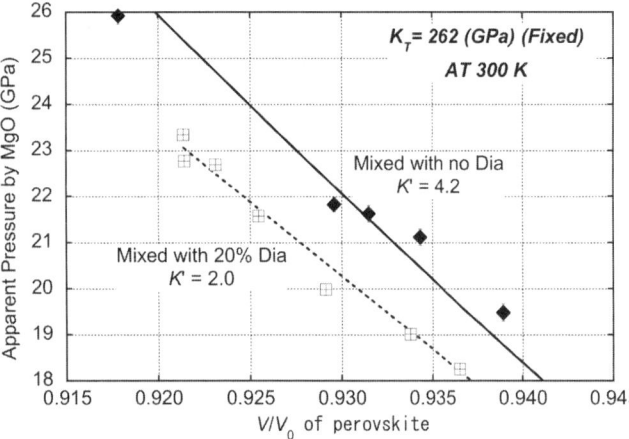

Figure 11. Apparent compression curves of $MgSiO_3$ perovskite at room temperature. Pressures are obtained from the compression of the pure MgO standard (solid diamonds) and that mixed with 20 wt% diamond (open squares). The samples and pressure standards were heated above 1400 K in order to minimize deviatoric stresses. The experiments with the MgO pressure standard mixed with 20 wt% diamond (Dia) gives a smaller K′ value with the fixed K_T value (262 GPa). The MgO pressure standard mixed with 20 wt% diamond would give a 9% smaller pressure than the pure MgO standard.

they reported difficulty in the transition from olivine to wadsleyite by simply increasing the pressure at temperatures higher than 1270 K. In the cases of the akimotoite-perovskite transition in $MgSiO_3$ and dissociation of ringwoodite to perovskite + periclase, the normal and reversed experiments were found to be very difficult to execute.

CONCLUDING REMARKS

In order to study the structure of Earth's mantle, comparison of pressures of the important phase transitions with seismic observation is essential. For this purpose, the phase relations have to be determined as a function of temperature with an accuracy of 0.2 GPa or less. High P-T in situ X-ray diffraction in a multianvil apparatus in combination with highly brilliant X-rays by synchrotron radiation can be a very powerful technique to precisely determine the boundaries of the important phase transitions and reactions. We can study the phase relations at pressures up to 30 GPa (depth of 800 km) using tungsten carbide anvils. Pressures exceeding 50 GPa (depth of 1200 km) can be generated using sintered diamond anvils. The temperature conditions can be studied up to 2500 K.

Energy-dispersive X-ray diffraction is adopted to obtain sufficient diffraction patterns for a limited data acquisition time. Sample pressures are estimated from volume of pressure standards such as NaCl, MgO, and Au, and temperature is indicated by a thermocouple. In order to calculate pressure from these data, the following equations of state are recommended: Brown (1999) for NaCl, Matsui et al. (2000) for MgO, and Fei et al. (2004a) for Au. Among these pressure standards, the use of MgO is strongly recommended, partially because of the relatively slow grain growth rate, and also because of the available high-temperature thermoelastic data that can be used to construct its equation of state.

So far, the following phase transition and reaction boundaries have been studied: coesite-stishovite transition in SiO_2, olivine-ringwoodite transition in Fe_2SiO_4, olivine-wadsleyite-ringwoodite transition in Mg_2SiO_4, akimotoite-perovskite transition in $MgSiO_3$, dissociation of ringwoodite to perovskite + periclase in Mg_2SiO_4, and dissociation of garnet to perovskite + corundum in $Mg_3Al_2Si_3O_{12}$. The binary olivine-wadsleyite transition in $(Mg,Fe)_2SiO_4$ has also been studied. The following studies are recommended because of their achievement of pairs of normal and reversal runs: coesite-stishovite transition in SiO_2 by Zhang et al. (1996), olivine-ringwoodite transition in Fe_2SiO_4 by Yagi et al. (1987), and olivine-wadsleyite transition in Mg_2SiO_4 by Morishima et al. (1994).

The akimotoite-perovskite transition in $MgSiO_3$ and dissociation of ringwoodite to perovskite + periclase have been repeatedly studied by different groups. The results agree within 1 GPa, but they show discrepancy to some extent. In the case of the dissociation of ringwoodite, all studies show disagreement of their phase boundary at the depth of the 660 km discontinuity. Although the disagreement becomes smaller by improvements in the equations of state of the pressure standards, there are still significant disagreements. In addition to the location of the boundary, its slope seems less steep than previously believed.

Accurate determinations of phase boundaries are difficult because of: (1) uncertainties in temperature distributions in a high-pressure cell, (2) unknown pressure effects on the electromotive force, (3) unreliable construction of equations of state suitable for relatively low-pressure (~30 GPa) and high-temperature experiments, (4) pressure intensification and reduction of pressure standards mixed with other materials, and (5) sluggish kinetics, especially for the phase transition at relatively high pressures.

ACKNOWLEDGMENTS

The in situ, high pressure-temperature, X-ray diffraction experiments on the binary phase relations between ringwoodite and perovskite + periclase were made at the SPring-8 with the approval of the Japan Synchrotron Radiation Research Institute (JASRI) (proposal no. 2003A0087-ND2-np, 2003B0638-CD2b-np, 2004A0368-ND2b-np, and 2004B0497-ND2b-np). These experiments were conducted with the following colleagues: N. Tomioka, K. Saito, S. Yokoshi, K. Kawabe, and M. Sugita. I would like to thank I. Getting for providing information about his Johnson noise measurement. I also would like to thank Y. Wang and H. Muller for providing information on the facilities at APS and HASYLAB, respectively. This research was supported by the 21st Century COE program of the Japan Society for the Promotion of Science. S.-M. Zhai and G. Mantilake are acknowledged for their comments.

REFERENCES CITED

Akaogi, M., and Ito, E., 1993, Refinement of enthalpy measurement of $MgSiO_3$ perovskite and negative pressure-temperature slopes for perovskite-forming reactions: Geophysical Research Letters, v. 20, p. 1839–1842.

Akaogi, M., and Navrotsky, A., 1984, The quartz-coesite-stishovite transformations: New calorimetric measurements and calculation of phase diagrams: Physics of the Earth and Planetary Interiors, v. 36, p. 124–134, doi: 10.1016/0031-9201(84)90013-X.

Akaogi, M., Ito, E., and Navrotsky, A., 1989, Olivine-modified spinel-spinel transitions in the system Mg_2SiO_4-Fe_2SiO_4: Calorimetric measurements, thermochemical calculation, and geophysical application: Journal of Geophysical Research, v. 94, p. 15,671–15,685.

Akaogi, M., Yusa, H., Shiraishi, K., and Suzuki, T., 1995, Thermodynamic properties of α-quartz, coesite, and stishovite and equilibrium phase-relations at high-pressures and high-temperatures: Journal of Geophysical Research, v. 100, p. 22,337–22,347, doi: 10.1029/95JB02395.

Akaogi, M., Tanaka, A., and Ito, E., 2002, Garnet-ilmenite-perovskite transitions in the system $Mg_4Si_4O_{12}$-$Mg_3Al_2Si_3O_{12}$ at high pressures and high temperatures: Phase equilibria, calorimetry and implications for mantle structure: Physics of the Earth and Planetary Interiors, v. 132, p. 303–324, doi: 10.1016/S0031-9201(02)00075-4.

Akimoto, S., 1987, High-pressure research in geophysics: Past, present and future, in Manghnani, M.H., and Syono, Y., eds., High-Pressure Research in Mineral Physics: Tokyo, Terra Scientific Publishing, p. 1–13.

Akimoto, S., Fujisawa, E., and Katsura, T., 1965, The olivine-spinel transition in Fe_2SiO_4 and Ni_2SiO_4: Journal of Geophysical Research, v. 70, p. 1969–1977.

Anderson, O.L., Isaak, D.G., and Yamamoto, S., 1989, Anharmonicity and the equation of state for gold: Journal of Applied Physics, v. 65, no. 4, p. 1534–1543.

Boehler, R., Ross, M., and Boercker, D.B., 1997, Melting of LiF and NaCl to 1 Mbar: Systematics of ionic solids at extreme conditions: Physical Review Letters, v. 78, p. 4589–4592, doi: 10.1103/PhysRevLett.78.4589.

Brown, J.M., 1999, The NaCl pressure standard: Journal of Applied Physics, v. 86, p. 5801–5808, doi: 10.1063/1.371596.

Chudinovskikh, L., and Boehler, R., 2004, $MgSiO_3$ phase boundaries measured in the laser-heated diamond cell: Earth and Planetary Science Letters, v. 219, no. 3–4, p. 285–296.

Decker, D.L., 1971, High-pressure equation of state for NaCl, KCl, and CsCl: Journal of Applied Physics, v. 42, p. 3239–3244, doi: 10.1063/1.1660714.

Dubrovinsky, L.S., and Saxena, S.K., 1997, Thermal expansion of periclase (MgO) and tungsten (W) to melting temperatures: Physics and Chemistry of Minerals, v. 24, p. 547–550, doi: 10.1007/s002690050070.

Fei, Y.W., Li, H., Hirose, K., Minarik, W., Van Orman, J., Sanloup, C., van Westrenen, W., Komabayashi, T., and Funakoshi, K., 2004a, A critical evaluation of pressure scales at high temperatures by $in~situ$ X-ray diffraction measurements: Physics of the Earth and Planetary Interiors, v. 143–144, p. 515–526, doi: 10.1016/j.pepi.2003.09.018.

Fei, Y.W., Van Orman, J., Li, J., van Westrenen, W., Sanloup, C., Minarik, W., Hirose, K., Komabayashi, T., Walter, M.J., and Funakoshi, K., 2004b, Experimentally determined postspinel transformation boundary in Mg_2SiO_4 using MgO as an internal pressure standard and its geophysical implications: Journal of Geophysical Research, v. 109, B02305, doi: 10.1029/2003JB002562.

Flanagan, M.P., and Shearer, P.M., 1998, Global mapping of topography on transition zone velocity discontinuities by stacking SS precursors: Journal of Geophysical Research, v. 103, p. 2673–2692, doi: 10.1029/97JB03212.

Getting, I.C., and Kennedy, G.C., 1970, Effect of pressure on the EMF of chromel-alumel and platinum-platinum 10% rhodium thermocouples: Journal of Applied Physics, v. 41, p. 4552–4562, doi: 10.1063/1.1658495.

Hirose, K., Fei, Y.W., Ono, S., Yagi, T., and Funakoshi, K., 2001a, In situ measurements of the phase transition boundary in $Mg_3Al_2Si_3O_{12}$: Implications for the nature of the seismic discontinuities in the Earth's mantle: Earth and Planetary Science Letters, v. 184, p. 567–573, doi: 10.1016/S0012-821X(00)00354-X.

Hirose, K., Komabayashi, T., Murakami, M., and Funakoshi, K., 2001b, In situ measurements of the majorite-akimotoite-perovskite phase transition boundaries in $MgSiO_3$: Geophysical Research Letters, v. 28, p. 4351–4354, doi: 10.1029/2001GL013549.

Irifune, T., Nishiyama, N., Kuroda, K., Inoue, T., Isshiki, M., Utsumi, W., Funakoshi, K., Urakawa, S., Uchida, T., Katsura, T., and Ohtaka, O., 1998, The postspinel phase boundary in Mg_2SiO_4 determined by $in~situ$ X-ray diffraction: Science, v. 279, p. 1698–1700, doi: 10.1126/science.279.5357.1698.

Isaak, D.G., Anderson, O.L., and Goto, T., 1989, Measured elastic moduli of single crystal MgO up to 1800 K: Physics and Chemistry of Minerals, v. 16, p. 704–713, doi: 10.1007/BF00223321.

Ito, E., and Takahashi, E., 1989, Postspinel transformations in the system Mg_2SiO_4-Fe_2SiO_4 and some geophysical implications: Journal of Geophysical Research, v. 94, no. B8, p. 10,637–10,646.

Jamieson, J.C., Fritz, J.N., and Manghnani, M.H., 1982, Pressure measurement at high temperature in X-ray diffraction studies: Gold as a primary standard, in Akimoto, S., and Manghnani, M.H., eds., High-Pressure Research in Geophysics: Tokyo, Center for Academic Publishing, p. 27–48.

Karato, S.-I., Toriumi, M., and Fujii, T., 1980, Dynamic recrystallization of olivine single crystals during high-temperature creep: Geophysical Research Letters, v. 7, p. 649–652.

Kato, T., Ohtani, E., Morishima, H., Yamazaki, D., Suzuki, A., Suto, M., Kubo, T., Kikegawa, T., and Shimomura, O., 1995, In situ X-ray observation of high-pressure phase transitions of $MgSiO_3$ and thermal expansion of $MgSiO_3$ perovskite at 25 GPa by double-stage multi-anvil system: Journal of Geophysical Research, v. 100, p. 20,475–20,481, doi: 10.1029/95JB01688.

Katsura, T., and Ito, E., 1989, The system Mg_2SiO_4-Fe_2SiO_4 at high pressures and temperatures: Precise determination of stabilities of olivine, modified spinel, and spinel: Journal of Geophysical Research, v. 94, p. 15,663–15,670.

Katsura, T., Yamada, H., Shinmei, T., Kubo, A., Ono, S., Kanzaki, M., Yoneda, A., Walter, M.J., Ito, E., Urakawa, S., Funakoshi, K., and Utsumi, W., 2003, Post-spinel transition in Mg_2SiO_4 determined by high P-T $in~situ$ X-ray diffractometry: Physics of the Earth and Planetary Interiors, v. 136, p. 11–24, doi: 10.1016/S0031-9201(03)00019-0.

Katsura, T., Funakoshi, K., Kubo, A., Nishiyama, N., Tange, Y., Sueda, Y., Kubo, T., and Utsumi, W., 2004a, A large-volume high-pressure and high-temperature apparatus for $in~situ$ X-ray observation, 'SPEED-Mk.II': Physics of the Earth and Planetary Interiors, v. 143–144, p. 497–506, doi: 10.1016/j.pepi.2003.07.025.

Katsura, T., Yamada, H., Nishikawa, O., Song, M.S., Kubo, A., Shinmei, T., Yokoshi, S., Aizawa, Y., Yoshino, T., Walter, M.J., Ito, E., and Funakoshi, K., 2004b, Olivine-wadsleyite transition in the system $(Mg,Fe)_2SiO_4$: Journal of Geophysical Research, v. 109, p. B02209, doi: 10.1029/2003JB002438.

Kubo, A., and Akaogi, M., 2000, Post-garnet transitions in the system $Mg_4Si_4O_{12}$-$Mg_3Al_2Si_3O_{12}$ up to 28 GPa: Phase relations of garnet, ilmenite and perovskite: Physics of the Earth and Planetary Interiors, v. 121, p. 85–102, doi: 10.1016/S0031-9201(00)00162-X.

Kuroda, K., Irifune, T., Inoue, T., Nishiyama, N., Miyashita, M., Funakoshi, K., and Utsumi, W., 2000, Determination of the phase boundary between ilmenite and perovskite in $MgSiO_3$ by $in~situ$ X-ray diffraction and quench experiments: Physics and Chemistry of Minerals, v. 27, p. 523–532, doi: 10.1007/s002690000096.

Li, J., Hadidiacos, C., Mao, H.K., Fei, Y.W., and Hemley, R.J., 2003, Behavior of thermocouples under high pressure in a multi-anvil apparatus: High-Pressure Research, v. 23, p. 389–401, doi: 10.1080/0895795031000088269.

Matsui, M., Parker, S.C., and Leslie, M., 2000, The MD simulation of the equation of state of MgO: Application as a pressure calibration standard at high temperature and high pressure: The American Mineralogist, v. 85, p. 312–316.

Morishima, H., Kato, T., Suto, M., Ohtani, E., Urakawa, S., Utsumi, W., Shimomura, O., and Kikegawa, T., 1994, The phase boundary between α- and β-Mg_2SiO_4 determined by $in~situ$ X-ray observation: Science, v. 265, p. 1202–1203, doi: 10.1126/science.265.5176.1202.

Mueller, H.J., Lauterjung, J., Schilling, F.R., Lathe, C., and Nover, G., 2002, Symmetric and asymmetric interferometric method for ultrasonic compressional and shear wave velocity measurements in piston-cylinder and multi-anvil high-pressure apparatus: European Journal of Mineralogy, v. 14, p. 581–589, doi: 10.1127/0935-1221/2002/0014-0581.

Nishiyama, N., Katsura, T., Funakoshi, K., Kubo, A., Kubo, T., Tange, Y., Sueda, Y., and Yokoshi, S., 2003, Determination of the phase boundary between the B1 and B2 phases in NaCl by $in~situ$ X-ray diffraction: Physical Review, v. 68, p. 134109, doi: 10.1103/PhysRevB.68.134109.

Ono, S., Katsura, T., Ito, E., Kanzaki, M., Yoneda, A., Walter, M.J., Urakawa, S., Utsumi, W., and Funakoshi, K., 2001, In situ observation of ilmenite-perovskite phase transition in $MgSiO_3$ using synchrotron radiation: Geophysical Research Letters, v. 28, p. 835–838, doi: 10.1029/1999GL008446.

Sato, Y., Akimoto, S., and Inoue, K., 1973, Pressure intensification in the composite material: High Temperatures–High Pressures, v. 5, p. 289–297.

Serghiou, G., Zerr, A., Chudinovskikh, L., and Boehler, R., 1995, The coesite-stishovite transition in a laser-heated diamond cell: Geophysical Research Letters, v. 22, p. 441–444, doi: 10.1029/94GL02692.

Shim, S.-H., Duffy, T.S., and Shen, G.Y., 2001, The post-spinel transformation in Mg_2SiO_4 and its relation to the 660-km seismic discontinuity: Nature, v. 411, p. 571–574, doi: 10.1038/35079053.

Shim, S.-H., Duffy, T.S., and Takemura, K., 2002, Equation of state of gold and its application to the phase boundaries near 660 km depth in Earth's mantle: Earth and Planetary Science Letters, v. 203, p. 729–739, doi: 10.1016/S0012-821X(02)00917-2.

Shimomura, O., Yamaoka, S., Yagi, T., Wakatsuki, M., Tsuji, K., Fukunaga, O., Kawamura, H., Aoki, K., and Akimoto, S., 1985, Multi-anvil type high pressure apparatus for synchrotron radiation: Solid state physics under pressure, in Minomura, S., ed., Recent Advance with Anvil Devices: Tokyo and Dordrecht, KTK and Reidel, p. 351–356.

Shimomura, O., Utsumi, W., Taniguchi, T., Kikegawa, T., and Nagashima, T., 1992, A new high pressure and high temperature apparatus with sintered diamond anvils for synchrotron radiation use, in Syono, Y., and Manghnani, M.H., eds., High-Pressure Research: Application to Earth and Planetary Sciences: Tokyo, Terra Scientific Publishing Company, p. 3–11.

Speziale, S., Zha, C.S., Duffy, T.S., Hemley, R.J., and Mao, H.K., 2001, Quasi-hydrostatic compression of magnesium oxide to 52 GPa: Implications for the pressure-volume-temperature equation of state: Journal of Geophysical Research, v. 106, p. 515–528, doi: 10.1029/2000JB900318.

Suito, K., 1977, Phase relations of pure Mg_2SiO_4 up to 200 kilobars, in Manghnani, M.H., and Akimoto, S., eds., High-Pressure Research: Application to Geophysics: London, Academic Press, p. 255–266.

Suzuki, A., Ohtani, E., Morishima, H., Kubo, T., Kanbe, Y., Kondo, T., Okada, T., Terasaki, H., Kato, T., and Kikegawa, T., 2000, *In situ* determination of the phase boundary between wadsleyite and ringwoodite in Mg_2SiO_4: Geophysical Research Letters, v. 27, p. 803–806, doi: 10.1029/1999GL008425.

Takahashi, E., Yamada, H., and Ito, E., 1982, An ultrahigh-pressure furnace assembly to 100 kbar and 1500 °C with minimum temperature uncertainty: Geophysical Research Letters, v. 9, p. 805–807.

Takemura, K., 2001, Evaluation of the hydrostaticity of a helium-pressure medium with powder X-ray diffraction techniques: Journal of Applied Physics, v. 89, p. 662–668, doi: 10.1063/1.1328410.

Utsumi, W., Funakoshi, K., Urakawa, S., Yamakata, M., Tsuji, K., Konishi, H., and Shimomura, O., 1998, SPring-8 beamlines for high pressure science with multi-anvil apparatus: Review of High-Pressure Science and Technology, v. 7, p. 1484–1486.

Wang, Y.B., Rivers, M., Uchida, T., Murray, P., Shen, G.Y., Sutton, S., Chen, J.H., Xu, Y.Q., and Weidner, D.J., 2000, High pressure research using large-volume presses at GeoSoilEnviroCARS, Advanced Photon Source, *in* Manghnani, M.H., and Nicol, M.F., eds., Science and Technology of High Pressure, Proceedings of AIRAPT-17: Hyderabad, Universities Press, p. 1047–1052.

Weidner, D.J., Vaughan, M.T., Ko, J., Wang, Y., Liu, X., Yeganeh-Haeri, A., Pocalo, R.E., and Zhao, Y., 1992, Characterization of stress, pressure, and temperature in SAM85, a DIA type high pressure apparatus, *in* Syono, Y., and Manghnani, M.H., eds., High-Pressure Research: Application to Earth and Planetary Sciences: Tokyo, Terra Scientific Publishing Company, p. 13–17.

Weidner, D.J., Wang, Y.B., and Vaughan, M.T., 1994, Yield strength at high-pressure and temperature: Geophysical Research Letters, v. 21, p. 753–756, doi: 10.1029/93GL03549.

Woodland, A.B., and Angel, R.J., 2000, Phase relations in the system fayalite-magnetite at high pressures and temperatures: Contributions to Mineralogy and Petrology, v. 139, p. 734–747, doi: 10.1007/s004100000168.

Yagi, T., and Akimoto, S., 1976, Direct determination of coesite-stishovite transition by *in situ* X-ray measurements: Tectonophysics, v. 35, p. 259–270, doi: 10.1016/0040-1951(76)90042-1.

Yagi, T., Akaogi, M., Shimomura, O., Suzuki, T., and Akimoto, S., 1987, *In situ* observation of the olivine-spinel phase transformation in Fe_2SiO_4 using synchrotron radiation: Journal of Geophysical Research, v. 92, p. 6207–6213.

Zhang, J.Z., Liebermann, R.C., Gasparik, T., Herzberg, C.T., and Fei, Y.W., 1993, Melting and subsolidus relations of SiO_2 at 9–14 GPa: Journal of Geophysical Research, v. 98, p. 19,785–19,793.

Zhang, J.Z., Li, B., Utsumi, W., and Liebermann, R.C., 1996, *In situ* X-ray observations of the coesite stishovite transition: Reversed phase boundary and kinetics: Physics and Chemistry of Minerals, v. 23, p. 1–10, doi: 10.1007/BF00202987.

MANUSCRIPT ACCEPTED BY THE SOCIETY 14 AUGUST 2006

Multianvil techniques in conjunction with synchrotron radiation at Deutsches ElektronenSYnchrotron (DESY) – HAmburger SYnchrotron LABor (HASYLAB)

Hans J. Mueller[†]
Frank R. Schilling
Christian Lathe

GeoForschungsZentrum Potsdam in der Helmholtz-Gemeinschaft, Department 5, D-14473 Potsdam, Germany

ABSTRACT

During the early 1980s, geoscientists worldwide realized synchrotron radiation was a highly valuable tool for in situ experiments, i.e., experiments under simulated Earth mantle conditions. MAX80 at Deutsches ElektronenSYnchrotron (DESY) – HAmburger SYnchrotron LABor (HASYLAB), Hamburg, a single-stage multianvil DIA system at a synchrotron beam line was among the high-pressure pioneer apparatus designed in Japan. MAX80 is equipped to perform ultrasonic interferometry in conjunction with synchrotron radiation measurements, i.e., X-ray diffraction (XRD) and X-radiography. The maximum conditions are ~12 GPa at 2000 K.

To make transition-zone conditions accessible and to achieve bigger specimen volumes, the sister apparatus MAX200x, a double-stage DIA system, was recently installed at the HASYLAB HARWI-II beam line. MAX2000x is designed to reach 25 GPa and 2400 K, simultaneously. MAX200x is driven by a hydraulic ram with a maximum load of 1750 tons. Derived from the successful equipment of MAX80 and adapted to the new task, MAX200x is equipped for XRD with a Ge-solid-state detector, for transient ultrasonic interferometry, and it has a radiography system to measure change of volume and shape of the sample under in situ conditions. A stepper motor–driven slits system allows the X-ray beam size and shape to be optimized for experiments.

Parallel to the installation of MAX200x, some experiments were carried out to improve the potentials of multianvil apparatus in terms of maximum pressure and limitation of stress inside the sample and the anvils. Some recent results of these experiments as well as the data from the first experiments with the new double-stage system are reported here.

Keywords: multianvil high-pressure cell, DIA, ultrasonic interferometry, ultrasonic data transfer function, XRD, X-radiography, quartz to coesite transition, clinoenstatite.

[†]E-mails: hjmuel@gfz-potsdam.de; hans-joachim.mueller@gfz-potsdam.de.

Mueller, H.J., Schilling, F.R., and Lathe, C., 2007, Multianvil techniques in conjunction with synchrotron radiation at Deutsches ElektronenSYnchrotron (DESY) – HAmburger SYnchrotron LABor (HASYLAB), *in* Ohtani, E., ed., Advances in High-Pressure Mineralogy: Geological Society of America Special Paper 421, p. 207–226, doi: 10.1130/2007.2421(12). For permission to copy, contact editing@geosociety.org. ©2007 Geological Society of America. All rights reserved.

INTRODUCTION

In 1969, the first human put his foot on the surface of the moon. Now 35 yr later, mankind is going to fly to our outer neighbor planet—Mars. Geoscience and planetary science approach each other more and more. Satellite-based Earth observation has become indispensable for geoscience. Probably, in the not too distant future, we will record detailed seismic data from planets and planetary bodies. To interpret them and to compare them with existing terrestrial data will be a tremendous challenge and the biggest opportunity geoscience ever had at the same time. But we have to realize that our current understanding of the ongoing dynamic processes deep inside our home planet is still quite limited.

In situ studies of Earth's materials are the necessary complement to the progress of global seismology in general and of the tomographic method in particular. Understanding and modeling of mantle dynamics require more detailed insights into the structural and physical properties of materials relevant for great depths. Ultradeep subduction, penetrating into the lower mantle, and the interpretation of slab recycling in plumes require a fundamental understanding of the water and energy budget of our planet Earth (Bolfan-Casanova et al., 1997, 1998, 2002; Kohlstedt, 1994a, 1994b, 1996; Kohlstedt et al., 1994, 1996; Litasov et al., 2003, 2004; Litasov and Ohtani, 2003a, 2003b, 2005; Ohtani and Maeda, 2001; Ohtani et al., 2000, 2001a, 2001b, 2003, 2004; Rubie and Ross-II, 1994; Rubie et al., 1990; Suzuki et al., 2001). Following this line, the transition zone and the core-mantle boundary are regions of special interest. To face this scientific challenge, we have to improve and develop in situ high-pressure tools and techniques. By combining structural examinations and measurements of physical properties, e.g., elastic constants, large-volume presses can significantly enhance precision. To enhance the pressure range of multianvil apparatus, new materials are required, e.g., polycrystalline diamond with ceramic binder, big synthetic single-crystal diamonds grown from the gas phase, composite materials using carbon fiber. New techniques, e.g., high brilliance X-ray and high-intensity neutron sources, 3-D simulation software, and methods for transient measurements in mineral physics, significantly improve the potentials of this high-pressure technique, and they are available now or will be in the near future. High-pressure technology seems to be on the way to becoming a key technology for geosciences, material sciences, physics, and chemistry.

EXPERIMENTAL TECHNIQUES

Single-Stage Multianvil MAX80

Pressure Generation

MAX80 consists of a 2500 N hydraulic ram and two load frames that drive four reaction bolsters for the lateral anvils (Fig. 1). The lateral bolsters together with the top and bottom ones push six tungsten carbide (TC) anvils, which can compress a cubic sample volume of 512–216 mm^3 (Mueller et al., 2005b). The total weight is ~1.5 tons. The original MAX80 was a Japanese design (Shimomura et al., 1984, 1985; Liebermann et al., 1985; Yagi et al., 1987a, 1987b; Yagi, 1988). It started opera-

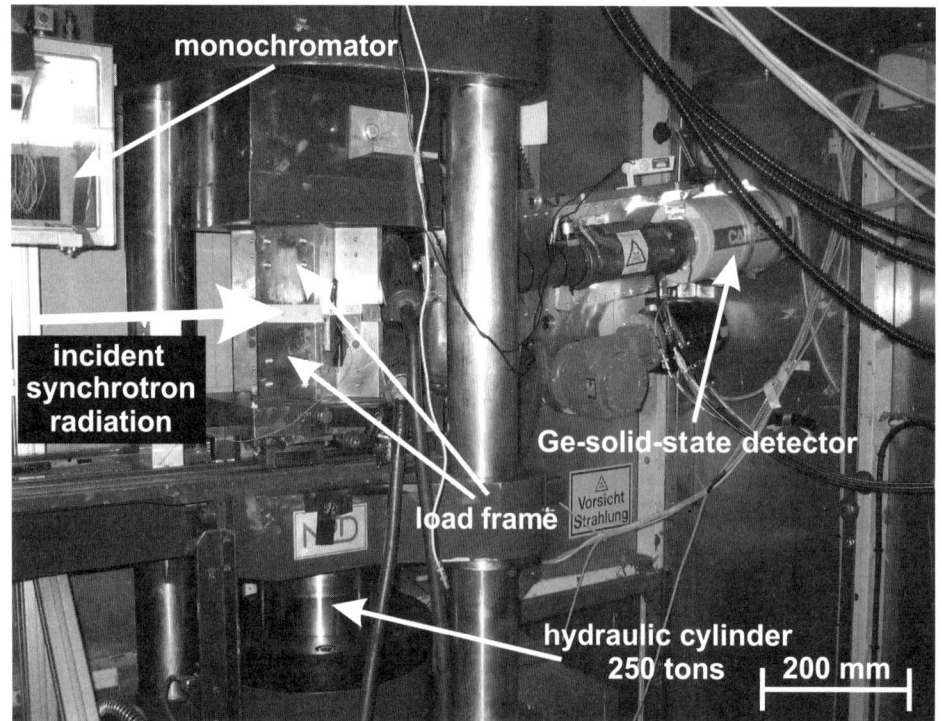

Figure 1. MAX80 showing load frame, Ge-solid-state detector, and monochromator (not in working position); white synchrotron radiation is used. The bottom hydraulic cylinder pushes the bottom anvil against the top anvil. At the same time, both load frames push the four lateral anvils toward the central boron epoxy cube. The incident synchrotron beam is guided through the gaps between the tungsten carbide anvils to the sample. The Ge-solid-state detector records the diffracted beam at a fixed angle. For angle-dispersive X-ray diffraction (XRD), a double-crystal, fixed-offset monochromator with silicon (311) single crystals can be moved to the white synchrotron beam.

tions at the Photon Factory, the synchrotron radiation laboratory at the National Laboratory for High-Energy Physics in Tsukuba, Japan. MAX80 stands for MultiAnvil-type high-pressure X-ray system. It was designed to obtain powder diffraction patterns using the energy-dispersive technique under high-pressure and high-temperature conditions. Shortly after the original press, a similar sister apparatus was installed in Europe in 1991 at HASYLAB (HAmburger SYnchrotron LABoratory) at DESY (Deutsches ElektronenSYnchrotron), Hamburg, Germany (Hinze et al., 1991, 1992). Today MAX80 at HASYLAB is operated by GeoForschungsZentrum Potsdam. Other similar apparatuses, with slight modifications, have come into use worldwide.

The press operated at beam-line F2.1 at HASYLAB, DESY, is equipped with three anvil sets with different truncations: 6 mm, 4 mm, and 3.5 mm. The maximum attainable pressures using the tungsten carbide anvils with the smallest front face reach 12 GPa at 2000 K produced by an internal graphite heater. The biggest truncation limits the maximum pressure to ~7 GPa. To enhance the lifetime of the anvils, we usually use cubic boron nitride anvils instead of the classical tungsten carbide anvils (Mueller et al., 2002). Use of sintered diamond anvils as second stage in similar-sized multianvil apparatus (MA8 type apparatus MAX80, MAX90) results in maximum pressures of more than 30 GPa (Utsumi et al., 1992; Shimomura et al., 1992; Kondo et al., 1993a, 1993b; Funamori et al., 1996). Recent studies using the stronger DIA-type press SPEED MkII (Katsura et al., 2004) at SPring-8, Japan, have reached 63 GPa with a load of ~10,000 kN (Ito et al., 2005).

Compression of the boron epoxy cubes results in a nearly hydrostatic (quasi-hydrostatic) pressure at the center. Depending on the given load, some part of the cubes floats out between the gaps of the anvils forming the gaskets (Fig. 2). This common technique is simple for the user, but it results in high lateral tension in the outer parts of the cube accompanied by strong plastic deformation, and it limits the "survival" of the anvils and the maximum pressure.

Prefabricated gasket insets are one way to increase the lateral anvil support at the expense of pressure efficiency, i.e., the relation between applied ram load and resulting pressure. We successfully used gasket strips made from Klinger SIL C-4400, an industrial sealing material made from NBR (Nitrile Butadiene Rubber) tied p-aramide fibers, for tungsten carbide and cubic boron nitride anvils. The experiments demonstrated that maximum pressures at least 25% higher compared to the standard MAX80 configuration are easily achievable while using prefabricated gaskets. Further benefits are a reduced number of blow-outs, i.e., decompression of the cell assembly by gasket failure, and higher X-ray intensity at peak pressures due to the larger gap width between the anvils (Mueller et al., 2005a).

X-Ray Diffraction (XRD)

Beam-line F2.1 is a bending magnet beam line at the storage ring DORIS III (DOuble RIng Storage) that has a circumference of 300 m, converted to a pure synchrotron in 1993. It works with a maximum energy of 4.5 GeV in five bunch mode for ~500 h/yr. DORIS III works in two bunch mode 20% of the time.

Most of the experiments are carried out using energy-dispersive X-ray diffraction. The synchrotron beam is guided between the tungsten carbide anvils through the sample. A Ge-solid-state detector (resolution: 135 eV for 6.3 keV and 450 eV for 122 keV) records the diffracted beam at a fixed angle. The maximum available 2θ range is 30°. MAX80 can be adjusted in three directions: (1) vertically to focus the X-ray beam on different parts of the setup, e.g., specimen and pressure standard

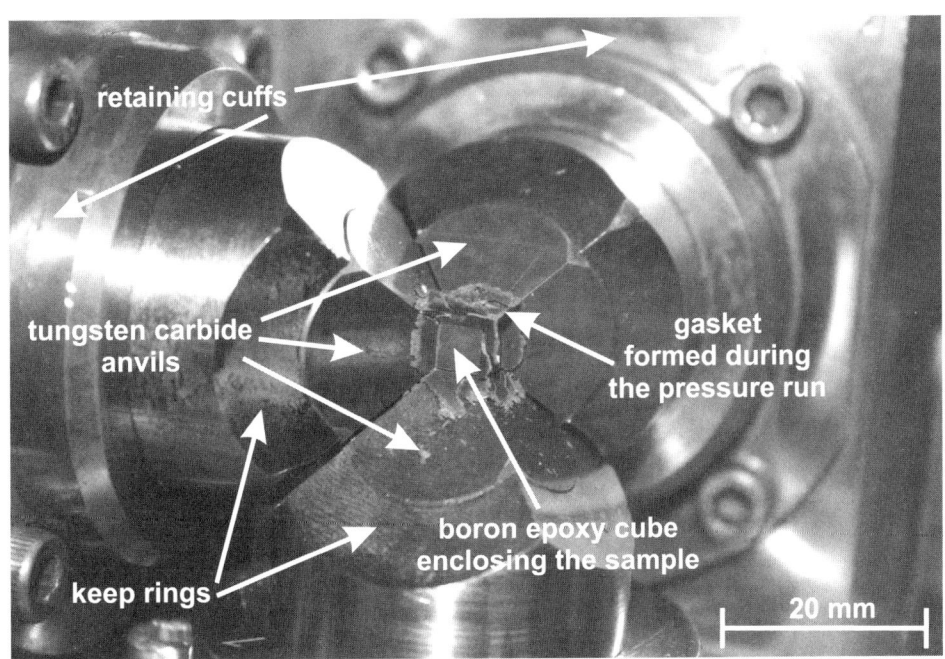

Figure 2. MAX80 showing boron epoxy cube with gaskets after a high-pressure run. Two lateral anvils and the top anvil have already been removed after the high-pressure run. Due to the load of high confining pressure, part of the boron epoxy cube flows into the gaps between the anvils forming gaskets. The gaskets "seal" the volume under pressure.

as well as to compensate for the pressure-induced deformation; (2) horizontally, perpendicular to the synchrotron beam, to adjust the synchrotron beam to the slots between the tungsten carbide anvils with an accuracy of 10 μm using stepper motors; and (3) the stepper motor–driven rotational table allows adjustment of the opposite-facing slots between the anvils parallel to the synchrotron beam with an accuracy of 0.01°. The diffractometer is linked with the high-pressure apparatus and has four degrees of freedom: (1) rotation around the vertical axis of the sample, (2) horizontal translation perpendicular to the beam, (3) vertical translation, and (4) 2θ rotation of the detector. The primary beam size and the diffracted beam size are limited by stepper motor–driven crossed-slit systems to 50×50 μm and 500×500 μm, respectively. Energy-dispersive X-ray diffraction may also be performed to measure the pressure inside the boron epoxy cube using the high-pressure equation of state for NaCl (Decker, 1971). All pressure measurements during the experiments described in this paper were performed this way.

For angle-dispersive X-ray diffraction, a double-crystal, fixed-offset monochromator with silicon (311) single crystals is installed additionally, in order to get higher resolution and to allow structure refinements under in situ conditions. The monochromator is calibrated in the wavelength range between 0.4 and 0.6 Å. For online data acquisition, a Stoe linear position sensitive detector (7° at 15 cm distance) is available. The detector is equipped with a conical slit system. Further progress concerning the resolution could be reached by using the imaging plate.

X-Radiography

X-radiography allows us to measure variations of sample length at high pressure-temperature (*P-T*) conditions (Li et al., 2001). It is a prerequisite for ultrasonic experiments to determine elastic properties. Viscosity measurement of melts by the falling sphere method is a further application for X-radiography under high-pressure conditions (Tinker et al., 2004).

To establish an X-radiography system at MAX80, the original fixed double-slit unit was exchanged with a four-blade high-precision slit system of Advanced Design Consulting USA, Inc. (ADC). The maximum slit opening is 1 inch. The motion reproducibility is 1 μm with a motion resolution of 0.4 μm remote controlled by a computer. Because the four blades can be moved independently from each other, the slits system is not only able to control the beam size, but it can also define the X-ray beam position. For X-radiography, the blades are opened far enough that the X-ray beam covers the whole sample length, including a part of the adjacent buffer and reflector rods. Using anvils absorbing the synchrotron radiation, e.g., made from tungsten carbide, the maximum horizontal opening of the beam is adapted to the maximum available gap between the lateral anvils, ~1.5 mm at normal pressure and <0.5 mm at maximum pressure conditions to limit the scattered radiation inside the hutch and to enhance the resolution of the diffraction patterns.

After penetrating the high-pressure cell assembly, the X-ray shadow graph is partially converted by fluorescence of a 0.1-mm-thick Ce: YAG (Cerium doped Yttrium Aluminum Garnet) crystal ([by courtesy of IKZ (Institut für Kristallzüchtung)]) to an optical image of ~540 nm wavelength (green). An aluminum-coated mirror decouples the optical from the non-converted X-ray image, which is absorbed in the beam-stop (Figs. 3 and 4). Otherwise the intense X-rays would destroy the CCD (Charge-Coupled Device) chip of the camera and would cause radiation damage in the optical glass of the lenses. The limit of the optical resolution is determined by the wavelength of green light of ~0.5 μm and the aperture of the objective of less than 0.5 to a total of ~1 μm (Mueller et al., 2005a). The minimum object distance is a strongly limiting factor for the

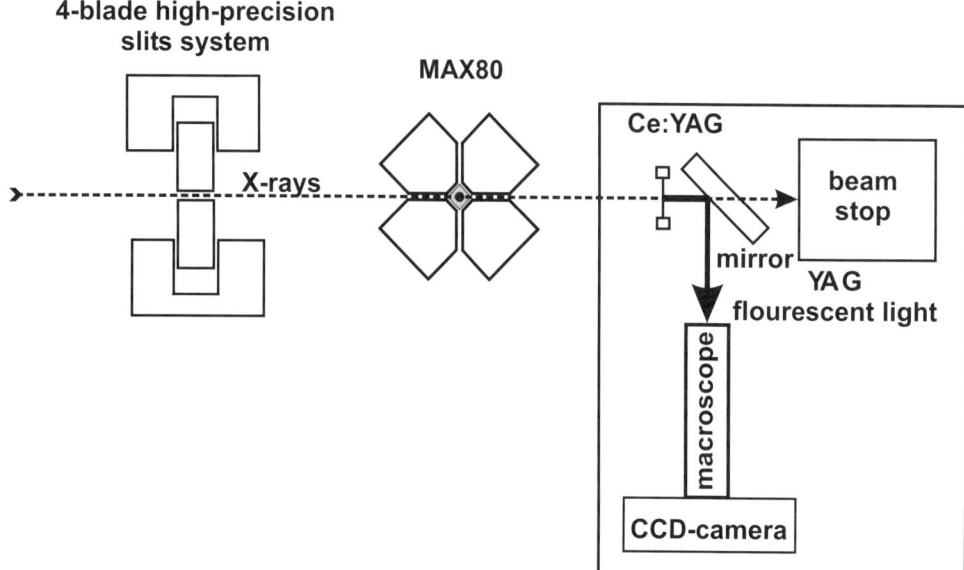

Figure 3. Scheme of X-radiography. The slits system shapes the incident synchrotron beam to a size slightly larger than the sample dimensions. After passing the sample inside MAX80, an X-ray shadow graph of the sample is projected on the Ce: YAG (Cerium doped Yttrium Aluminum Garnet) crystal. It converts the X-ray image to a visible one by fluorescence. The mirror redirects this optical image to a microscope with a CCD (Charge-Coupled Device) camera. The direct X-ray beam is absorbed by the beam stop.

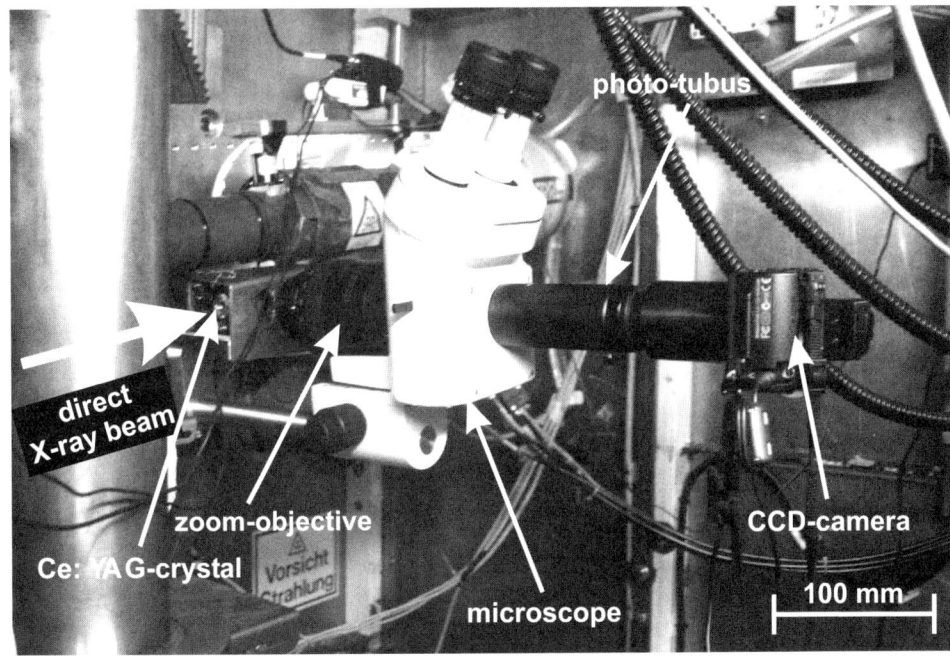

Figure 4. MAX80 showing X-radiography equipment assembled below the Ge-solid-state detector for energy-dispersive X-ray diffraction. The lead shielding limits the scattered radiation. The objective of the microscope is required to magnify the shadow graph of the sample to the chip size of the CCD (Charge-Coupled Device) camera to get an optimum resolution of the saved image.

selection of useful objectives. We assembled the objective of a microscope and a 5 megapixels color CCD camera able to take images and videos as well (Fig. 4).

The evaluation of the shadow graphs in terms of sample length measurement is performed by densitometry profiling, i.e., image processing software analyzes the brightness of the image along a predefined line. Therefore the image is transformed into a black and white gray-scale image. Because most of the minerals under investigation do not have a distinct difference in X-ray absorption with respect to the other parts of the setup, 5-μm-thick gold foils are used as markers. Figure 5 shows the shadow graph of a clinoenstatite specimen in between an Al_2O_3 part (top) and the NaCl pressure standard (bottom) and the corresponding image-processing evaluation. The dark shadow inside the NaCl standard close to the clinoenstatite interface corresponds to the thermocouple.

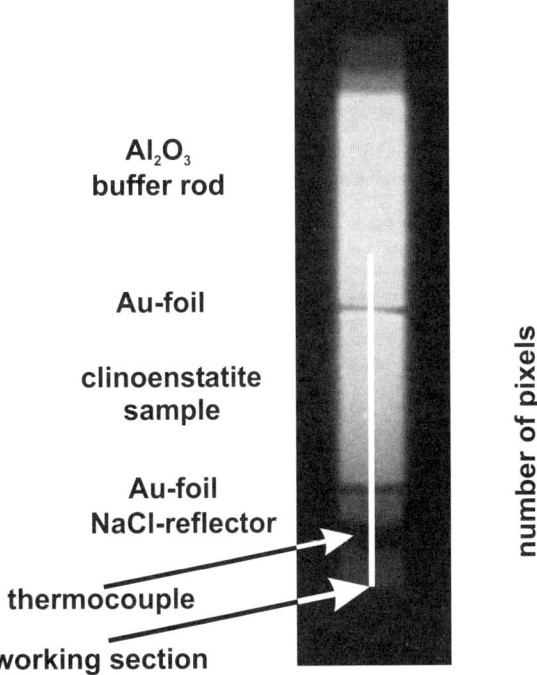
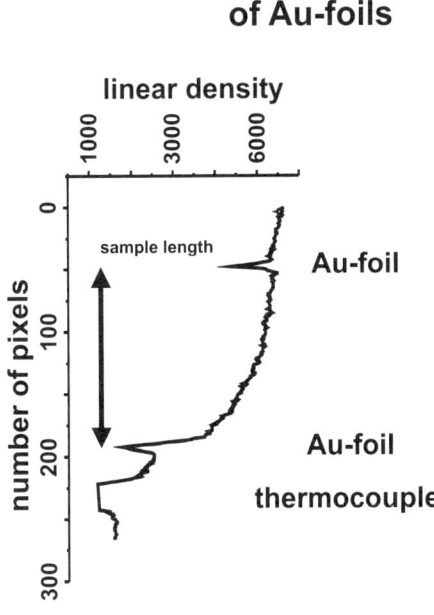

Figure 5. Sample length measurement under in situ conditions by image processing of X-ray shadow graphs. The X-ray shadow graph (left) shows a clinoenstatite sample between an Al_2O_3 buffer rod (top) and the NaCl reflector (bottom). The black part inside the reflector is the thermocouple. The density difference between the three parts is too small for a secure sample length determination. Gold foils serving as ultrasonic bonds at the same time are placed between the adjacent parts as markers. The white line inside the shadow graph is the gauge length evaluated by the image-processing system. The resulting sample length measured in pixels is transformed to millimeters by the initial calibration image of the sample with a given length. Because the pixel size corresponds to 0.0044 mm, the sample shown in the shadow graph has a length of 0.635 mm.

Ultrasonic Interferometry

Frequency sweep. Corresponding to the crucial importance seismology has for the exploration of Earth's deep interior, elastic-wave velocity measurements under simulated mantle conditions allow us to "better" interpret seismic observations. Because the sample size has to be larger than the ultrasonic wavelength, high-pressure measurements have to be performed in the ultrasonic frequency range—lower MHz range for multi-anvils, upper MHz to GHz range for diamond anvil cells (DAC). The classical traveltime method, common in rock physics, is not useful because the ultrasonic transducers are too brittle and too temperature sensitive to put them inside the high-pressure cell. Assembling them outside this cell, e.g., at the rear side of the anvils (see Fig. 6), results in a very long entire travel path in comparison to the sample length. For the configuration shown in Figure 6, the sample length is only 5% of the whole travel path. The resulting accuracy would be insufficient. Contrary to the direct traveltime measurement, ultrasonic interferometry uses the interference between the incident and reflected waves inside the sample. This way, the technique becomes relatively independent of the length of the travel path but requires a careful optimization of the reflection coefficients at all interfaces. Otherwise, the level of the ultrasonic energy reflected after passing the sample becomes too low for evaluation. To evaluate the reflected signals, the parts representing the reflection from the sample's front and rear face are selected and digitally interfered. The result is a periodically and equidistantly changing energy level (constructive and destructive interference) as a function of frequency. The periodicity of the interference pattern results in a much higher precision (0.2%–1.5%), ~1–3 times higher (Schreiber et al., 1973; Li et al., 1998) than classical traveltime methods (Birch, 1960, 1961).

The technique was first described by McSkimin (1950). Kinoshita et al. (1979) and Fujisava and Ito (1984, 1985) took the first steps to adapt the method to cubic anvil apparatus and uniaxial split-sphere apparatus. In the 1990s, Li et al. (1994, 1995a, 1995b, 1996a, 1996b, 1996c) optimized the method for multi-anvil devices and pushed pressures and temperatures to transition-zone conditions.

To adapt MAX80 to ultrasonic interferometry, the spacer located between upper and lower anvil and corresponding bolster had to be replaced by a redesigned part that had a cavity at one side to keep the ultrasonic transducers free of any stress (Fig. 6). In principle, two types of ultrasonic setups were used in MAX80—symmetrical and asymmetrical ones. Symmetrical setup means that the buffer and reflector (the parts above and below the sample, see Fig. 6) are made from the same material and have the same shape. Consequently, the measurements can be performed from both sides of the sample. Asymmetrical setups permit optimization of the reflection coefficients in the cell assembly to a great extent, but they require the transducers for generation and detection of elastic compressional (P) and shear (S) waves to be attached to the same anvil. Using single-mode transducers for generating pure P or S waves, respectively, four transducers in a relatively large cavity have to be cemented to the anvil slightly eccentric. For symmetrical setups, one single transducer (P or S) is cemented at the top and bottom anvil. Because the transducers act sequentially as generator and receiver of ultrasonic waves, a directional bridge is needed to prevent the strong excitation wavelet from hitting the sensitive input of the receiving amplifiers. The transducers are overtone-polished LiNbO disks with a natural frequency of 33.3 MHz. The usual sweep range is 5–65 MHz with a frequency step of 100 kHz, i.e., a sinusoidal wavelet of frequency x is transmitted to the transducer, the reflected signal is received, digitized, and saved in a computer, next the wavelet with the frequency $x + 100$ kHz is transmitted to the transducer, and so on (Mueller et al., 2003, 2005a).

Data transfer function technique. The major problem of all classical ultrasonic interferometry is that it is very time-consuming, e.g., a 60 MHz frequency sweep with 100 kHz steps lasts for more than 30 min. Acquisition of the raw data for both

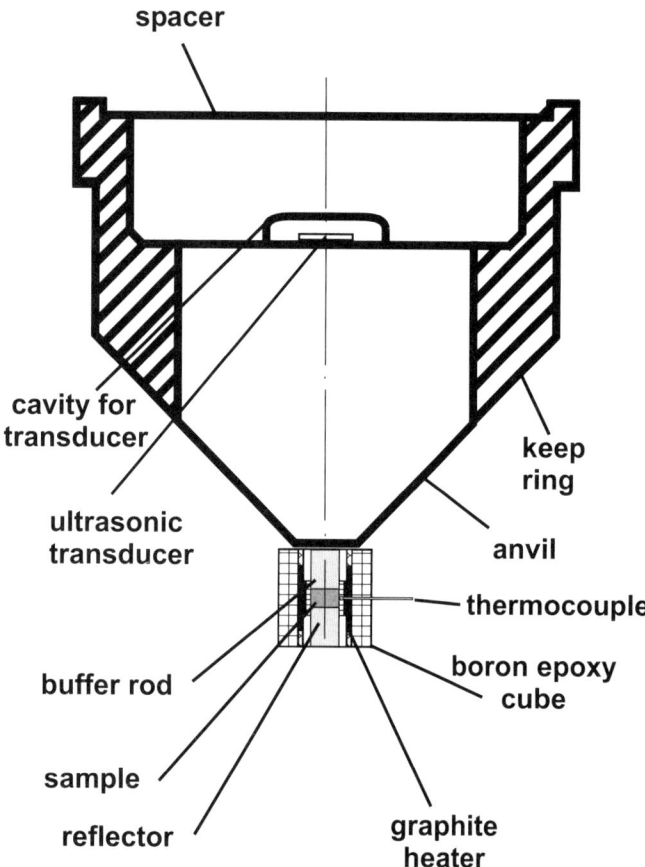

Figure 6. MAX80 sketch of ultrasonic top anvil and high-pressure cell assembly. For ultrasonic interferometry, the spacer needs a cavity to keep the brittle ultrasonic transducer free of any mechanical stress. The generated ultrasonic waves travel along the anvil axis, enter the buffer rod and the sample, and are finally reflected at the sample-reflector interface. Ultrasonic interferometry evaluates the superposition of the waves reflected from the buffer-sample and sample-reflector interface. The NaCl reflector is also used as pressure standard measured by X-ray diffraction.

velocities (V_p and V_s) requires ~1 h. This is much slower than the parallel-performed XRD, X-radiography, temperature measurement. Transient measurements are impossible. One solution for this time problem with the sweep technique is to perform all the single measurements of the sweep technique at the same time. This has become available only recently. It is the data transfer function (DTF) technique. The first mineral physics use of this technique was published by Li et al. (2002). Independent from that, a similar method was developed for MAX80 at HASYLAB (Mueller et al., 2005a). For that purpose, an excitation function is calculated stimulating all the frequencies inside the whole needed frequency range. This is the Fourier transform of the frequency range in time domain. To further improve the DTF technique, the transient response of the ultrasonic transducer, its resonance curve, and the parameters of the arbitrary waveform generator, which transforms the calculated function to an electrical signal, are taken into account. This is a modified "raised cosine function" produced by a digital filter. The response of the working section to this excitation is the data transfer function (DTF), which includes, in analogy to the excitation function, the responses of all monochromatic waves between the upper and lower cutoff frequencies. Saving the received DTF completes the measurement. This process lasts 2–30 s depending on resolution and equipment. Using triple-mode transducers (see a following section on ultrasonic interferometry), the saved DTF contains the same information content as the results of a 1-h-long measurement with the sweep technique. To reproduce the response for each single frequency, the data transfer function is convolved with the single frequencies. The convolution theorem states that the convolution of two functions corresponds to the multiplication of its Fourier transforms.

The expenditure of time is shifted from the experiment to the adjacent mathematical evaluation. The further evaluation corresponds to that of the sweep technique. The DTF technique requires an extremely high resolution of the saved DTF. This function (DTF) is the superposition of an infinite quantity of monochromatic responses between the upper and lower cutoff frequencies. Consequently, the resolution has to be high enough so that even the monochromatic frequency with the lowest amplitude can be resolved after the convolution of the DTF. Otherwise, the reproduced monochromatic response is distorted. Because the Fourier transform is performed numerically by fast discrete Fourier transform (FFT), the signal has to be saved with three times the duration of the evaluation range, at a minimum, to prevent its distortion by breakoff effects. For a bandwidth of 60 MHz and a travel time of ~15 μs, the DTF should be saved with a duration of 50 μs and a resolution of 200,000 data points at a minimum (Fig. 7). Only the DTF technique makes ultrasonic interferometry adequate for synchrotron facilities without restrictions and suitable for transient measurements (Li et al., 2002; Mueller et al., 2005a).

Double-Stage DIA-Type Multianvil MAX200x

Pressure Generation

MAX200x was planned as a supplement to and continuation of the development of MAX80 in order to be able to simulate Earth's transition-zone conditions with similar or even bigger-sized samples. All experiences gathered at visits to the multianvil apparatus at NSLS (National Synchrotron Light Source), APS (Advanced Photon Source) and SPring-8 in 2002 and 2003 were taken under consideration. In order to minimize the deviatoric stress inside the second-stage anvil cubes, MAX200x was also designed as a DIA-system. A maximum load of 1750 tons will guarantee pressures of 25 GPa in a large sample volume. All techniques described for MAX80-XRD, X-radiography, and ultrasonic interferometry are available with MAX200x. The hydraulic ram was designed and manufactured by Max Voggenreiter GmbH (mavo) as a classical four-column

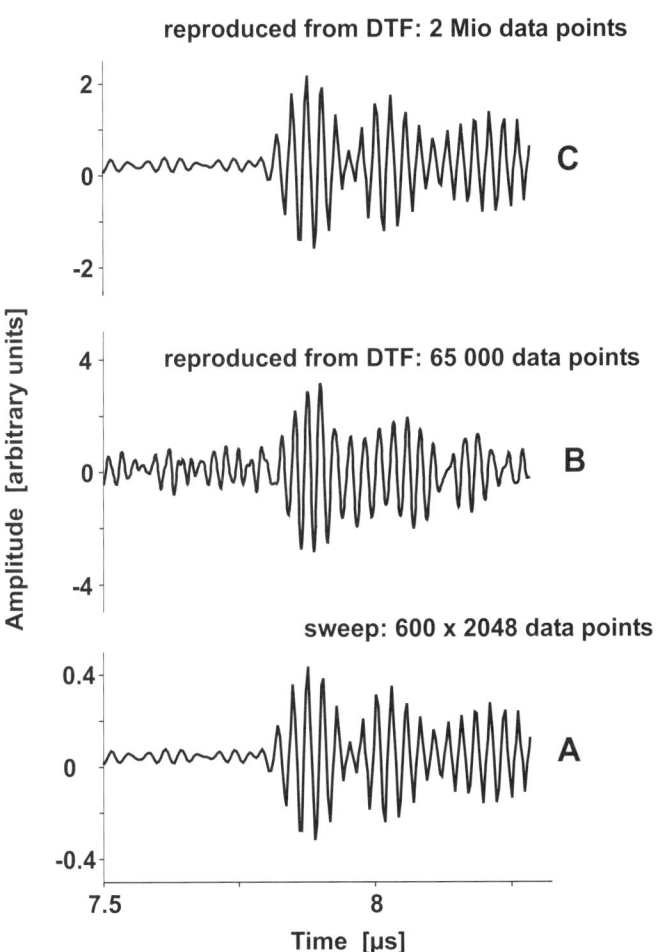

Figure 7. Reproduction of monochromatic ultrasonic signals from the data transfer function (DTF) by convolution in dependence on DTF resolution. The DTF technique requires a very high resolution of the saved DTF, because the DTF represents the superposition of an infinite quantity of monochromatic responses between the upper and lower cutoff frequencies. Otherwise, the reproduced single-frequency response does not correspond to the signal saved at monochromatic excitation.

Figure 8. MAX200x rotation table assembled in the pit at HARWI-II hall at Deutsches ElektronenSYnchrotron (DESY) – HAmburger SYnchrotron LABor (HASYLAB). Adjusting MAX200x to the synchrotron beam requires three different machine tables: rotation, lateral displacement, and lifting. Because the synchrotron beam has a position of only 1200 mm above the ground of the experiment hall, a pit of 3700 mm length, 3220 mm width, and 1400 mm depth had to be constructed.

press frame with oil-pressure storage device, i.e., the oil pressure can be held without continuous pumping. The maximum load is reached by applying an oil pressure of 700 bars in the working cylinder suspended from the trestle bridge. To be able to align the specimen to the fixed synchrotron beam, the whole press can be positioned by a combination of three machine tables for rotation, horizontal shift, and lifting with a precision of 0.01 mm and 0.01°, respectively. As the synchrotron beam has a position of only 1200 mm above the ground of the experiment hall, a pit of 1400 mm depth with a 500-mm-thick concrete foundation had to be constructed to support the total weight of ~30 tons. Figure 8 shows the assemblage of the bottom rotation table in the pit.

As described in the previous section on pressure generation, MAX200x also needs two load frames to drive the four lateral anvils. The design of the module, in principle an enlarged version of MAX80's parts, had to be adapted to bigger size and weight. The bottom load frame together with the lateral reaction bolsters can be mechanically pushed to the front of the press using a motor-driven thrust chain for the replacement of the high-pressure cell. Figure 9 shows MAX200x inside the hutch with the module in working position. The second stage consists of eight 32-mm-long tungsten carbide anvils, each with a truncation directed to the center of the arrangement that will compress an octahedron equipped with specimen, pressure standard, internal

Figure 9. MAX200x assembled at HARWI-II hall inside the hutch with module in working position. For high-pressure cell replacement, the whole module can be mechanically pushed to the assembly table at the front side of the press.

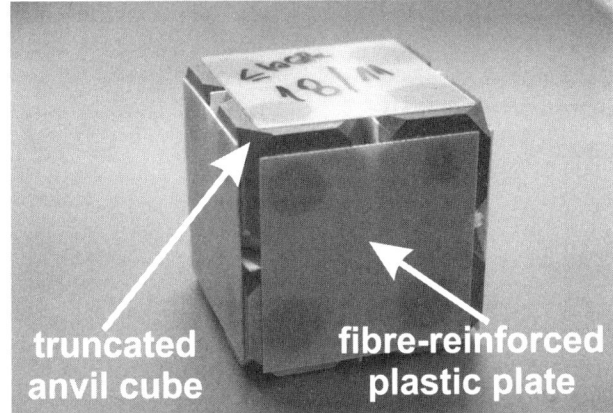

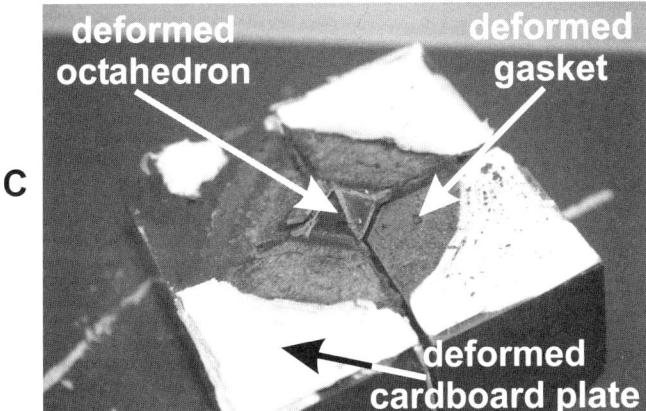

Figure 10. MAX200x second-stage setup: (A) ready for use, where six fiber-reinforced plastic plates keep the second-stage anvils together; (B) after the high-pressure experiment, where the plastic plates are broken because of the compression of the setup; and (C) high-pressure setup dismounted in part after the run, where the upper four anvils have been removed. The deformed octahedron with strongly deformed gaskets is visible. The white cardboard plates support the gaskets and limit their deformation.

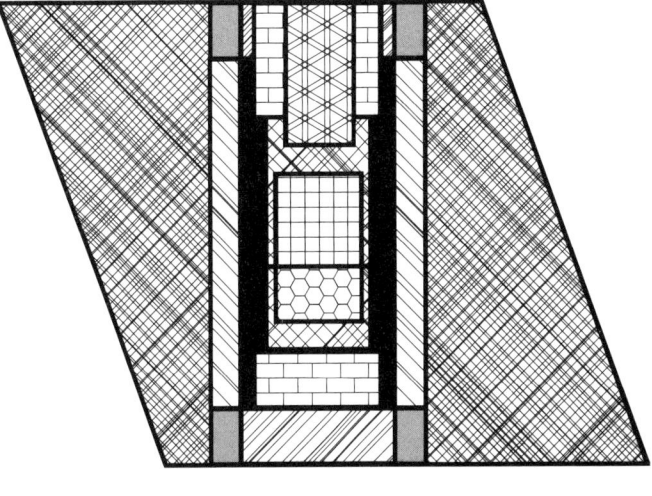

▨ MgO (chrome stabilized)

◩ Al$_2$O$_3$

▢ MgO

■ graphite / LaCrO$_3$

▧ molybdenum

▦ MgO

⊠ BN

⬡ Au / BN standard

▦ sample

▩ four-hole alumina sleeve

Figure 11. MAX200x showing internal (10/5) octahedron-cell assembly for X-ray diffraction (XRD). The Al$_2$O$_3$ sleeve is used for thermoinsulation. The molybdenum parts lead the heating current to the stepped graphite heater. Stepping the heater results in a decreased temperature gradient along the sample axis, because the stronger heat production at the ends compensates for the heat losses to the anvils in part. An alumina sleeve with four holes carries the thermocouple. The Au/BN standard is used as X-ray diffraction pressure marker.

electrical heater, and further parts if necessary. The octahedron is surrounded by prefabricated gaskets made from pyrophyllite. The setup is held together by glue on plates of fiber-reinforced plastic. Figures 10A–10C show a second-stage setup before and after a high-pressure experiment with the maximum load of 1750 tons, as well as the setup dismounted in part after the run. The setup inside the octahedron (Fig. 11) is derived and adapted from that of MAX80 (Huppertz, 2003; Leinenweber, 2005; Mueller et al., 2005a, 2005b).

The pressure calibration was performed for different high-pressure cell assemblies: 11/8, 10/5, and 7/3. The octahedra were made from sintered MgCr$_2$O$_4$ using 95% high-purity MgO. The pressure was measured by energy-dispersive X-ray diffraction using the high-pressure equation of state for MgO (Jamieson et al., 1982). Figure 12 shows the results for the first series of experiments.

Synchrotron Radiation Facilities and Measurements

The three centers of the Helmholtz-Gemeinschaft DESY, GKSS (Gesellschaft für Kernenergieverwertung in Schiffbau und Schiffahrt mbH), and GFZ (GeoForschungsZentrum Potsdam) jointly operate the new high-energy beam-line HARWI-II at the storage ring DORIS III at HASYLAB. To penetrate deeply into materials, the optics of the beam line were designed to provide a white beam of 0.5 mm diameter and a 50 mm × 10 mm monochromatic X-ray beam with an energy range of 20–250 keV.

The design goal for the new HARWI-wiggler was a maximized flux at 100 keV at an aperture of ~1 cm. The new device delivers a factor of 20 more energy than the old wiggler at the HARWI-beam line. Figure 13 shows the flux graphs calculated for 1 × 1 mm. In a wide range, the flux scales linearly with the area of aperture. The total power is 28 kW, and the on-axis power density reaches 41 kW/mrad. Right after the exit chamber of the wiggler-beam a horizontal slit is positioned that defines the height of the beam. A 1.5 mm C-filter is permanently installed as a high pass filter to limit the heat generation by X-ray absorption in the cell assemblies. An additional Cu-filter is available for experiments using high-energy radiation (above 60 keV). The Be-window at the front end has a distance of ~38 m from the center of the wiggler. The white beam for MAX200x has a diameter of 0.7 mm (Fig. 14). Two types of monochromators are operated in vacuum. A double Laue monochromator in horizontal geometry delivers beams of 10 × 10 mm in size with an energy range of 50–250 keV and an offset of 3 cm of the diffracted beam. The required monochromator tank is 3 m long, 2 m broad, and 1 m high. It is designed and operated by GKSS. The second monochromator produces a beam of 50 mm in width and 10 mm in height in a vertical diffraction geometry optimized for imaging techniques like tomography and Diffraction-Enhanced Imaging (DEI) performed by the partner institution GKSS at the joint beam line. The energy ranges from ~20 keV up to 200 keV. The beam offset is ~4 cm (Beckmann et al., 2003).

The primary slit is a SL-V-TU-200–10 from Advanced Design Consulting Inc. (ADC), a high-precision slits system with four tungsten blades driven by a stepper motor with a maximum aperture of 25.4 mm. Because the distance between the beam parts (see Fig. 14) exceeds the slits' aperture, the slits system is tied to the vacuum tube by metal bellows and can be shifted. The high-energy white beam is sent through a vacuum cube as long as possible to limit the energy loss of the incident beam by scattering to air molecules and to decrease the resulting scattered radiation inside the hutch at the same time.

The horizontal goniometer (mavo) is mounted on the optical Table 117-0000-4 (ADC) equipped with eleven stepper motors.

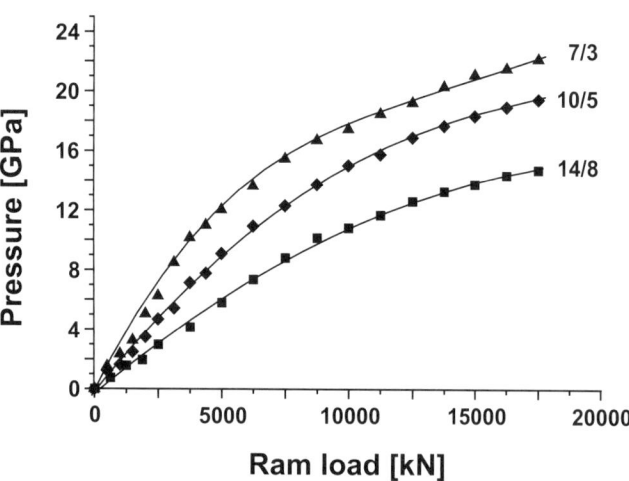

Figure 12. MAX200x pressure calibration for 14/8, 10/5, and 7/3 cell assemblies using X-ray diffraction (XRD) and the equation of state for MgO (Jamieson et al., 1982). The octahedra were made from sintered $MgCr_2O_4$ using 95% high-purity MgO. The first number of the cell specification marks the edge length of the octahedron; the second number (after the slash) marks the truncation edge length of the second-stage anvils.

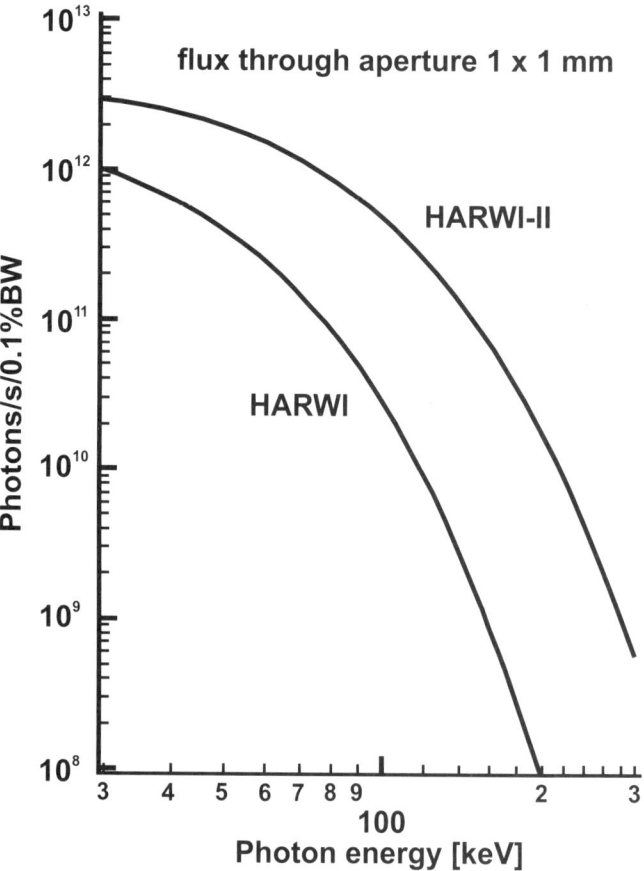

Figure 13. MAX200x flux comparison through a small aperture of 1 × 1 mm for the old (HARWI) and the new designed HARWI-II-wiggler. Over a broad photon energy range, the new wiggler produces a one order of magnitude higher flux (modified from Beckmann et al., 2003).

HARWI-II - beams

output of monochromator

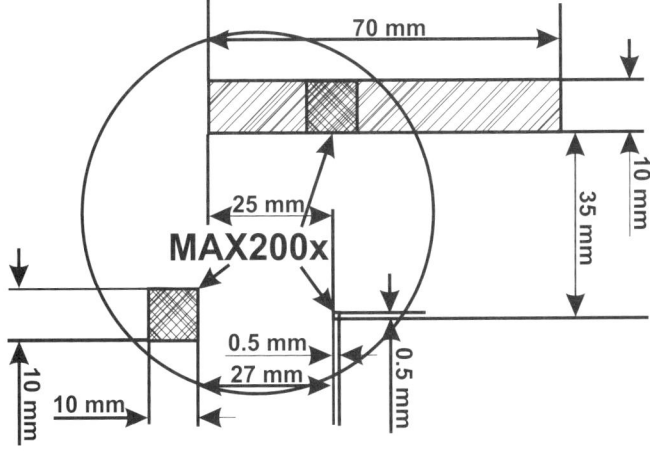

Figure 14. MAX200x X-ray beam size, shape, and location at the exit of the optics hutch of the HARWI-II beam line. The small white beam is used for energy-dispersive X-ray diffraction (XRD). Contrary to MAX80, X-radiography has to be performed using monochromatic beams because of the higher flux and higher energy range. In addition, monochromatic radiation results in higher resolution of the saved shadow graphs (modified from Beckmann et al., 2003).

The Ge-coaxial-detector GLP-06165/05P by ORTEK (resolution at 5.9 keV = 170 eV; resolution at 122 keV = 504 eV) is tipped with two high-precision slits SL-A-TU-100–10 (ADC) at a distance of ~1500 mm from each other to block scattered radiation and diffracted X-rays from nonsampled parts of the high-pressure cell.

For converting the X-ray shadow graphs and decoupling the optical images, the same design as that for MAX80 was adopted. The optical system consists of the macroscope Z16 APO A and the 1.4 megapixels black-and-white CCD camera DFC 350FX (Fig. 15). Transmitted using a Fire-Wire controller and four repeaters because of the long cables, the images are saved to a computer hard drive outside the hutch. The optimum macroscope-camera combination was selected by an experiment using a double-blind study from 13 options.

Two ionization chambers, IONIKA DN160 ISO-K, are available to measure X-ray absorption.

Ultrasonic Interferometry

Triple-mode transducers. The way to perform ultrasonic measurements with a double-stage multianvil apparatus such as MAX200x is to attach a transducer to the anvil cubes of the second stage. The only option to do this is the small triangular truncation at the outer side of the whole setup opposite to the same truncation that compresses the octahedron. With a 10/5 setup, the available space is a triangle with an edge length of 5 mm suitable for cementing a transducer disk of 3.8 mm in diameter.

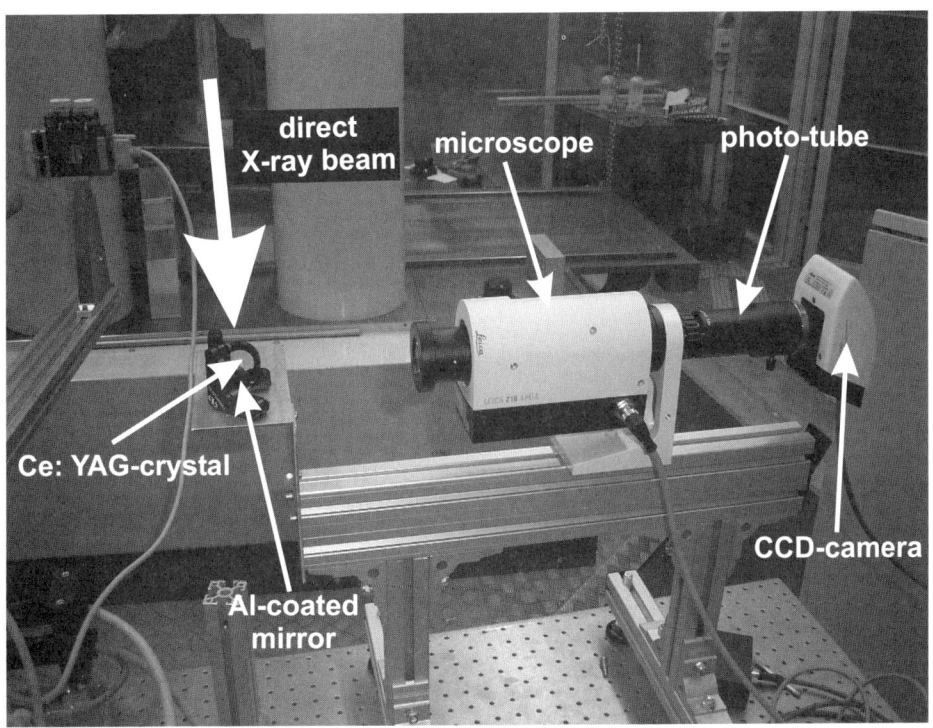

Figure 15. MAX200x X-radiography equipment. A Ce:YAG (Cerium doped Yttrium Aluminum Garnet) crystal ([by courtesy of IKZ (Institut für Kristallzüchtung)]) converts the X-ray shadow graph to an optical image. An Al-coated mirror decouples the optical from the nonconverted X-ray image. The optical system consists of the microscope Z16 APO A and the 1.4 megapixel black-and-white CCD (Charge-Coupled Device) camera DFC 350FX.

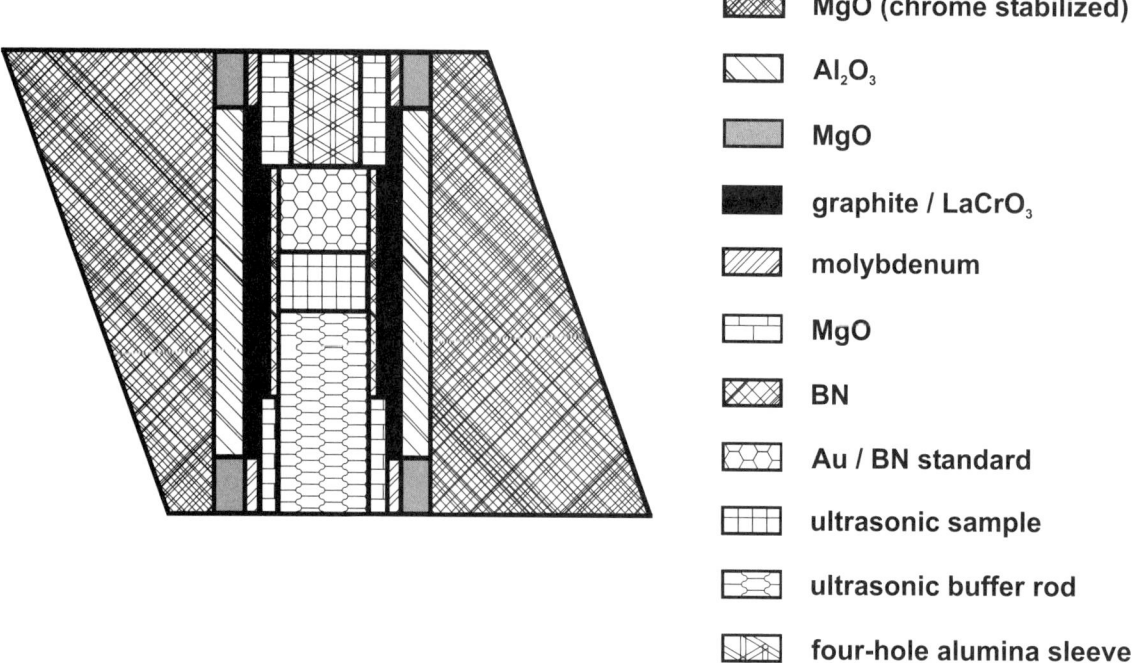

Figure 16. MAX200x internal (10/5) octahedron-cell (asymmetrical) for ultrasonic interferometry in conjunction with X-ray diffraction. Ultrasonic interferometry requires the exchange of some parts. The hot-isostatic pressed sample is located between the reflector used as pressure standard at the same time (top) and the buffer rod (bottom).

Another option is to use anvil cubes made with one additional greater truncation for ultrasonic experiments. If an asymmetrical ultrasonic configuration (Fig. 16) (see also the previous section on Ultrasonic Interferometry) is used, triple-mode transducers made from $LiNbO_3$ are needed to generate compressional, shear, and dual-mode waves at the same time (Fig. 17), because there is no further space available to attach more than one transducer (Li et al., 2004; Li et al., 2005). At the same time, this method also accelerates transient measurements because all the needed information to measure the velocities of P and S waves is saved within one single file. Figure 18 shows the electronic equipment for DTF ultrasonic interferometry for MAX200x.

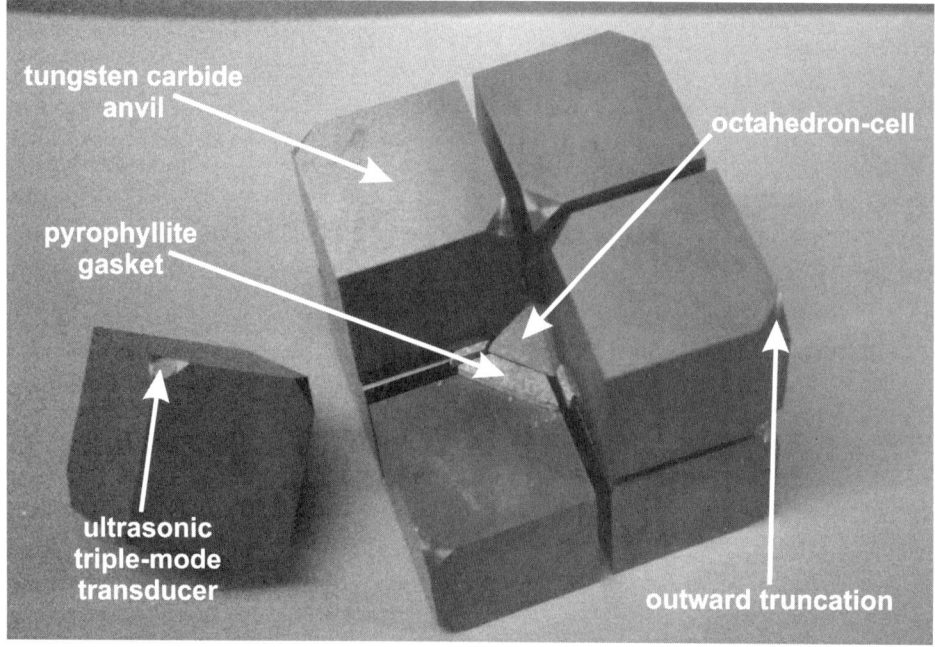

Figure 17. MAX200x second-stage cell with ultrasonic transducer cemented on one outward anvil truncation. For demonstration, the gaskets are not glued on the anvils. The cardboard plates, the self-adhesive Teflon tape, and the glass-fiber reinforced epoxy sheets are also removed. Use of a triple-mode transducer (10° rotated Y-cut $LiNbO_3$) allows the elastic-wave velocity measurement of compressional and shear waves to be performed at the same time.

RESULTS AND DISCUSSION

Multistaging Experiments: NaCl

A promising way to increase the maximum pressure of multianvil devices is "multistaging," i.e., implementing an additional set of "subanvils" between the anvils and sample, which results in a better distribution and limitation of the stress inside the anvils. Contrary to the common opinion of overshooting the maximum crushing strength, most of the anvils fail in high-pressure experiments because the maximum tensile stress is exceeded as a result of lateral deformation. Utsumi et al. (1986) published a technique to reach 60 GPa pressure by using sintered diamond anvils as second stage in a single-stage DIA-type multianvil apparatus. Wang and Utsumi (2005, personal commun.) reported similar experiments performed in a deformation-DIA (D-DIA). There was a strong indication that more than 70 GPa can be reached by additional uniaxial stress starting from 65 GPa under hydrostatic conditions.

As our first approach, we performed similar experiments using fine-grained tungsten carbide as anvil material inside an 8 mm boron epoxy cube. The anvils had a conical shape with a cylindrical shaft. The specimen, a 1:1 mixture by volume of NaCl and BN placed between both front faces, was enclosed

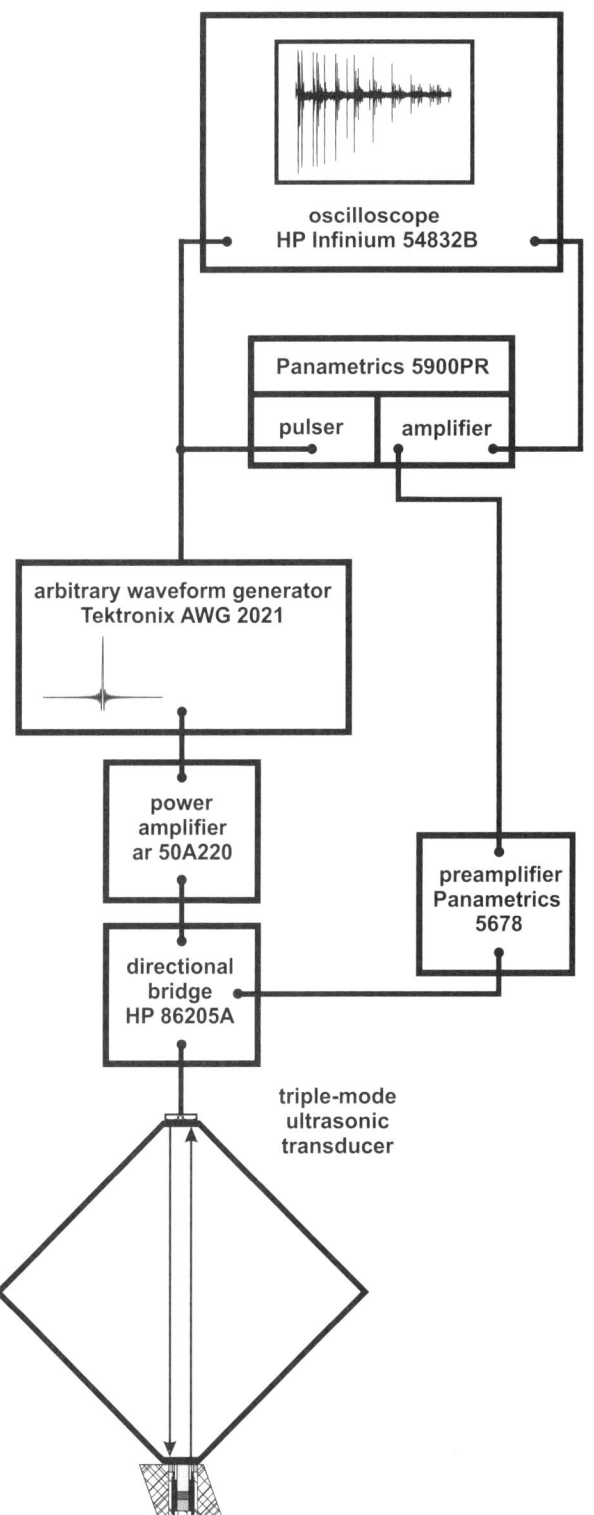

Figure 18. MAX200x electronic equipment for data transfer function (DTF) ultrasonic interferometry. The arbitrary waveform generator transforms the calculated excitation function into an electric signal, which is applied to the transducer via a power amplifier. The directional bridge prevents the sensitive pre-amplifier from being destroyed by these strong signals.

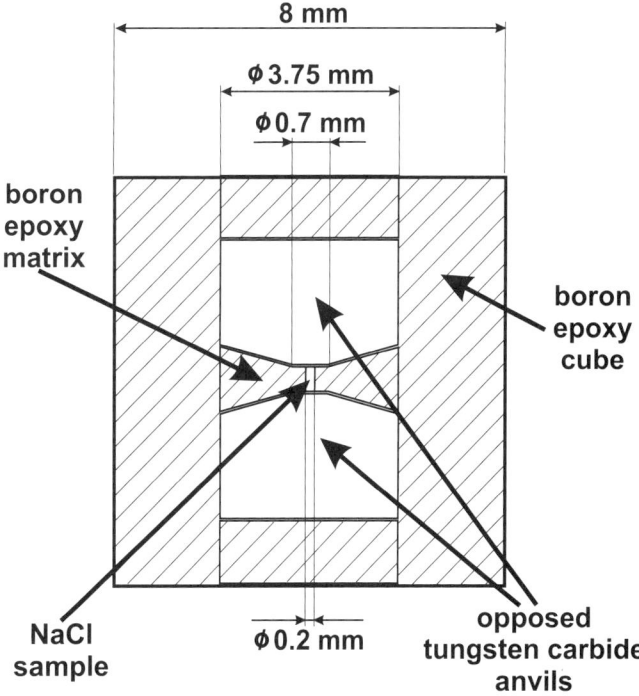

Figure 19. MAX80 opposed-anvil–type setup inside a single-stage multianvil 8 mm boron epoxy cube (based on Utsumi et al., 1986). The advance of MAX80's top and bottom anvils compresses the sample between the opposed anvils. The advance of the four lateral anvils prevents the boron epoxy matrix from being squeezed out.

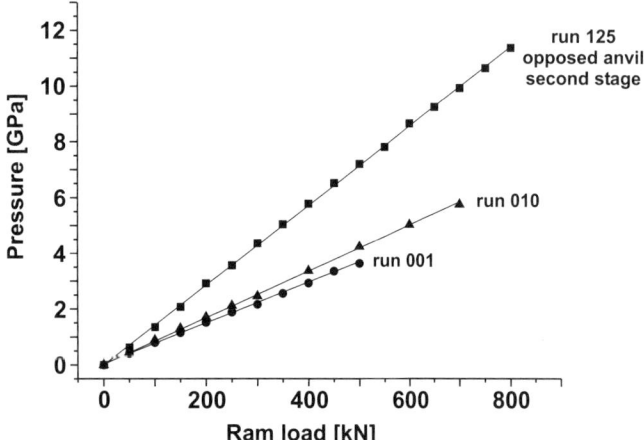

Figure 20. MAX80 pressure-doubling by additional opposed-anvil configuration (run 125) in comparison to two single-stage runs using the same anvil truncation of 6 mm. The pressure inside the NaCl sample was measured by X-ray diffraction (XRD) using the equation of state by Decker (1971).

in a boron epoxy matrix with an internal diameter of 0.2 mm (Fig. 19). The experimental results showed a pressure doubling in comparison to the original 8 mm cell assembly. Figure 20 compares the pressure data for the double-stage experiment measured by XRD using the Decker scale (Decker, 1971) with the results for an original 8 mm cell. The lower maximum pressure in comparison to Utsumi et al. (1986) is due to the porosity of the manually compressed specimen and the nonoptimum conical surface of the boron epoxy matrix used in this first experiment. Despite the pressure doubling, small gaskets were formed during the run. Normally, higher loads result in higher pressures inside the cube and more material flows out to form long thin gaskets. Consequently, it is surprising that higher loads and much higher pressures result in much smaller and shorter gaskets at first glance. The key to the observation is the change of the whole cube deformation because of the internal second-stage anvil configuration.

We also performed an ultrasonic interferometry measurement using "multistaging." To increase the reflection area, the diameter of the sample space in the matrix and the diameter of the subanvil's front face was increased to 1.7 mm. The top subanvil was acoustically coupled to MAX80's upper anvil by a platinum disk of the same diameter as the cylindrical shaft of the subanvil. From an acoustical point of view, the travel path has three buffer rods—top MAX80 anvil, platinum disk, upper internal opposed anvil—and the bottom internal opposed anvil as reflector (compare Fig. 19). As a result of preparatory experiments, we had found that the upper cutoff frequency for the existing transducer–tungsten carbide anvil system was ~300 MHz. An excitation function representing the bandwidth of 100–300 MHz was calculated and used in the experiment (Fig. 21). Under high-pressure conditions, we were able to detect ultrasonic energy reflected at the interfaces of the NaCl sample

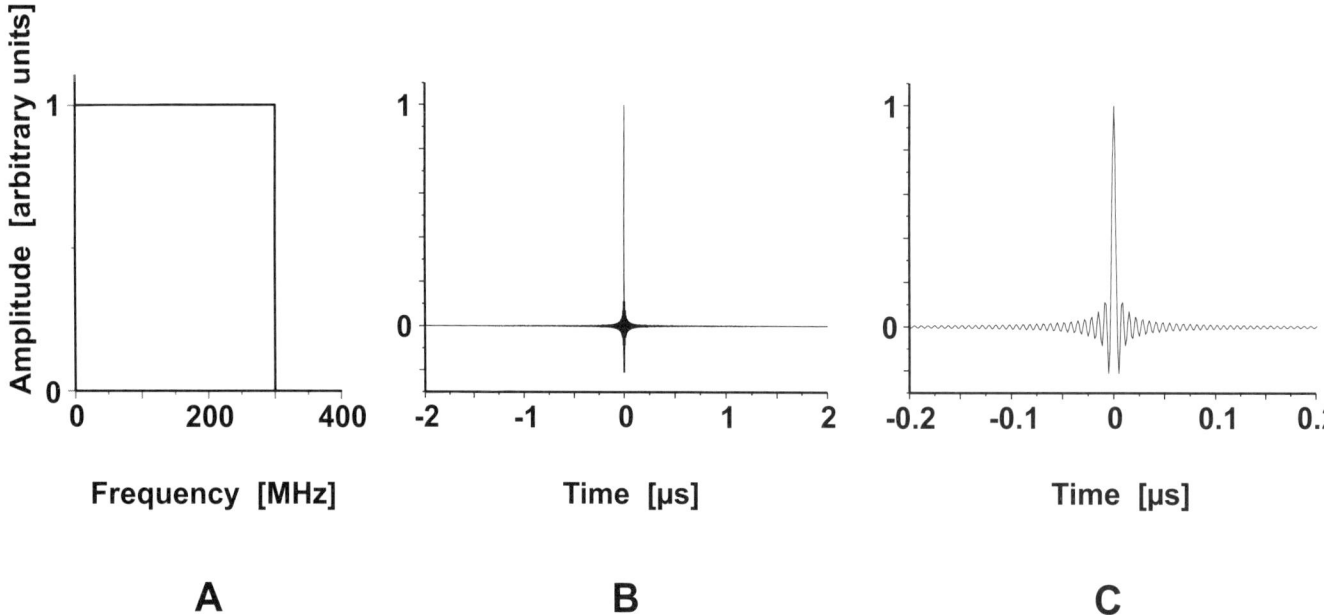

Figure 21. Excitation function representing the frequency range 100–300 MHz for "multistaging" ultrasonic interferometry in time and spatial domain. The smaller sample requires higher frequencies. (A) Frequency band pass of the calculated excitation function in frequency domain. (B) Calculated excitation function of 4 μs duration in time domain. (C) Central part of the excitation function shown in B displayed with 10× higher time resolution (0.4 μs) in time domain.

as the fundamental condition for interferometric data evaluation, which is in accordance with previously published data for single-stage experiments (Mueller et al., 2005a, 2005b).

Transient Measurements: Quartz to Coesite Transition

The high-pressure SiO_2 polymorph coesite is an important mineral in the subduction process, including in crustal material as has been observed in rocks of the Dora Maira Massif (western Alps) by Chopin (1984), Gillet et al. (1984), and also in other subducted continental rocks (Schreyer, 1995). The quartz to coesite transition is thus of fundamental importance in the effort to understand the processes within subducting lithosphere. Furthermore, the nature of the quartz to coesite transition is controversial, because high-pressure XRD studies suggest an intermediate phase during the transformation process (Bourret et al., 1986; Heaney and Veblen, 1991; Zinn et al., 1997a, 1997b; Prakapenka et al., 2004).

We used natural fine-grained quartzite with a SiO_2 content of more than 99 wt% (Schilling, 1999) for the experiments. The samples were shaped to cylinders of 2.4 mm diameter and 1.6 mm length and 2.0 mm diameter and 1.2 mm length, respectively, using a high-precision cylindrical grinding machine. Both end faces were polished for optimum ultrasonic coupling. Figure 22 shows the results for the elastic-wave velocities using 8 mm as well as 6 mm boron epoxy cubes. A minimum pressure of 4.5 GPa was applied to observe the phase transition from quartz to coesite. At 800 °C, it took place in less than 2 min, much too fast for the classical sweep technique. If the frequency span is reduced and the number of steps is minimized, the variation in elastic properties during the phase transition could be observed at temperatures between 720 °C and 750 °C. At lower temperatures, the transition did not take place at all or stopped at ~50% transformation. The temperature-induced weakening of the boron epoxy cube resulted in a stress release in the whole high-pressure cell. Together with the phase transition–induced volume reduction of the sample, a dramatic pressure decrease occurred that could not be compensated by increasing the load of the press. The pressure measurement was performed using the X-ray diffraction from the NaCl pressure marker also used as ultrasonic reflector.

To perform experiments at higher pressures, we used a setup for 6 mm boron epoxy cubes and 4 mm anvil truncation. Furthermore, we used gasket insets (Mueller et al., 2003, 2005a) to support the cubes. In Figure 22, the variation of compressional and shear-wave velocities, V_p and V_s, during the quartz to coesite phase transformation is shown.

Figure 23 compares the V_p data and the results of the X-ray diffraction measurements performed simultaneously. It demonstrates the first transient ultrasonic interferometry technique. The measurement was accelerated by limiting the frequency span to a little more than 2 MHz and using a frequency step of 300 kHz. The observed peak shift corresponds to an increase of V_p from ~6.5 km/s to ~7.5 km/s with some indication of an intermediate state. Corresponding DTF measurements are in progress.

Multi-Cycle Experiments: Clinoenstatite

One of the most challenging recent tasks for mineral physics has been the investigation of unquenchable phase transitions. As an interesting example, we measured the elastic-wave velocities at the high-pressure clinoenstatite ($MgSiO_3$, HCEn) to low-pressure clinoenstatite transition (LCEn) under high-pressure and high-temperature conditions in conjunction with in situ XRD. For ultrasonic interferometry experiments, low-pressure clinoenstatite powder synthesized at ambient pressure was hot-isostatic-pressed (HIP) (Liebermann et al., 1975) at 0.4 GPa and 1400 °C for 2 h in MAX80 to obtain minimum-porosity samples. The elastic-wave velocities, V_p and V_s, of the clinoenstatite sample were measured in situ using the sweep as well as the DTF technique (Mueller et al., 2005b).

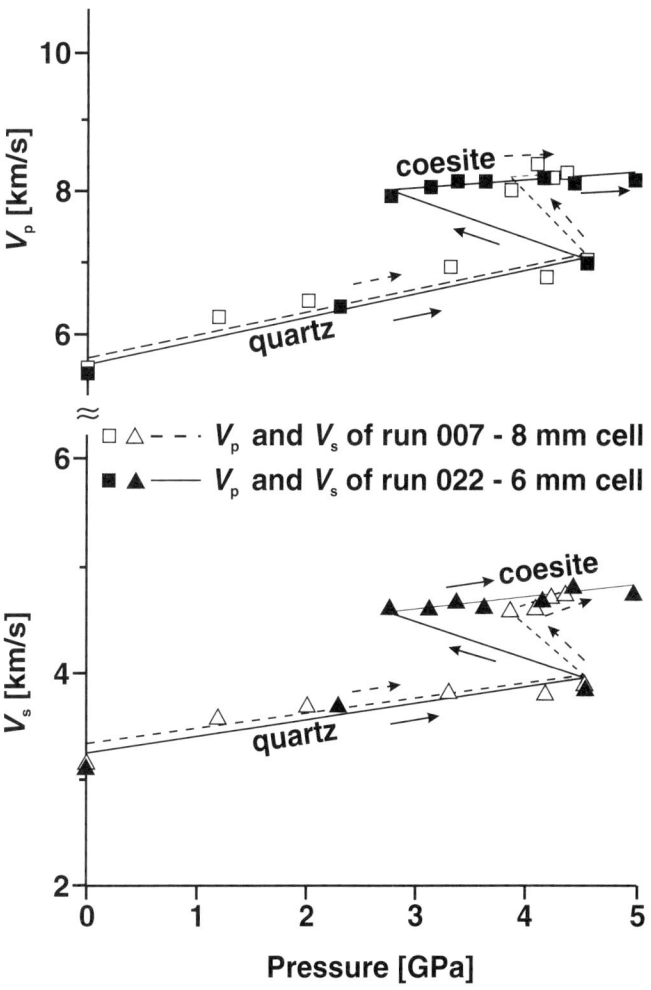

Figure 22. MAX80 V_p and V_s for quartz and coesite using 8 mm and 6 mm cubes. The pressure drop during the phase transformation to coesite is the result of stress release in the cell assembly by increasing temperature and also of the volume reduction of the sample. The uncertainty of the interferometric V_p and V_s measurement is <1%, i.e., smaller than the symbol size.

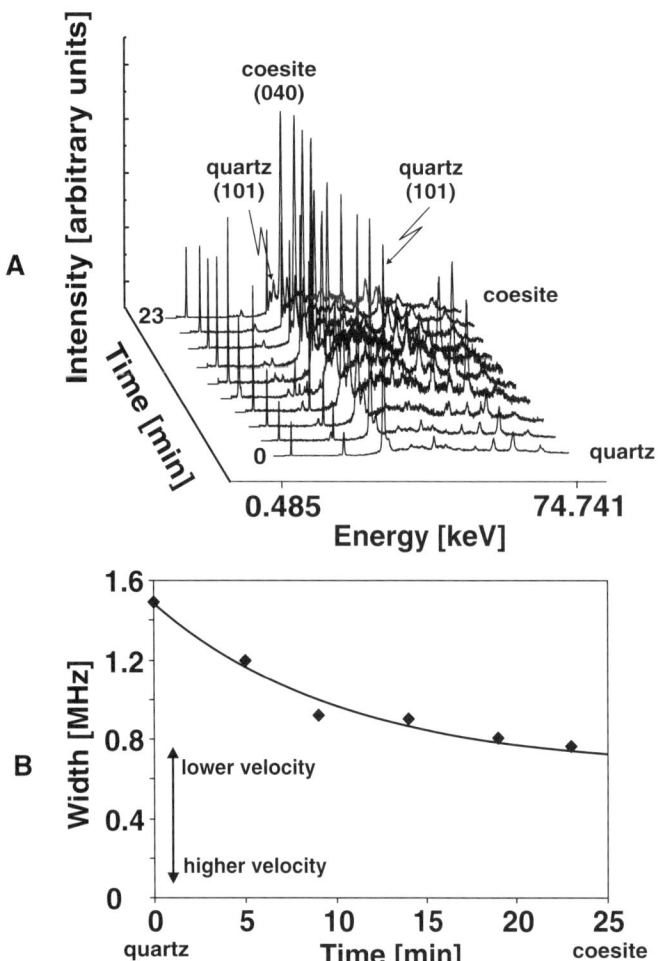

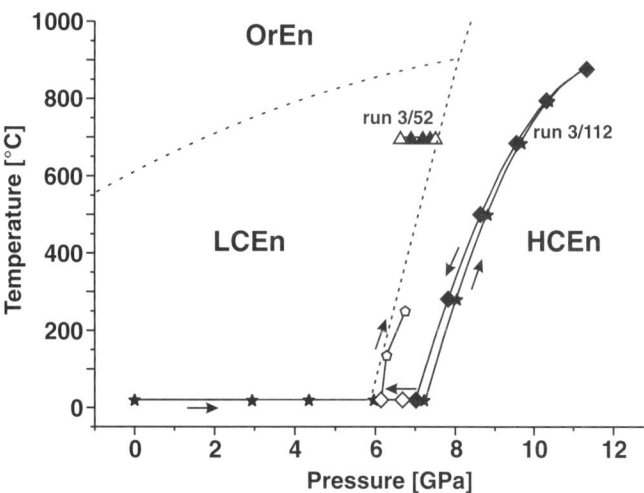

Figure 24. MAX80 pressure-temperature (P-T) paths for ultrasonic measurements with clinoenstatite (MgSiO$_3$, high-pressure clinoenstatite [HCEn] and low-pressure clinoenstatite [LCEn]). First hot-isostatic-pressed low-pressure clinoenstatite is transformed to high-pressure clinoenstatite (solid stars) by increasing the pressure. The temperature is increased by 1 K/min to release the stress. The thermal expansion further increases the pressure (solid stars). Next, the temperature is decreased by 1 K/min (solid diamonds). The next phase is the pressure decrease down to ~6 GPa at normal temperature (open diamonds). Now, the next temperature cycle is started (open pentagons). Run 3/52 aims at the velocity dependence of low-pressure clinoenstatite on pressure at constant temperature of 700 °C performed with sweep (solid triangles) and data transfer function (DTF) (open triangles) techniques. OrEn—orthoenstatite.

Figure 23. MAX80 transient measurements at the quartz to coesite transition. (A) Transient X-ray diffraction (XRD) patterns. (B) Transient measurement of compressional wave traveltime. The width is the distance between adjacent maxima and minima of the interference pattern of the echoes from both sample front faces. These maxima and minima represent the frequencies of constructive and destructive interference. The width is a measure of the traveltime inside the sample.

First we carried out ultrasonic experiments in a prograde manner, i.e., pressure and temperature were increased during the measurements. Kung et al. (2004) described and explained the opposite strategy, because otherwise the measured phase boundaries were apparently shifted due to the unreleased stress inside minimum-porosity samples as used in ultrasonic experiments. Therefore, we also performed an experiment in retrograde manner. Figure 24 illustrates the P-T paths of our experiments in the phase diagram of MgSiO$_3$ (Angel and Hugh-Jones, 1994; Mueller et al., 2005b).

Figure 25 summarizes the measured ultrasonic wave velocities of two experiments. After prograde crossing the phase boundary at 6.4 GPa, the temperature was further increased up to 700 °C to rule out crossing the phase boundary to high-pressure clinoenstatite during the following pressure increase up to 7.5 GPa. Run 3/52 aimed at the velocity dependence on pressure at constant temperature of 700 °C. For V_p and V_s velocity determinations, the pressure derivatives at 700 °C of 0.8 km/(s GPa) and 0.7 km/(s GPa) were determined for low-pressure clinoenstatite, respectively. The measurements were performed using the sweep technique at both pressure levels and the DTF technique also at these points and in between. The results demonstrate a good correspondence of both techniques (Mueller et al., 2005b).

Run 3/112 was a multicycle retrograde experiment in the high-pressure clinoenstatite stability field. After increasing the pressure up to 7.2 GPa, the temperature was increased by 1 K per minute up to 875 °C. As a result of thermal expansion, the pressure increased up to 11.3 GPa ("thermal pressure"). After this, the temperature was reduced with the same low negative temperature increment. The ultrasonic measurements were performed during this second segment of the experiment. The results show the combined P-T dependence along the path shown in Figure 24. After reaching room temperature, the pressure was released down to 6.1 GPa. During this segment V_p and V_s dependence on pressure at normal temperature was measured for a stress-released sample. The pressure derivatives for high-pressure clinoenstatite at normal temperature were determined to be 0.089 km/(s GPa) for V_p and 0.02 km/(s GPa) for V_s, respectively. The results correspond well with the data of Kung et al. (2004) for high-pressure clinoenstatite. After this segment, the next temperature cycle with

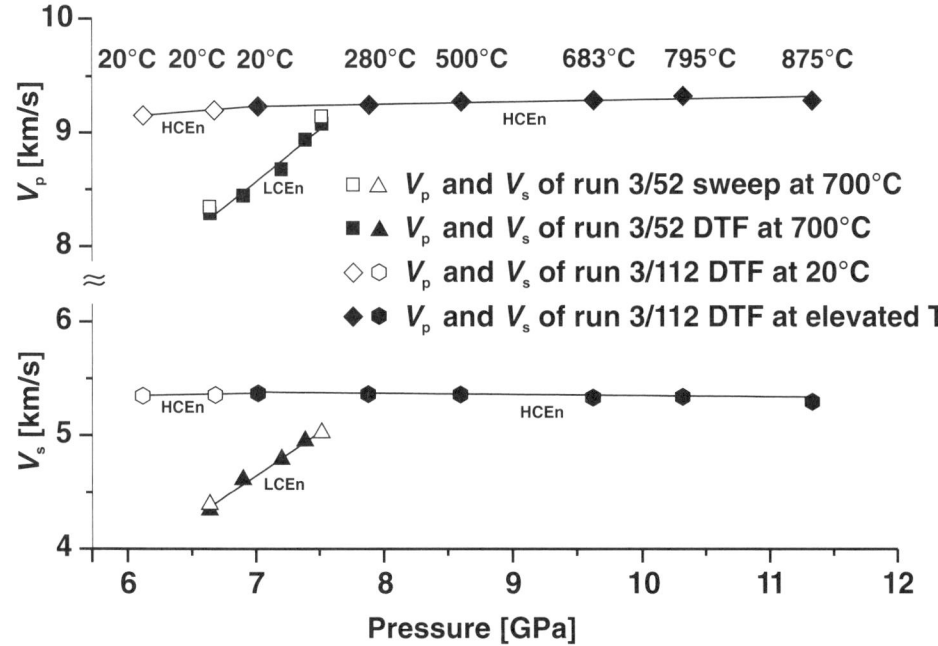

Figure 25. MAX80 compressional and shear-wave velocities, V_p and V_s, in clinoenstatite at various pressure and temperature conditions (hot isostatic pressed [HIP] sample). Because of the pressure-temperature (P-T) path shown in Figure 24, the results for high-pressure clinoenstatite represent the combined P-T effect on velocities, contrary to the pure P influence in the low-pressure clinoenstatite data. The uncertainty of the interferometric V_p and V_s measurement is <1%, i.e., smaller than the symbol size. DTF—data transfer function; HCEn—high-pressure clinoenstatite; LCEn—low-pressure clinoenstatite.

2 K per minute was launched. The slightly higher temperature increment was acceptable for all further cycles because the sample was almost completely stress-released after the first cycle. The slow pressure release did not indicate remarkable new stress inside the sample. Figure 26 shows the pressure-induced length decrease of a clinoenstatite sample monitored by X-radiography.

The low-pressure to high-pressure clinoenstatite phase transition might be an important reaction in deeper parts of cold, fast subducting slabs. Because of high subduction velocity, the temperature increase is retarded, i.e., high-pressure clinoenstatite keeps stable at much deeper levels than under temperature-equilibrium conditions. If the phase transition takes place at greater depths yet, the higher V_p and V_s pressure derivatives of low-pressure clinoenstatite might result in a greater velocity drop than the published 0.5% (Kung et al., 2004; Mueller et al., 2005a). Another aspect is the influence on the rheology of downgoing slabs by transformation plasticity (e.g., Poirier, 1982; Schmidt et al., 2003) of clinoenstatite-bearing rocks (Mueller et al., 2005a).

SUMMARY

The data demonstrate the potentials and technical details of two multianvil apparatus for the experimental simulation of Earth's mantle conditions installed at HASYLAB–DESY,

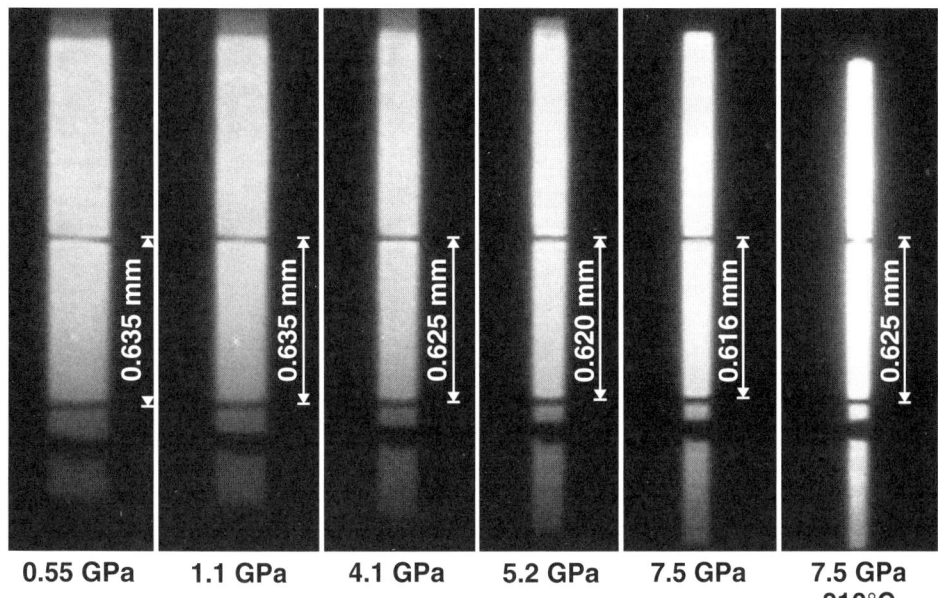

Figure 26. Pressure-induced length decrease of a clinoenstatite sample monitored by X-radiography under in situ conditions. The sample length is measured by image processing of the shadow graphs, i.e., the number of pixels between the two dark markers (Au foil) is measured. The pixel size is determined from a calibration with known sample length at ambient conditions. For further details, compare with Figure 5. The decreasing width of the shadow graphs is the result of the decreasing gap between the anvils at increasing pressure.

Hamburg—MAX80 and MAX200x. Besides the synchrotron radiation techniques, XRD and X-radiography, different ultrasonic interferometry methods are described. With respect to future scientific challenges and present-day technological feasibility, we described approaches to new techniques and presented the most recent results.

ACKNOWLEDGMENTS

We would like to express our special thanks to the editor E. Ohtani for his initiative, guidance, and patience. Our special thanks are due to Y. Wang and B. Li, without their helpful and fruitful discussions, the demonstrated development would have been much less successful. The constructive reviews of two unknown referees, which substantially improved the paper, are gratefully acknowledged. We also acknowledge the kind guidance and the support of B. Liebermann, D. Weidner, and M. Vaughan. We are especially grateful to S. Ganschow for the Ce:YAG crystal, manufactured and put at our disposal by IKZ, and J. Kulesza (ADC) for his untiring support. The authors are particularly indebted to B. Wunder for sample courtesy, S. Speziale for looking through the manuscript, M. Kreplin for the special preparation, N. Heidbrook, H. Zink, J. Novak, and O. Beimgraben (DESY) for their kind assistance, as well as all colleagues of the high-pressure mechanical workshop for their dedicated support.

REFERENCES CITED

Angel, R.J., and Hugh-Jones, D.A., 1994, Equations of state and thermodynamic properties of enstatite pyroxenes: Journal of Geophysical Research, v. 99, p. 19,777–19,783, doi: 10.1029/94JB01750.

Beckmann, F., Lippmann, T., Metge, J., Martins, R.V., Dose, T., and Schreyer, A., 2003, HARWI-II Material Science Beamline: HASYLAB Annual Report 2003, p. 139–142.

Birch, F., 1960, The velocity of compressional waves in rocks to 10 kilobars. Part 1: Journal of Geophysical Research, v. 65, p. 1083–1102.

Birch, F., 1961, The velocity of compressional waves in rocks to 10 kilobars. Part 2: Journal of Geophysical Research, v. 66, p. 2199–2224.

Bolfan-Casanova, N., Keppler, H., and Rubie, D.C., 1997, Distribution of water between lower mantle phases; $MgSiO_3$-ilmenite, $MgSiO_3$-perovskite and stishovite: Eos (Transactions, American Geophysical Union), v. 78, p. 736.

Bolfan-Casanova, N., Keppler, H., and Rubie, D.C., 1998, Partitioning of water between high pressure phases in the system $MgSiO_3$ + 0.5 wt.% H_2O: Min. Magazine, v. 62A, Part 1, p. 185–186.

Bolfan-Casanova, N., Mackwell, S., Keppler, H., McCammon, C., and Rubie, D.C., 2002, Pressure dependence of H solubility in magnesiowuestite up to 25 GPa; implications for the storage of water in the Earth's lower mantle: Geophysical Research Letters, v. 29, 1449, doi: 10.1029/2001GL014457.

Bourret, A., Hinze, E., and Hochheimer, H.D., 1986, Twin structure in coesite studied by high resolution electron microscopy: Physics and Chemistry of Minerals, v. 13, p. 206–212, doi: 10.1007/BF00308163.

Chopin, C., 1984, Coesite and pure pyrope in high-grade blueschists of the western Alps; a first record and some consequences: Contributions to Mineralogy and Petrology, v. 86, p. 107–118, doi: 10.1007/BF00381838.

Decker, D.L., 1971, High-pressure equation of state for NaCl, KCl, and CsCl: Journal of Applied Physics, v. 42, p. 3239–3244, doi: 10.1063/1.1660714.

Fujisava, H., and Ito, E., 1984, Measurements of ultrasonic wave velocities in solids under high pressure, in Proceedings of the 4th Symposium on Ultrasonic Electronics: Tokyo, Japan (1983): Journal of Applied Physics, Supplement 23-1, p. 51–53.

Fujisava, H., and Ito, E., 1985, Measurements of ultrasonic wave velocities of tungsten carbide as a standard material under high pressure up to 8 GPa, in Proceedings of the 5th Symposium on Ultrasonic Electronics: Tokyo, Japan (1984): Journal of Applied Physics, Supplement 24-1, p. 103–105.

Funamori, N., Yagi, T., Utsumi, W., Kondo, T., Uchida, Y., and Funamori, M., 1996, Thermoelastic properties of $MgSiO_3$ perovskite determined by in-situ X-ray observations up to 30 GPa and 2000 K: Journal of Geophysical Research, B, Solid Earth and Planets, v. 101, p. 8257–8269, doi: 10.1029/95JB03732.

Gillet, P., Ingrin, J., and Chopin, C., 1984, Coesite in subducted continental crust; P-T history deduced from an elastic model: Earth and Planetary Science Letters, v. 70, p. 426–436, doi: 10.1016/0012-821X(84)90026-8.

Heaney, P.J., and Veblen, D.R., 1991, Observations of the α–β phase transition in quartz: A review of imaging and diffraction studies and some new results: The American Mineralogist, v. 76, p. 1018–1032.

Hinze, E., Kremmler, J., and Lauterjung, J., 1991, Viel-Stempel-Hochdruckapparatur für Pulverdiffraktometrie mit Synchrotronstrahlung (in German): Deutsches ElektronenSYnchrotron (DESY), Jahresreport 1991.

Hinze, E., Kremmler, J., and Lauterjung, J., 1992, Mehrstempel-Hochdruckapparatur für Pulverdiffraktometrie unter geowissenschaftlich relevanten Bedingungen, MAX-80 (in German), Förderung der Grundlagenforschung durch den Bundesminister für Forschung und Technologie, Ergebnisberichte 1989–1992, Erforschung kondensierter Materie und Atomphysik im Verbund mit Großgeräten: Physik-Chemie-Biologie, Band II, Festkörperphysik und Materialforschung, p. 84–88.

Huppertz, H., 2003, Multianvil High-Pressure/High-Temperature Syntheses in Solid State Chemistry: Dresden, Hochdruck-Workshop.

Ito, E., Katsura, T., Aizawa, Y., Kawabe, K., Yokoshi, S., Kubo, A., Nozawa, A., and Funakoshi, K., 2005, High-pressure generation in the Kawai-type apparatus equipped with sintered diamond anvils: Application to the wurtzite-rocksalt transformation in GaN, in Chen, J., Wang, Y., Duffy, T., Shen, G., and Dobrzhinetskaya, L., eds., Advances in High Pressure Technology for Geophysical Application: Amsterdam, Elsevier Besloten Vennootschap, p. 451–460.

Jamieson, J.C., Fritz, J.N., and Manghnani, M.H., 1982, Pressure measurement at high temperature in X-ray diffraction studies: Gold as a primary standard, in Akimoto, S., and Manghnani, M.H., eds., High Pressure Research in Geophysics: Tokyo, Center for Academic Publications, p. 27–47.

Katsura, T., Funakoshi, K., Kubo, A., Nishiyama, N., Tange, Y., Sueda, Y., Kubo, T., and Utsumi, W., 2004, A large-volume high P-T apparatus for in situ X-ray observation "SPEED-MkII": Physics of the Earth and Planetary Interiors, v. 143–144, p. 497–506, doi: 10.1016/j.pepi.2003.07.025.

Kinoshita, H., Hamaya, N., and Fujisava, H., 1979, Elastic properties of single-crystal NaCl under high pressures to 80 kbar: Journal of Physics of the Earth, v. 27, p. 337–350.

Kohlstedt, D.L., Keppler, H., and Rubie, D.C., 1994a, Solubility of water in α–, β– and γ–$(Mg,Fe)_2SiO_4$ at high pressures: Eos (Transactions, American Geophysical Union), v. 75, p. 652.

Kohlstedt, D.L., Rubie, D.C., and Keppler, H., 1994b, Experimental constraints on the solubility of water in upper mantle minerals: Terra Abstracts, v. 6, suppl. 1, p. 29.

Kohlstedt, D.L., Keppler, H., and Rubie, D.C., 1996, Solubility of water in the α, β and γ phases of $(Mg,Fe)_2SiO_4$: Contributions to Mineralogy and Petrology, v. 123, p. 345–357, doi: 10.1007/s004100050161.

Kondo, T., Sawamoto, H., Yoneda, A., Kato, M., Matsumuro, A., Yagi, T., and Kikegawa, T., 1993a, The use of sintered diamond anvils in the MA8 type high-pressure apparatus: PAGEOPH, v. 141, p. 601–611, doi: 10.1007/BF00998347.

Kondo, T., Sawamoto, H., Yoneda, A., Kato, M., Matsumuro, A., and Yagi, T., 1993b, Ultrahigh-pressure and high temperature generation by use of the MA8 system with sintered diamond anvils: High Temperatures, High Pressures, v. 25, p. 105–112.

Kung, J., Li, B., Uchida, T., Wang, Y., Neuville, D., and Liebermann, R.C., 2004, In situ measurements of sound velocities and densities across the orthopyroxene → high-pressure clinopyroxene transition in $MgSiO_3$ at high pressure: Physics of the Earth and Planetary Interiors, v. 147, p. 27–44, doi: 10.1016/j.pepi.2004.05.008.

Leinenweber, K., 2005, Designing multi-anvil assemblies: The COMPRES cell assembly project: Argonne, Advanced Photon Source, COMPRES Multi-Anvil Workshop.

Li, B., Liebermann, R.C., and Jackson, I., 1994, Measurements of elastic wave velocity of polycrystal Al_2O_3 to 10 GPa: Eos (Transactions, American Geophysical Union), v. 75, p. 596.

Li, B., Liebermann, R.C., Gwanmesia, G.D., and Jackson, I., 1995a, Elastic wave velocities of mantle minerals to 10 GPa in multi-anvil apparatus by in situ ultrasonic techniques: Eos (Transactions, American Geophysical Union), v. 76, p. 277.

Li, B., Gwanmesia, G.D., and Liebermann, R.C., 1995b, Elastic wave velocity of olivine and beta polymorphs of Mg_2SiO_4 at transition zone pressures: Eos (Transactions, American Geophysical Union), v. 76, p. 619.

Li, B., Rigden, S.M., and Liebermann, R.C., 1996a, Elasticity of stishovite at high pressure: Physics of the Earth and Planetary Interiors, v. 96, p. 113–127, doi: 10.1016/0031-9201(96)03144-5.

Li, B., Jackson, I., Gasparik, T., and Liebermann, R.C., 1996b, Elastic wave velocity measurement in multi-anvil apparatus to 10 GPa using ultrasonic interferometry: International Union of Geodesy and Geophysics XXI Special Volume: Physics of the Earth and Planetary Interiors, v. 98, p. 79–91, doi: 10.1016/S0031-9201(96)03173-1.

Li, B., Gwanmesia, G.D., and Liebermann, R.C., 1996c, Sound velocities of olivine and beta polymorphs of Mg_2SiO_4 at Earth's transition zone pressures: Geophysical Research Letters, v. 23, p. 2259–2262, doi: 10.1029/96GL02084.

Li, B., Chen, G., Gwanmesia, G.D., and Liebermann, R.C., 1998, Sound velocity measurements at mantle transition zone conditions of pressure and temperature using ultrasonic interferometry in a multianvil apparatus, in Manghnani, M.H., and Yagi, T., eds., Properties of Earth and Planetary Materials at High Pressure and Temperature: American Geophysical Union Geophysical Monograph 101, p. 41–61.

Li, B., Vaughan, M.T., Kung, J., and Weidner, D.J., 2001, Direct length measurement using X-radiography for the determination of acoustic velocities at high pressure and temperature: NSLS (National Synchrotron Light Source) Activity Report 2001, p. 2-103–2-106.

Li, B., Chen, K., Kung, J., Liebermann, R.C., and Weidner, D.J., 2002, Sound velocity measurement using transfer function method: Journal of Physics Condensed Matter, v. 14, p. 11,337–11,342, doi: 10.1088/0953-8984/14/44/478.

Li, B., Kung, J., and Liebermann, R.C., 2004, Modern techniques in measuring elasticity of Earth materials at high pressure and high temperature using ultrasonic interferometry in conjunction with synchrotron X-radiation in multi-anvil apparatus: Physics of the Earth and Planetary Interiors, v. 143–144, p. 559–574, doi: 10.1016/j.pepi.2003.09.020.

Li, B., Kung, J., Uchida, T., and Wang, Y., 2005, Simultaneous equation of state, pressure calibration and sound velocity measurements to lower mantle pressures using multi-anvil apparatus, in Chen, J., Wang, Y., Duffy, T., Shen, G., and Dobrzhinetskaya, L., eds., Advances in High Pressure Technology for Geophysical Application: Amsterdam, Elsevier Besloten Vennootschap, p. 49–66.

Liebermann, R.C., Ringwood, A.E., Mayson, D.J., and Major, A., 1975, Hot-pressing of polycrystalline aggregate at very high pressure for ultrasonic measurements, in Osugi, J., ed., Proceedings of the 4th conference on High Pressure: Tokyo, Physico-Chemical Society of Japan, p. 495–502.

Liebermann, R.C., Prewitt, T.C., and Weidner, D.J., 1985, Large-volume high-pressure mineral physics in Japan: Eos (Transactions, American Geophysical Union), v. 66, p. 138.

Litasov, K., and Ohtani, E., 2003a, Stability of various hydrous phases in CMAS pyrolite-H_2O system up to 25 GPa: Physics and Chemistry of Minerals, v. 30, p. 147–156, doi: 10.1007/s00269-003-0301-y.

Litasov, K., and Ohtani, E., 2003b, Hydrous solidus of CMAS-pyrolite and melting of mantle plumes at the bottom of the upper mantle: Geophysical Research Letters, v. 30, p. 2143, doi: 10.1029/2003GL018318.

Litasov, K., and Ohtani, E., 2005, Phase relations in hydrous MORB at 18–28 GPa: Implications for heterogeneity of the lower mantle: Physics of the Earth and Planetary Interiors, v. 150, p. 239–263, doi: 10.1016/j.pepi.2004.10.010.

Litasov, K., Ohtani, E., Langenhorst, F., Yurimoto, H., Kubo, T., and Kondo, T., 2003, Water solubility in Mg-perovskites and water storage capacity in the lower mantle: Earth and Planetary Science Letters, v. 211, p. 189–203, doi: 10.1016/S0012-821X(03)00200-0.

Litasov, K., Ohtani, E., Suzuki, A., Kawazoe, T., and Funakoshi, K., 2004, Absence of density crossover between basalt and peridotite in the cold slabs passing through 660 km discontinuity: Geophysical Research Letters, v. 31, L24607, doi: 10.1029/2004GL021306.

McSkimin, H.J., 1950, Ultrasonic measurement techniques applicable to small solid specimens: The Journal of the Acoustical Society of America, v. 22, p. 413–418, doi: 10.1121/1.1906618.

Mueller, H.J., Lauterjung, J., Schilling, F.R., Lathe, C., and Nover, G., 2002, Symmetric and asymmetric interferometric method for ultrasonic compressional and shear wave velocity measurements in piston-cylinder and multi-anvil high-pressure apparatus: European Journal of Mineralogy, v. 14, p. 581–589, doi: 10.1127/0935-1221/2002/0014-0581.

Mueller, H.J., Schilling, F.R., Lauterjung, J., and Lathe, C., 2003, A standard free pressure calibration using simultaneous XRD and elastic property measurements in a multi-anvil device: European Journal of Mineralogy, v. 15, p. 865–873, doi: 10.1127/0935-1221/2003/0015-0865.

Mueller, H.J., Lathe, C., and Schilling, F.R., 2005a, Simultaneous determination of elastic and structural properties under simulated mantle conditions using multi-anvil device MAX80, in Chen, J., Wang, Y., Duffy, T., Shen, G., and Dobrzhinetskaya, L., eds., Advances in High Pressure Technology for Geophysical Application: Amsterdam, Elsevier Besloten Vennootschap, p. 67–94.

Mueller, H.J., Schilling, F.R., Lathe, C., and Lauterjung, J., 2005b, Calibration based on a primary pressure scale in a multi-anvil device, in Chen, J., Wang, Y., Duffy, T., Shen, G., and Dobrzhinetskaya, L., eds., Advances in High Pressure Technology for Geophysical Application: Amsterdam, Elsevier Besloten Vennootschap, p. 427–449.

Ohtani, E., and Maeda, M., 2001, Density of basaltic melt at high pressure and stability of the melt at the base of the lower mantle: Earth and Planetary Science Letters, v. 193, p. 69–75, doi: 10.1016/S0012-821X(01)00505-2.

Ohtani, E., Mizobata, H., and Yurimoto, H., 2000, Stability of dense hydrous magnesium silicate phases in the systems Mg_2SiO_4-H_2O and $MgSiO_3$-H_2O at pressures up to 27 GPa: Physics and Chemistry of Minerals, v. 27, p. 533–544, doi: 10.1007/s002690000097.

Ohtani, E., Toma, M., Litasov, K., Kubo, T., and Suzuki, A., 2001a, Stability of dense hydrous magnesium silicate phases and water storage capacity in the transition zone and lower mantle: Physics of the Earth and Planetary Interiors, v. 124, p. 105–117, doi: 10.1016/S0031-9201(01)00192-3.

Ohtani, E., Litasov, K., Suzuki, A., and Kondo, T., 2001b, Stability field of new hydrous phase, δ-AlOOH, with implications for water transport into the deep mantle: Geophysical Research Letters, v. 28, p. 3991–3993, doi: 10.1029/2001GL013397.

Ohtani, E., Toma, M., Kubo, T., Kondo, T., and Kikegawa, T., 2003, In situ X-ray observation of decomposition of superhydrous phase B at high pressure and temperature: Geophysical Research Letters, v. 30, p. 1029, doi: 10.1029/2002GL015549.

Ohtani, E., Litasov, K., Hosoya, T., Kubo, T., and Kondo, T., 2004, Water transport into the deep mantle and formation of a hydrous transition zone: Physics of the Earth and Planetary Interiors, v. 143–144, p. 255–269, doi: 10.1016/j.pepi.2003.09.015.

Poirier, J.P., 1982, On transformation plasticity: Journal of Geophysical Research, v. 87, p. 6791–6797.

Prakapenka, V.P., Shen, G., Dubrovinsky, L.S., Rivers, M.L., and Sutton, S.R., 2004, High pressure induced phase transformation of SiO_2 and GeO_2: Difference and similarity: Journal of Physics and Chemistry of Solids, v. 65, p. 1537–1545, doi: 10.1016/j.jpcs.2003.12.019.

Rubie, D.C., and Ross, C.R., II, 1994, Kinetics of the olivine-spinel transformation in subducting lithosphere: Experimental constraints and implications for deep slab processes: Physics of the Earth and Planetary Interiors, v. 86, p. 223–241.

Rubie, D.C., Webb, S.L., and Brearley-Adrian, J., 1990, Mechanisms and kinetics of the olivine-spinel transformation in subducting slabs: Eos (Transactions, American Geophysical Union), v. 71, p. 966.

Schilling, F.R., 1999, A transient technique to measure thermal diffusivity at elevated temperatures: European Journal of Mineralogy, v. 11, p. 1115–1124.

Schmidt, C., Bruhn, D., and Wirth, R., 2003, Experimental evidence of transformation plasticity in silicates: Minimum of creep strength in quartz: Earth and Planetary Science Letters, v. 205, p. 273–280, doi: 10.1016/S0012-821X(02)01046-4.

Schreiber, E., Anderson, O.L., and Soga, N., 1973, Elastic Constants and Their Measurement: New York, McGraw-Hill, 196 p.

Schreyer, W., 1995, Ultradeep metamorphic rocks: The retrospective viewpoint: Journal of Geophysical Research, v. 100, p. 8353–8366, doi: 10.1029/94JB02912.

Shimomura, O., Yamaoka, S., Yagi, T., Wakatsuki, M., Tsuji, K., Fukunaga, O., Kawamura, H., Aoki, K., and Akimoto, S., 1984, Multianvil type X-ray apparatus for synchrotron radiation: Proceedings of Material Research Society Symposium Proceedings, v. 22, p. 17.

Shimomura, O., Yamaoka, S., Yagi, T., Wakatsuki, T., Tsuji, K., Fukunaga, O., Kawamura, H., Aoki, K., and Akimoto, S., 1985, Multi-anvil type high-pressure apparatus for synchrotron radiation, in Minomura, S., ed., Solid State Physics under Pressure: Recent Advance with Anvil Devices: Tokyo and Dordrecht, KTK Science Publishers, Tokyo and Reidel, p. 351–356.

Shimomura, O., Utsumi, W., Taniguchi, T., Kikegawa, T., and Nagashima, T., 1992, A new high-pressure and high-temperature apparatus with sintered diamond anvils for synchrotron radiation use, in Syono, Y., and Manghnani, M.H., eds., High-Pressure Research: Application to Earth and Planetary Sciences: Tokyo, Terra Scientific Publishing Company, p. 3–11.

Suzuki, A., Ohtani, E., Kondo, T., Kuribayashi, T., Niimura, N., Kurihara, K., and Chatake, T., 2001, Neutron diffraction study of hydrous phase G: Hydrogen in the lower mantle hydrous silicate, phase G: Geophysical Research Letters, v. 28, p. 3987–3990, doi: 10.1029/2001GL013260.

Tinker, D., Lesher, C.E., Baxter, G.M., Uchida, T., and Wang, Y., 2004, High-pressure viscosimetry of polymerized silicate melts and limitations of the Eyring equation: American Mineralogy, v. 89, p. 1701–1708.

Utsumi, W., Toyama, N., Endo, S., Fujita, F.E., and Shimomura, O., 1986, X-ray diffraction under ultrahigh pressure generated with sintered diamond anvils: Journal of Applied Physics, v. 60, p. 2201–2204, doi: 10.1063/1.337178.

Utsumi, W., Yagi, T., Leinenweber, K., Shimomura, O., and Taniguchi, T., 1992, High-pressure and high-temperature generation using sintered diamond anvils, in Syono, Y., and Manghnani, M.H., eds., High-Pressure Research: Application to Earth and Planetary Sciences: Tokyo, Terra Scientific Publishing Company, p. 37–42.

Wang, Y., and Utsumi, W., 2005, 6/2: A possible marriage between muti-anvil and DAC (oral paper): COMPRES Annual Meeting 2005, Mohonk Mountain House, New Palz, June 16–19.

Yagi, T., 1988, MAX80: Large-volume high-pressure apparatus combined with synchrotron radiation: Eos (Transactions, American Geophysical Union), v. 69, no. 12, p. 18–27.

Yagi, T., Akaogi, M., Shimomura, O., Suzuki, T., and Akimoto, S., 1987a, In situ observation of the olivine-spinel transformation in Fe_2SiO_4 using synchrotron radiation: Journal of Geophysical Research, v. 92, p. 6207.

Yagi, T., Akaogi, M., Shimomura, O., Tamai, H., and Akimoto, S., 1987b, High pressure and high temperature equations of state of majorite, in Manghnani, M.H., and Syono, Y., eds., High-Pressure Research in Mineral Physics: American Geophysical Union Geophysical Monograph 39, p. 141–147.

Zinn, P., Lauterjung, J., and Wirth, R., 1997a, Kinetic and microstructural studies of the crystallisation of coesite from quartz at high pressure: Physics of the Earth and Planetary Interiors, v. 212, p. 691–698.

Zinn, P., Hinze, E., Lauterjung, J., and Wirth, R., 1997b, Kinetic and microstructural studies of the quartz-coesite phase transition: Physics and Chemistry of the Earth, v. 22, p. 105–111, doi: 10.1016/S0079-1946(97)00089-X.

MANUSCRIPT ACCEPTED BY THE SOCIETY 14 AUGUST 2006

X-ray microtomography under high pressure

Takeyuki Uchida[†]
Yanbin Wang
Frank Westferro
Center for Advanced Radiation Sources, University of Chicago, Chicago, Illinois 60637, USA

Mark L. Rivers
Center for Advanced Radiation Sources, University of Chicago, Chicago, Illinois 60637, USA, and Department of Geophysical Sciences, University of Chicago, Chicago, Illinois 60637, USA

Jeff Gebhardt
Center for Advanced Radiation Sources, University of Chicago, Chicago, Illinois 60637, USA

Stephen R. Sutton
Center for Advanced Radiation Sources, University of Chicago, Chicago, Illinois 60637, USA, and Department of Geophysical Sciences, University of Chicago, Chicago, Illinois 60637, USA

ABSTRACT

We present a new technique for X-ray microtomography under high pressure. By modifying an opposed-anvil high-pressure cell known as the Drickamer cell, monochromatic X-ray radiographs can be collected through the entire cell assembly and a thin-walled containment ring. We designed a rotation mechanism to rotate the Drickamer cell from 0 to 180° under hydraulic load, and examined pressure-generation efficiencies of the Drickamer cell up to 8 GPa at room temperature using the energy-dispersive technique through the containment ring, which allowed us to conduct a detailed evaluation of effects of geometric factors of the Drickamer anvils for tomography application. The maximum attainable load supported by the containment ring is proportional to the anvil diameter. Cells with larger anvil diameters are less pressure efficient, although they can reach higher pressures with much higher loads. Pressure efficiency generally increases with the tapering angle and decreases with tip diameter of the anvils. However, cells with larger tapering angles are more unstable, causing blowouts beyond a certain pressure. We evaluated the quality of X-ray images using the optical setup for conventional tomography at the GSECARS (GeoSoilEnviroCARS [Consortium for Advanced Radiation Sources]) beam line, the Advanced Photon Source. Noise level in the images depends on the material used for the containment ring. Containment rings made of either cubic boron nitride or silicon carbide allow us to better observe the images, but these materials are brittle and prone to mode-1 failure and are not suitable for high-pressure generation. The noise level of

---

[†]E-mail: uchida@cars.uchicago.edu.

aluminum-alloy rings is somewhat higher, but the material is much more ductile, and hence it is capable of supporting higher loads. Using the aluminum-alloy containment ring, we conducted a commissioning run of tomography up to 3 GPa. We demonstrate that the high-pressure tomography setup is useful for studying internal structure of objects and density of melt and fluid under pressure.

Keywords: tomography, Drickamer cell, high pressure, synchrotron radiation.

INTRODUCTION

X-ray computed microtomography (CMT) has various applications in geophysics, geology, environmental and material sciences, because it is capable of imaging internal structures of objects, especially when samples are precious, fragile, or time-consuming to prepare (e.g., Rivers et al., 1999). Data collection is usually conducted at ambient condition or at high temperatures by attaching a furnace. Pressure, which is another important parameter in many applications, has not been fully explored in previous efforts of tomography studies. If high-pressure and high-temperature conditions can be applied simultaneously to the sample, then the tomography imaging technique may open a new window to many opportunities in various scientific and engineering fields.

One of the most critical issues in performing tomography experiments under high pressure is the limited X-ray access to the sample because of the highly absorbing materials, such as tungsten carbide and tool steel, typically used in multianvil apparatus for pressure generation. The diamond anvil cells (DAC), on the other hand, generally compress samples too small to be useful for imaging internal structures of bulk samples in many applications. Our choice of high-pressure apparatus is an opposed-anvil high-pressure cell, sometimes referred to as the Drickamer cell, which has several unique advantages over other high-pressure apparatus, such as the multianvil systems and the DAC. The sample is compressed between a pair of opposed anvils similar to the DAC. The relatively large volume of the sample chamber in the Drickamer cell permits the incorporation of a resistive heater for high-temperature generation, with stable control of temperature, unlike the laser-heated DAC, in which temperature control may vary depending on the laser-absorbing material used. The Drickamer cell is also more load efficient than most multianvil apparatuses, which typically require several hundred tons, making it difficult to rotate the sample under pressure. Furthermore, by employing X-ray transparent containment rings, a 360° X-ray accessibility to the sample can be obtained in the plane perpendicular to the loading axis. This feature is essential for our tomographic imaging application.

Two sets of Harmonic Drive™ gear reducers and thrust bearings are employed on the top and bottom sides of the Drickamer cell, in order to rotate the high-pressure module continuously and precisely from 0 to 180° under load for tomography observations. The designed mechanism has benefited from the rotational deformation apparatus (Yamazaki and Karato, 2001) and is described elsewhere (Wang et al., 2005).

In the present study, the development and characterization of the Drickamer cell is discussed in the context of examining X-ray imaging and diffraction capabilities for applications of X-ray microtomography under high pressure. We present results from a series of tests for optimization of the modified Drickamer anvil cell, in order to reach the desired pressure within the load capacity of the rotation module and to collect better X-ray images. We also present results obtained from a commissioning test performed under pressure with monochromatic synchrotron radiation and finally discuss the possibility of the Drickamer cell for other physical property measurements such as radial distribution function and density measurements of melts.

EXPERIMENTAL METHODS

Figure 1 shows a schematic view of a typical tomography setup at the GSECARS (GeoSoilEnviroCARS [Consortium for Advanced Radiation Sources]) beam line at the Advanced Photon Source (APS) (Rivers et al., 1999). A large uniform white X-ray beam is monochromatized by a Si(111) monochromator. Incident beam stability is controlled by the feedback system for the Si monochromator. At the APS, the X-ray beam is delivered to each station at so-called "top up" mode so that the ring current is virtually constant at 100 mA, and thus the brightness of incident beam remains unchanged. Temperature of the Si monochromator is constantly monitored and adjusted by a water-cooling system. Position and tilt of the monochromator is continuously adjusted depending on the beam intensity. Thus, the incident beam intensity is practically constant during the entire image-data-collection process. The available energy range of the monochromatic beam is up to 65 keV. As the photon energy increases, the number of photons passing through both the objects of interest and the surrounding material increases, reducing the contrast. On the other hand, decreasing the photon energy too much would also lead to poor contrast because attenuation increases. Thus, there is an optimal energy range for imaging contrast, which depends on the sample of interest. The energy is carefully selected in our experiments by maximizing imaging contrast. The transmitted X-rays are converted to visible light with a single-crystal YAG (yttrium aluminum garnet) scintillator (0.2 mm thick). The visible light signals are reflected by an optical mirror and detected by a 1242 × 1152 pixel fast-charge coupled device (CCD) detector through a microscope objective, through which magnification of the image can be adjusted.

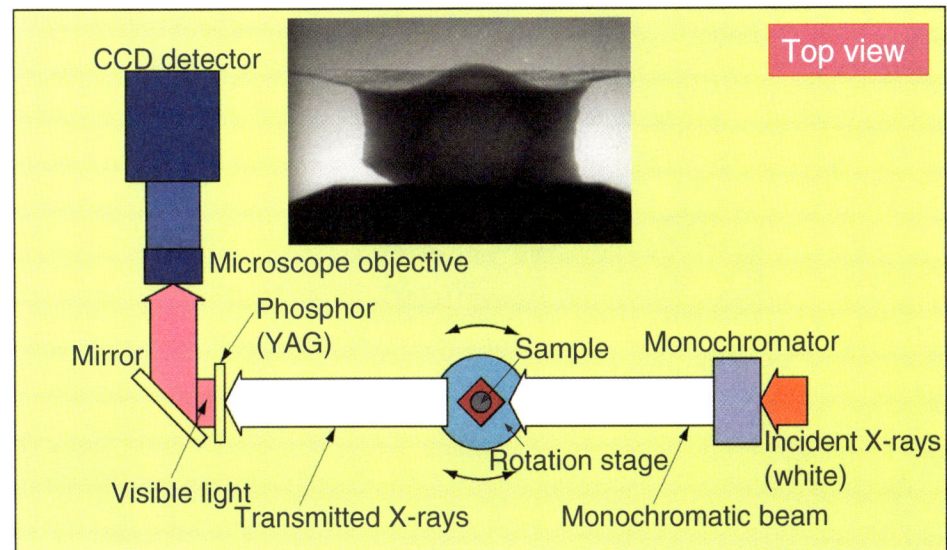

Figure 1. Schematic diagram of X-ray microtomography setup at the GeoSoilEnviroCARS (GSECARS) beam line. Incident white X-rays are monochromatized by Si(111) monochromator. Transmitted monochromatic X-ray is converted to visible light by a YAG (yttrium aluminum garnet) single-crystal scintillator; the visible light image is reflected by a 90° by mirror, and then it is detected by a visible light charge coupled device (CCD) detector through a microscope objective.

The sample is rotated so that X-ray radiographs are collected as a function of rotation angle with a typical step size of 0.25°. In order to minimize effects of beam nonuniformity and optic imperfection, (e.g., intensity variation of incident X-ray beam, distortion of monochromator, dust particles on the mirror, the scintillator, or the objective, etc.), flat-field images are taken by driving the sample out of the beam path after every 50 images. All of the images are then treated by dividing the corresponding flat-field image for background correction. A CMT software package, which can be downloaded from our web page at http://cars.uchicago.edu, is used to perform tomography reconstruction and visualization of the three-dimensional sample image (Rivers et al., 1999).

In order to conduct microtomography under high pressure, the sample rotation stage is replaced with a high-pressure module (Fig. 2; Wang et al., 2005), which is capable of rotating the sample by at least 180° under hydraulic loads up to ~50 tons. Figure 3 shows the Drickamer cell that we adopted, which consists of a pair of tungsten carbide (WC) anvils (top and bottom) and a containment ring. Various X-ray transparent materials have been considered and tested as candidates for the containment ring, including an aluminum alloy, cubic boron nitride (cBN), or silicon carbide (SiC). Anvil geometry (Fig. 3C) is also important in pressure generation, and several geometric parameters have been examined. The parameters tested include (1) anvil diameter (10 and 20 mm), (2) diameter of truncation tip surface (2.5, 3.0, and 3.5 mm), and (3) tapering angle from the tip surface (10° and 20°). The overall design of the cell assembly is quite simple: The gap between the two anvils is filled with a pressure medium made of amorphous boron powder mixed with epoxy resin (BE), which is transparent to X-rays and exhibits no diffraction peaks. The sample is loaded into the sample chamber (~2 mm diameter and 2 mm height, i.e., the initial gap between anvils is 2 mm) at the center of the BE pressure medium with (for pressure efficiency test) or without (for imaging test) a graphite tube.

We conducted three different sets of experiments to examine three major functions of the high-pressure tomography apparatus: (1) diffraction capability and pressure generation, (2) imaging capability, and (3) overall system performance. For the first test set, the Drickamer cell was loaded, without the rotating mechanism, into the 250 ton press installed at Sector 13 bending magnet beam line at the Advanced Photon Source (Wang et al., 1998; Uchida et al., 2002). Powdered NaCl was filled in a graphite tube (1.4 mm inner diameter), which would be used as the heater for future high-temperature experiments, inside the sample chamber (2.0 mm diameter and 2.0 mm height) in the middle of the BE pressure medium (Fig. 3). White X-radiation was used with a Ge single-element solid-state detector (SSD) to measure lattice parameters of NaCl at a fixed 2θ angle of 5.5° based on the energy-dispersive method. Using the equation of state of NaCl from Decker (1971), pressures were computed from the lattice parameters of NaCl during compression and decompression cycles.

For the second test set (imaging capability), tomography data were collected using the entire Drickamer cell assembly (including the outer containment ring and the pressure medium) shown in Figure 3A, utilizing the ambient tomography setup at the GSECARS beam line (Fig. 1), without applying any load to increase pressure. The sample consisted of a sapphire sphere (1.0 mm diameter) embedded in a mixture of Fe and FeS_2 powders (containing 9 wt% S in total). This assemblage was packed in the sample chamber in the middle of the BE pressure medium. The energy of the monochromatic X-ray beam used was 40 keV, based on optimization of image contrast between the sapphire sphere and surrounding Fe-S mixture. Total field of view was ~9 mm × 8 mm, which resulted in the spatial resolution of ~13.3 µm/pixel, after 2 × 2 binning. Overall, 720 X-ray projections were collected by rotating the sample from 0° to 179.75° at 0.25° steps, and then tomographic reconstruction was performed using filtered back-projection.

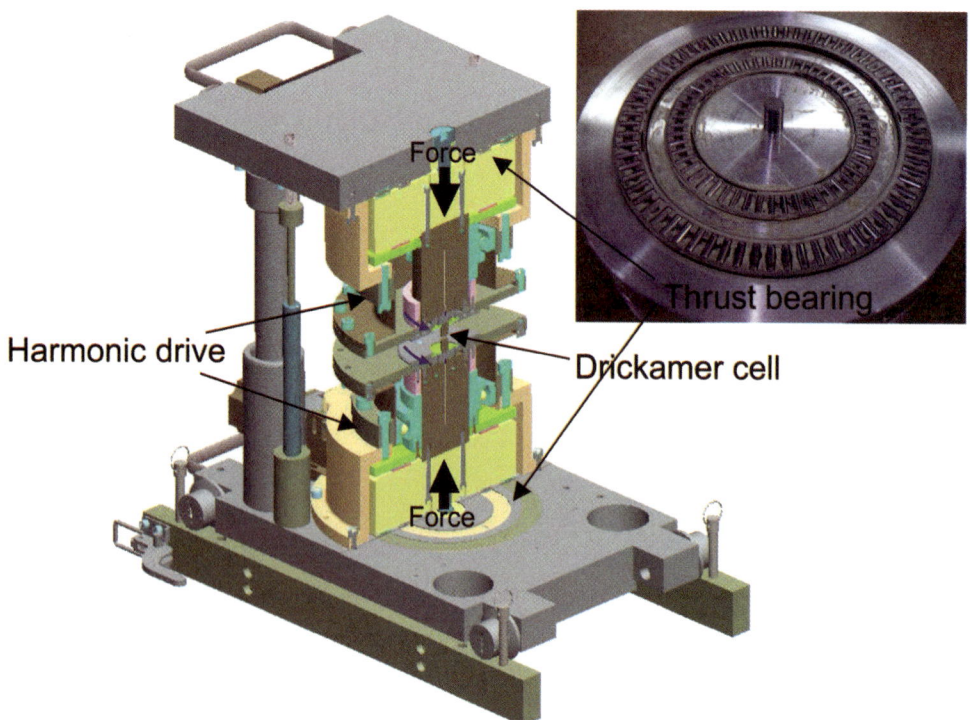

Figure 2. A cut-away view of the high-pressure tomography module. Two pairs of thrust bearings (picture inset) support an applied load up to 50 tons. Two sets of the Harmonic Drive™ gear reducers and stepper motors rotate independently, and the upper and lower devices are synchronized for rotating the cell to collect images for tomography reconstruction. The outer diameter of the Harmonic Drives is 300 mm.

For the third test set, a commissioning run was performed using the same sapphire/Fe-S sample and Al-alloy containment ring used in the second test to evaluate performance of the rotation mechanism under pressure (Fig. 2; Wang et al., 2005). The field of view was ~7.6 mm × 4.9 mm with 5.9 μm/pixel after 2 × 2 binning. Under pressure, the entire Drickamer cell was rotated to collect radiographs of the sample at various angles from 0° to 179.5° with 0.5° step size. Computational reconstruction of these projections was conducted for the data collected at 0, 3, 6, 9, and 12 tons. No X-ray diffraction was attempted in this test. However, using the elastic constants of sapphire, we can estimate pressure from the volume change of the bulk sapphire sphere measured in high-pressure tomography.

EVALUATION OF THE MODIFIED X-RAY TRANSPARENT DRICKAMER CELL

Pressure Generation

Figure 4 shows a representative NaCl X-ray diffraction pattern. Excellent counting statistics were obtained for a 200 s counting time. In addition to NaCl diffraction peaks, the strongest graphite 002 peak was also present in the pattern. No extra peak contamination was observed.

Figure 5 compares the pressure efficiency among various anvil geometric parameters. Cells with larger anvil diameters (20 mm; solid circles) show lower efficiency than those with smaller diameters (10 mm; diamonds), but they are stable to higher loads. As a result, the maximum attainable pressure for 20-mm-diameter anvils is higher than that for 10-mm-diameter anvils. However, larger anvils contain thicker pressure media in the X-ray beam path, more X-ray absorption, and may raise noise levels in X-ray images. The high loads required for generating high pressures may also become a limiting factor, since the rotation mechanism for the high-pressure tomography module is designed for a maximum load of 50 tons.

With the same anvil diameter, anvils with smaller tip surface (truncation) and/or steeper tapering angles show higher pressure efficiency (Fig. 5). However, steeper tapering angle results in larger pressure gradient and the pressure condition is less stable—experiments often ended up with a blowout by fracturing the containment ring (Fig. 5, inset). The pressure, not the load, at which blowout occurred appeared reproducible (at 6.6 GPa) despite the different anvil-geometric parameters. Therefore, these blowouts may be related to the strength of the Al-alloy containment ring. The inset of Figure 5 shows that blowouts caused brittle failure of the containment ring rather than ductile deformation. Out of the three materials tested for anvil containment ring, Al-alloy appears to be the most practical. Other containment rings made of cBN or SiC do not affect pressure efficiency; however only a few tons of load lead to failure of the ring. These materials, cBN and SiC, are good for X-ray transparency, but they are too brittle to maintain structural integrity. They are also much more expensive.

These results demonstrate that the modified Drickamer cell works well when collecting diffraction data through a thin, X-ray transparent containment ring. Using Al-alloy containment rings, pressures can be generated up to 8 GPa using larger anvils. Using smaller anvils (10 mm diameter), the current pressure range comfortably achievable is below 6 GPa.

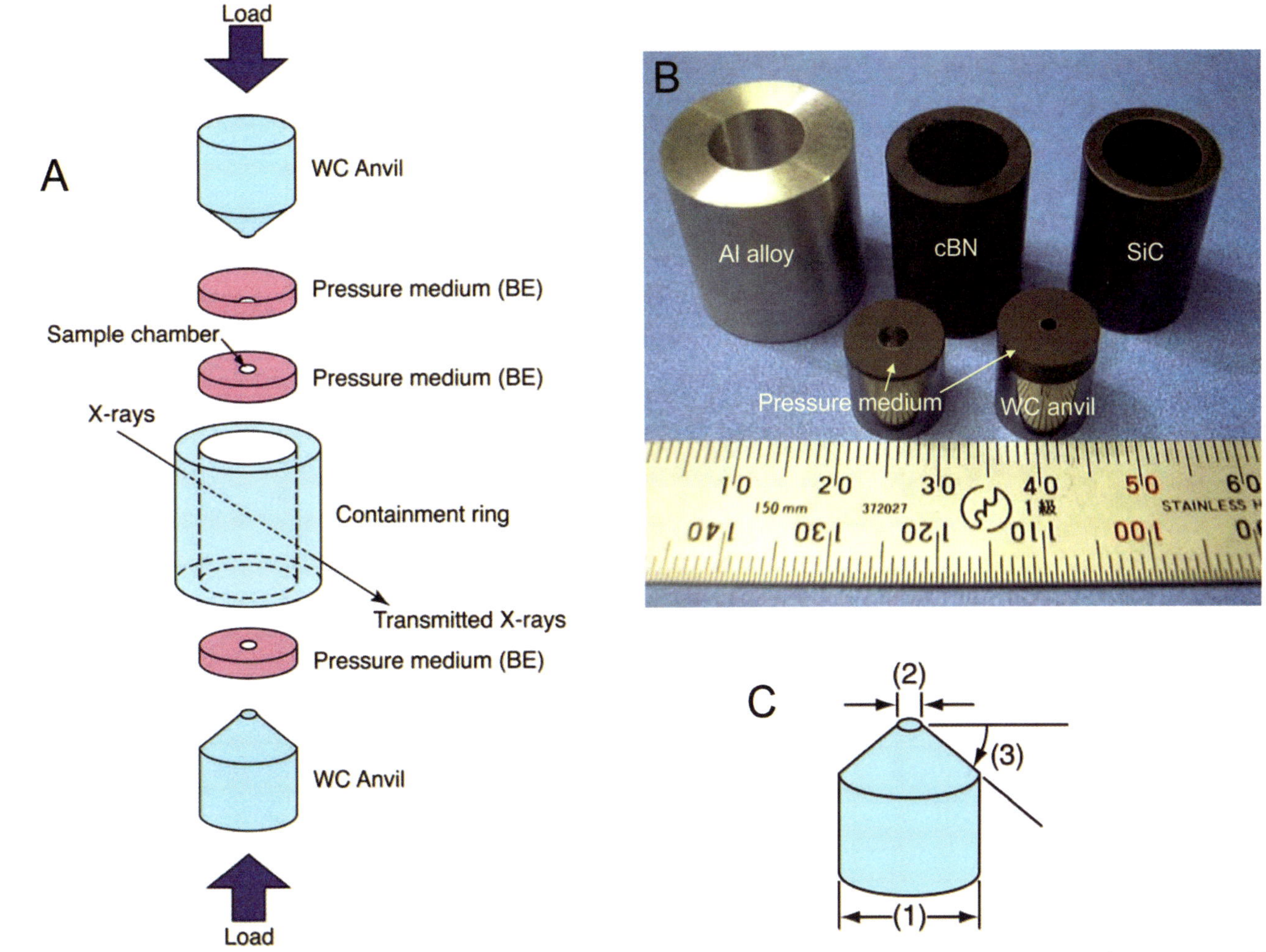

Figure 3. (A) Schematic configuration of the modified Drickamer cell, (B) photograph of its components, and (C) geometric parameters of the Drickamer anvil. Two opposed tungsten carbide (WC) anvils compress the sample through the pressure medium, which is made of amorphous boron powder solidified by epoxy resin (BE). Variation of containment ring includes aluminum alloy, cubic boron nitride (cBN), and silicon carbide (SiC). The low X-ray absorption of the pressure medium and containment ring allows a wide 2θ range for X-ray access, as well as imaging capability. Efficiency of pressure generation was examined by changing (1) anvil diameter, (2) tip diameter, and (3) tapering angle of the geometric parameters. By using harder material such as sintered diamond as anvil material, the pressure range could be extended to over 30 GPa.

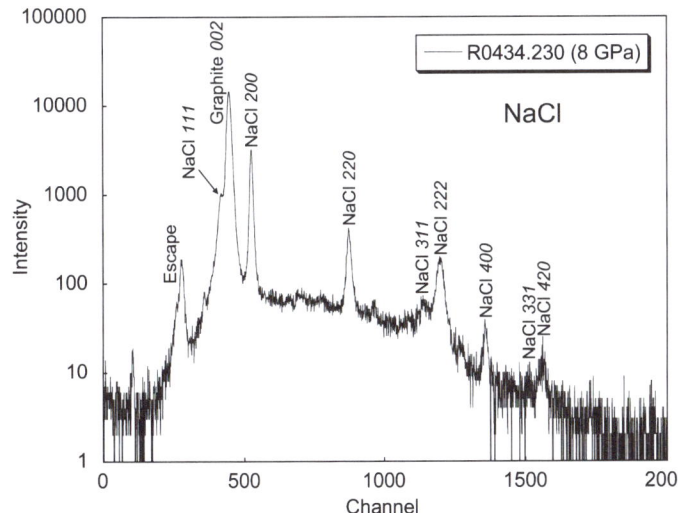

Figure 4. Representative energy-dispersive X-ray diffraction pattern of NaCl sample collected at ~8 GPa. Exposure time was 200 s. Horizontal axis is channel, which is proportional to X-ray energy; vertical axis is intensity (note log scale) in arbitrary units. Five diffraction peaks (200, 220, 222, 400, and 420) from the NaCl sample are identified for pressure measurements. The main graphite diffraction peak (200) overlaps with the NaCl (111) peak. Other graphite peaks are too weak to identify.

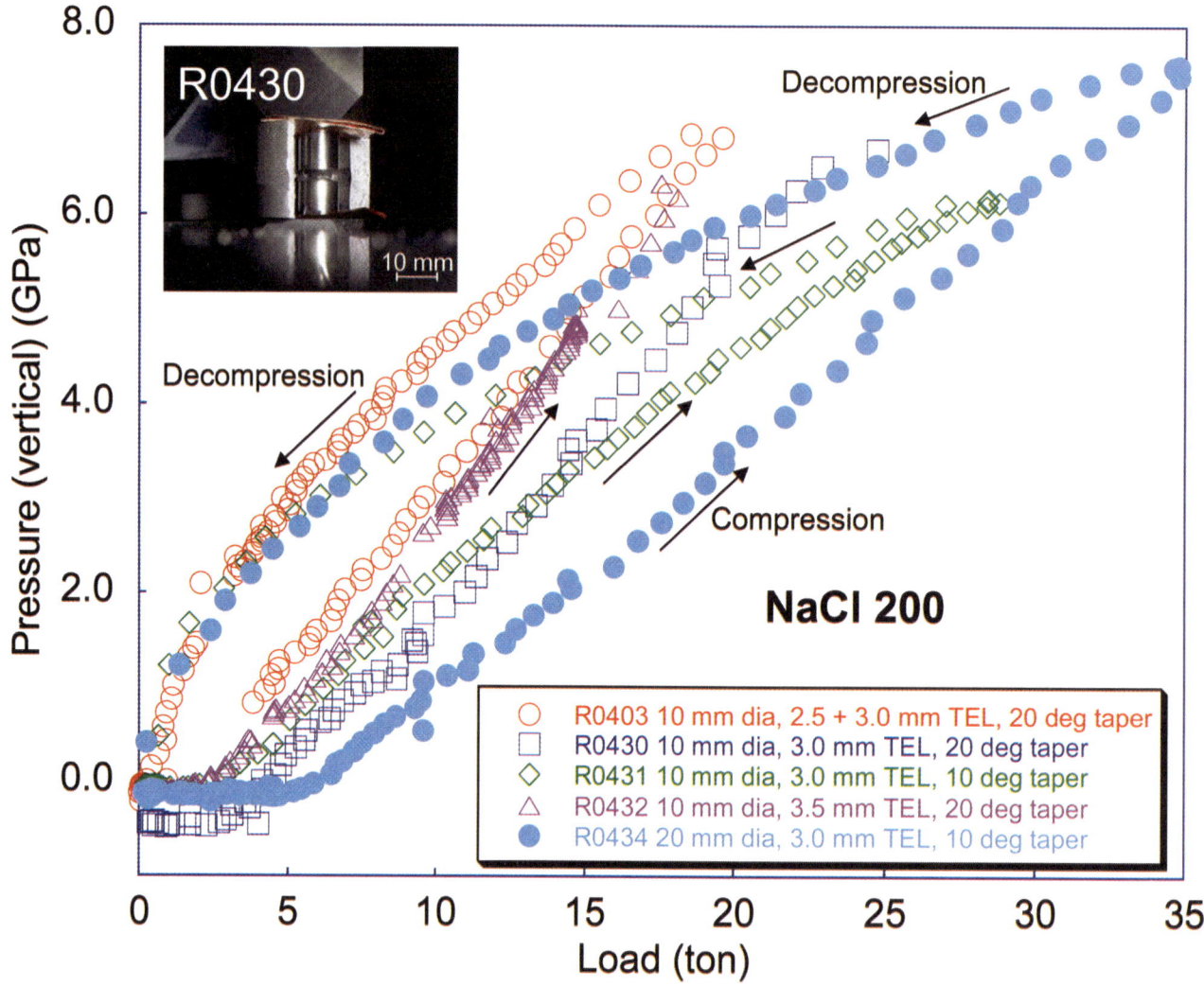

Figure 5. Pressure-generation efficiency of the tomographic Drickamer cell. Larger-diameter anvils have lower efficiency but allow application of more load. Smaller tip surface and steeper tapering angle increase efficiency, but experiments often end with a blowout (photo of fractured ring inserted).

Image Contrast

Figure 6 shows selected reconstructed image slices of the sapphire sphere and Fe-S mixture in the horizontal direction collected within various containment rings at ambient conditions using the conventional tomography setup. The contrast is reversed: Brighter areas correspond to higher density and/or higher atomic number (less X-ray transparency). The inner boundaries of the containment rings cannot be observed in the images because of the field of view we chose. All slices were produced at the same height, in the same horizontal plane, allowing direct comparison of the effect of different containment ring materials tested.

Because imaging through the Drickamer cell required longer exposure time, due to intensity loss by the absorption of the containment ring and the pressure medium, a flat-field collected through air, with the same exposure time, would saturate the CCD used in our test. For data collection with containment rings, flat-field was obtained through both the containment ring and the pressure medium with exposure time of 40 s. We are still working to find the best approach for flat-field correction. The current treatment introduces additional noise to the images and a slight systematic background variation to the images.

The noise level varies depending on the material from which the containment ring is made. The fuzziest images were those obtained within the Al-alloy ring (Fig. 6B). With SiC (Fig. 6C) and cBN (Fig. 6D), contrasts of the boundaries between the sapphire sphere and Fe-S mixture and between Fe-S mixture and pressure medium were better imaged. To

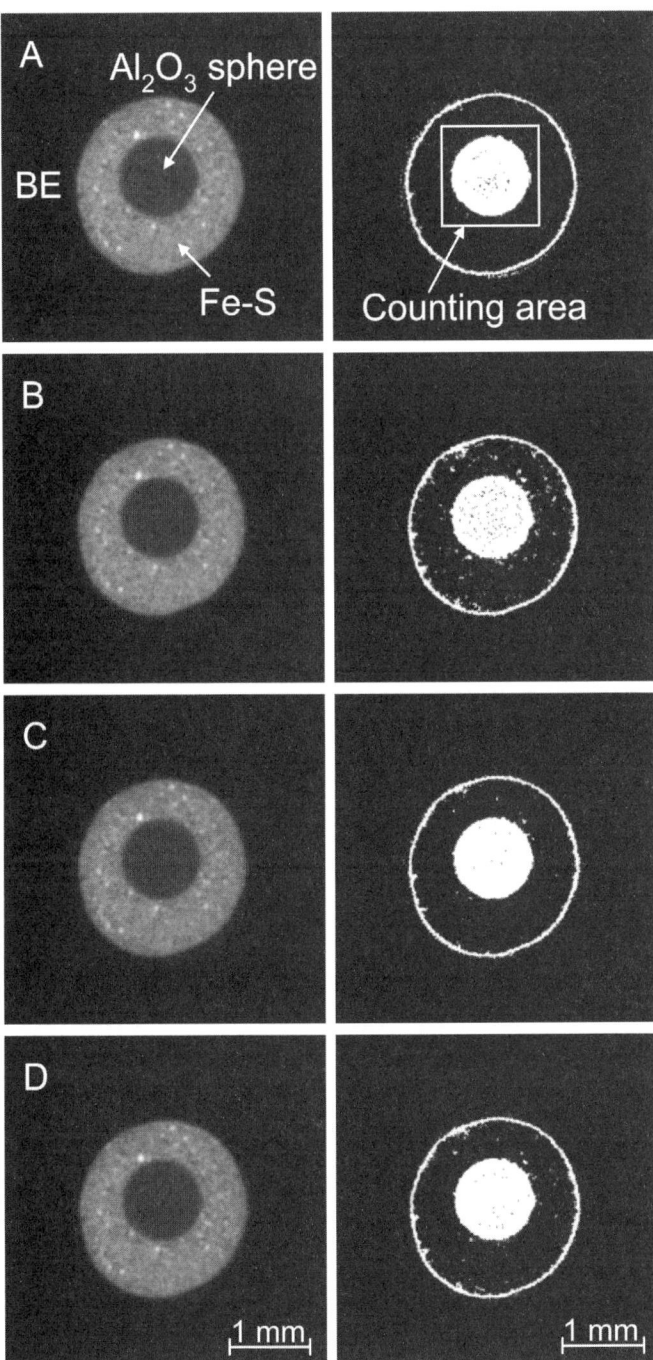

Figure 6. Comparison of reconstructed tomography slices: (A) without any containment ring, (B) with a 5-mm-thick Al-alloy ring, (C) with a 2-mm-thick SiC ring, and (D) with a 2.5-mm-thick cBN ring. The contrast is reversed: less absorption material corresponds to darker pixels. The left column shows reconstructed slices, while the right column shows filtered images to count the numbers of pixels (white dots). Counting area, which is indicated by the white rectangle, is defined to count the number of pixels of the sphere. Depending on the containment ring, the noise level is different. BE—boron powder solidified by epoxy resin pressure medium.

quantify the fuzziness, these images were filtered by setting the upper and lower intensity thresholds (Fig. 6, right column). The number of pixels in this intensity range was counted, simulating a volume measurement of sapphire. The filtering process could not remove ring artifacts caused at the boundary between Fe-S mixture and pressure medium (Fig. 6, right column). To remove the pixels counted from this artifact, we constrained a counting area by a rectangular parallelepiped shown in Figure 6A. The Al-alloy caused ~10% of error in pixel counting, which was mainly due to the noise slightly outside the sphere (Fig. 6B). Compared to Al-alloy, the errors with SiC and cBN rings were relatively small (~5%).

The pixel counting is also affected by the selection of intensity threshold. The choice of upper and lower bound intensity is somewhat arbitrary. When the upper bound is high, the image picks up noise outside the sample. When the upper bound is reduced, the outside noise level decreases, which may reduce volume of the sapphire sphere as well. A careful selection of intensity threshold is therefore of fundamental importance.

To address this issue, we developed a more quantitative method to determine the threshold: Some reconstructed slices were selected as samples and processed by a curve-fitting routine. Figure 7 summarizes the procedure. A reconstructed image is first decomposed into a series of 1-pixel stripes (Fig. 7A). The intensity profile of the 1-pixel stripes is characterized by a U-shaped trough with two sharp edges marking the ends of the sample (Fig. 7B). By taking the absolute values of the derivatives of the intensity distribution with respect to the pixels, the end positions are defined by the two peaks in Figure 7C. A curve-fitting routine is then applied to the 1-pixel stripes to obtain the boundary position between the sphere and Fe-S mixture in terms of pixels. This process is applied to several samples of reconstructed slices to determine sphere-surrounding material boundary. Based on the information of these boundary conditions, intensity threshold is reset to extract the sample pixels from other materials, corresponding to the conversion from the left column to the right column in Figure 6.

Typical curve-fitting error is comparable to a pixel, and therefore the uncertainties in diameter could be about the resolution of the image (5.9 μm/pixel) if the errors in the peak fit are directly transferred to the errors in diameter. In our case, however, results of peak fitting constrain the intensity threshold, which further constrains the volume of the sphere. If the number of sampled reconstructed slices is increased, the intensity threshold can be more accurately defined, reducing the error in volume measurement. Typical error in diameter currently achievable was less than a micron. Because the diameter of the sphere is only ~1 mm, the error in volume measurement was still larger than that using the diffraction technique. One way to reduce the error is to increase resolution of the image. Larger magnification microscope objective and higher-resolution CCD detector will help us to achieve such high-resolution imaging, leading to more accurate volume measurements.

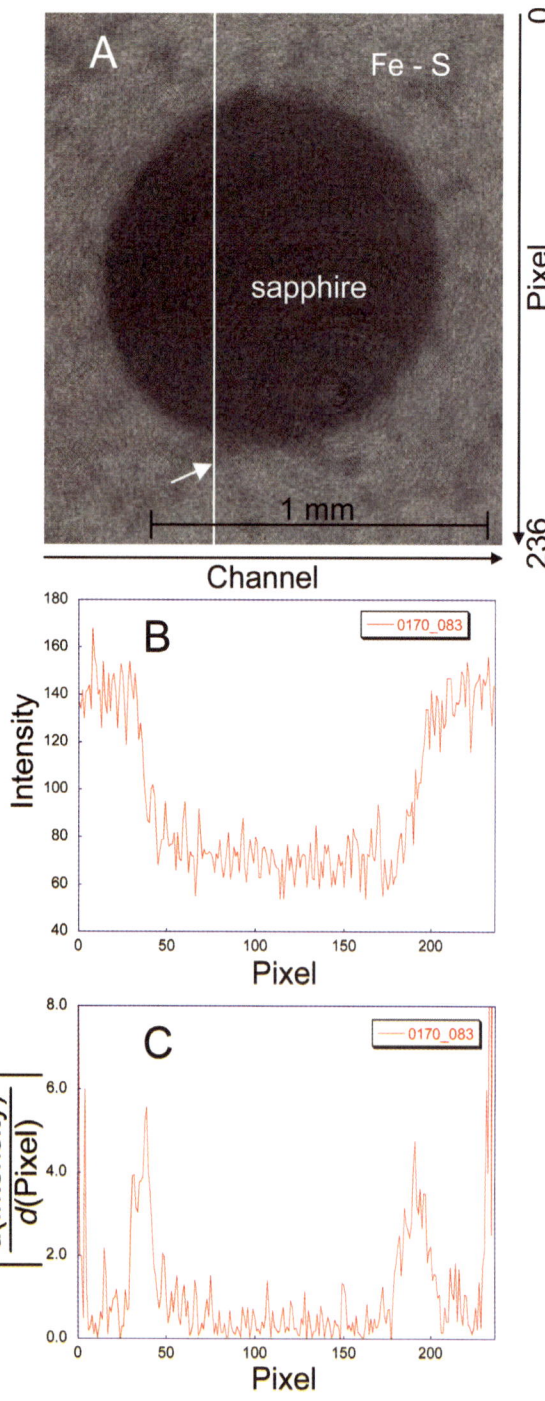
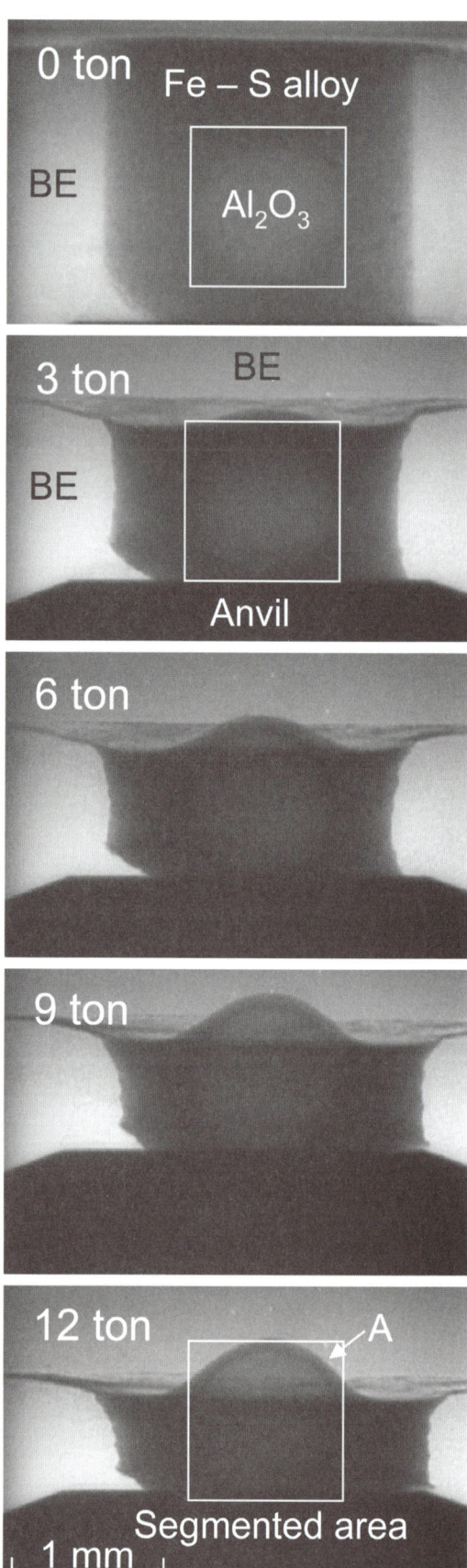

Figure 7. Quantitative threshold determination for locating the sphere-matrix interface. (A) An example of a reconstructed slice. The contrast is reversed (brighter contrast corresponds to higher density or absorption coefficient). In the vertical direction, a series of 1-pixel stripes is created at each channel (horizontal axis). (B) From a 1-pixel stripe, indicated by vertical white line, an intensity profile is created against pixel number (vertical axis in A), which features a U-shaped trough with two sharp edges. (C) Absolute values of the derivatives of the intensity distribution in B are then plotted with respect to vertical distance; these show the end positions as the two peaks. A curve-fitting routine determines the peak position for all of the 1-pixel stripes.

TEST RESULT OF TOMOGRAPHY CELL UNDER PRESSURE

Figure 8 shows X-ray projections during compression. At ambient pressure, the sphere is completely embedded in the surrounding Fe-S mixture. With increasing load (and hence pressure), the BE pressure medium pushes the Fe-S mixture down, forcing it to flow away from the top of the sapphire sphere, which does not deform so much due to its high strength. Note also that compression reduces porosity, homogenizing the intensity from BE and providing better contrast between the sample and BE.

To measure the volume of the sapphire sphere, three-dimensional (3-D) tomography images were reconstructed. Similar to the method described in Image Contrast, reconstruction was performed in a segmented area shown as a rectangle in Figure 8. By identifying the intensity threshold between the sphere and the Fe-S mixture, the boundary between the sphere and the mixture is defined, thus allowing extraction of the sphere volume. However, this spatial limitation method worked fine only up to 3 tons. As the Fe-S mixture was pushed down, the BE at the top of the sample began to approach the sphere. The X-ray intensity of BE, indicated by "A" in Figure 8, was close to that of sapphire, and the two materials were indistinguishable in terms of intensity in reconstructed 3-D image. This problem was solved by using the program "Blob3d." Figure 9A shows a reconstructed 3-D image for the segmented area shown in Figure 8. Because the BE outside the sample has similar intensity to that of the sapphire sphere, both areas are identified as one material in Blob3d. However, Blob3d allows separation of these areas into two sub-blobs (Figs. 9B and 9C). Volumes of the sphere can then be accurately and automatically determined from the extracted image (Fig. 9C). To be consistent, the volume measurements were conducted using Blob3d throughout the run for measurement.

The errors in the volume measurement ranged from 0.3% to 0.7% (0.002–0.004 mm^3), mostly due to the shadowing of the deformed anvil underneath the sphere. Figure 10 shows pressure estimated from the volume change of sapphire sphere, based on the equation of state of sapphire (Sato and Akimoto, 1979).

These test results, although performed using a solid sample, demonstrate the potential to measure the volume of noncrystalline materials. Previous melt-density measurements using X-ray radiography with only one-dimensional data assumed that the geometry of the molten sample was perfectly symmetric and remained unchanged throughout the experiment (e.g., Katayama et al., 1998). In our new technique, this assumption is no longer required, and density of melts can be inferred directly from the sample volume even when the molten sample is distorted.

DISCUSSION AND SUMMARY

For X-ray diffraction, the modified thin-walled Drickamer anvil cell is useful to obtain robust data. Exposure time was 200 s, similar to that in typical multianvil experiments. Although the diffraction test was conducted using energy-dispersive technique, our results indicate that angle-dispersive diffraction experiments are also feasible using monochromatized X-rays because no extra diffraction peak was observed except for the peak from the sample and surrounding graphite heater in energy-dispersive diffraction. However, if the angle-dispersive diffraction system does not have any collimating mechanism, a significant contribution from containment ring will be expected. Either a small access hole in the containment ring or a solar slit system (Yaoita et al., 1997) will reduce this contribution. Because of the X-ray visibility through the Drickamer cell, diffraction data can be collected through the containment ring with large 2θ angles. This capability enables many kinds of experiments, including measurement of radial distribution functions for melts, which requires wide reciprocal space range. On the other hand, unit-cell volume measurements using the energy-dispersive diffraction may experience significant error due to nonhydrostatic stress induced by uniaxial compression in the Drickamer cell. The pressure measured from the energy-dispersive technique will depend on the direction in which the diffraction pattern is collected, because the lattice strain measured from the distortion of the X-ray Debye ring may vary significantly for high strength materials (e.g., Uchida et al., 2005). A flexible system combining energy- and angle-dispersive methods will satisfy a variety of experimental needs efficiently.

Pressure can be comfortably generated up to 6 GPa using tungsten carbide anvils and an Al-alloy containment ring. Because some physical properties change dramatically with pressure (e.g., solubility of gas), this capability opens a new window of opportunities for both the tomography community and the high-pressure community.

X-ray imaging tests indicate that volume measurements of the sample are viable as long as the comparison of volume is conducted in the same run, where systematic errors can be minimized. For example, measurement of volume change between glasses and liquids can be carried out with the current setup. Under pressure, background due to pressure medium becomes more homogeneous, enabling better analysis of microstructure. On the other hand, detailed observation of the sample texture requires further improvements in the reduction of the noise contrast and enhancement of imaging resolution. The current noise level makes it difficult to distinguish the

Figure 8. Radiographic projections collected during compression. At ambient pressure, the sphere is completely embedded in the surrounding Fe-S mixture. With increasing load (pressure), the boron epoxy (BE) pressure medium on top of the sample forces the Fe-S mixture to flow away from the sapphire sphere, which does not deform so much due to its hardness. Note that compression homogenizes the intensity from BE, providing better contrast between the sample and the pressure medium. The rectangles indicate the segmented area, where reconstruction was performed to estimate the volume of the sphere.

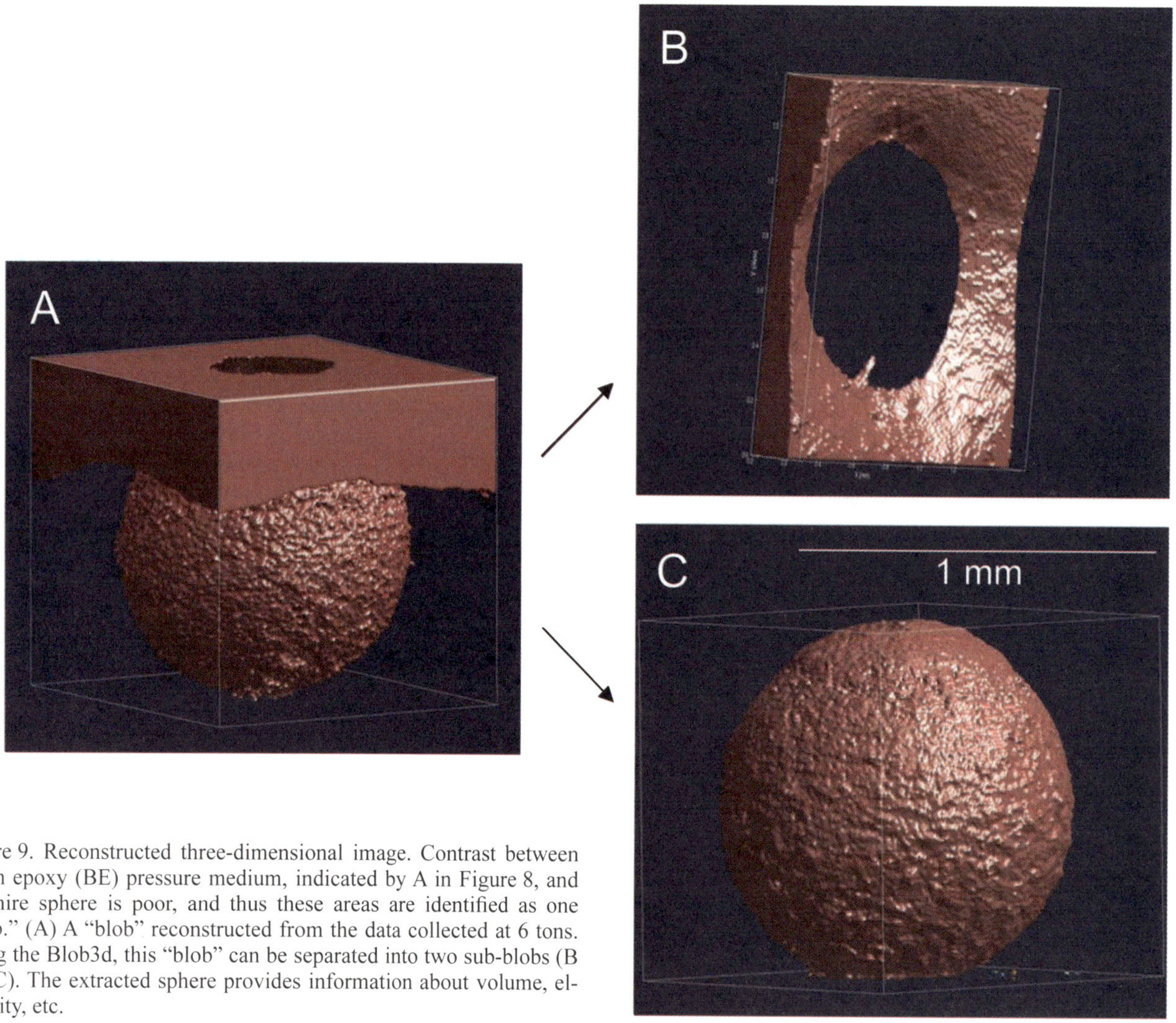

Figure 9. Reconstructed three-dimensional image. Contrast between boron epoxy (BE) pressure medium, indicated by A in Figure 8, and sapphire sphere is poor, and thus these areas are identified as one "blob." (A) A "blob" reconstructed from the data collected at 6 tons. Using the Blob3d, this "blob" can be separated into two sub-blobs (B and C). The extracted sphere provides information about volume, ellipticity, etc.

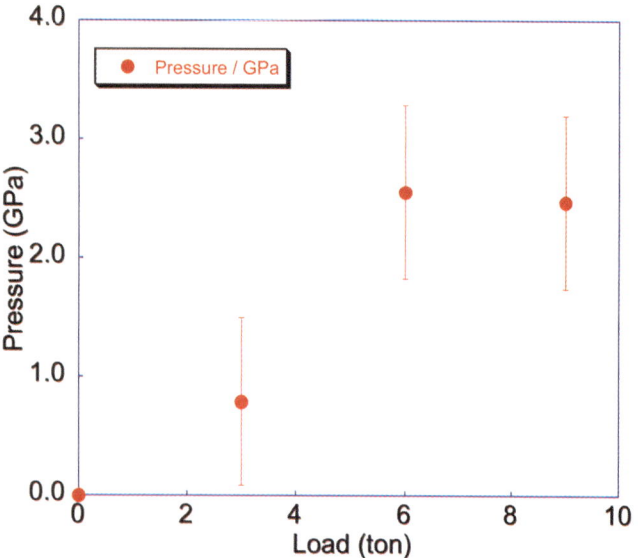

Figure 10. Pressure efficiency obtained from volume change in sapphire sphere. The datum at the highest load (12 ton) is significantly affected by anvil shadowing effects and is not plotted in this figure.

sample from surrounding material quantitatively. A better approach will need to be developed for flat-field correction. Image analysis by curve-fitting routine provides quantitative insight about intensity range in order to extract the objects of interest.

ACKNOWLEDGMENTS

We thank N. Lazarz, F. Sopron, M. Jagger, C. Pullins, N. Nishiyama, and GSECARS (GeoSoilEnviroCARS [Consortium for Advanced Radiation Sources]) personnel for their support during the design and testing of the high-pressure tomography module. The apparatus design benefited from rotational Drickamer apparatus developed at Minnesota University. We thank S. Karato and D. Yamazaki for useful information about the rotational Drickamer apparatus. We thank E. Ohtani and two reviewers for careful reviews. The synchrotron work was performed at GeoSoilEnviroCARS (GSECARS), Sector 13, Advanced Photon Source at Argonne National Laboratory. GSECARS is supported by the National Science Foundation–Earth Sciences, Department of Energy–Geosciences, W. M. Keck Foundation, and the U.S. Department of Agriculture. Use of the Advanced Photon Source was supported by the U.S. Department of Energy, Basic Energy Sciences, Office of Energy Research, under contract no. W-31-109-Eng-38. This work was supported by National Science Foundation grant EAR-001188.

REFERENCES CITED

Decker, D.L., 1971, High-pressure equation of state for NaCl, KCl, and CsCl: Journal of Applied Physics, v. 42, p. 3239–3244, doi: 10.1063/1.1660714.

Katayama, Y., Tsuji, K., Shimomura, O., Kikegawa, T., Mezouar, M., Martinez-Garcia, D., Besson, J.M., Häusermann, D., and Hanfland, M., 1998, Density measurement of liquid under high pressure and high temperature: Journal of Synchrotron Radiation, v. 5, p. 1023–1025, doi: 10.1107/S0909049597015239.

Rivers, M.L., Sutton, S.R., and Eng, P., 1999, Geoscience applications of X-ray computed microtomography, in Bonse, U., ed., Developments in X-Ray Tomography II: Washington, D.C., The International Society for Optical Engineering, p. 78–86.

Sato, Y., and Akimoto, S., 1979, Hydrostatic compression of four corundum-type compounds: α-Al_2O_3, V_2O_3, Cr_2O_3, and α-Fe_2O_3: Journal of Applied Physics, v. 50, p. 5285–5291, doi: 10.1063/1.326625.

Uchida, T., Wang, Y., Rivers, M.L., Sutton, S.R., Weidner, D.J., Vaughan, M.T., Chen, J., Li, B., Secco, R.A., Rutter, M.D., and Liu, H., 2002, A large-volume press facility at the Advanced Photon Source: Diffraction and imaging studies on materials relevant to the cores of planetary bodies: Journal of Physics Condensed Matter, v. 14, p. 11,517–11,523, doi: 10.1088/0953-8984/14/44/509.

Uchida, T., Wang, Y., Rivers, M.L., and Sutton, S.R., 2005, Stress and strain measurements of polycrystalline materials under controlled deformation at high pressure using monochromatic synchrotron radiation, in Chen, J., Wang, Y., Duffy, T.S., Shen, G., and Dobrzhinetskaya, L.F., eds., Advances in High-Pressure Technology for Geophysical Applications: Amsterdam, Elsevier B.V., p. 137–166.

Wang, Y., Rivers, M., Sutton, S., Eng, P., Shen, G., and Getting, I., 1998, A multi-anvil, high-pressure facility for synchrotron radiation research at GeoSoilEnviroCARS at the Advanced Photon Sources: Review of High Pressure Science and Technology, v. 7, p. 1490–1495.

Wang, Y., Uchida, T., Westferro, F., Rivers, M.L., Nishiyama, N., Gebhardt, J., Lesher, C.E., and Sutton, S.R., 2005, High-pressure X-ray tomography microscope: Synchrotron computed microtomography at high pressure and temperature: The Review of Scientific Instruments, v. 76, p. 073709, doi: 10.1063/1.1979477.

Yamazaki, D., and Karato, S., 2001, High-pressure rotational deformation apparatus to 15 GPa: The Review of Scientific Instruments, v. 72, p. 4207–4211, doi: 10.1063/1.1412858.

Yaoita, K., Katayama, Y., Tsuji, K., Kikegawa, T., and Shimomura, O., 1997, Angle-dispersive diffraction measurement system for high-pressure experiments using a multichannel collimator: The Review of Scientific Instruments, v. 68, p. 2106–2110, doi: 10.1063/1.1148103.

MANUSCRIPT ACCEPTED BY THE SOCIETY 14 AUGUST 2006

Index

A
achondrites, 58–59, 69–72
activation enthalpy, perovskite transport and, 29
akimotoite
   in harzburgite, 137
   Martian meteorites and, 70
   seismic discontinuities and, 117
   shocked meteorites and, 66–67
   Tenham chondrite and, 75
   water solubility in, 137
   Zagami meteorite and, 76
akimotoite-perovskite transitions, 189, 198–199
ALH84001, Martian meteorites and, 69
alkali-chloride melts, 89, 93–94, 100
alphabet phases, phase relations in hydrous peridotite and, 122–123
aluminosilicates, unshocked meteorites and, 61
aluminous phases, stability of in lower mantle, 8–10
aluminum
   effect of on equation of state calculations, 21, 31
   magnesium rich perovskite and, 8–9
   mechanisms of substitution of into perovskite, 8, 17–18
   O vacancies and, 21
   storage of minor elements in perovskite and, 19
aluminum-alloy rings, 227–228
amorphization, pressure-induced of solids, 112
amphibole, 122, 125, 127
Anderson scale, 199
anharmonic potentials, 161, 181
anorthosite, 72–73
antiferromagnets, wüstite as, 48
anvil-with-hole high pressure apparatus, diamond synthesis and, 86–87
apatite, behavior of at high pressures, 68
arc magmatic systems, water flux in, 139–140
atomic diffusivity, perovskite and, 29–30
augite, water content in, 135

B
backing plates, diamond anvil cell analysis and, 177
basalts
   melting phase relations in hydrous, 127
   peridotite vs., 62
   subsolidus phase relations of hydrous, 125–127
   water and, 116
   water content of, 140, 141–142, 144
   water transport to deep mantle and, 115
   See also Mid-ocean ridge basalt (MORB)
Birch's law, 165
Blob3d, 235
Botswanian diamond inclusions, 84, 86
Brillouin-zone boundary, 163
brine inclusions, 130
Brown scale, 195, 197
bulk rocks, 6–7, 69–70
bulk sound velocities, post-perovskite and, 44

C
carbonaceous chondrites, 58, 73
carbonate-carbon systems, 85–86
carbonate-silicate systems, 85–86
carbonates, 84
carbonatite melts, 88–90, 130
cell patterns, ice VII and, 109
Chagatai carbonatite rocks, 86, 94–99
chalcopyrite, diamond synthesis and, 84
chassignite, Martian meteorites and, 69
Chassigny meteorite, 70
chemical vapor deposition, diamonds and, 73
chlorite, water transport to deep mantle and, 115
choke points, chlorite breakdown and, 122
chondrites
   carbonaceous, 58, 73
   petrology and mineralogy of, 58
   shocked meteorites and, 63–69, 75–76
   as source of planet material, 23
chondrules, petrology and mineralogy of, 58
Christoffel equation, 163
chromite, shocked meteorites and, 69
clinoenstatite, 59, 221–223
clinopyroxene, 4, 94, 135
Cocconino sandstones, 72
coesite, 9–10, 60, 136, 221
coesite-stishovite transitions, 9–10, 195, 196
compound mixtures, nuclear resonance spectroscopy and, 166
compression mechanisms, 175, 179–180
compression-wave velocities, nuclear resonance spectroscopy and, 165–166
computed microtomography (CMT). *See* X-ray microtomography at high pressure
conductivity, electrical, 30–32
CONUSS software, 162
convection, mantle, 4
core, hydrogen storage potential of, 115, 142–144
corundum, 60, 200
cotunnite, rutile and, 62
covalency, admixture of in stishovite under pressure, 182
creep and viscosity, perovskite and, 31
cristobalite, unshocked meteorites and, 60–61
crystal structures, 41, 62–63, 176–178

D
D″ seismic discontinuity, post-perovskite and, 37, 38, 43, 117
Debye sound velocity measurements, 163–165, 166
deep mantle, fluid/melt miscibility in, 130
deep water cycle, 141–142
defect density, kinetics and, 202
defect population, 8, 29, 31
degassing, 115, 116, 141–142
dehydration states, 115, 140, 142–144
density of states (DOS), 160, 162–163
depth, packing efficiency and, 15–16
Deutsches ElektronenSYnchotron (DESY), 207, 209, 216
DHMS, metastability of in mantle, 123
diamond
   alkali-chloride melts and, 89–94
   carbonatitic melts similar to fluid-bearing multiphase inclusions and, 88–89
   Chagati carbonatite rocks and, 94
   conditions for formation of in high-pressure experiments, 97–98
   experimental data for formation of, 85–86
   fluid inclusions in, 129–130
   garnet-pyrrhotite melting equilibrium and, 94–97
   genesis of in mantle conditions, 99
   high-pressure Ca-silicate inclusions in, 6
   high-pressure genesis of, 83–84, 100–101
   Kokchetav massif and, 94
   natural data for syngenetic inclusions in, 84–85
   origins of carbonatitic parental melts and, 99
   origins of in meteorites, 73
   sulfur-graphite melts and, 89
   See also Kimberlitic diamonds
diamond anvil cells
   ice VII and, 105–107
   periclase, wüstite studies and, 48
   phase transition experimentation and, 2, 10, 190
   single-crystal structural analysis and, 175, 177–178
diffraction intensity measurements, 177, 180–181, 229, 235
diopside, 6, 59–60, 135
directional dependence, nuclear resonance spectroscopy and, 161–162
dissociations, X-ray diffraction analysis of, 198–200
distortion, wüstite and, 48
Dora Maira Massif, 221
double-stage KAWAI-type apparatuses, 190
Drickamer cells
   advantages of, 228, 235–236
   evaluation of, 230–234
   experimental methods using, 228–230
   modification of, 227

E
eclogites, diamond synthesis and, 84, 86, 94, 96, 97
elastic parameters, 23–24, 43, 48–49, 180
electrical conductivity, perovskite and, 30–32
electromagnetic radiation, nuclear resonance spectroscopy and, 158–159
electromotive force, 189, 200
electroneutrality, storage of minor elements in perovskite and, 19
enstatite, 59, 135. *See also* Clinoenstatite
enthalpy, perovskite transport properties and, 29
entropy, 175, 184–187
epidote, melting phase relations of hydrous basalt and, 127

F
fayalite-ringwoodite transitions, 195–196
Fei scale, 195
feldspars, unshocked meteorites and, 61
ferropericlase
   Mössbauer spectroscopy and, 48, 51
   seismic discontinuities and, 123–125
   ultrasonic interferometry and, 49
   water solubility in, 138
   X-ray diffraction analysis and, 48–49, 51–52
   See also Periclase
Fourier transform infrared spectroscopy (FTIR), 131
frequency sweeps, MAX80 and, 212

fullerenes, 73
furnace assemblies, 192

G

garnet
  experimental phase transitions of, 4–6
  garnet-pyrrhotite melting equilibrium and, 94–97
  hydrogen incorporation to, 131, 136
  Martian meteorites and, 71
  melting of, 125
  NAL (new aluminous) phase and, 9
  transition zone and, 2, 9
  unshocked meteorites and, 59, 60
  water solubility in, 127, 135–136
  X-ray diffraction analysis of dissociation of, 189, 200
gold, as X-ray diffraction pressure standard, 193, 195
grains, 19, 25–26, 29, 32, 85–86
Grüneisen parameters, nuclear resonance spectroscopy and, 157, 166–167
GSECARS beam line, 227, 228

H

HAmburger SYnchotron LABor (HASYLAB), 207, 209
harmonic model, 161
HARWI-wiggler, 216–217
harzburgite, akimotoite in, 137
hexaluminosilicates, Martian meteorites and, 70
high-pressure minerals, incorporation of hydrogen in, 131
H-meteorites, 58
hollandite, 9, 61, 68, 70–71, 75–76
hotspots, deep water cycle and, 141
hydrogarnet substitution, 131, 136
hydrogen, 8, 115, 131, 142–144
hydrohyl forms, 131
hydrostatic conditions, structure conditions under, 179–180
hydrous olivine, 133

I

ice VII, 105–106, 108–112
ideal entropy. See Maximum entropy method
ilmenite, 4, 59–60, 66–67, 175, 184–186
image contrast, microtomography and, 232–234
impact events, 58. See also Shocked meteorites
impedance contrasts, mantle olivine and, 3
inclusions, 19–20, 83–86, 129–130
incompatible elements, MORB water content and, 140
in situ X-ray diffraction. See X-ray diffraction analysis
interatomic distance, determination of, 182
iron, 8, 18, 24–26, 29–30, 32
island arcs, water content of, 140

K

kadeite, water content in, 135
KAWAI-type anvil apparatuses, 190–191
KFMASH, 127–129
kimberlitic diamonds, 4, 85–86
kimberlitic pyropes, 136
kinetics, X-ray diffraction analysis and, 189, 202–203
Kokchetav rock, diamond synthesis and, 86, 94, 95

komatiite, melting of, 125
kyanite, water solubility in, 137

L

Lamb-Mössbauer factor, 161, 162, 166
larnite, unshocked meteorites and, 60
lattice vibrations, nuclear resonance spectroscopy and, 160
lawsonite, 115, 128
light rare earth elements (LREEs), MORB water content and, 140
Lindemann law, perovskite melting and, 29
lingunite, Tenham chondrite and, 75
lithosphere, 2, 7, 9, 117–118, 140–141
L-meteorites, 58
lower mantle
  diamond synthesis and, 84
  elastic parameters for refining composition of, 23–24
  equation of state of perovskite in, 21
  iron dismutation in, 25–26, 32
  periclase, wüstite and, 44–48, 53–54
  perovskite in, 8, 15–16, 31–32
  phase transitions in, 37–38
  problems with using perovskite as analog for, 31–32
  stable aluminous and silica phases in, 8–10
  water solubility in minerals in, 138–139
  water storage in, 144

M

magnesiowüstite, 2–5, 10, 59, 75
magnesium oxide, as X-ray diffraction pressure standard, 193–195
magnetic ordering, wüstite and, 48, 52–53
magnetism, 48, 52–53, 167–168
majorite
  Martian meteorites and, 70
  melting of, 125
  phase transformations of, 4–6
  shocked meteorites and, 66
  subduction of oceanic lithosphere and, 7
  Tenham chondrite and, 75
  transition zone and, 2
  unshocked meteorites and, 59, 60
majorite-perovskite transitions, 3, 5–7
Manicouagan impact crater, 72–73
mantle
  diamond genesis in, 99–101
  fluid/melt miscibility in, 130
  mobility of hydrous fluid in, 130
  subduction zones and, 116–120
  water partitioning between phases of, 139
  water storage capacity of, 142–144
  See also Lower mantle; Upper mantle
mantle convection, postspinel transformation boundary and, 4
mantle fluids, 129–130
Martian meteorites, shock-induced phase transformations and, 58, 69–72
maskelynite, 69–70, 71, 72
Matsui scale, 193–195
MAX80, 207, 208–213
MAX200x, 207, 213–218
maximum entropy method, electron-density distribution by, 175, 184–187
melt pockets, mineralogy of, 71–72
merrillite, 68–69

metagraywackes, metasediment phase relations and, 127
metals, diamond synthesis and, 84
metapelites, 127
metasediments, 127–129
metastability, 123, 179–180
meteorites, 58, 69–72, 73. See also Shocked meteorites; Unshocked meteorites
microtomography. See X-ray microtomography at high pressure
mid-ocean ridge basalt (MORB), 7, 62, 127, 140, 144
minor elements, perovskite structure and, 16, 18–20, 31
miscibility, 130
Mössbauer spectroscopy, 48–51, 168–170. See also Synchotron Mössbauer spectroscopy
multianvil apparatus, 2, 10, 189–190, 207–208, 223–224
multistaging, sodium chloride and, 219–221
muscovite, metasediment phase relations and, 128

N

nakhalites, Martian meteorites and, 69
NAL (new aluminous) phase, 9, 139
nonstoichiometric substitution, of Al ions into perovskite, 8, 17–18
NRIXS measurements, overview of, 160–161
nuclear resonance spectroscopy
  compound mixtures and, 166
  compression- and shear-wave velocities and, 165–166
  debye sound velocity measurements and, 163–165
  directional dependence and, 161–162
  experimental methods using, 159–160
  Grüneisen parameters and, 166–167
  NRIXS measurements and, 160–161
  nuclear resonance basics and, 158–159
  overview of, 131, 157–158, 170
  quasi-harmonic model and, 161
  SMS measurements and, 162
  sound velocity measurements and, 162–163
  temperature and, 167
  valence and spin state and, 168–170

O

oceanic lithosphere, 2, 7, 9, 117–118
OH-stretching bands, 135
olivine-spinel transitions, 15
olivine-wadsleyite transitions, 3, 139, 197–198
olivines
  experimental phase transitions of, 2–4
  hydrous, 133
  impedance contrasts and, 3
  seismic discontinuities and, 139
  shocked meteorites and, 64–66
  Tenham chondrite and, 75–76
  unshocked meteorites and, 59
  upper mantle and, 2
  water in polymorphs of, 123–125
  water solubility in, 116, 131–133
  water transport to deep mantle and, 115
  X-ray diffraction analysis of, 196
omphacite, Zagami meteorite and, 76
opposed-anvil high-pressure cells. See Drickamer cells
orthopyroxene, 4, 135
orthorhombic perovskite, 16–17, 37, 38, 60

O vacancies, 21, 30, 31
oxygen fugacity, 24–26, 29–30, 87, 98–99

P

packing efficiency, 15–17
parental media, diamond formation and, 83–84, 99–101
partitioning, of water in mantle phases, 139
pentlandite, diamond synthesis and, 84
periclase
    Mössbauer spectroscopy and, 48, 49–50, 51
    ultrasonic interferometry and, 49
    water solubility in, 138
    wüstite and, 47–48, 53–54
    X-ray diffraction analysis and, 48–52, 193
    See also Ferropericlase
peridotites
    alphabet phases and, 122–123
    bulk rock phase transitions and, 6
    diamond synthesis and, 84, 86
    melting of, 125
    melting relations of hydrous, 125
    olivine polymorphs, seismic discontinuities and, 123–125
    phase relations to 6 GPa and, 121–122
    unshocked meteorites and, 62
    water solubility in, 116
    water transport to deep mantle and, 115, 142
perovskite
    akimotoite-perovskite transitions and, 189, 198–199
    aluminum and, 8–9, 17–18, 19, 20–23
    crystal chemistry of Ca form of, 16–20
    elastic parameters and, 23–24
    in lower mantle, 8, 15–16, 31–32
    majorite-perovskite transitions and, 3, 5–7
    Martian meteorites and, 70
    melting of, 28–29
    oxygen fugacity and, 24–26, 29–30
    phase transformation in PBNM silicate perovskite and, 26–28
    properties of lower-mantle phase and, 31–32
    seismic discontinuities and, 117, 123–125
    shocked meteorites and, 4, 66–67
    spin transitions and, 10
    stability of, 38
    subduction of oceanic lithosphere and, 7
    Tenham chondrite and, 75
    transition zone and, 2
    transport properties of, 16, 29–31
    unshocked meteorites and, 59, 60
    water solubility in, 138
    X-ray diffraction analysis and, 193, 198–199, 200
    Zagami meteorite and, 76
    See also Post-perovskite
perovskite-majorite transitions, 7
phase transitions
    atomic diffusion and, 29
    in bulk rocks, 6–7
    effect of water on, 116
    hydrostatic condition and, 179–180
    of iceVII to HPHTA ice, 109–112
    Martian meteorites and, 69–72
    multianvil experiments and, 190
    olivine system and, 2–4
    perovskite in lower mantle and, 8, 38–39
    to post-perovskite, 37–39
    pressure-induced, 176, 179–180
    pressure-induced amorphization of solids, 112
    of pyroxenes and garnet, 4–6
    stable aluminous and silica phases in lower mantle and, 8–10
    in subducting slabs, 118
    in transition zone and lower mantle, 1–2, 10
    unquenchable, 221–223
    wüstite and, 48
    X-ray diffraction analysis and, 189–193, 203
    See also Shocked meteorites; Specific transitions
phases, alphabet, 122–123
phengite, 115, 127, 128
phlogopite, 122, 125
phonon excitation, nuclear resonance spectroscopy and, 160
phonon lifetime concept, 161
phonon scattering length, 161
plagioclases, 61, 75
plate tectonics, dehydration sites in lower mantle and, 115, 142–144
polarization anisotropy, post-perovskite and, 43–44
polyhedral stacking, 15, 28
Popigai impact crater, 73
post-perovskite
    anticorrelation between S-wave and bulk sound velocities and, 44
    crystal structure of, 39–41
    discovery of phase transition to, 38–39
    isostructural compounds of, 41–42
    lower mantle and, 117
    overview of, 37–38
    seismic anomalies in lowermost mantle and, 43–44
    stability of perovskite and, 38
    structure of, 28, 32
    understanding of, 158
    unshocked meteorites and, 59
pressure
    activation enthalpy and, 29
    atomic diffusivity and, 30
    effect of on crystal structure, 176–177
    MAX80 and, 208–209
    MAX200x and, 213–216
    maximum entropy method and, 175, 184–187
    microtomography and, 228–229, 235
    multistaging and, 219–221
    phase relations in hydrous peridotite and, 121–122
    X-ray diffraction analysis and, 201–202
pressure standards, X-ray diffraction analysis and, 189, 192, 193–195
pseudo-potential model, 181
pVT equation, lower-mantle perovskite phases and, 22–23
pyrite, diamond synthesis and, 84
pyrolite, 6, 7, 62, 124
pyrolite model, mantle composition and, 116
pyrope, 60, 135–136
pyroxenes, 4–6, 70
pyrrhotite, 71–72, 84, 94, 96–97

Q

quartz, 9–10, 60, 136, 221
quartz-coesite transitions, 9–10, 221
quasi-harmonic model, overview of, 161
quenching, lack of, 221–223

R

Raman spectroscopy, ice VII and, 105–106, 107–108, 109, 110
recoilless absorption of X-rays, 161
redox conditions, diamond synthesis and, 98–99
richterite, phase relations in hydrous peridotite and, 122
Ries crater, 73
ringwoodite
    experimental phase transitions of, 2–4
    fayalite-ringwoodite transitions and, 195–196
    seismic discontinuities and, 123–125
    shocked meteorites and, 64–66
    Tenham chondrite and, 75–76
    transition zone and, 2
    unshocked meteorites and, 59
    water solubility in, 116, 137
    water transport to deep mantle and, 115
    X-ray diffraction analysis of, 189, 193, 195–196, 197, 199–200
Rodinia supercontinental breakup, 140
rutile, 61–62, 73, 137

S

scattering strength, nuclear resonance spectroscopy and, 158–159
scheelite, rutile and, 62
sclorite, phase relations in hydrous peridotite and, 122
secondary ion mass spectrometry (SIMS), hydrogen abundance and, 131
seismic discontinuities
    mantle composition, subduction zones and, 116–118
    olivine-wadsleyite transition and, 3, 139
    post-perovskite and, 43–44
    of transition zone, 2
    water in olivine polymorphs and, 123–125
    water transport to deep mantle and, 115, 139–141
    See also D″ seismic discontinuity
serpentine, 115, 122
serpentinized peridotite, water transport to deep mantle and, 142
shear properties, 20–21, 23–24
shear-wave velocities, nuclear resonance spectroscopy and, 165–166
shergottites, 69–70, 76–77
Shergotty meteorite, 69–70
Shim scale, 195
shock melt veins, 70–72, 74–75
shocked meteorites
    achondrites and, 69–72
    chondrites and, 63–69, 75–76
    hollandite, CAS phases and, 9
    ilmenite and perovskite and, 4
    overview of, 57–58, 76–78
    pressure-temperature-time characteristics of, 74–76
    shock-wave propagation and, 62–63
    Tenham chondrite and, 75–76
    terrestrial rock phase transformations and, 69–72
    Zagami shergottite and, 76
shocked terrestrial rocks, 72–74
shock-wave experiments, 29, 62
silicates
    diamond synthesis and, 94, 96, 97
    Martian meteorites and, 70

stability of in lower mantle, 8–10
    water solubility in, 136–137
    water solubility in olivines and, 133
Simon equation, ice VII and, 105, 108
single-stage cubic anvil apparatuses, 190
sintered diamond anvils, 2
Sixiangkou chondrite, 76
SMS measurements. *See* Synchotron Mössbauer spectroscopy
sodium chloride, 193, 219–221
solubility. *See* Water solubility
sound velocity measurements, nuclear resonance spectroscopy and, 162–165
SPEED-Mk.II, 190, 191
speeds of sound, modeling of for lower-mantle, 23
Speziale scale, 193–195
spin transitions, 10, 27–28, 168–170
stacking, 15, 28
standards, X-ray diffraction analysis and, 189, 192, 193–195
staurolite, metasediment phase relations and, 127
stishovite
    coesite-stishovite transitions and, 9–10, 195, 196
    diffraction intensity measurements of, 175
    Martian meteorites and, 69–70
    rutile and, 61–62
    structure of under high pressure, 181–184
    subduction of oceanic lithosphere and, 7
    transition zone and, 9
    unshocked meteorites and, 60–61
    water solubility in, 136
    X-ray diffraction analysis and, 195, 196
    Zagami meteorite and, 76
stoichiometric substitution, of Al ions into perovskite, 8, 17–18
subduction, oceanic lithosphere and, 2, 7, 9
subduction zones
    dehydration sites and, 142
    majorite phase transitions in, 5
    mantle composition and, 116–120
    phase relations in basaltic systems and, 125
    thermal modeling of, 142
    water transport to deep mantle and, 115, 129–130, 139–142
sulfides, diamond synthesis and, 84–86, 89, 97
superionic water, 112
sursassite, metasediment phase relations and, 128
S-wave velocities, post-perovskite and, 44
symmetry, decomposition of perovskite and, 27
symmetry elements, structural changes of, 15
synchrotron Mössbauer spectroscopy, 158–159, 162, 168–170
synchrotron radiation
    analysis of Earth's interior using, 175–177, 187
    high-pressure structure study using, 178–179
    MAX200x and, 216–217
    single-crystal structural analysis and, 180–184
    structure transitions, hydrostatic condition and, 179–180
    use of multianvil apparatus with, 207–208, 223–224
syngenetic inclusions, 83–85
synthesis conditions, effect of on perovskite equation of state, 21–22

T

talc, phase relations in hydrous peridotite and, 122
temperature
    activation enthalpy and, 29
    atomic diffusivity and, 30
    effect of on equation of state calculations, 20–21
    nuclear resonance spectroscopy and, 167
    phase transitions in bulk rocks and, 7
    shocked meteorites and, 74–76
    X-ray diffraction analysis and, 189, 201
Tenham meteorite, 66–67, 75–77
terrestrial rocks, new shock-induced phase transformations in, 72–74
thermal displacement parameters, increasing pressure and, 175, 181
Thomson-scattering cross section, 158
time-discrimination tricks, 159
titanite, unshocked meteorites and, 60
tomography. *See* X-ray microtomography at high pressure
topaz, 115, 128
trace elements, 16, 18–20, 31, 61–62
transition pressure, AL-insertion into perovskite and, 17
transition zone
    overview of phase transitions of minerals in, 1–2, 10
    seismic discontinuities of, 2
    transport of water to, 115–116, 144
    water solubility in, 116
    water solubility in minerals in, 137
    water storage in, 142–143, 144
transitions. *See* Phase transitions
trapped high-density melt, existence of, 116
travel time method, 212
tridymite, unshocked meteorites and, 60
triple-mode transducers, MAX200x and, 217–218
twins, perovskite and, 31

U

Udachnaya pipe, 86
ultrasonic interferometry
    clinoenstatite and, 221–223
    MAX80 and, 212–213
    MAX200x and, 217–218
    multistaging and, 220–221
    periclase, wüstite studies and, 49, 50
unshocked meteorites, 58–62
upper mantle, 2, 115, 131–136, 141–142, 144
ureilites, diamond and, 73

V

valence electrons, 168–170, 175, 181
velocity jumps, mantle olivine and, 3
viscosity, 31
volcanism, degassing of upper mantle and, 115, 141–142

W

wadsleyite
    experimental phase transitions of, 2–4
    olivine-wadsleyite transitions and, 3, 139, 197–198
    seismic discontinuities and, 123–125
    shocked meteorites and, 64–66
    transition zone and, 2
    unshocked meteorites and, 59
    water solubility in, 116, 137
    water transport to deep mantle and, 115
    X-ray diffraction analysis of transition to, 189, 197–198
walstromite, unshocked meteorites and, 60
water
    lower-mantle minerals and, 10
    role of in Earth's interior, 116
    storage of minor elements in perovskite and, 19–20
    superionic, 112
    transport and storage of in mantle, 139–144
    See also Ice VII
water solubility
    hydrogen incorporation to high-pressure minerals and, 131
    in lower mantle minerals, 138–139
    partitioning in mantle phases and, 139
    in silica polymorphs in upper mantle, 136
    in transition zone minerals, 137
    in upper mantle minerals, 131–136
Whitlockite, behavior of at high pressures, 68
wollastonite, unshocked meteorites and, 60
wüstite
    Mössbauer spectroscopy and, 48, 49–50
    periclase and, 47–48, 53–54
    ultrasonic interferometry and, 49, 50
    X-ray diffraction analysis and, 48–49, 50–51

X

X-radiography, 18, 210–212
X-ray diffraction analysis
    high-pressure structure study using, 178
    ice VII and, 105–106, 108, 111
    kinetics and, 202–203
    MAX80 and, 209–210
    overview of, 190–193
    periclase, wüstite studies and, 48–52
    phase relations of mantle minerals and, 189–190, 203
    phase relations studied with, 195–201
    post-perovskite and, 37–38, 39–42
    pressure measurement and, 201–202
    pressure standards and, 193–195
    techniques for, 190–193
    temperature measurement and, 201
X-ray microtomography at high pressure
    evaluation of equipment for, 230–234
    methods for, 228–229
    overview of, 227–228, 235–237
    test results of, 235

Y

Yamato chondrite, 76

Z

Zagami shergottite, 76–77
Zeeman splitting, nuclear resonance spectroscopy and, 167–168
zircon, 61–62, 137
zoisite, 115, 127